Doubled Haploid Production in Crop Plants

Doubled Haploid Production in Crop Plants
A Manual

Edited by

M. Maluszynski
*Joint FAO/IAEA Division of Nuclear Techniques in Food and Agriculture,
Vienna, Austria*

K.J. Kasha
*University of Guelph,
Department of Plant Agriculture,
Guelph, Canada*

B.P. Forster
*Scottish Crop Research Institute,
Dundee, U.K.*

and

I. Szarejko
*University of Silesia,
Department of Genetics,
Katowice, Poland*

KLUWER ACADEMIC PUBLISHERS
DORDRECHT / BOSTON / LONDON

A C.I.P. Catalogue record for this book is available from the Library of Congress.

ISBN 1-4020-1544-5

Published by Kluwer Academic Publishers,
P.O. Box 17, 3300 AA Dordrecht, The Netherlands.

Sold and distributed in North, Central and South America
by Kluwer Academic Publishers,
101 Philip Drive, Norwell, MA 02061, U.S.A.

In all other countries, sold and distributed
by Kluwer Academic Publishers,
P.O. Box 322, 3300 AH Dordrecht, The Netherlands.

Printed on acid-free paper

All Rights Reserved
© 2003 IAEA
No part of this work may be reproduced, stored in a retrieval system, or transmitted
in any form or by any means, electronic, mechanical, photocopying, microfilming,
recording or otherwise, without written permission from the Publisher, with the exception
of any material supplied specifically for the purpose of being entered
and executed on a computer system, for exclusive use by the purchaser of the work.

Printed in the Netherlands.

Editorial

The views expressed in the papers, nomenclature used and way of presentation remain the responsibility of the authors concerned, who are also responsible for any reproduction of copyright material. All published papers were reviewed by the following group of 'COST 851' Programme experts: S.B. Andersen, B. Bohanec, P. Devaux, M. Hansen and M. Wedzony.

The excellent assistance of Mrs. Katayon Allaf and Kathleen Weindl, Joint FAO/IAEA Division, Plant Breeding and Genetics Section for editorial help, retyping some manuscripts and making corrections in English, is highly appreciated.

Contents

Preface vii
Abbreviations ix
Colour Figures xi

1. Production of doubled haploids in crop plants. An introduction
   K.J. Kasha and M. Maluszynski... 1

2. Protocols for major crops
   Barley
   2.1. Doubled haploid production in barley using the *Hordeum bulbosum* (L.) technique
         P. Hayes, A. Corey and J. DeNoma.. 5
   2.2. The *Hordeum bulbosum* (L.) method
         P. Devaux... 15
   2.3   Anther culture in barley
         C. Jacquard, G. Wojnarowiez and C. Clément... 21
   2.4. Barley anther culture
         L. Cistué, M.P. Vallés, B. Echávarri, J.M. Sanz and A. Castillo............................. 29
   2.5. Anther culture for doubled haploid production in barley (*Hordeum vulgare* L.)
         I. Szarejko... 35
   2.6. Barley isolated microspore culture protocol
         K.J. Kasha, E. Simion, R. Oro and Y.S. Shim... 43
   2.7. Barley isolated microspore culture (IMC) method
         P.A. Davies... 49
   Wheat
   2.8. Doubled haploid production in wheat through wide hybridization
         M.N. Inagaki.. 53
   2.9. Protocol of wheat (*Triticum eastivum* L.) anther culture
         J. Pauk, R. Mihály and M. Puolimatka... 59
   2.10. Protocol for producing doubled haploid plants from anther culture of wheat
         (*Triticum aestivum* L.)
         B. Barnabás... 65
   2.11. Wheat anther culture
         S. Tuvesson, R. von Post and A. Ljungberg.. 71
   2.12. Haploid wheat isolated microspore culture protocol
         K.J. Kasha, E. Simion, M. Miner, J. Letarte and T.C. Hu...................................... 77
   2.13. Production of doubled haploids in wheat (*Triticum aestivum* L.) through
         microspore embryogenesis triggered by inducer chemicals
         M.Y. Zheng, W. Liu, Y. Weng, E. Polle and C.F. Konzak...................................... 83
   Maize
   2.14. Isolated microspore culture in maize (*Zea mays* L.), production of doubled-haploid
         via induced androgenesis
         M.Y. Zheng, Y. Weng, R. Sahibzada and C.F. Konzak... 95
   2.15. Anther culture of maize (*Zea mays* L.)
         B. Barnabás... 103
   Rice *indica/japanica*
   2.16. Laboratory protocol for anther culture technique in rice
         F.J. Zapata-Arias... 109
   Triticale
   2.17. Triticale anther culture
         S. Tuvesson, R. von Post and A. Ljungberg... 117
   2.18. Protocol for anther culture in hexaploid triticale (x *Triticosecale* Wittm.)
         M. Wędzony... 123

2.19. Protocol of triticale (x *Triticosecale* Wittmack) microspore culture
J. Pauk, R. Mihály, T. Monostori and M. Puolimatka 129

2.20. Protocol for doubled haploid production in hexaploid triticale (x *Triticosecale* Wittm.) by crosses with maize
M. Wędzony 135

Rye

2.21. Protocol for rye anther culture
S. Immonen and T. Tenhola-Roininen 141

2.22. Microspore culture of rye
S. Pulli and Y.-D. Guo 151

Oat

2.23. Oat haploids from wide hybridization
H.W. Rines 155

Durum wheat

2.24. Haploid and doubled haploid production in durum wheat by wide hybridization
P.P. Jauhar 161

2.25. Haploid and doubled haploid production in durum wheat by anther culture
P.P. Jauhar 167

Timothy

2.26. Anther culture and isolated microspore culture in timothy
S. Pulli and Y.-D. Guo 173

Ryegrass and other grasses

2.27. Doubled haploid induction in ryegrass and other grasses
S.B. Andersen 179

Rapeseed

2.28. Microspore culture in rapeseed (*Brassica napus* L.)
J.B.M. Custers 185

Broccoli

2.29. Protocol for broccoli microspore culture
J.C. da Silva Dias 195

Other Brassicas

2.30. Microspore culture of *Brassica* species
A. Ferrie 205

2.31. Protocol for microspore culture in Brassica
M. Hansen 217

Tobacco

2.32. Anther and microspore culture in tobacco
A. Touraev and E. Heberle-Bors 223

Potato

2.33. Haploid production of potatoes by anther culture
G.C.C. Tai and X.Y. Xiong 229

2.34. Anther culture through direct embryogenesis in a genetically diverse range of potato (*Solanum*) species and their interspecific and intergeneric hybrids
V.-M. Rokka 235

2.35. Potato haploid technologies
M.J. De,Maine 241

Linseed/flax

2.36. Anther culture of linseed (*Linum usitatissimum* L.)
K. Nichterlein 249

Sugar beet
- 2.37. Doubled haploid production of sugar beet (*Beta vulgaris* L.)
  E. Wremerth Weich and M.W. Levall 255

Asparagus
- 2.38. Asparagus microspore and anther culture
  D.J. Wolyn and B. Nichols 265

Onion
- 2.39. *In vitro* gynogenesis induction and doubled haploid production in onion (*Allium cepa* L.)
  L. Martínez 275
- 2.40. Haploid induction in onion *via* gynogenesis
  M. Jakše and B. Bohanec 281

Apple
- 2.41. *In vitro* androgenesis in apple
  M. Höfer 287

Aspen
- 2.42. Doubled haploid production in poplar
  S.B. Andersen 293

Cork oak
- 2.43. Oak anther culture
  M.A. Bueno and J.A. Manzanera 297

Citrus
- 2.44. Haploids and doubled haploids in *Citrus* spp.
  M.A. Germanà 303

3. Published protocols for other crop plant species
   M. Maluszynski, K.J. Kasha and I. Szarejko 309

4. Application of doubled haploid production techniques
   - 4.1. Doubled haploids in breeding
     W.T.B. Thomas, B.P. Forster and B. Gertsson 337
   - 4.2. Doubled haploid mutant production
     I. Szarejko 351
   - 4.3. Barley microspore transformation protocol by biolistic gun
     Y.S. Shim and K.J. Kasha 363
   - 4.4. Doubled haploids in genetic mapping and genomics
     B.P. Forster and W.T.B. Thomas 367

5. Protocols for chromosome counting
   - 5.1. Cytogenetic tests for ploidy level analyses – chromosome counting
     J. Maluszynska 391
   - 5.2. Ploidy determination using flow cytometry
     B. Bohanec 397

6. Major media composition and basic equipment for DH laboratory
   - 6.1. Major media composition
     compiled by I. Szarejko 405
   - 6.2. Basic equipment for maize microspore culture laboratory
     compiled by M.Y. Zheng, Y. Weng, R. Sahibzada and C.F. Konzak 415

List of contributors 417

Preface

The production of doubled haploids has become a necessary tool in advanced plant breeding institutes and commercial companies for breeding many crop species. However, the development of new, more efficient and cheaper large scale production protocols has meant that doubled haploids are also recently being applied in less advanced breeding programmes. This Manual was prepared to stimulate the wider use of this technology for speeding and opening up new breeding possibilities for many crops including some woody tree species. Since the construction of genetic maps using molecular markers requires the development of segregating doubled haploid populations in numerous crop species, we hope that this Manual will also help molecular biologists in establishing such mapping populations.

For many years, both the Food and Agriculture Organization of the United Nations (FAO) and the International Atomic Energy Agency (IAEA) have supported and coordinated research that focuses on development of more efficient doubled haploid production methods and their applications in breeding of new varieties and basic research through their Plant Breeding and Genetics Section of the Joint FAO/IAEA Division of Nuclear Techniques in Food and Agriculture. The first FAO/IAEA scientific network (Coordinated Research Programme - CRP) dealing with doubled haploids was initiated by the Plant Breeding and Genetics Section in 1986. This was followed by a CRP on doubled haploids in cereals involving countries in Latin America and by other CRPs connecting doubled haploids with molecular marker techniques or with their application for developing stress tolerant mutant germplasm in crop species. Also, numerous FAO/IAEA training courses were organized by the Joint Division in association with CRP's and Technical Cooperation Projects.

Requests from researchers and trainees for systematised protocols dealing with the production of doubled haploids in various species were instrumental in initiating work on this Manual, and the idea to prepare it was strongly supported by European scientists working under the EU large scale program 'COST 851' on 'Gametic cells and molecular breeding for crop improvement'. Many of the authors who provided protocols for the manual participated in these activities, thereby helping to transfer doubled haploid technology in countries and over continents. Also, five leading experts from the EU-COST 851 were invited to review the suitability of the manuscripts presented for publication in the Manual.

The Manual focuses on efficient protocols – laboratory recipes – for producing doubled haploids in major crop species. Additionally, references on about 200 doubled haploid protocols for other plant/crop species together with chapters illustrating the current status of the use of doubled haploids in plant breeding and basic research are included. However, the Manual does not include reviews on doubled haploids in particular crops or groups of crops as these were recently the subject of another series of books.

James D. Dargie
Director, Joint FAO/IAEA Division
of Nuclear Techniques in Food and Agriculture
Vienna, Austria

Abbreviations

2,4,5-T	2,4,5-Trichlorophenoxyacetic acid
2,4-D	2,4-Dichlorophenoxyacetic acid
2-HNA	2-Hydroxynicotinic acid
2iP	6-(γ,γ-Dimethyloamino)purine
ABA	Abscisic acid
AFLP	Amplified fragment length polymorphism
Ancymidol	α-Cyclopropyl-α-(4-methoxyphenyl)-5-pyrimidinemethanol
BA (BAP)	6-Benzylmaminopurine
BaYMV	Barley yellow mosaic virus
BaYDV	Barley yellow dwarf virus
BSA	Bulk segregant analysis
CPA	p-Chlorophenoxyacetic acid
DAPI	4'6-Diamidino-2-phenylindole dihydrochloride
DH	Doubled haploid(s); doubled haploidy
Dicamba	3,6-Dichloro-o-anisic acid (or 3,6-Dichloro-2-methoxybenzoic acid)
DMSO	Dimethyl sulphoxide
EDTA	Ethylenediaminetetraacetic acid
EI	Ethyleneimine
ELS	Embryo-like structure
EMS	Ethyl methanesulfonate
ENU	N-nitroso-N-ethylurea
GA	Gibberelic acid
GA$_3$	Gibberelin A$_3$
IAA	Indole-3-acetic acid
IBA	Indole-3-butric acid

MES	2-(*N*-Morpholino)ethanesulphonic acid
MNH (MNU)	*N*-nitroso-*N*-methylurea
MNNG	*N*-methyl-*N*'-nitro-*N*-nitrosoguanidine
NAA	1-Naphtalenacetic acid
NaN$_3$	Sodium azide
N$_f$	Fast neutrons
PAA	Phenylacetic acid
Picloram	4-Amino-3,5,6-trichloropicolinic acid
RAPD	Random amplified polymorphic DNA
RFLP	Restriction fragment length polymorphism
RYMV	Rye yellow mosaic virus
SSD	Single seed descent
SSR	Simple sequence repeat (microsatelite)
Thidiazuron	1-Phenyl-3-(1,2,3-thidiazol-5-yl) urea
TIBA	2,3,5-Triiodobenzoic acid

Colour Figures

Figure 2.2-1. a) Vernalization of *H. bulbosum* plants in a cold-frame; b) Emasculation of a barley spike; c) *H. bulbosum* spike sheading pollen (*); d) Collecting *H. bulbosum* pollen; e) Pollination of a barley spike; f) Spray a barley spike 1-2 days after pollination. Figures originated from Florimond Desprez and from Crop and Food Research (*), Christchurch, NZ (courtesy of Dr. Paul Johnson).

Figure 2.2-2. a) Print of a bar coded label; b) Bagged spikes after spraying; c) Embryo excision from seeds (*); d) Embryo culture on medium (*); e) Developing embryos on medium (*); f) Plantlets ready to be transferred into potting mix. Figures originated from Florimond Desprez and from Crop and Food Research (*), Christchurch, NZ (courtesy of Dr. Paul Johnson).

Figure 2.2-3. a) Transfer of plantlets into potting mix; b) Plants at 2-3 tiller stage; c) Haploid plants of barley prepared to be treated with colchicine; d) Colchicine treatment of haploid plants of barley; e) Plants in soil after colchicine treatment; f) Doubled-haploid plants covered with a bag; g) Doubled-haploid plants after self-pollination; h) Manual collection of fertile spikes from doubled-haploid plants; i) Gathering the fertile spikes of each doubled-haploid in a bag; j) Identification of each bag with a label. Figures originated from Florimond Desprez.

Figure 2.4-1. Barley anther culture. a) Anthers after four days of pre-treatment in mannitol; b) Embryos after 12-14 days in induction medium; c) Embryos after 18-20 days in induction medium; d) Well-developed embryos before transferring to regeneration medium; e) Green and albino plants in regeneration medium; f) Transfer of plants to soil; g) Doubled haploid plants cultivated in the greenhouse; h) Field trials of doubled haploid lines.

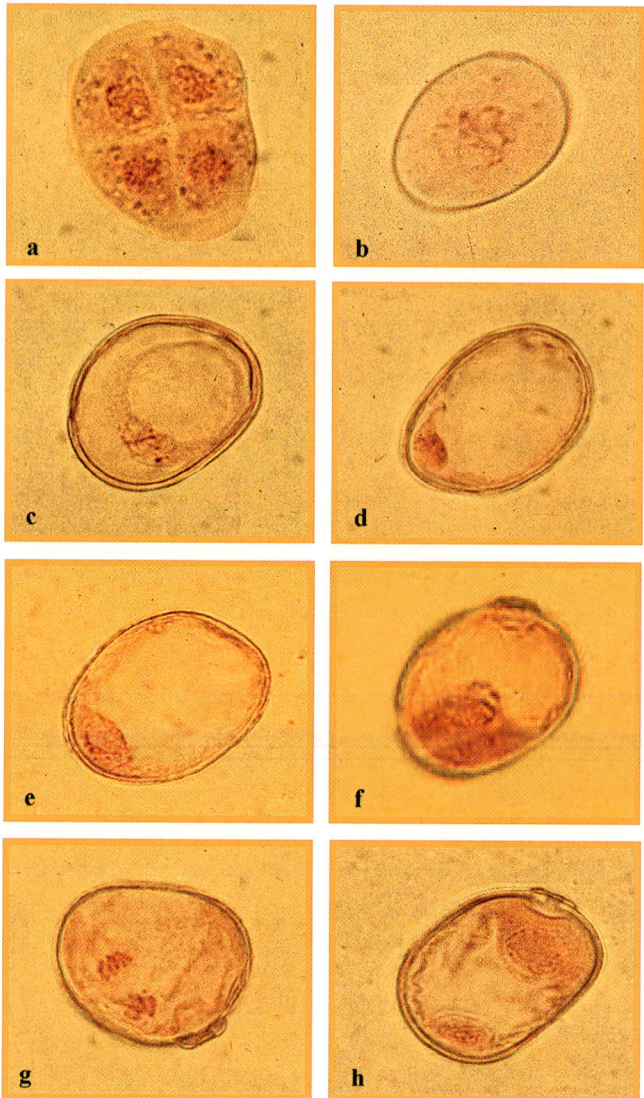

Figure. 2.5-1. Stages of microspore development in barley: a) tetrads; b) early uninucleate; c) early-mid uninucleate; d) mid uninucleate; e) mid-late uninucleate; f) late uninucleate; g) anaphase of the first microspore division; h) bi-nucleate.

Fig. 2.5-2. Production of barley doubled haploids through anther culture: a) tillers containing spikes at the proper developmental stage (the arrow indicates a distance between the flag and the penultimate leaf); b) a microspore at the mid uninucleate stage; c) a microspore at the mid-late uninucleate stage; d) spikes selected for cold pre-treatment; e) induction plates 30 days after anther inoculation; f) microspore-derived calli/embryos on Ficoll medium; g) an anther with embryo-like structures on induction medium; h) an androgenic embryo 35 days after anther inoculation; i) regenerating calli/embryos 14 days after transfer onto regeneration medium; j) a plantlet developing from callus on regeneration medium; k) regenerants on rooting medium; l) young plants transferred into soil; m) spontaneously doubled, fertile microspore derived barley plants (DH1); n) field evaluation of DH2 lines.

Figure 2.7-1. a) Eberbach blender model 8580; b) 100 μm sieves; c) Band of viable microspores at the interface of maltose and mannitol in the neck of the volumetric flask.

Figure 2.7-2. a) Microspore colonies in liquid medium prior to subculture to solidified medium; b) Regenerants growing on solidified medium; c) Regenerants on rooting medium; d) Regenerant plants in soil; e) Regenerants transferred to greenhouse.

Figure. 2.8-1. Stages in producing wheat doubled haploids through ultra-wide crosses. a) Wheat spikes at ear emergence (*left*), emasculation (*center*) and seed setting (*right*). b) Germination of maize pollen on wheat stigma. Bar 0.1 mm. c) Injection of 2,4-D into wheat spike using a syringe. d) Pearl millet pollen at collection (*upper*) and after being dried for two hours (*lower*). Bar 0.1 mm. e) Detached-tiller culture of wheat at emasculation (*left*), pollination (*center*) and seed setting (*right*). f) Wheat seeds obtained from self-pollination, and from crosses with maize, pearl millet and sorghum (*from left to right*). Bar 2 mm. g) Apparatus used for embryo rescue on a laminar flow bench.

Figure 2.8-1 (continuation). h) Wheat embryos obtained from self-pollination, and from crosses with maize pearl millet and sorghum (*from left to right*), showing differences in size and shape among crosses. Bar 2 mm. i) Plant developing from cultured embryos (*from left to right*; 0, 1, 2 and 3 weeks after incubation). j) Culture room accommodating test tubes and plastic dishes. k) Wheat haploid plants transferred to potted soil for further development. l) Somatic chromosomes ($2n=3x=21$) of a wheat haploid plant. Bar 10μm. m) Colchicine treatment of wheat haploid plants at tillering. n) Wheat plants transferred to potted soil after colchicine treatment. o) Wheat spikes of colchicine-treated plants, covered with glassine bag (*left*), and producing doubled haploid grains (*right*). p) Doubled haploid plants grown for seed multiplication. q) Doubled haploid lines grown in the field, showing uniformity within lines.

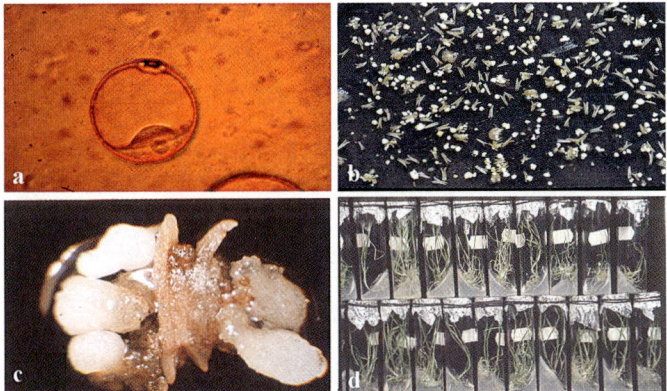

Figure 2.9-1. a) Vacuolated, mid uninucleated wheat microspore after two weeks in cold pretreatment. Don't use older stages for wheat anther isolation; b) Float anther culture using P-4mf culture medium with embryo-like-structures (ELS) at the 6th week of culture; c) A responsive anther with five individual ELS; d) Rooting and tillering of androgenic haploids.

Figure 2.9-2. a) Root tip cytology from a wheat haploid plant with 21 chromosomes ($2n=3x=21$); b) Colchicine treatment of haploids under a well-controlled greenhouse condition; c) Acclimatisation of transplanted colchicine treated plantlets; d) Microspore-derived DH breeding lines in greenhouse propagation.

Figure 2.9-3. Yield test and propagation of DH lines in Szeged, Hungary. Seed and bread of the first anther culture-derived DH variety: GK Délibáb (mirage), released and patented in 1992 (upper right corner).

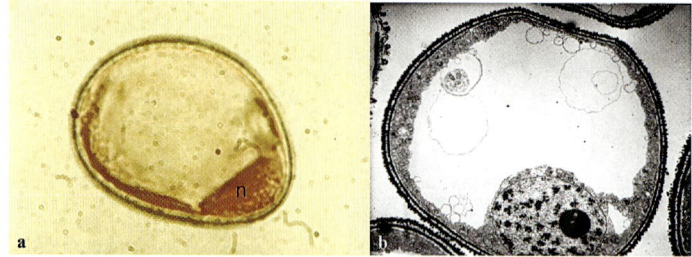

Figure 2.10-1. a) *Triticum aestivum* microspore in the late uninuclear stage of development, n - nucleus; b) doubled chromosome set in a microspore nucleus (TEM picture).

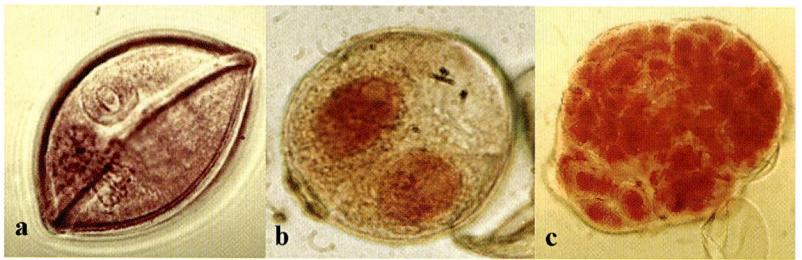

Figure 2.10-2. Effect of colchicine on microspore androgenesis *in vitro*; a) C-mitosis; b) symmetrical division; c) microspore derived embryo.

Figure 2.10-3. Pollen embryos (e) and callus (c) in anther culture.

Figure 2.10-4. a) Regeneration of pollen plants; b) Doubled haploid plants grown in the growth chamber.

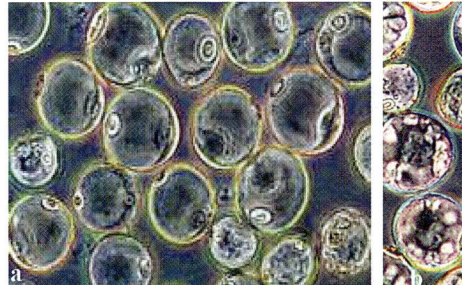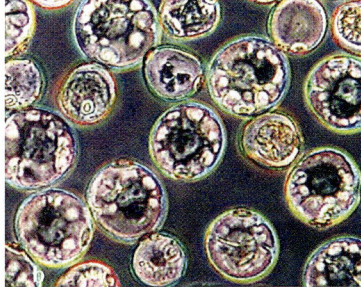

Figure 2.13-1. Wheat microspores are induced by an inducer chemical to develop toward embryoid formation. a) Uni-nucleate microspores, naturally programmed for gametophytic pathway toward pollen formation; b) Triggered by inducer chemical, microspores are reprogrammed for sporophytic development toward embryoid formation.

Figure 2.13-2. a) Pretreatment apparatus in "flask" system; b) Disinfected spikes.

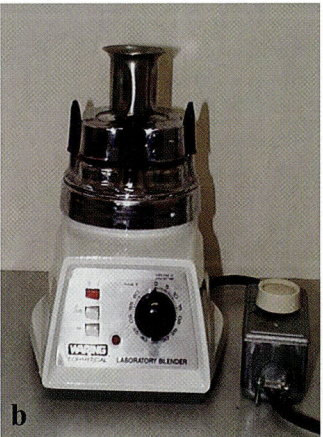

Figure 2.13-3. a) Cut florets in blender cup ready to be blended; b) Blender cup assembled to the blender for blending to release microspores.

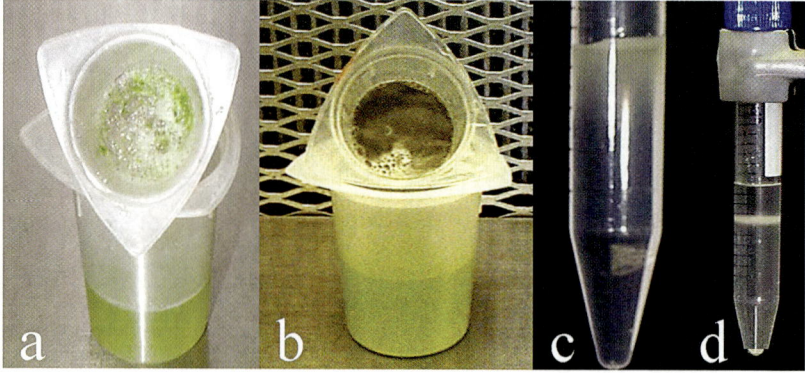

Figure 2.13-4. From left to right: 100 μm (a) and 38μm (b) mesh filter in beakers, overlay of 0.3 M mannitol + microspores over 0.58 M maltose (c), and band with androgenic wheat microspores after gradient centrifugation (d).

Figure 2.13-5. Wheat microspore embryogenesis was triggered by an inducer chemical and nursed by optimal culture condition *in vitro;* a) Dividing microspores at day 7 in liquid culture medium; b) Developing pre-embryos at day 21; c) Mature embryos at day 30; d) Embryos geminated on solid medium.

Figure 2.13-6. a) Wheat microspore-derived plants in trays; b) Plants in trays with plastic cover.

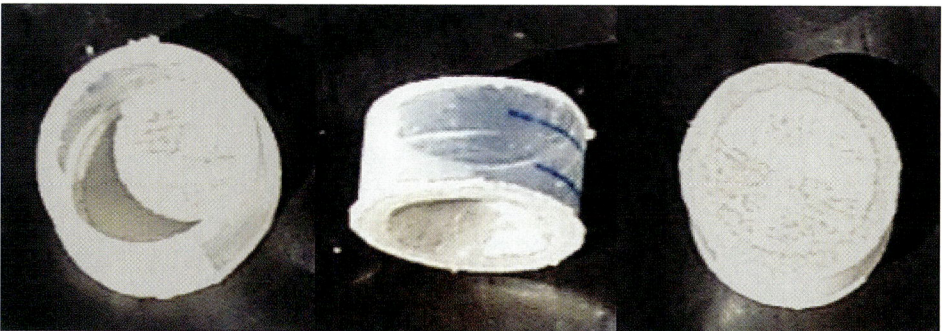

Figure 2.13-7. Tube filter, which can be placed inside 30x10 mm or larger culture Petri-dishes.

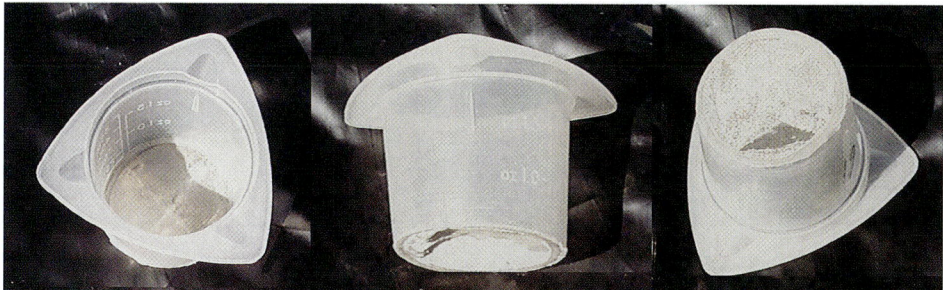

Figure 2.13-8. Beaker filter, which can be placed inside another beaker and/or beaker filter.

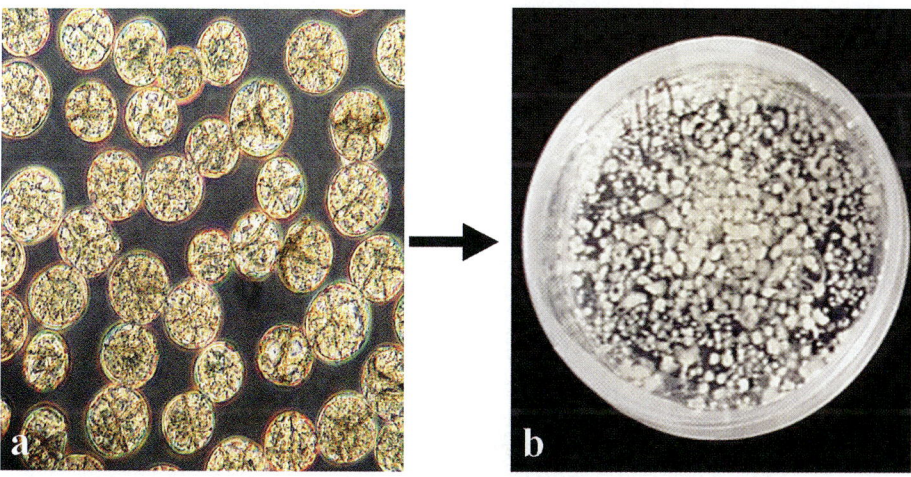

Figure 2.13-9. a) Near 100% dividing microspores; b) Mature wheat embryoids from embryogenic microspores.

Figure 2.14-1. a) Tassel following sampling; b) Tassels or florets ready for disinfection, c) Pretreated florets ready for microspore isolation; d) Blender and blender cup.

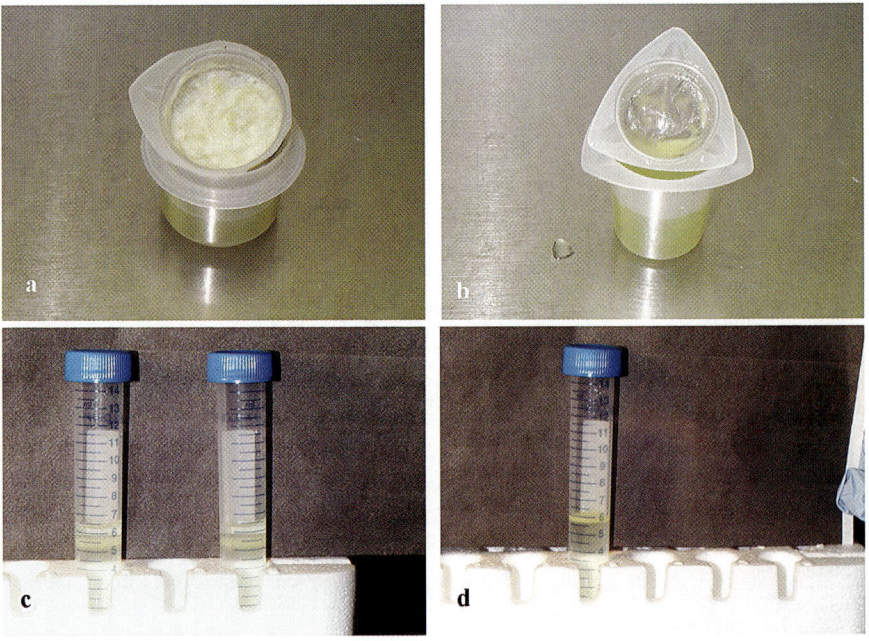

Figure 2.14-2. a) Filtering through 100 μm filter; b) Microspores trapped on 50 μm mesh filter; c) Viable microspores in band(s); d) Potentially embryogenic microspore band.

Figure 2.14-3. a) Maize microspores ready for induction culture; b) Cell divisions in 3d culture; c) Multicellular structures (11th day); d) Pro-embryoids emerging from the exine.

Figure 2.14-4. a) Some mature embryoids/calli; b) Plantlets on Reg-II medium; c) Plantlets on Reg-III (for rooting); d) Plantlets ready for transfer to GH.

Figure 2.14-5. a) Plants 4 d after the transplant; b) Plants 10 d after the transplant; c) Plants 5 weeks after the transplant; d) Mature ears from DH plants.

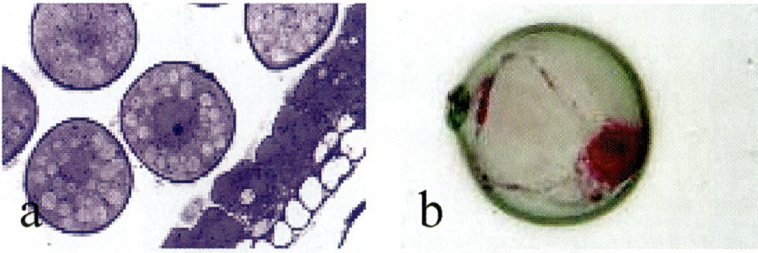

Figure 2.15-1. Development of the microspores during cold pre-treatment. a) mid uninuclear microspore; b) late uninuclear microspore.

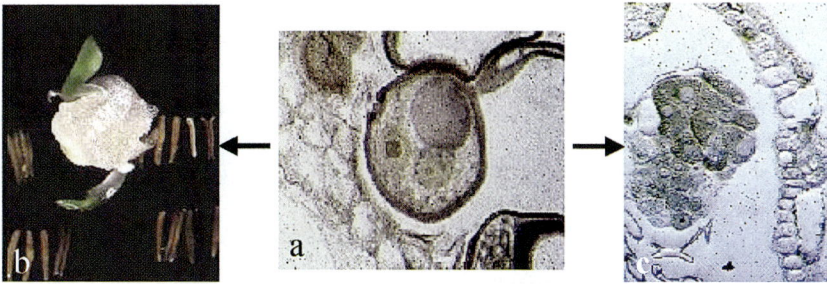

Figure 2.15-2. Androgenic development of maize microspores. a) microspore divided asymmetrically; b) callus; c) embryo formation.

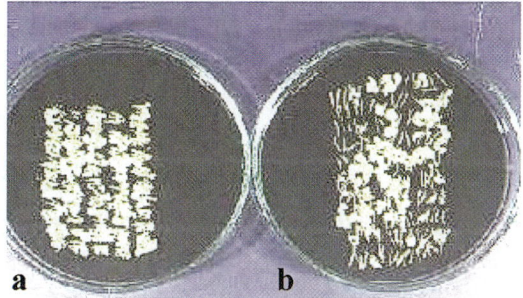

Figure 2.15-3. Androgenic response of the anthers of the DH105×HMv5405 hybrid induced without (a) and in the presence of colchicine (b).

Figure 2.15-4. Plantlet differentiation (a) and growth (b).

Figure 2.15-5. DH line (No.109) selected from anther culture of the exotic genotype Chi 592.

Figure 2.15-6. DH line No.105 selected from anther culture of an SC hybrid SR88×Chi 592 of partly exotic origin

Figure 2.16-1. a) Selected rice seeds; b) Growing of donor plants; c) Panicles showing different flag leave length; d)Wrapped panicles in moisten paper towel ready for cold shock; e) Florets containing anthers. Anthers containing microspores at the right stage of development should not exceed half of the floret length; f) Microspore at the early uninucleate stage of development; g) Microspore at the late uninucleate stage of development; h) Microspore at the early binucleate stage of development; i) Microspore at the late binucleate stage of development

Figure 2.16-2. a) Plating of anthers under sterile conditions; b) Close up of figure a; c) Callus induction; d) Embryogenic globular structures developed from microspores; e) Calli being plated for plant regeneration; f) Shoot development in regeneration medium; g) Regenerated plants; h) Regenerated plants placed in Yoshida's culture solution for a good root development; i) Haploid plant ready to be diploidized by immersing the plant in a colchicine solution.

Figure 2.18-1. Stages of hexaploid triticale anther culture, (a-d) induction phase; a) Longitudinally broken anthers of variety Vero show embryo-like structures at the end of the fourth week of the induction phase. b) 5th week of anther culture of variety Bogo conditioned with ovaries (one can be noticed in the centre), androgenic structures of various shapes are visible; c) and d) Variability of shapes and sizes of androgenic structures produced from the anthers of variety Salvo after 5 weeks of the induction phase; e and f) Regeneration phase; e) Green and albino plants regenerated 3 weeks after placing androgenic structures onto the regeneration medium. f) Green plants were rooted for two weeks in magenta boxes. At this stage they are most sensitive to the colchicine treatment according to the protocol 1.

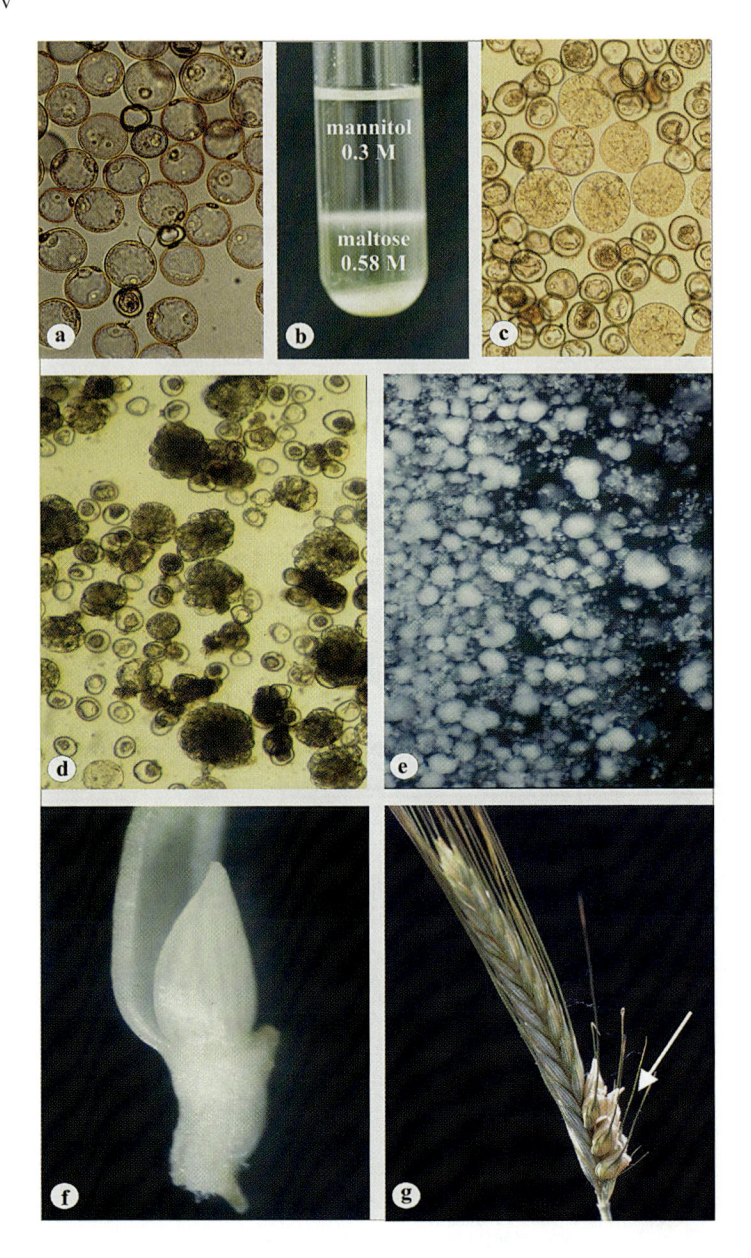

Figure 2.19-1. *In vitro* androgenesis of isolated triticale microspores in directly isolated microspore culture; a) Freshly isolated late-stage uninucleated microspores in 0.3 M mannitol; b) Separation of viable (band between 0.58 M maltose and 0.3 M mannitol) and dead (at the bottom of the tube) microspores from blendor macerated crude head suspension on the maltose/mannitol cushion; c) Dividing multicellular (dense cells) and non-responsive (dead) microspores 1-week after isolation; d) Well-growing microspore-derived aggregates 2-week after isolation; e) ELS in the 4th week of culturing while non-divided microspores (little white points) are visible in background; f) Germinated bipolar embryo on regeneration medium with first and second leaf initiatives (up) and at the opposite side (down) little root is visible at 6-7 week after isolation; g) Microspore-derived colchicine treated androgenic triticale spike with sectorial fertility (arrow).

XXXV

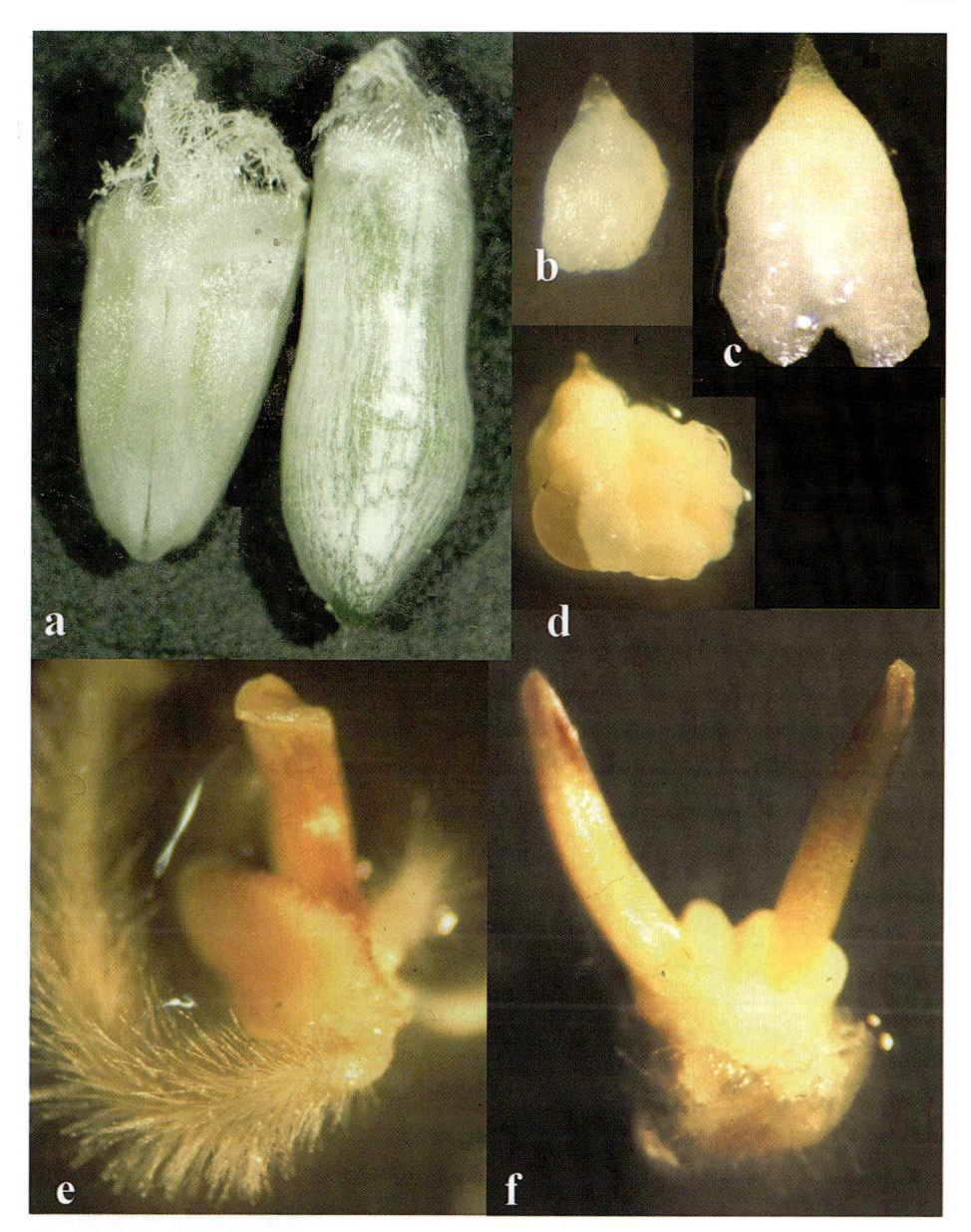

Figure 2.20-1. a) At the left, kernel-like structure (enlarged ovary) obtained after pollination of triticale variety Salvo with maize variety Gama and treatment with 100 mg/L dicamba applied one day later. At the right kernel developing after the selfpollination of the same cultivar of the same age, i. e. 18 day after pollination; b-d) Triticale haploid embryos at time of isolation for culture; e and f) Triticale haploid embryos germinating *in vitro*.

Figure 2.21-1. a,b) Optimal microspore developmental stages for rye anther culture, (a) late uninucleate stage, (b) first pollen mitosis (anaphase); c) Responding rye anthers on induction medium; d,e) Proliferation of ELS and callus on induction medium; f) Green plant regeneration through gametic embryogenesis on induction medium; g) anther culture derived green rye regenerants on regeneration medium; h) Anther culture derived green plants from F_1 cross.

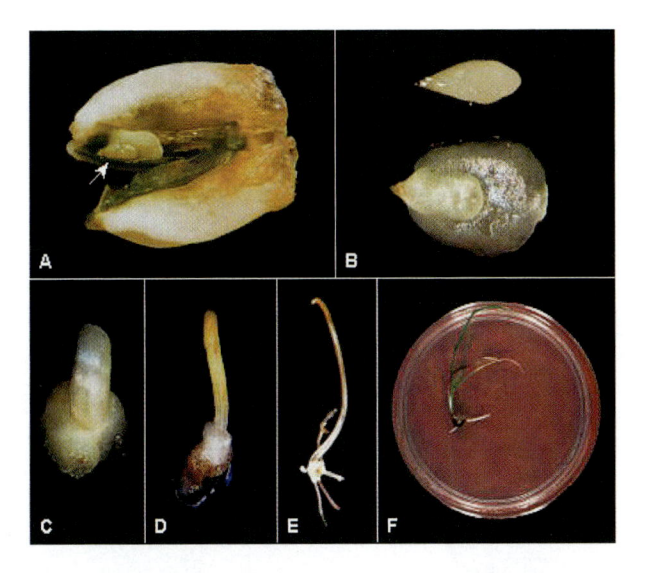

Figure 2.24-1. Stages of embryo germination and development of haploid durum plants: a) A small embryo (*arrow*) in the empty karyopsis (without endosperm), 16 days after pollination and daily post-pollination treatments with 3 mg/L 2, 4-D + 120 mg/L $AgNO_3$; b) *Lower*: the embryo shown in A is rescued onto hormone-free MS medium placed in an incubator in the dark. *Upper*: a torpedo-shaped, small embryo rescued from a different karyopsis. Some of these small embryos are also viable; c,d) Stages of embryo sprouting on the medium in the dark; e) A young plantlet (without chlorophyll) with well developed coleoptile and several primary roots, transferred to a lighted incubator; f) Green plantlet with small primary roots. Note chlorophyll that developed under light.

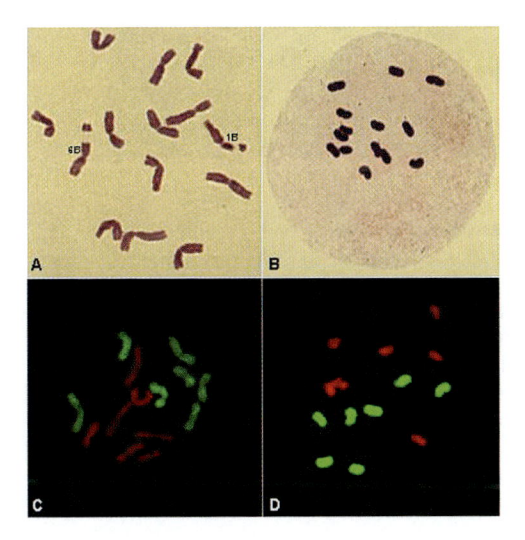

Figure 2.24-2. Somatic and meiotic chromosomes of haploid durum plants: a) Conventionally stained 14 somatic chromosomes from root tips. Note the satellited chromosomes 1B and 6 B; 1 B has a smaller satellite; b) Conventionally stained meiotic chromosomes from pollen mother cells, note 14 unpaired chromosomes (univalents). Pairing between the A-genome and the B-genome chromosomes is suppressed by *Ph1* (see also Fig. d); c) Somatic chromosomes (after fluorescent GISH) from root tips of haploid plantlets. Note 7 A-genome chromosomes (green) and 7 B-genome chromosomes (red). Labeled (biotinylated) total genomic DNA from A-genome donor, *T. urartu*, was used as a probe; d) Meiotic chromosomes from pollen mother cells of immature anthers of haploid plants. Fluorescent GISH reveals the 7 univalents of the A genome (green) and 7 univalents of the B genome (red). Note *Ph1*-induced suppression of pairing.

Figure 2.25-1. Different stages of durum anther culture, green and albino regenerants, and chromosomes of haploid durum plantlet. a) Mid uninucleate microspore; note the nucleus and germpore; b) Appropriately staged yellowish green anthers, cultured on liquid medium; note that both lobes are in contact with the medium; c) An anther on BAD-3 induction medium, three weeks after culture showing the formation of several calli; d) Callus initiated on BAC-1 medium transferred to BAD-1 differentiation medium 4 weeks after initial anther culture; e) Differentiating callus showing embryoid formation with a shoot initial. Note chlorophyll that developed in the dark; f) Embryoid one day after transfer under light. Note well-developed leaves with chlorophyll; g) Healthy, greed plantlets on the regeneration medium, 8 weeks after initial culturing of the anther; h) A well-developed albino plantlet, 8 weeks after initial anther culture. Such plantlets do not survive; i) 14 somatic chromosomes (photographed under phase lens) from root tips of an albino haploid.

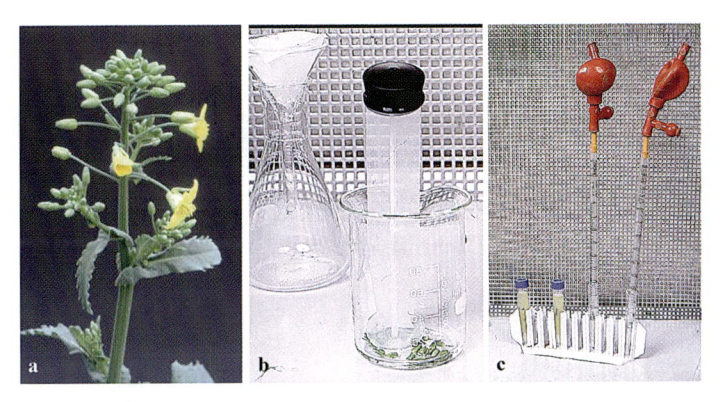

Figure 2.28-1. Materials for *Brassica napus* isolated microspore culture. a) Young inflorescence to serve flower buds for microspore culture; b) Plunger for squeezing the flower buds, i.e. the backside of the piston of a disposable 35 ml syringe; c) Holder with centrifuge tubes with microspore suspensions (two tubes at the left).

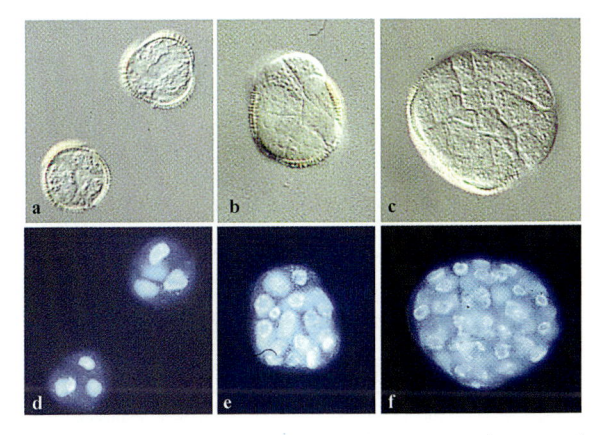

Figure 2.28-2. Successive stages of *Brassica napus* microspore embryogenesis. a,b,c) Differential interference contrast micrographs (DIC); d,e,f) DAPI staining to visualize nuclei. a,d) Microspores with four vegetative nuclei, after 2 days of culture; b,e) Multi-cellular microspore at the rupture stage, with at least 24 nuclei, after 5 days of culture; c,f) Early globular embryo, after 6 days of culture.

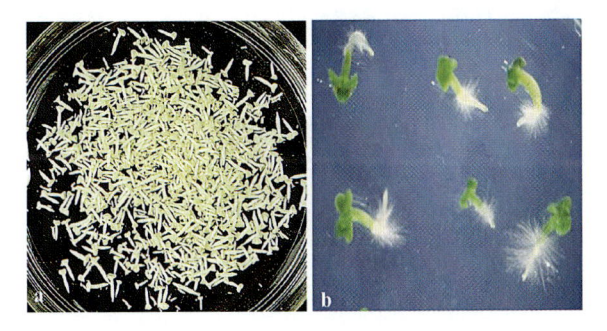

Figure 2.28-3. *Brassica napus* microspore culture. a) Torpedo and cotyledonary embryos after 21 days of culture; b) Embryos after 10 days on germination medium.

Figure 2.30-1. Stages of microspore culture from donor plants to field evaluation of doubled haploids.

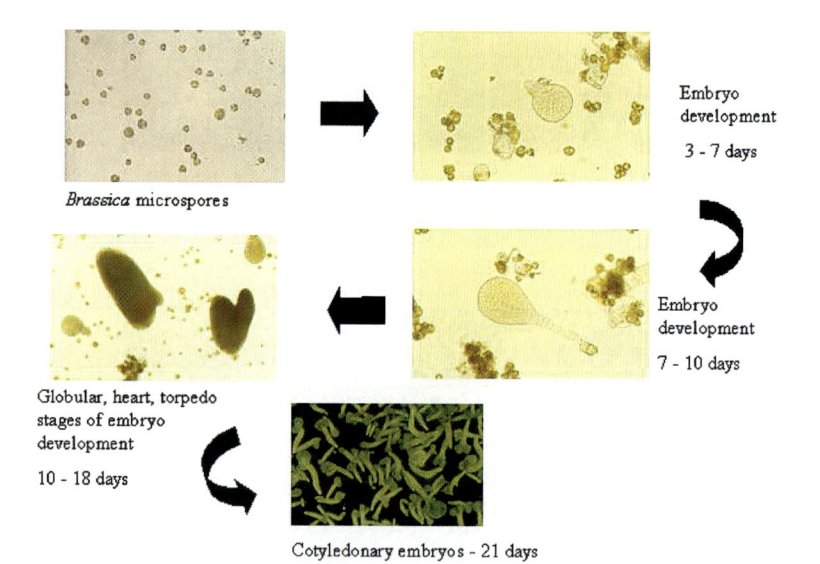

Figure 2.30-2. Embryo development from microspores of *Brassica*.

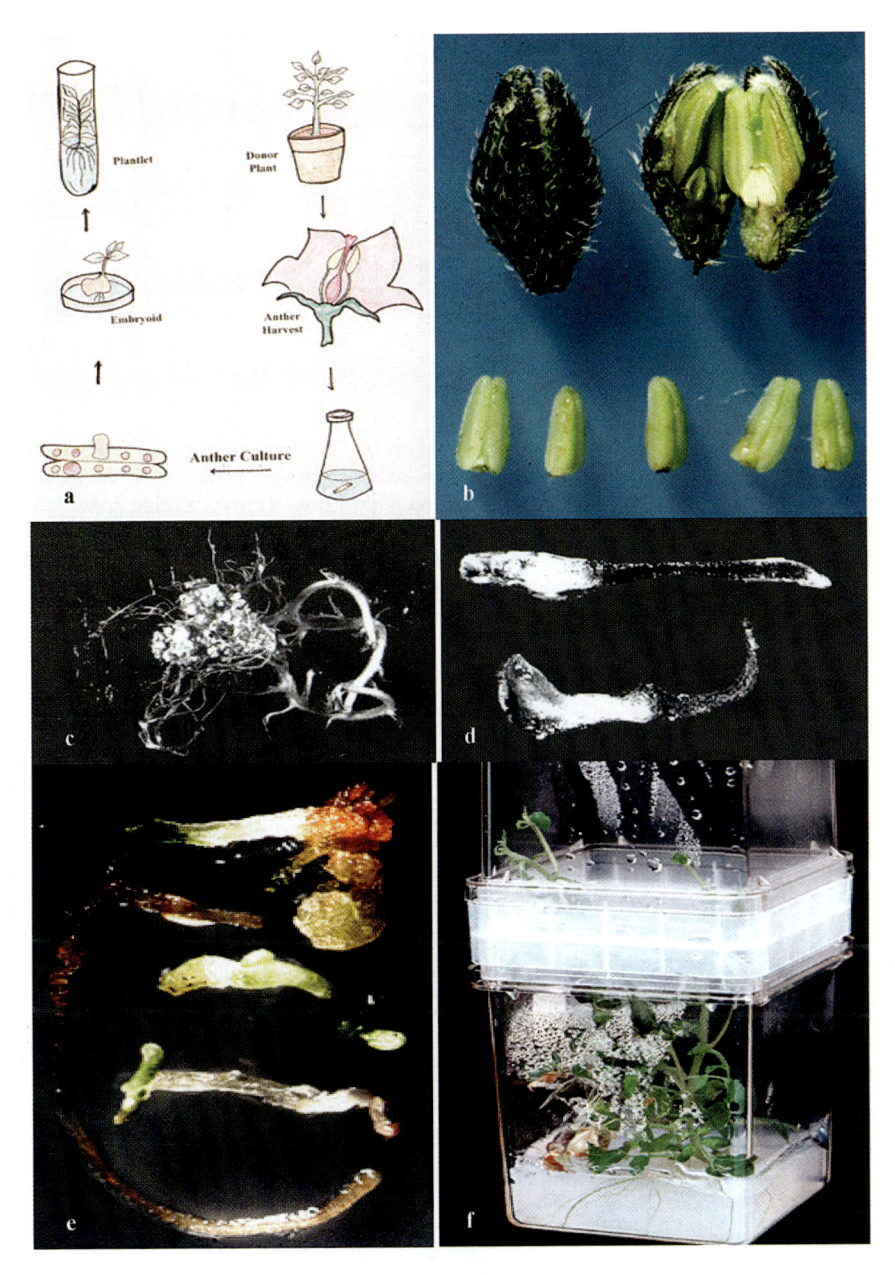

Figure 2.33-1. Illustration on developmental landmarks *in vitro* in anther culture of potatoes. a) Anther culture procedure; b) Anthers; c) Callus; d) Embryoids; e) Plantlets; f) Plants.

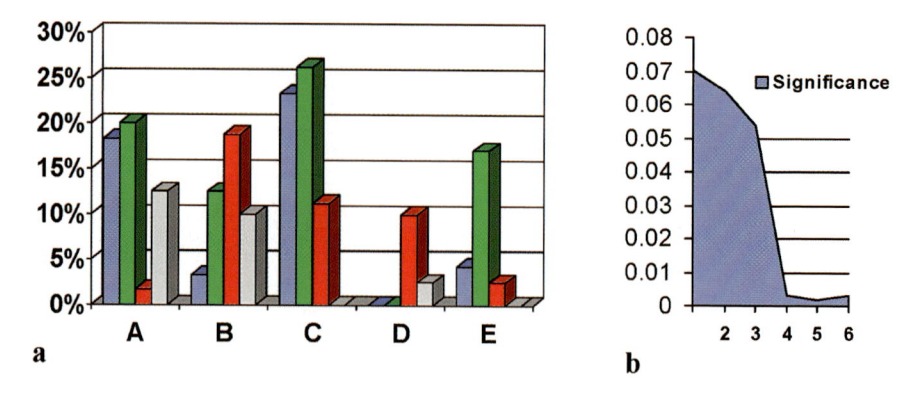

Figure 2.37-1. a) The within-line variation. Variation in embryo response between plants within lines (lines A-E including four plants per line). The figure also depicts the between-line variation; b) The within-line variation imposes a risk that ovules are collected from a low-producing donor. The figure shows the rapid risk-reduction as the number of donor plants increase. Thus, to reduce the effect imposed by the within-line variation, ovules should be collected from 4 or more plants, e.g. 20 ovules per Petri dish, 4 Petri dishes per plant and ≥4 plants. This procedure also allows a fair estimation of the line response.

Figure 2.37-2. a) Section through *Beta vulgaris* flower bud uncovering anthers and ovule; b) Petri dishes enclosed by plastic bags facilitate the handling of large numbers; c) Haploid embryo emerging from the brown ovule; d) Haploid shoot culture, characterised by narrow leaves and slender appearance.

Figure 2.39-1. Inoculated onion flowers after two days of culture.

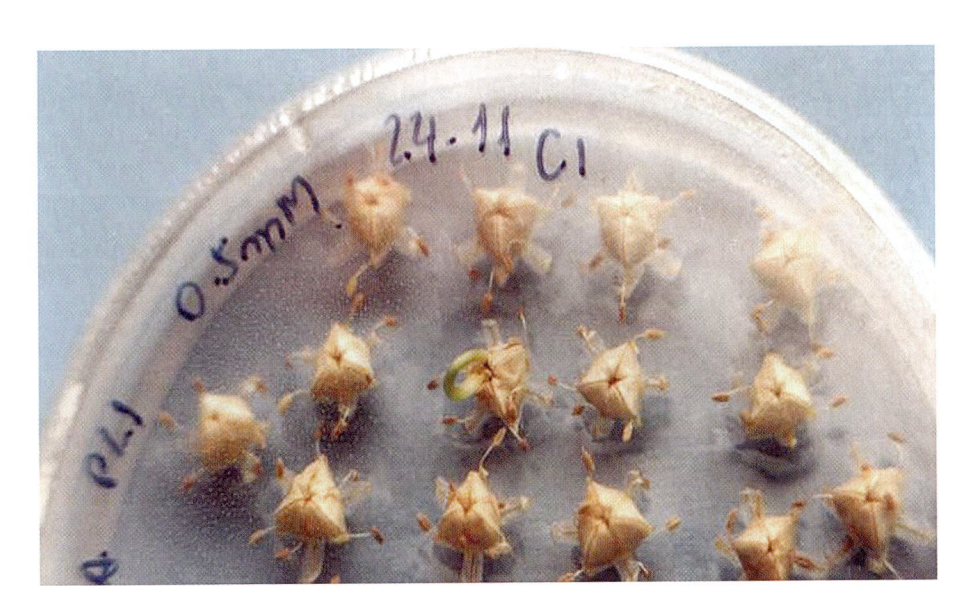

Figure 2.39-2. Gynogenic embryos breaking through the ovary wall.

Figure 2.41-1. Anther culture in apple. a) Flower buds at the picking stage for *in vitro* androgenesis; b) Two anther cycles with ten anthers out of each cycle; c) Embryo at the torpedo stage; d) Formation of secondary embryos; e) Adventitious shoot formation; f) Rooted androgenic shoots.

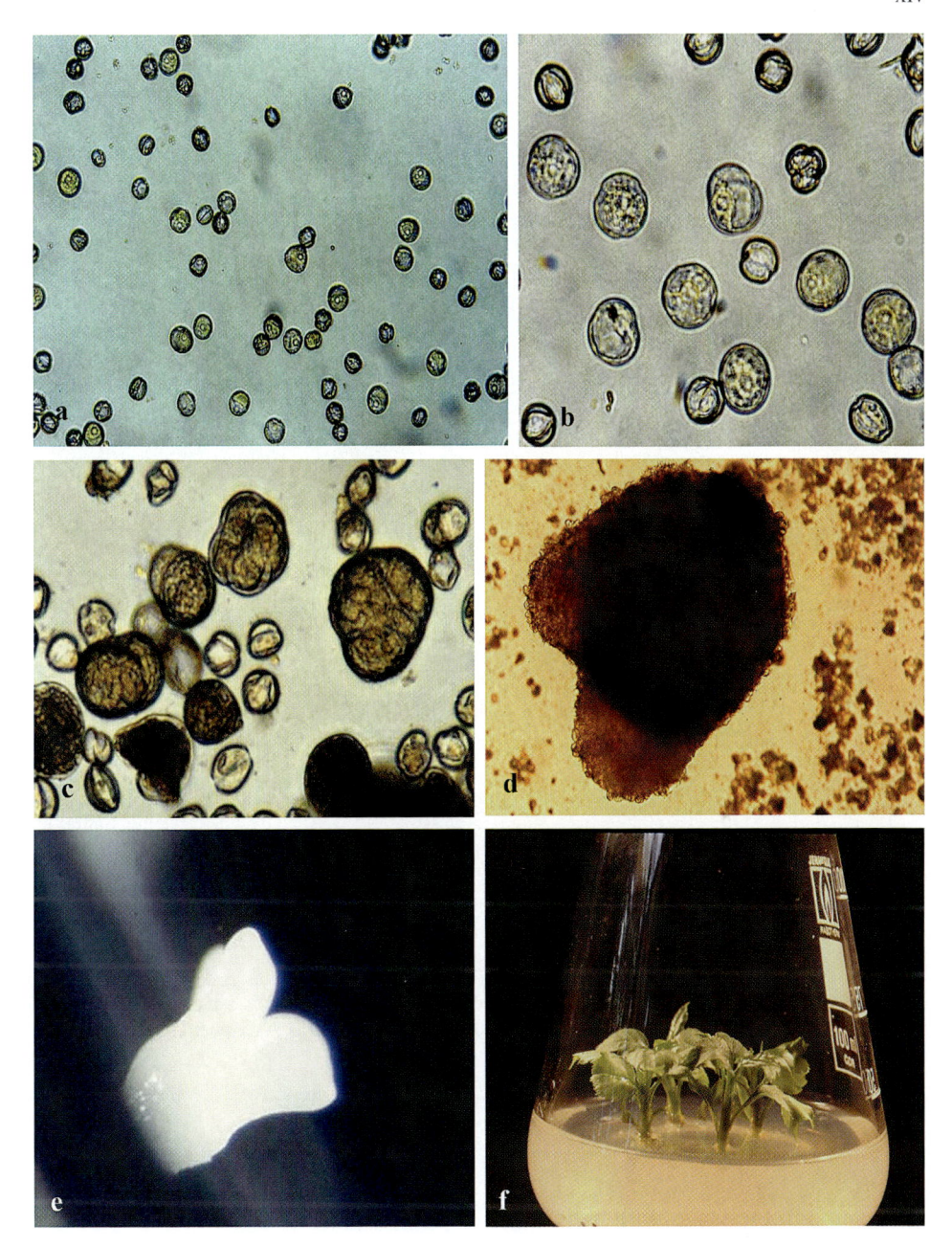

Figure 2.41-2. Microspore culture in apple. a) Freshly isolated microspores; b) Population of microspores in medium B after starvation; c) Induction of sporophytic divisions after 3-4 weeks in induction medium; d) Proembryo formation after 5-8 weeks; e) Torpedo-shaped embryo after 12 weeks of culture; f) Regenerated shoots in the phase of micropropagation.

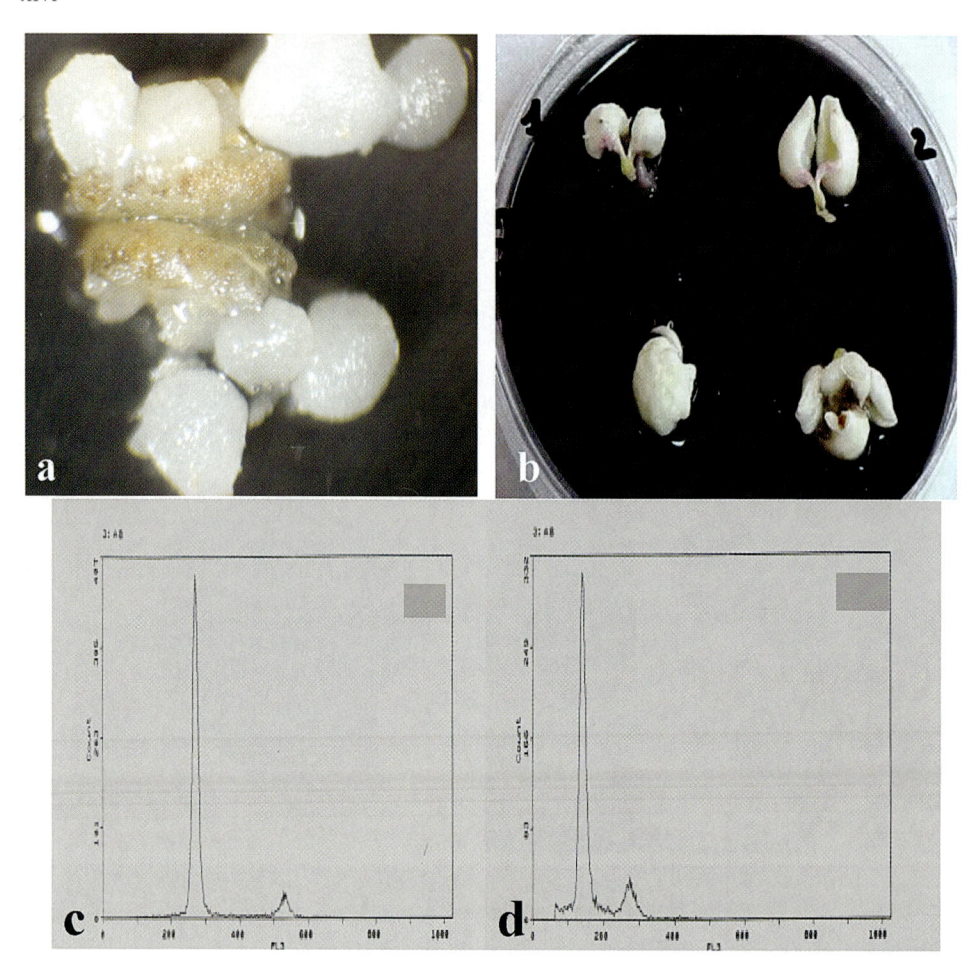

Figure 2.43-1. a) Oak embryos induced in anther culture; b) Anther embryos in development medium; c) Flow cytometry graphic of a diploid control with the G1 DNA peak (2C) set at channel number 250; d) Flow cytometry graphic of an anther embryo showing a G1 DNA peak at channel 125 (1C).

1
Production of doubled haploids in crop plants.
An introduction

K.J. Kasha and M. Maluszynski[1]
Department of Plant Agriculture, Biotechnology Division, University of Guelph, Guelph, Ontario Canada N1G 2W1; [1]*Plant Breeding and Genetics Section, Joint FAO/IAEA Division, P.O. Box 100, Vienna, Austria*

Terminology

Haploid is the general term for plants (sporophytes) that contain the gametic chromosome number (n). Thus, in a diploid sporophytic ($2n$) species the haploid could also be called *monoploid (x)* as they have only one set of chromosomes. In polyploid species, the haploids (n) have more than one set of chromosomes and are *polyhaploids*. The haploid plant from an *autotetraploid* ($4x$) with four sets of one genome was originally called a *dihaploid* (because $2n=2x$). It is important to note that when the chromosome number of a haploid is doubled, it should be called a *doubled haploid* (DH) and not a dihaploid. The dihaploid is not homozygous as it represents two chromosome sets selected from four sets in the autotetraploid, whereas the doubled haploid from a monoploid or an *allohaploid* should be completely homozygous. Since there has always been concern about the terminology used in haploidy, R. A. de Fossard was asked to formulate guidelines on terminology to be presented at the 1st International Symposium on *Haploids in Higher Plants* in 1974 (Kasha, 1974). A group of scientists met to discuss this terminology guideline and the consensus was adopted by the Symposium and published in its Proceedings (de Fossard, 1974). Thus "x" or multiples of "x" should be used to denote the number of sets of chromosomes while "n" and "$2n$" are used for the gametic and sporophytic chromosome numbers respectively. For example, a bread wheat plant (*Triticum aestivum* L.) could be described as $2n=6x$. Its usual haploid gamete would be $n=3x$ and the sporophytic plant from the gamete would be $2n=3x$. In addition to *euhaploids* with complete sets of chromosomes, one could also produce or find *aneuhaploid* plants where the chromosomes do not represent balanced complete sets of chromosomes (e.g. $2n=3x$-1 or $2n=3x$+1). For the many other possible variations please see de Fossard (1974). Other valuable terms are *androgenetic* haploids that are derived from the male gamete and *gynogenetic* haploids that are derived from the female gamete.

History

Riley (1974) provided an excellent introduction to haploidy for the 1st International Symposium that also covered some terminology but included a historical perspective for those working in this area. Riley noted the first recognition of haploids in plants was by A.D. Bergner in 1921 working with *Datura stramonium* L.and this was reported in Science by Blakeslee *et al.* (1922). Subsequently haploids were reported in many species but at low and variable frequencies. Kimber and Riley (1963) in a review pointed out that the two major problems limiting the use of haploids were their low frequencies and genotype differences in response. Thus, research was directed at finding methods of producing haploids in crop species to address these problems.

In the diploids of cross-pollinating species like maize or rye, the haploids may uncover deleterious genes, possibly resulting in abnormal plants and lower viability. In diploid inbreeding species, these deleterious factors have been bred or selected out so that the haploids are usually quite normal in morphology and vigorous. These trends also exist in polyploids where outbreeding species tend to be autoploids while inbreeding species have evolved towards alloploidy. The haploids from autopolyploids suffer from inbreeding depression and some sterility whereas the haploids from allopolyploids are more vigorous but sterile until their chromosome number is doubled, then they are fertile.

The 1st International Symposium in 1974 on Haploids in Higher Plants requested authors to provide papers that summarized the literature on haploidy prior to that time. There are still a few copies of these Proceedings available and one may be obtained by writing or e-mailing kkasha@uoguelph.ca

Haploids systems

The first attempt to use haploidy in breeding appears to be Chase (1952) who selected the low frequency of *parthenogenic* haploids (egg cell develops into an embryo without fertilization) in maize and then applied chromosome doubling treatments to produce inbred lines. Many methods were tried in various labs to improve the frequency of parthenogenesis. These include pollen irradiation prior to use in pollination, selection of seeds with twin embryos, sparse pollination, alien cytoplasm, wide hybridization and genetic stocks that influenced haploid frequencies or carried visible dominant markers for coloured embryos or roots.

The remarkable discovery that haploid embryos and plants can be produced by culturing anthers of *Datura* (Guha and Maheshwari, 1964, 1966) brought renewed interest to haploidy. This was quickly attempted in many species but the frequencies were very low, relative to the large numbers of pollen per floret. Then in 1970, Kasha and Kao reported haploid production in barley following wide hybridization and the subsequent preferential elimination of the wild species chromosomes during early embryogenesis. This provided a system that produced larger numbers of haploids across most genotypes. This method quickly entered into breeding programs and showed promise when field evaluated in comparison to traditional barley breeding methods (Choo *et al.*, 1985). To-date, well over 60 barley cultivars have been produced around the world using that method. Subsequently, the pollination of wheat with maize (Laurie and Bennett, 1988), sorghum or millet pollen has enabled the chromosome elimination method of haploid production to be used in many wheat breeding and research programs. However, this procedure so far has been restricted to just a few crops.

Chase (1962) proposed the *Analytic* breeding method for autopolyploid species where the chromosome number of an autotetraploid would be reduced to the diploid level for breeding and selection and then returned to the tetraploid level for production. This system employed haploids and 2*n* gametes (unreduced chromosome number) in various ways. Hougas *et al.* (1958) demonstrated the use of wide crosses between wild 2*x* and cultivated 4*x* potato to produce *dihaploids* of potato in quite adequate quantities, apparently through parthenogenesis. Through the use of systems to produce 2*n* gametes, chromosome numbers could be returned to the tetraploid level following breeding at the diploid level (Mendiburu *et al.*, 1974).

The potential for higher frequencies of haploids from immature pollen culture has enticed extensive research and some successes. Today, isolated microspore culture is being utilized in breeding programs in many crop species as will be seen in the protocols listed in this manual. One advantage is that a high frequency of chromosome doubling can take place early in the culture phase, resulting in completely fertile doubled haploid plants. Crops where such an system is quite extensively used are rapeseed (Canola), tobacco, wheat and barley.

Other systems for haploid production are *ovule culture* as used in sugar beets and onions and *semigamy* used in cotton. Semigamy represents an incomplete fertilization process where the sperm enters the egg cell, and embryo development is induced. However, the nuclei do not fuse and the resulting embryo develops as a chimera with some sectors haploid from the male and others haploid from the female gamete. By the use of visible marker genes, these sectors can be distinguished and the haploids isolated. Ovule culture has been possible in some species and has been utilized in sugar beet and onion breeding. Since there is usually only one egg cell per floret, ovule culture has much less potential than microspore culture. However in the onion, there are three carpels per floret and each with two egg cells which makes ovule culture more attractive.

Efficiency of haploid systems and applications

Of the two major systems in use today for breeding, namely wide hybridization and anther/microspore cultures, there does not appear to be much difference in cost of haploid production or in the time and amount of labour required to produce them. The limitation of wide hybridization is that thus far, it is restricted to the cereals where the chromosome elimination system appears to operate or to the potato. Microspore culture should be feasible in most species but it has taken a long time to develop an efficient system in a limited number of crops. Each crop has different requirements and thus, the need for extensive research to develop a system. Never the less, microspore culture has more potential in view of the large numbers of microspores per floret versus one or a few egg cells. In barley and wheat where both systems have been fairly well developed, the yield of green plants from isolated microspore culture can be up to 100 times higher than from wide hybridisation in the most responsive genotypes. Wide hybridization has the advantage of being quite effective across genotypes and produces little or no induced variation from culturing the embryos. However, as the microspore culture systems have improved in crops, plants are being derived through embryogenesis rather than from callus and this should reduce the extensive culture induced variability found in earlier reports on anther and microspore cultures. With improvement of frequencies from microspores, the genotype variability in response has also been reduced so that the system is being used in breeding programs. The feature of a high frequency of spontaneous chromosome doubling in some crops which results in completely fertile doubled plants without subsequent doubling treatments is another advantage of microspore culture.

In breeding, the instant production of true breeding lines in diploid or allopolyploid species saves a number of generations in the breeding program. In addition, selection is much more effective on pure lines than on early generation segregating lines and quantitative trait evaluation can also begin earlier in the program, saving time and space. Barley was the first crop where extensive studies and comparisons of breeding methods were conducted. Being a self-pollinated diploid species it was also possible to study gene interactions and quantitative traits. Cho *et al.* (1985) summarized the work on barlcy breeding with haploids produced from the wide hybridisation/chromosome elimination method and provides a good review of the genetics and breeding with doubled haploids.

Besides for breeding, haploids have been advantageous in research areas such as mutation studies, gene mapping, genomics and as targets for transformation. In transformation, haploid embryos may be produced and used as targets for bombardment procedures or with

Agrobacterium. Of great interest is the targeting of haploid immature pollen (microspores) for transformation to obtain directly doubled haploid plants homozygous for the transgene(s). In cereals, the doubling of the chromosome number occurs by nuclear fusion after the first nuclear division when using mannitol pretreatments (Kasha *et al.*, 2001). Thus, it is important to get the foreign genes incorporated prior to this stage in order to obtain doubled haploids that are homozygous for the transgenes.

In this manual on haploid protocols we thought it would be of value to include summary chapters on applications of haploids in the various areas that we have just mentioned. Some information on applications will also be found in the protocol chapter introductions. For more extensive reviews on haploids, one may consult the series of five volumes on "*In Vitro* Haploid Production in Higher Plants" edited by Jain, Sopory and Veilleux (1996-1997).

References

Blakeslee, A. F., J. Belling, M. E. Farnham and A. D. Bergner, 1922. A haploid mutant in the Jimson weed, *Datura stramonium*. Science **55**: 646-647.
Chase, S. S., 1963. Analytic breeding of *Solanum tuberosum* L. Can.J.Genet.Cytol. **5**: 359-363.
Choo, T. M., E. Reinbergs and K. J. Kasha, 1985. Use of haploids in breeding barley. Plant Breed.Rev. **3**: 219-252.
de Fossard, R. A., 1974. Terminology in "Haploid" research. In: Haploids in Higher Plants: Advances and Potential. Kasha, K.J. (Ed.), Univ. of Guelph, Guelph, pp.403-410.
Guha, S. and S. C. Maheshwari, 1964. *In vitro* production of embryos from anthers of *Datura*. Nature **204**: 497.
Guha, S. and S. C. Maheshwari, 1966. Cell division and differentiation of embryos in the pollen grains of *Datura in vitro*. Nature **212**: 97-98.
Hougas, R. W., S. J. Peloquin and R. W. Ross, 1958. The potential of haploid potatoes for research and breeding. Amer.J.Potato Research **35**: 442.
Jain, S.M., S.K. Sopory and R.E. Veilleux (Eds). 1996-1997. *In Vitro* Haploid Production in Higher Plants. Vols. 1-5. Kluwer Academic Publ. Dordrecht.
Kasha, K. J. (Editor) 1974. Proc. 1st International Symposium on Haploids in Higher Plants: Advances and Potential. Univ. of Guelph, Guelph, pp.421.
Kasha, K. J., T. C. Hu, R. Oro, E. Simion and Y. S. Shim, 2001. Nuclear fusion leads to chromosome doubling during mannitol pretreatment of barley (*Hordeum vulgare* L.) microspores. J.Exp.Bot. **52**(359): 1227-1238.
Kasha, K. J. and K. N. Kao, 1970. High frequency haploid production in barley (*Hordeum vulgare* L.). Nature **225**: 874-876.
Kimber, G. and R. Riley, 1963. Haploid angiosperms. Botanical Rev. **29**: 480-531.
Laurie, D. A. and M. D. Bennett, 1988. The production of haploid wheat plants from wheat x maize crosses. Theor.Appl.Genet. **76**: 393-397.
Mendiburu, A. O., S. J. Peloquin and D. W. S. Mok, 1974. Potato breeding with haploids and *2n* gametes. In: Haploids in Higher Plants. Kasha, K.J. (Ed.) University of Guelph, Guelph, pp.249-258.
Riley, R., 1974. The status of haploid research. In: Haploids in Higher Plants: Advances and Potential. Kasha,K.J. (Ed.) Univ. of Guelph, Guelph, pp.3-9.

2.1
Doubled haploid production in barley using the *Hordeum bulbosum* (L.) technique

P. Hayes, A. Corey and J. DeNoma
Department of Crop and Soil Science, Oregon State University, Corvallis, OR 97331 USA

Introduction

In 1989 we reported modifications of the *Hordeum bulbosum* technique that, in our lab, led to greater efficiencies and less germplasm specificity (Chen and Hayes, 1989). Over the years, we continued to refine these extra steps (tiller culture, floret culture, and *in vitro* vernalization of haploid plantlets, regardless of growth habit) because they gave us better results. When the *H. bulbosum* system is operating well, we can predict with relative certainty that about half of our emasculations will lead to fertile green plants. We offer the following protocols to the barley research community with the caveat that they worked well for us, but that additional modifications will probably be necessary in other environments. A simple factor such as ambient relative humidity – which is greater than 80% for seven months of the year in western Oregon - can have a profound effect on DH production and could emerge as a principal constraint in drier environments.

Protocol

Donor plants and growth conditions

Healthy plants are prerequisite for success. Our greenhouses are antiquated and adhere in only the most general way to our target settings. We strive for temperatures of 16°C night and 18°C day. However, temperature control is so poor that we produce DH lines only between October and May, as greenhouse temperatures soar in spring and summer. As mentioned earlier, ambient humidity is high, but we further supplement with a fog system in our greenhouse, which enhances the survivability of haploids when they are first taken out of culture, and later, after they are treated with colchicine. We provide supplemental illumination with high intensity lights, set for a 16 h light/8 h dark cycle. Over the years, we have used a range of different types of fixtures and bulbs, with no apparent differences. We take every precaution to prevent insect infestations and fungal disease epidemics, as chemical controls for these pests often lead to sterility or abnormalities in floral structure or development.

We control aphids by the use of Marathon, a soil applied, systemic insecticide (Olympic Horticultural Products, Marathon 1% Granular, Imidacloprid). We treat at label rates once plants are transplanted to soil and well established. Plants at this stage usually have 5-6 leaves. In the event that Marathon is not applied, we use a nicotine smoke fumigant (Plant Products Corporation) at the label rate. For control of thrips, we use a predatory mite biological control agent (*Amblyseius cucumeris*) (Evergreen Growers Supply). The principal disease we have is powdery mildew, which we have controlled by the use of Strike fungicide (Olympic Horticultural Products, Strike 25 WDG, Triazole), following label rates.

H. bulbosum plants

During peak DH production periods we would devote an entire greenhouse bench to pollen production. We use bulk pollen from three clones. Clones 2921-1 and 2920-4 were obtained from Dr. Duane Falk, University of Guelph. Clone CRD was obtained from Dr. Richard Pickering, Crop and Food Research, New Zealand. We divide the *H. bulbosum* clones annually, and vernalize the divided plant material annually at 6°C for eight weeks, with 8 h light/16 h dark cycle.

The DH production process

As shown in Table 2.1-1, the DH production system involves nine steps. We use the vernalization step for all plant material, both winter and spring habit, as it gives stronger plants that are better able to withstand the colchicine treatment.

Table 2.1-1. Overview of doubled haploid production using the *Hordeum bulbosum* technique, as applied at Oregon State University

Steps	Duration	Procedure
Emasculation	48 h	pollination
Pollination	24 h (may go 48 h)	foret culture
Floret culture	9-11 days	embryo rescue
Place in dark	7-10 days	first growth check
Transfer tube to light	2 leaves/top of tube	vernalization
Vernalization	6-8 weeks	transplant to greenhouse
Transplant	2-3 weeks	colchicine
Colchicine	2 months	harvest seed

Emasculation

We use both the scissor emasculation technique (floret cut perpendicular to floral axis, removing top ¾ of anthers) and the full emasculation technique (floret cut perpendicular to floral axis and anthers removed). We found no adverse effects from scissor emasculations, which are much faster than the full emasculation technique. This technique is not recommended for critical *H. vulgare* x *H. vulgare* crosses, as some self-pollination may occur. With the *H. bulbosum* crosses, the endosperm is watery whereas the endosperm of the rare seeds resulting from *H. vulgare* self-pollination is hard. Florets appeared to have less damage and there was no increase in contamination with the scissor technique. There was no apparent increase in diploid embryos. Tag the stem and cover each emasculated head with glassine bag.

Media Recipes

Table 2.1-2. Summary of media and corresponding stock solutions for doubled haploid production of barley, using the *H. bulbosum* technique as applied at Oregon State University

Media	Stock solutions
Barley Spike Medium	A - F
Barley Floret Culture Medium	G - J
Barley Embryo Rescue Medium	K - M

Tiller culture

After emasculation, cut the tiller at its base and place it in a clean container of distilled water. Surface sterilize the stems and place tillers in spike medium. Trim the tiller, leaving two or more nodes below the head. Remove excess leaf material. Surface sterilize the stem with a 10 minute wash in 20% Clorox bleach solution with Tween 20 wetting agent. Rinse in sterile distilled water. Place the tillers in a beaker of sterile distilled water. About an inch above the stem's base, make a 45-degree cut across the stem, pressing the culm against the side of the beaker with the scalpel. Dip scalpel in 70% ETOH between each severing event to prevent cross contamination with bacteria which may be present inside the stem. Place the cut tiller in a tube of spike medium and close the top tightly with parafilm. Return the spike to the greenhouse to wait 24 to 48 hours before pollinating with *H. bulbosum*.

Barley Spike Medium - Stock Solutions
Stock Solution 'A'
    $CaCl_2$ x $2H_2O$ 3,000 mg
    $CoCl_2$ x $6H_2O$ 1 mg
    (make solution of 100mg $CoCl_2$ x $6H_2O$ in 100 ml dH_2O and add 1 ml to stock)
    KI   10 mg
    Bring volume to 500 ml with dH_2O (deionized H_2O)

Stock Solution 'B'
    KH_2PO_4 3,000 mg
    K_2HPO_4 500 mg
    Na_2MoO_4 x $2H_2O$ 10 mg
    Bring volume to 500 ml with dH_2O

Stock Solution 'C'
    MgSO$_4$ x 7H$_2$O   3,000 mg
    CuSO$_4$ x 5H$_2$O   1 mg
    (make solution of 100 mg CuSO$_4$ x 5 H$_2$O in 100 ml dH$_2$O and add 1 ml to stock)
    MnSO$_4$ 110 mg
    ZnSO$_4$ x 7H$_2$O   100 mg
    Bring volume to 500 ml with dH$_2$O

Stock Solution 'D'
    H$_3$BO$_3$ 100 mg
    Bring volume to 500 ml with dH$_2$O

Stock Solution 'E'
    KNO$_3$ 20 g
    NH$_4$NO$_3$ 10 g
    Bring volume to 500 ml with dH$_2$O

Stock Solution 'F'
    FeSO$_4$ x 7H$_2$O   557 mg (+ 20 ml H$_2$O)
    Na$_2$EDTA   745 mg (+ 100 ml H$_2$O)
    Heat Na$_2$ EDTA
    Mix with FeSO$_4$ over stirrer and heat until fully dissolved and turns yellow
    Bring volume to 500 ml with dH$_2$O
    Store in dark bottle

Preparation of Barley Spike Medium - 1000 ml
    – Use:
        Stock Sol. 'A'   25 ml
        Stock Sol. 'B'   25 ml
        Stock Sol. 'C'   25 ml
        Stock Sol. 'D'   25 ml
        Stock Sol. 'E'   25 ml
        Stock Sol. 'F'   25 ml
    – Bring volume to 1 L with dH$_2$O
    – pH to 5.7
    – Dispense into tubes (25 ml/tube)
    – Autoclave

Pollination of *H. vulgare* spp. *vulgare* with *H. bulbosum*
Collect *H. bulbosum* pollen in the morning by gently tapping mature heads over a glass Petri dish. Using a paintbrush, gently brush pollen over each floret. Record the pollination date on the glassine bag and place it back over the head. Leave the pollinated tiller in greenhouse for another 24 h before bringing it into the lab to establish it in floret culture. In some crosses, embryo numbers were increased by repeatedly pollinating heads at 24, 48 and 72 h following emasculation.

Floret culture
Supplies
Laminar flow hood setup and running; 70% ethanol for disinfecting hood surface; 1 non-sterile 100 ml graduated cylinder with 20% Clorox bleach solution and Tween 20; sterile 100 ml graduated cylinders (one for each set of plants); 1+ liters sterile distilled water; 30 or 50

ml sterile beakers with sterile water; waste water container; sterile Petri dishes; sterile instruments, including: scissors, forceps, scalpel with blade; floret medium.

Sterilization
Gentle agitation for 10 minutes in 20% Clorox bleach solution with 2 drops of Tween. Then use multiple sterile water rinses.

Procedure
1. Trim: Bring the pollinated spikes into laboratory. Away from the laminar flow hood, carefully trim off all leaf sheaths, awns, anthers and other unnecessary plant parts, which may harbor contaminants. Cut the stem to about 15 cm length. If you must keep spikes individually labeled, color code each by wrapping a small piece of colored tape around the stem at the base of the head. (Writing may bleach off of the tape during washing).

2. Surface sterilization: Place inflorescence head-down into the 100 ml graduated cylinder containing bleach solution. Immediately start the timer. Cover the top of cylinder with parafilm and gently agitate continuously for 10 minutes.

3. Rinse: Since the plant material is now assumed to be sterile, from this point on work in the laminar flow hood using good sterile technique. Immediately following the 10 minutes wash, use sterile forceps to transfer the spikes, head-down, into a sterile graduated cylinder of sterile water. Gently agitate spikes. Pour off the rinse water and add fresh sterile rinse water. Rinse 5 to 6 times in this manner.

4. Dry: Stand the inflorescence upright in the hood with the base of their stems in small beakers of sterile water, like little barley bouquets. Put no more than three heads in each beaker. Allow the heads to dry in the hood's air flow for about two hours.

5. Place on floret medium: Once dry, place individual florets on medium. Trim off the stem and lay the inflorescence in a sterile Petri dish. Cut each floret such that a piece of the rachis remains attached to the floret. Stand the florets upright in medium. Place 10 to 20 florets per dish, fewer if contamination is a problem. Label and date each dish. Seal the dishes with Parafilm and place in the growth chamber at 18°C, with a 16 h light/ 8 h dark photoperiod regime.

6. Check for contamination: Check plates daily for contamination. If contamination appears, transfer the uncontaminated florets to a Petri dish of fresh medium. Use care to avoid spreading the contamination.

Barley Floret Culture Medium – Stock Solutions
Stock Solution 'G'
  $MnSO_4 \times 4H_2O$ 440 mg
  $ZnSO_4 \times 7H_2O$ 150 mg
  H_3BO_3 160 mg
  KI   80 mg
  Bring volume to 100 ml with dH_2O

Stock Solution 'H'
    Glycine   200 mg
    Thiamine   100 mg
    Pyridoxine   50 mg
    Nicotinic Acid   50 mg
    Bring volume to 100 ml with dH_2O
    To store: freeze vitamins in pre-measured amounts

Stock Solution 'I'
    2,4-D   120 mg; (dissolve in 1 ml 95% ethanol)
    Bring volume to 100 ml with dH_2O (Best to make fresh or store for < 1 week)

Stock Solution 'J'
    Kinetin (6-fufurylaminopurine)   50 mg
    (dissolve in 95% ETOH or 1 ml of HCl and slowly add water).
    Bring volume to 100 ml with dH_2O (Best to make fresh or store for < 1 week)

Preparation of Barley Floret Culture Medium – 1000 ml
- Weigh out the six macro salts listed below:
    $(NH_4)_2SO_4$ 463 mg
    KNO_3 2,830 mg
    KH_2PO_4 400 mg
    $MgSO_4 \times 7 H_2O$ 185 mg
    $CaCl_2 \times 2 H_2O$ 166 mg
    FeNaEDTA   40 mg
- Bring to volume in 900 ml of H_2O
- Mix thoroughly
- Add Stock Sol. 'G'   1 ml
- Add sucrose   90 g
- Adjust pH to 5.8
- Add agar   8.5 g
- Autoclave
- Prepare the following addition:
    Stock Sol. 'H'   1 ml
    Myo-Inositol   2 g
    L-glutamic acid   160 mg
    Stock Sol. 'I'   1 ml
    Stock Sol. 'J'   1 ml
    Gentamicin sulfate   75 ul
- Bring volume to 100 ml
- Filter sterilize
- Allow autoclaved solution to cool to about 65°C
- Add filter-sterilized solution to autoclaved medium and stir
- Pour about 10 ml per 60x15mm Petri plate

Embryo rescue
Embryos are rescued after 9 to 11 days in floret culture.

Supplies
Laminar flow hood setup and running; dissecting microscope; 70% ethanol for disinfecting hood surface; sterile Petri dishes; embryo medium; sterile instruments, including fine forceps and scalpels.

Procedure
1. Set up: Allow embryo medium to warm to room temperature on laboratory bench. Surface sterilize the hood and dissecting microscope with 70% ethanol. Do not get ethanol on the microscope lenses. Flame instruments three times and then set on sterile Petri dishes at the back of the hood. Check each floret culture dish for contamination that could be spread during embryo rescue.

2. Embryo dissection: Using good sterile technique, work one dish of florets at a time. Use fine forceps to remove the seed from the tissue covering it by giving the seed a sharp jerk. Place seeds in a sterile Petri dish. Move the dish to the dissecting microscope stage. Carefully dissect out each embryo and place it on the embryo medium, three embryos per tube. Alternatively, one can use cell wells (see Comment at the end of this section). If possible, lay the scutellum side of the embryo on the medium.

Barley Embryo Regeneration Medium - Stock Solutions
Stock Solution 'K'

$NaH_2PO_4 \times H_2O$	3,000 mg
KNO_3	50,000 mg
$(NH_4)_2SO_4$	2,680 mg
$CaCl_2 \times 2H_2O$	3,000 mg
$Na_2\ EDTA \times 2H_2O$	820 mg
$FeSO_4 \times 7H_2O$	548 mg
$MgSO_4 \times 7H_2O$	5,000 mg

Bring volume to 1 L with dH_2O

Stock Solution 'L'

$MnSO_4 \times 4H_2O$	1,000 mg
$ZnSO_4 \times 7H_2O$	200 mg
H_3BO_3	300 mg
$Na_2MoO_4 \times 2H_2O$	25 mg
KI	75 mg
$CuSO_4 \times 5H_2O$	2.5 mg
$CoCl_2 \times 6H_2O$	2.5 mg

For copper and cobalt, make a stock solution of 250 mg of each in 100 ml of water and then add 1 ml of this stock to Stock Solution 'L'.
Bring volume to 500 ml with dH_2O

Stock Solution 'M'

Nicotinic Acid	100 mg
Thiamine HCl	100 mg
Pyridoxine HCl	100 mg

Bring volume to 500 ml with dH_2O
Freeze vitamins in individual pre-measured quantities

Preparation of Barley Embryo Regeneration Medium (after Jensen's modified B5) – 1000 ml
- Use:

Stock Sol. 'K'	50 ml
Stock Sol. 'L'	5 ml
Stock Sol. 'M'	5 ml
Myo-inosotol	100 mg
Sucrose	30 g

- Bring volume to 1 L
- Adjust pH to 5.8
- Add agar   7.5 g
- Bring to full boil to dissolve agar
- Dispense into 10 ml/25 mm x 100 mm tubes
- Autoclave
- Cool on slant
- If wells are used dispense 2 ml/well

3. Germination: Place tubes or cell wells in boxes (dark) in the growth chamber at 18°C, with a 16 h light/8 h dark photoperiod regime. Check the embryos after seven days and move embryos that have germinated to the light in the same growth chamber.

4. Transfers: If the embryos are germinated in cell wells, they must be transferred to test tubes as they germinate. Use long forceps, so your fingers never come near the mouth of the test tube. Transfer each embryo into a tube. Slow growing plantlets that have been on the same medium for longer than one month seem to obtain a boost by being transferred to a tube of fresh medium. Allow the transferred plant to re-establish growth for at least one week before moving it into vernalization.

5. Vernalization: When haploid plants have two leaves and are pushing the lid of the test tube, move the tube to the vernalization chamber for six to eight weeks.

Comment
Embryo rescue medium may be in either test tubes or cell wells. If contamination is a problem or if the individual is uncertain of their sterile technique, test tubes offer fewer chances for spreading contamination. Some operators find that embryo rescue is quicker when using cell wells. Cell wells may also allow recovery of more late-germinating embryos and weaker plantlets. If using test tubes, place up to three embryos per tube. If using cell wells, place two embryos per well. Once embryos germinate, they must be transferred from the cell well to a tube, which adds an extra step.

Haploid plant transfer
After vernalization, haploid plantlets are transferred to vermiculite. Gently pull plantlet from medium and rinse medium from roots. Transfer plantlets to moistened fine vermiculite and water with a dilute complete liquid fertilizer.

Colchicine doubling
Prepare colchicine solution

Colchicine	1 g
DMSO	20 ml
10 ppm GA_3	10 ml
Tween 20	10 drops

H_2O	1 L

Comment
Mix colchicine solution in a fume hood. Do not inhale colchicine powder. Wear gloves. Use only designated glassware. We obtain colchicine in 1 g containers from Sigma, so does not need to be weighed. Dump the container's entire contents into solution, rinsing the container twice. We prepare the GA_3 beforehand in a stock solution of 10 mg GA_3 per 1 L dH_2O.

Treatment
Submerge crowns in colchicine solution and place under growth lights for 5 hours. Rinse 15 minutes in running tap water.

Chromosome doubling procedure
1. Establish plants in greenhouse: Following vernalization, remove the plantlets from tissue culture into peat pots with vermiculite in the greenhouse. Cover the plants with netting and place them under mist. The covers can be removed after about a week. Colchicine the plants when they become established and hardy enough to survive the treatment, but before they become too large.

2. Colchicine treatment: Make sure you have a system to keep plants properly labeled throughout the treatment process. We label plastic stakes with cross codes and maintain groups of five plants with an appropriately labeled stake throughout the trimming, colchicine treatment, and washing phases. Remove plants from the vermiculite and wash and trim their roots to about 1.5 cm. Cut away excess leaves and trim tips of leaves. Fill test tubes about 1/3-full with the colchicine solution. About five plants can fit into each tube. Submerge the crowns in solution and leave them for five hours. During treatment, plants should be under growth lights and temperature should be approximately 16°C.

3. Rinse: After five hours, remove plants from colchicine and stand them in the holes of the test tube rack used for washing. Set the rack in the designated washing dish. (We use a 5 cm deep, white porcelain tray.) Rinse plants in the sink under running tap water for at least 15 minutes. Pour used colchicine into a properly labeled waste container. Wear gloves while handling treated plants. Replace plants into vermiculite and place them back in the greenhouse. Cover the plants with netting and place under mist.

4. Safety precautions: Because of the toxicity of colchicine, a separate and well-labeled work area should be set aside for colchicine treatments only. All containers used in the process should be designated for this use only. Take special care not to contaminate the work area with spills, etc. Pour the used colchicine into clearly labeled bottles for hazardous waste disposal (use the orange labels provided by OSU). Used bench paper and latex gloves should be sealed in garbage bags before placing them into the dumpster. Always wear gloves and protective clothing, including long sleeves and a plastic apron. Remove and properly discard your gloves before touching things like doorknobs and telephones. If colchicine contacts your skin, immediately flush skin with running water.

Efficiency and Applications

The *H. bulbosum* technique for producing doubled haploids of barley has been a cornerstone of the Oregon State University Barley Breeding and Genetics Program. We have released three varieties developed by this technique ('Strider', a winter habit six-row; 'Orca', a spring habit two-row; and 'Tango', a spring habit six-row). We have also used the technique to produce an array of genetic stocks for structural and functional genomics research. These include the principal mapping populations of the North American Barley Genome Project (http://www.css.orst.edu/barley/nabgmp/nabgmp.htm) and stocks developed through marker-assisted selection (Toojinda et al., 1998; Zhu et al., 1999).

We have made this extensive use of the *H. bulbosum* technique due to its functionality across a broad range of germplasm and its relative simplicity. Large-scale production of DH lines by the *H. bulbosum* technique is expensive and time-consuming. At our peak, we were producing up to two thousand DH lines per year, with a full-time skilled technician dedicated to the endeavor.

Acknowledgement
The presented protocols and procedures were prepared by Jeanine DeNoma, based on the contributions of many colleagues who spent long hours in the greenhouse and laboratory. Special thanks to Dr. Fuqiang Chen, who pioneered the floret culture procedure, Aihong Pan, who further modified the procedure achieved truly high throughput, Kurt Farris, who produced some very important doubled haploids as an undergraduate student, Mary Jo Lundsten, who managed the laboratory with grace and serenity, and Claudio Jobet, who turned a special project into a cornerstone of our program by producing the Oregon Wolfe Barley mapping population (http://www.css.orst.edu/barley/WOLFEBAR/WOLFNEW.HTM).

References

Chen, F.Q. and P.M. Hayes, 1989. A comparison of *Hordeum bulbosum*-mediated haploid production efficiency in barley using *in vitro* floret and tiller culture. Theor.Appl.Genet. **77**: 701-704.

Toojinda, T., E. Baird, A. Booth, L. Broers, P. Hayes, W. Powell, W. Thomas, H. Vivar and G. Young, 1998. Introgression of quantitative trait loci (QTLs) determining stripe rust resistance in barley: an example of marker-assisted line development. Theor.Appl.Genet. **96**: 123-131.

Zhu, H., G. Briceno, R. Dovel, P.M. Hayes, B.H. Liu, C.T. Liu and S.E. Ullrich, 1999. Molecular breeding for grain yield in barley: an evaluation of QTL effects in a spring barley cross. Theor.Appl.Genet. **98**: 772-779.

2.2
The *Hordeum bulbosum* (L.) method

P. Devaux
Florimond Desprez, Biotechnology Laboratory, P.O. Box 41, 59242 Cappelle en Pévèle, France

Introduction

The development of superior barley cultivars has been a continuing dynamic breeding process for many different characters. Being primarily self-pollinated, barley landraces and modern cultivars have been mostly homozygous and several methods of breeding including pedigree, bulk, backcross, single-seed descent have been successful for many years. The newest and most important development in breeding methods of barley has been the recovery of high frequencies of doubled haploids (DH) i.e. homozygous lines in a single generation. Several methods are available for DH production in barley. The first to be reported in the early 1970s was the interspecific hybridization of cultivated barley (*Hordeum vulgare* L.) with *H. bulbosum* L. (Kasha and Kao, 1970). Haploid plantlets of *H. vulgare* are generated as a result of gradual elimination of *H. bulbosum* chromosomes from hybrid embryos between the two species. Scientists and breeders rapidly adopted the method because DH provide several advantages over standard selection and breeding methods. This chapter describes a protocol for the interspecific hybridization of *H. vulgare* with *H. bulbosum*, also called the *H. bulbosum* method that has been used at Florimond Desprez for twenty years to generate DH lines of barley. *H. bulbosum*. is a perennial outcrossing species found in the Mediterranean region. Although there are two cytotypes, diploid ($2n = 2x = 14$) and tetraploid ($2n = 4x = 28$), only the diploid form has been used to generate haploid plants of barley.

Protocol

Donor plants and growth conditions

The conditions in which donor plants have been raised are very critical to the success of the technique. Healthy, fast-growing and insect-free plants are a prerequisite to ensure fertilization, high seed set and embryo quality i.e. higher efficiencies. Although most predators can be easily controlled, some e.g. insects and especially California thrips (*Frankliniella occidentalis* Pergande), are almost impossible to eradicate once they have been introduced into a greenhouse. Consequently much care should be taken at the time of pollination (see below). Seeds of winter barley and the vegetatively propagated diploid *H. bulbosum* clones are planted individually in 2.5 and 5 L plastic containers respectively, containing commercial greenhouse potting mix (105D – Agrofino Products nv/sa, Arendonk, Belgium) and, whenever possible, left outside for vernalization from the end of September over winter with frost protection when necessary (Fig. 2.2-1a). When temperatures are getting too high for natural vernalization, winter barley and *H. bulbosum* plants are put in a cold room at 4°C for 8 weeks (8 h day length) and subsequently repotted in 2.5 and 5 L containers respectively, and transfer to the greenhouse. For spring barley, seeds are planted in the same conditions but the plants are raised directly in the greenhouse without any vernalization step. A basic fertilizer (Nitrophoska BASF 12:12:17) is applied at the beginning of the culture and a high nitrogen fertilizer (KEMIRA 33.5) about 3 weeks before heading. As our crossing period expands from March to October, the greenhouse has to be heated or cooled to keep the temperature inside at 21±4/17±2°C (day/night) with a 16 h photoperiod. Natural light is supplemented with artificial sodium lighting (Philips Agro SONT-400) as required, to maintain a photon flux density of 200-300 $\mu E\ m^{-2}\ s^{-1}$ at the soil surface.

Hybridization

Emasculation of barley florets is carried out by forming a slit with a forceps in the lemma, through which the three anthers are removed (Fig. 2.2-1b). A freshly collected *H. bulbosum* pollen mixture (Figs. 2.2-1c and d) from four genotypes is applied to receptive stigmas with a paintbrush 2-3 days later (Fig. 2.2-1e). Special care should be taken at this time not to bring larvae of California thrips along with pollen that could cause dramatic effects on caryopsis and embryo development. To facilitate the control of insect propagation, donor pollen and barley plants are raised in separate greenhouses. From emasculation to the gibberellic acid treatment, barley spikes are covered with glassine bags (15x8 cm).

Post-pollination treatment

Longevity of developing seeds in intergeneric and interspecific hybrids in the Poaceae has been improved by the application of gibberellic acid e.g. GA_3 and other plant growth regulators (Matzk, 1991; Pickering and Wallace, 1994) to florets after fertilization. Without GA_3 the developing seed degenerate before embryo rescue. GA_3 (Sigma G-7645) at 75 mg/L + 0.05% Tween 20 (Prolabo 28 829.296) is applied as a spray to florets (Fig. 2.2-1f) 1-2 days after pollination to enhance seed development and embryo size. Early in spring and in the fall, the same treatment is usually repeated one day after the first one. GA_3 + 2,4-D + Dicamba at 75 mg/L; 2 mg/L and 1 mg/L, respectively, has been used on the day after pollination to promote effectively seed longevity (Pickering, pers. comm.). Immediately after the post-pollination treatment, the glassine bag is replaced by a brown paper bag (15 x 4 cm) and identified by a bar coded label printed directly from a portable printer (Figs. 2.2-2a and b). Spikes are left on the plant until they are collected for seed dissection and embryo culture.

Spike collection, seed dissection and embryo culture

Eleven (summer) to 20 (early spring and late autumn) days after pollination, spikes covered by their labelled brown paper bags are cut off from the plants, immersed to a depth of 5-6 cm in a beaker containing tap water. If seeds are not dissected on the same day, the spikes can be stored up to 15 days in a fridge at 4°C in the dark. Non-shrivelled yellow to green seeds are removed from spikes and surface sterilized in a solution of calcium or sodium hypochlorite containing 2.5% (w/v) available chlorine for 5 minutes prior to three successive quick washes in sterile water. In summer, when *in vitro* contamination is much higher, calcium hypochlorite can be substituted by a 0.2% (w/v) aqueous solution of mercuric chloride (Prolabo 25.384.232) for 20 minutes. Although much care should be taken when manipulating and eliminating the highly toxic mercuric chloride, substantial efficiency has often been obtained when compared to the classical calcium hypochlorite. Five to 10 sterilized seeds are brought to one plate of a sterile Petri dish and embryos are excised under a dissecting microscope (x20) (Fig. 2.2-2c). Differentiated embryos are transferred to glass tubes (H82, 28 mm diameter) three per tube (Figure 2.2-2d), containing 8 ml of modified Gamborg's B-5 medium with 2% sucrose (Merck 7653), 0.8% agar (Sigma A-1296) but without 2,4-D (Table 2.2-1).

Table 2.2-1. Modified Gamborg's B-5 medium

Medium components	Concentration (mg/L)
Macro nutrients	
$NaH_2PO_4 \times H_2O$	150
KNO_3	2,500
$(NH_4)_2SO_4$	134
$MgSO_4 \times 7H_2O$	250
$CaCl_2 \times 2H_2O$	150
Micro nutrients	
$MnSO_4 \times H_2O$	10
H_3BO_3	3
$ZnSO_4 \times 7H_2O$	2
$Na_2MoO_4 \times 2H_2O$	0.25
$CuSO_4 \times 5H_2O$	0.039
$CoCl_2 \times 6H_2O$	0.025
KI	0.75
FeNa EDTA	40
Vitamins	
Nicotinic acid	1
Thiamine HCl	10
Pyridoxine HCl	1
Myo-Inositol	100
Miscellaneous	
Agar	8,000
Sucrose	20,000
pH	5.5

From labels on brown paper bags, new labels are printed out and pasted on corresponding vials. Embryos are incubated in darkness at 22±2°C and when coleoptiles have reached 1 cm in length (Fig. 2.2-2e) i.e. between 3 to 8 days later, the culture vials are put to a 16 h day length environment provided by a 1:1 mixture of Sylvania Gro-Lux F58W/GRO and Philips TLD58W/840 New Generation fluorescent lamps giving an irradiance of 100-150 $\mu E\ m^{-2}\ s^{-1}$ at a similar temperature. Plantlets at the two-leaf stage (Fig. 2.2-2f) are removed directly to potting mix (Fig. 2.2-3a), and the weaker ones covered with clear plastic beakers to maintain high humidity. A plastic label with a bar code is planted in each pot containing an individual plantlet.

Chromosome doubling

As most of the regenerated plants (>95%) are haploid, no ploidy level determination is carried out. Interspecific hybrids can be easily recognized by their growth habit, the presence of pubescent leaf sheaths and of course, the morphology of their spikes at heading time. They can therefore be discarded at an early stage to save room in the greenhouse. Among several anti-mitotic compounds tested, colchicine has remained the most efficient one for chromosome doubling of many plant species including barley. Reproducible and high chromosome doubling efficiency has been achieved with the 'Plant Cell Culture Tested' grade colchicine of Sigma (C-3915). For colchicine treatment, plants that have developed at least two tillers (Fig. 2.2-3b) are removed from the compost and roots are washed and cut back to about 1 cm (Fig. 2.2-3c). They are then immersed to a depth of 5 cm (Fig. 2.2-3d) in a 0.05% aqueous colchicine solution containing 2% dimethyl sulfoxide (Sigma D-4540) after having made single incisions into each tiller base. They are treated for 5 hours at 25°C under artificial light then rinsed in running tap water for a few minutes, potted in compost and placed to the greenhouse. After a re-establishment period of 15 days, plants of winter-type are vernalized in a controlled environment room at 4°C and 8 h day length for 6 weeks and left outside (Fig. 2.2-3e) or in a cold frame for several weeks to enhance tillering. Plants of spring-type can be sent directly to the greenhouse for further development. Chromosome doubling rates of >80% are commonly achieved, although there are significant differences in the proportions and fertility of DH plants obtained. To save space in the cold room, vernalization of winter barley haploids can be carried out while still in the culture vials and before colchicine treatment (Pickering, pers. comm.).

Seed collection

After they have developed several tillers, colchicine-treated haploid plants are transferred to a greenhouse kept at a moderate temperature (12±2°C) at the beginning, to a gradually higher temperature (20±2°C) at maturing time along with increasing photoperiod provided by artificial light when necessary. Just before reaching anthesis, all spikes of each plant are covered by a paper bag (Fig. 2.2-3f) to prevent cross-pollination. This phenomenon is rather frequent especially with winter-type of barley and particularly when plants are left outside. In this case, synthetic linen bags have been used. Immediately after seed set, the bags are removed from the plants to allow insect and pest control (Fig. 2.2-3g). At maturity, fertile spikes are collected manually from each plant (Fig. 2.2-3h) and put in a paper bag labeled with a bar code (Fig. 2.2-3i and j).

Multiplication and selection of DH plants

Mature seeds from DH are sown in the field for agronomic assessment. When limited numbers of seeds are available (<10), they are sown in shallow boxes in a greenhouse and at spring planted in the field for a first round of multiplication prior to selection. The most suitable lines are harvested and further large-scale trials and multiplication take place before the best selections are considered for registration.

Efficiency and Applications

The described *H. bulbosum* method, is a reliable technique that has been used for many years by researchers and breeders to generate thousands of DH lines of barley from which many cultivars have been derived and grown in several areas over the world (Devaux *et al.*, 1996). To generate these cultivars, DH lines have been derived from thousands of early hybrids, usually F_1. DH production efficiency depends on genetic factors including the barley and *H. bulbosum* genotypes and several environmental factors e.g. temperature (Pickering, 1984) and light. Because of these sources of variation, there are marked differences in the success rates, ranging from very low frequencies to as many as 130 fertile DH plants from a single three-way barley hybrid plant (Devaux unpublished). Overall, 5 to 10 DH plants per 100 florets of *H. vulgare* pollinated with *H. bulbosum* have been commonly reported by several researchers. Although genetic and environment factors influence the success rate, the technique has been improved at each stage of the process (reviewed by Pickering and Devaux, 1992). However to ensure production of DH lines from every possible barley hybrid in a breeding program, anther culture can be used in parallel, since recalcitrant genotypes to both methods have been identified (Devaux *et al.*, 1996). In addition, because of the different cultural conditions employed for these methods, it is possible to exploit both of them for optimum efficiency and minimum cost at different seasons of the year.

Acknowledgements
Many thanks to Dr. R.A. Pickering, Crop and Food Research, Christchurch (NZ) for his advice and useful comments and to C. Murez and A. Droulers and their co-workers for their excellent technical assistance.

References

Devaux, P., M. Zivy, A. Kilian and A. Kleinhofs, 1996. Doubled haploids in barley. In: Proc. V International Oat Conference and VII International Barley Genetics Symposium. Scoles, G. and B. Rossnagel (Eds.) Univ. of Saskatchewen, Saskatoon, pp. 213-222.
Kasha, K.J. and K.N. Kao, 1970. High frequency haploid production in barley (*Hordeum vulgare* L.). Nature **225**: 874-876.
Matzk, F., 1991. A novel approach to differentiated embryos in the absence of endosperm. Sex.Plant Reprod. **4**: 88-94.
Pickering, R.A., 1984. The influence of genotype and environment on chromosome elimination in crosses between *Hordeum vulgare* L. x *Hordeum bulbosum* L. Plant Sci.Lett. **34**: 153-164.
Pickering, R.A. and P. Devaux, 1992. Haploid production: approaches and use in plant breeding. In: Barley: Genetics, Biochemistry, Molecular Biology and Biotechnology. Shewry, P.R. (Ed.) CAB Int., Wallingford, pp. 519-547.
Pickering, R.A. and A.R. Wallace, 1994. Gibberelic acid + 2,4-D improves seed quality in *Hordeum vulgare* L. x *H. bulbosum* L. crosses. Plant Breed. **113**: 174-176.

2.3
Anther culture in barley

C. Jacquard, G. Wojnarowiez and C. Clément
Université de Reims Champagne Ardenne, UFR Sciences, Laboratoire de Biologie et Physiologie Végétales, BP 1039, 51687 Reims Cedex 2, France

Introduction

The production of doubled haploids through androgenesis represents a modern tool for the improvement of cultivated species enabling plant breeders to produce homozygous lines in a few months. In barley and other cereals, the use of androgenesis has generated a number of cultivars currently available and cultivated in many countries.

Androgenesis can be performed using either anther or isolated microspore culture. In both cases, it consists of the regeneration of haploid plantlets from microspores, which were initially destined to develop into pollen grains (Fig. 2.3-1). The microspore enters the androgenetic process following two possible pathways. In the first case, the microspore develops into a haploid callus from which haploid plantlets can be regenerated through indirect embryogenesis. In the second case the microspore develops directly into a haploid embryo (direct microspore embryogenesis) that further regenerates a haploid plantlet. The indirect microspore embryogenesis was mostly followed in the 70's but successive optimisations of the protocols progressively lead to direct embryogenesis in most cases.

The process of androgenesis can be separated into 3 steps: the pretreatment, or induction phase, destined to switch the pollen fate from the initial gametophytic program to the alternative sporophytic embryogenic program; the culture phase, which consists in the embryogenic development of the microspore; the regeneration phase allowing the development of androgenetic embryos into haploid plantlets (Fig. 2.3-1).

Protocol

Donor plants and growth conditions

The model genotype used is the 2-rowed winter cultivar 'Igri', though it is no longer used in breeding programmes. Spring lines of barley can be used following this protocol but the efficiency will be lower than winter types, especially regarding albinism.

The vigor of the donor plant is a crucial point for the success of androgenesis. During plant growth, any form of stress such as pesticide treatment, dehydration or wide temperature variations should be avoided. A minimal weekly preventive pesticide treatment can however be applied without significant changes in the yield of anther culture. Seeds are germinated on moistened filter paper in a closed Petri dish for a few days at room temperature and ambient light. After germination seedlings are planted in 20 cm diameter pots containing a mixture of sand, peat moss and soil (1:1:1). They are grown in a greenhouse at 25°C for a week under a 16 h photoperiod at approximately 80% relative humidity. For winter lines, seedlings are vernalized for 8 weeks at 4°C using a 12 h photoperiod at low light intensity (100 µE m^{-2} s^{-1}). Seedlings of vernalized winter or spring cultivars are next treated similarly. They are grown in the greenhouse at low temperature (12±2°C) and 80% relative humidity. The suitable photoperiod is 16 h and the light intensity should reach 1,000 µE m^{-2} s^{-1}. The quality of light can be a determinant and sodium vapour lamps are recommended. Addition of NPK granules during plant growth helps to increase plant vigour and the yield from androgenesis.

Spike sampling and sterilization

1. The developmental stage of microspores at sampling time is a determinant for their competence to androgenesis. Spikes are collected after approximately 8 weeks when the awns appear outside the upper leaf. This usually corresponds to the uninucleate stage.

2. The stage of development is checked by collecting an anther in the middle of the spike and squashing it in acetic carmine (5% in 45% acetic acid boiled for 1 h and filtered) on a glass slide. Acetic carmine binds to DNA and delineates the location of the nucleus in the microspore under the microscope. Spikes in which microspores have undergone mitosis are not suitable for androgenesis.

3. Suitable spikes are collected and sterilized in ethanol 70% for 5 min and rinsed in sterile distilled water for 5 min.

4. In each spike, the anthers from proximal and distal flowers are removed. The anthers from the central flowers are solely conserved for further pretreatment, representing approximately 30 anthers per spike.

Anther pretreatment

1. Collected anthers from the same spike are deposited in a 5 cm diameter Petri dish in 10 ml of P3 medium containing 2.5 mg/L $CuSO_4$ x $5H_2O$ and 62 g/L mannitol providing an osmotic pressure of 180 mosm/L. For sterilization, the pretreatment medium is autoclaved.

2. Anthers are pre-treated at 4°C in the dark at 80% relative humidity for 3-4 days.

Anther culture

1. After pretreatment, anthers are transferred, without rinsing, onto the C3 culture medium (Table 2.3-1).

2. Thirty anthers are plated per 5 cm diameter Petri dish. Dishes are sealed with Parafilm and maintained in the culture chamber at 26±2°C with 85% relative humidity, and kept in the dark for 21-28 days.

Table 2.3-1. Composition of the C3 and M1 media for anther culture in barley

Media components	C3 culture medium (mg/L)	M1 regeneration medium (mg/L)
Macro elements		
KH_4NO_3	166	166
KNO_3	1,900	1,900
$MgSO_4 \times 7H_2O$	374	374
KH_2PO_4	170	170
$CaCl_2$	22	22
Micro elements		
H_3BO_3	6.2	6.2
$ZnSO_4 \times 4H_2O$	8.6	8.6
KI	0.83	0.83
$Na_2MoO_4 \times 2H_2O$	0.25	0.25
$CuSO_4 \times 5H_2O$	2.5	0.025
$MnSO_4 \times 4H_2O$	22.3	22.3
FeNaEDTA	40	40
Vitamins		
myo-Inositol	0.1	0.1
Thiamine HCl	0.4	0.4
Growth regulators		
NAA	2	0.4
BAP	1	0.4
Others		
Glutamine	752	752
Maltose	60,000	-
Sucrose	-	30,000
Mannitol	32,000	-
Agarose	7,000	-
Agar washed	-	6,000
pH	5.6	5.6

Vitamins and growth regulators are filter sterilised

Haploid plantlet regeneration

1. When microspore derived embryos measure approximately 1-2 mm (Fig. 2.3-2), responding anthers are transferred onto the M1 regeneration medium (Table 2.3-1). The differences between the regeneration and the culture medium are the nature and concentration of the carbon source, the removal of mannitol, the replacement of agarose by agar washed (Sigma A8976) and the lower concentrations of plant growth regulators.

2. The Petri dishes are maintained in the culture chamber at 26±2°C and 85% relative humidity with a 16 h photoperiod, at 100 µE m^{-2} s^{-1}.

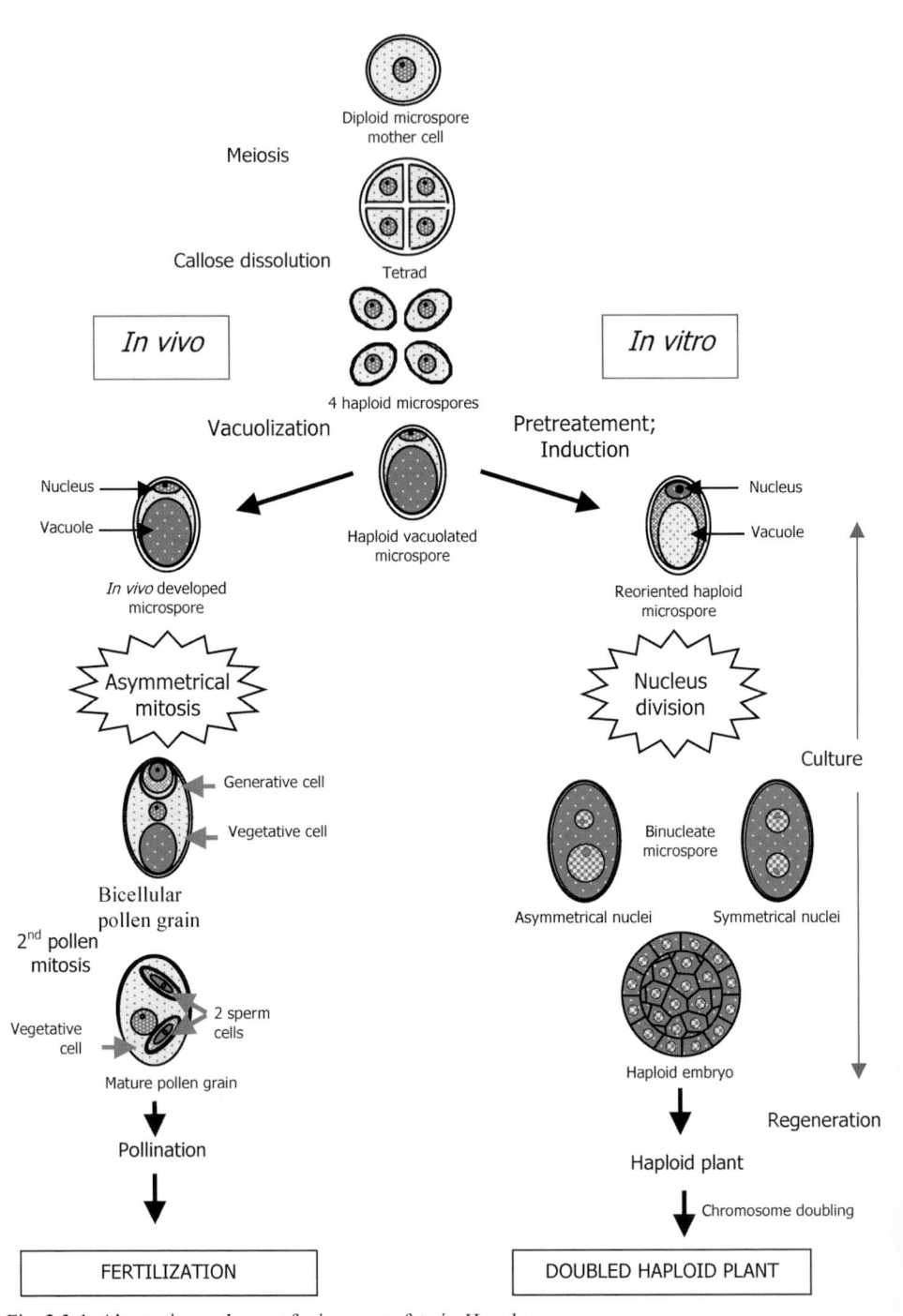

Fig. 2.3-1. Alternative pathway of microspore fate in *H. vulgare*.

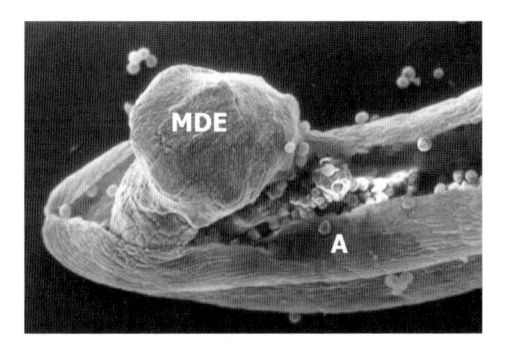

Fig.2.3-2. Microspore derived embryo (MDE) developing outside of the anther (A) prior to regeneration (x50).

3. After 2 weeks, green plantlets are transferred in 25 mm diameter culture tubes containing 25 ml of the regeneration medium described above and allowed to grow for 4 weeks in the culture chamber.

Plant ploidy evaluation

Doubled haploid plantlets are spontaneously recovered from anther culture as well as plants with other ploidy level. The former are cultivated as described above and grown for further agronomical evaluation. The latter are tested for their ploidy and haploid plantlets undergo chromosome doubling. Aneuploid, triploid or tetraploid plants are discarded.

1. In order to test the ploidy of an androgenetic plantlet, leaf tissue is homogenised in a buffer containing the Murashige and Skoog salts, 700 mM sorbitol and 1% (w/v) Triton X 100 at pH 6.6. For the preparation of samples, 40 mg of fresh tissue are homogenised in 2 ml of buffer in the presence of 16 µL filter sterilized DAPI (250 µg/ml).

2. The homogenate is used for the determination of cell ploidy using a CAII flow cytometer (Partec GmbH).

Chromosome doubling and doubled haploid plant obtaining

1. Regenerated green plantlets are transferred directly to 20 cm diameter pots containing a mixture of sand, peat moss and soil (1:1:1).

2. They are grown in the greenhouse at 25°C and 80% relative humidity. The suitable photoperiod is 16 h and the light intensity should reach 1,000 µE m^{-2} s^{-1}.

3. Plants are grown until they produce 3-4 tillers, then they are removed from the pot and washed in water.

4. Roots are trimmed at the stem bases (5 mm) and plantlets are treated with 0.05% colchicine in the presence of 2% dimethyl sulphoxide for chromosome doubling. Plants are immersed to a depth of 5 cm for 5 hours at 25-30°C.

5. Plants are washed for 5 min and then replanted into the pots and grown in the greenhouse.

Efficiency and Applications

In *H. vulgare* the androgenetic protocol has been widely improved for years in most cultivated lines (Pickering and Devaux, 1992; Jähne-Gärtner and Lörz, 1996) though some genoptypes remain recalcitrant. Indeed, the genotype is responsible for more than 60% of the variations obtained in anther cultures. The best results were obtained in the winter lines, especially in the winter variety Igri, which represents the model genotype for the improvement of the androgenesis protocol through anther culture. Another problem affecting the efficiency of androgenesis in barley is the production of albino plantlets in various proportions according to the cultivars (Caredda and Clement, 1999). Barley is particularly affected by this phenomenon as several genotypes cannot be used for improvement because of albinism. Both winter and spring lines are concerned, although the latter are more sensitive. For example, the winter variety Igri produces up to 89% of green haploid plantlets using the following protocol, whereas the spring variety Cork currently generates 99% albino plantlets.

This protocol has been optimized for the winter variety Igri and provides acceptable results (Table 2.3-2). The originalities of this protocol are mainly in the addition of both mannitol and copper sulfate during anther pretreatment and culture. Both of them have considerably increased the yield of this technique. The short 4 days pretreatment with mannitol instead of 4 weeks in cold water made it possible to remove the formation of callus during anther culture in favor of embryos. Besides, the addition of copper sulfate at 2.5 mg/L (10 µM) enhanced both the quantity and the quality of regenerated plantlets in the winter variety Igri.

Table 2.3-2. Yield of anther culture in the winter barley (*H. vulgare*) variety Igri

Parameters tested	RA (%)	RP/100 RA	GP/100 RA	GP (%)
Winter variety Igri	72.3	1245	1111.1	89.2

RA, responding anthers; RP/100 RA, regenerated plantlets per 100 responding anthers; GP/100 RA, green plantlets per 100 responding anthers; GP, percentage of green plantlets.

However, several parameters generate noticeable variations in the results obtained. For example, the season affects microspore behaviour in the anther. Actually, although the conditions described above are maintained rigorously throughout the year, the results fluctuate significantly. They are much better during the spring and the early summer, suggesting that endogenous rhythms affect barley plant physiology from the seed to the flowering plant, whatever the growth and *in vitro* culture conditions. It is likely that it will be difficult to remove this effect in the future.

The major obstacle to anther culture in numerous barley lines remains to be albinism. In some cultivars, 100% of microspore derived plantlets are albinos. This trait is not specific to anther culture and is currently studied in many other *in vitro* systems (Cho et al., 1998). At present, the optimization of the anther culture protocol did not provide considerable improvement of this parameter, suggesting that barley microspores and derived structures are particularly sensitive to albinism.

When albinism occurs, plastids have lost their internal membranes and accumulated lipids and prolamellar globules while chlorophyll is not synthesized. Comparing the winter variety Igri (producing 89% of green plantlets) and the spring variety Cork (producing 99% of albino plantlets), it was shown that microspore plastids are affected as early as the anther sampling stage, and that normal pattern of plastid development is not recovered during the androgenesis process (Caredda et al., 2000). Similar results were obtained in other lines, meaning that obtaining albino plants following anther culture may be correlated to the physiolgical state of plastids in the microspore at the time of sampling. Therefore, both fundamental and applied studies carried out in this respect may focus on the early stages of microspore development and on plant growth conditions. For example, the impact of copper during plant growth may provide interesting data, especially regarding albinism in concerned albino producing cultivars.

Acknowledgements
The authors would like to thank Dr. Pierre Devaux (Florimond Desprez Ind.) for his helpful advice in reading the manuscript.

References

Caredda, S. and C. Clement, 1999. Androgenesis and albinism in Poaceae: influence of genotype and carbohydrates. In: Anther and Pollen. From Biology to Biotechnology. Clement, C., E. Pacini and J.-C. Audran (Eds.) Springer-Verlag, Berlin, pp. 211-228.
Caredda, S., C. Doncoeur, P. Devaux, R.S. Sangwan and C. Clement, 2000. Plastid differentiation during androgenesis in albino and non-albino producing cultivars of barley (*Hordeum vulgare* L.). Sex. Plant Reprod. **13**: 95-104.
Cho, M.-J., W. Jiang and P.G. Lemaux, 1998. Transformation of recalcitrant barley cultivars through improvement of regenerability and decreased albinism. Plant Sci. **138**: 229-244.
Jähne-Gärtner, A. and H. Lörz, 1996. Protocols for anther and microspore culture of barley. In: Methods in Molecular Biology. Vol. 111: Plant Cell Culture Protocols. Hall, R.D. (Ed.) Humana Press Inc., Totowa, NJ. pp. 265-279.
Pickering, R.A. and P. Devaux, 1992. Haploid production: approaches and use in plant breeding. In: Barley: Genetics, Biochemistry, Molecular Biology and Biotechnology. Shewry, P.R. (Ed.) CAB Int., Wallingford, pp. 519-547.

2.4
Barley anther culture

L. Cistué, M.P. Vallés, B. Echávarri, J.M. Sanz and A. Castillo
Departamento de Genética y Producción Vegetal. Estación Experimental de Aula Dei, C.S.I.C. Apartado 202, 50080 Zaragoza, Spain

Introduction

The production of doubled haploid (DH) barley (*Hordeum vulgare* L.) plants has proven to be a highly valuable tool for plant breeding, allowing the release of a high number of new cultivars (Devaux *et al.*, 1996). Doubled haploids have also been fundamental for genetic analysis, linkage maps production and QTL analysis in barley. Several anther culture protocols have been established for an efficient production of DH lines by different public institutions and private companies. However, the low embryogenic capacity and the high albinism rate of some genotypes are still open fields for further improvements in these protocols. To be successfully used in a breeding program, any particular protocol should produce a large number of DH lines from all the genotypes. During the last years, our aim has been focused on the establishment of an efficient protocol for the production of doubled haploid plants from a wide range of genotypes, as well as from F_1 crosses of putative great agronomic performance. The results achieved over five years (1993-1997) in our laboratory resulted in a gradual increase in the numbers of green plants per 100 anthers (from 6.3 up to 17.4), as well as DH fertile lines per 100 anthers (from 1.2 to 10.3). Major changes of the initial protocol have been the increase of mannitol concentration in the pretreatment medium from 0.3 to 1.0 M, the introduction of Ficoll in the induction medium, the optimization of the regeneration medium composition, and the procedure for transferring the plants to soil (Cistué *et al.*, 1994, 1999; Castillo *et al.*, 2000).

Protocol

When establishing the protocol, attention was focused on the regular production of high number of doubled haploids plants from a wide range of genotypes.

Donor plants and growth conditions

The quality of the donor plants is a critical aspect regarding subsequent anther culture response. Donor plants must grown in optimal conditions, concerning photoperiod, intensity and quality of light, temperature and nutrition. Furthermore, plants must be free of pests and diseases, and the use of fungicides or pesticides should be avoided or reduced to a minimum. The phases of the donor plant growth take place in different growth chambers, each one with specific conditions in order to have a continuous production of spikes.

Seeds are sown in a paper-pot with a mixture of sand, vermiculite and peat in the same proportion. Vernalization is performed for one month at 4°C temperature, with 8/16 h light/dark photoperiod, and 100 $\mu E\ m^{-2}s^{-1}$ light intensity provided by fluorescent tubes (Mazdafluor 18W). Plantlets are transplanted to pots (30 cm diameter) with the soil mixture described above (2 plants/pot) and transferred to a growth chamber at 12°C, with 12 h photoperiod, and 500 $\mu E\ m^{-2}s^{-1}$ provided by high-pressure metal halide lamps (Phillips Powertone HPI-T Plus 400W). After one month, temperature is increased up to 18-21°C and the photoperiod to 16 h light, until the spikes are harvested. The relative humidity (RH) for both periods is 70-80%. The soil is fertilized with a combination of N:P:K (15:15:15) at the time of soil mixture preparation. Besides, the foliar fertilizer "Zelti foliage" (Zeneca-Agro S.A.) containing N:P:K (20:20:20) and micronutrients, is applied once a week during the whole growth cycle of the plants.

Harvest of spikes

Spikes are collected when most of the microspores are at the mid to late-uninucleate stage. Cytological examination is the most accurate method to identify this stage. Aceto-carmine staining squash, from fresh anthers from the central flowers of each spike, must be performed. Furthermore, it is recommended to check the proportion of viable microspores inside the anther by staining with fluorescein diacetate (FDA). Only spikes containing most of the microspores in mid- or late-uninucleate stage, and with higher than 70% microspore viability should be used for culture.

From a practical point of view, cytological examinations are performed on five spikes from each genotype and for each batch of plants. A correlation is established between the stages of development of the microspores and different morphological characters such as: the distance between the ligules of the flag leaf and the next lower leaf; colour and position of the upper part of the anthers with respect to the glumes; and the texture of the spikes.

Sterilization

Collected spikes are removed from sheath in a flow bench and sprayed with 70% ethanol. The strong morphological selection of the spikes should be completed at this moment using the above criteria. The selected spikes are placed in 9 cm diameter Petri dishes with a drop of water. The plates can be sealed with Parafilm and stored for 2 or 3 days in the refrigerator at 4°C in the dark until use.

Pretreatment

Out of the two mostly common pretreatments used, cold-shock and carbohydrate starvation, we are using the last one, which is based in the substitution of a metabolizable sugar by a non-metabolizable sugar like mannitol. Anthers are extracted from the spikes under a stereoscopic microscope, and inoculated immediately in a pretreatment medium (6 cm Petri dish), containing 0.7 M mannitol, 40 mM $CaCl_2$ and 8 g/L SeaPlaque agarose. When working with recalcitrant genotypes we increase the concentration of mannitol to 1-1.5 M. Normally 30 anthers are collected from each spike, 15 from each side of the 5 central flowers. Several spikes can be inoculated in each plate. These plates are sealed with Parafilm and incubated at 24 °C in the dark for 4 days.

Pretreatment medium
For preparation of 1000 ml (0.7 M mannitol):
1. Add 5.88 g of $CaCl_2$ x $2H_2O$ and 8 g of agarose to 500 ml of bi-destilled water and autoclave.
2. Dissolve 127.54 g of mannitol in 500 ml of bi-destilled water (heat at 37°C) and sterilise by passage through a 0.22 µm filter.
3. Mix both solutions and pour 8 ml in each Petri dish (6 cm diameter).

Induction

After the four days of pretreatment (Fig. 2.4-1a) fifteen anthers are inoculated in 3 cm diameter Petri dishes containing 1.5 ml FHG induction medium (Hunter, 1988). The FHG is a modified MS medium (Murashige and Skoog, 1962), where sucrose has been replaced by maltose in order to avoid the fast degradation of sucrose into glucose, and contains a high concentration of glutamine and a reduced amount of ammonium nitrate. This medium is supplemented with 4.4 µM of BAP (benzyl adenine). Two hundreds grams per litre of Ficoll Type-400 (Sigma F4375), a high molecular weight polysaccharide, is added to the medium in order to avoid the growth of microspores and embryos in anaerobic conditions due to their tendency to sink in liquid medium (Tab.2.4-1).

The plates are sealed with Parafilm and kept in a chamber at 24°C in the dark. After 12-14 days, 1.5 ml of the same medium containing 400 g/L Ficoll Type-400 is added for further development of the embryos (Fig 2.4-1b and c). Due to the high cost of Ficoll, plates with anthers of average or high responding genotypes that show a low number of dividing microspores are discarded.

Preparation of FHG medium - Ficoll 200 g/L
Prepare stock solutions (2x) containing macro nutrients + micro nutrients + Fe + thiamine HCl (Table 2.4-1), and freeze them.
For preparation of 1000 ml of culture medium:
1. Add 200 g of Ficoll Type-400 to 420 ml of bi-destilled water. Heat and stir vigorously on a magnetic stirrer. The Ficoll will not go into solution until it boils. Let it cold down at room temperature.
2. Defrost the stock solution (500 ml) and add the glutamine, myo-Iositol, maltose and growth regulators.

3. Mix both solutions, adjust the volume and the pH. Sterilize the solution by passage through a 0.22 μm filter.

Table 2.4-1. Culture media composition

Media components	FHGI embryo induction medium (mg/L)	FHGR plant regeneration medium (mg/L)	MSr rooting medium (mg/L)
Macro salts			
KNO_3	1,900	1,900	950
NH_4NO_3	165	165	825
KH_2PO_4	170	170	85
$CaCl_2 \times 2H_2O$	440	440	220
$MgSO4 \times 7H_2O$	370	370	185
Iron source			
$FeNa_2EDTA$	37.5	37.5	18.8
Micro salts			
$MnSO_4 \times 4H_2O$	22.3	22.3	11.2
$ZnSO_4 \times 7H_2O$	8.6	8.6	4.3
H_3BO_3	6.2	6.2	3.1
KI	0.83	0.83	0.42
$Na_2MoO_4 \times 2H_2O$	0.25	0.25	0.13
$CuSO_4 \times 5H_2O$	0.025	0.025	0.013
$CoCl_2 \times 6H_2O$	0.025	0.025	0.013
Vitamins			
Thiamine HCl	0.4	0.4	0.4
myo-Inositol	100	100	100
Other components			
Glutamine	730	-	-
Casein hydrolysate	-	-	1,000
Maltose	62,000	31,000	-
Sucrose	-	-	20,000
BAP	1	1	-
IAA	-	0.5	-
NAA	-	-	2
Ficoll	200,000 or 400,000	-	-
Phytagel	-	3,000	3,000
pH	5.8	5.8	5.8

Regeneration

One to two weeks after the refilling of the plates with medium (three to four weeks after culture initiation), the well-developed embryos (embryos with scutellum, coleoptile and coleorhiza) are transferred to regeneration medium for plant production (Fig. 2.4-1d). This medium (FHGR, Tab. 2.4-1) contains the macronutrients, micronutrients and vitamins from the FHG medium, but without organic nitrogen supply, a reduced amount of maltose, and 3 g/L Phytagel (Sigma P 8169) as a solidifying agent. Furthermore, it is supplemented with 4.4 µM of BAP and 2.5 µM of IAA (indole-3-acetic acid). This medium is not adequate for the regeneration of callus-like structures, which are discarded after 30 days of culture initiation.

Embryos in regeneration medium are placed in a growth chamber at 24°C, 16/8 h light/dark, with a light intensity of 200 µE $m^{-2}s^{-1}$ provided by fluorescent-incandescent lamps (Mazdafluor TF 58 W / LJ 54 – Philips 25W), and 70-80% RH. Germination of the embryos, and therefore plant production takes places in 10 to 18 days after transferring (Fig. 2.4-1e). At this moment, albino plants are identified and discarded.

Growth and vernalization of the plants
All green plantlets regenerated are transferred to Magenta boxes (9 plantlets per box) containing half-strength mineral salts MS medium, supplemented with 30 g/L sucrose, 2 mg/L NAA, and 8 g/L agar (Merck 1.01614), for development of a good root system (MSr, Table 2.4-1).

For vernalization of the plants, the Magenta boxes are kept in a growth chamber at 4°C, 8/16 h light/dark, at 100 µE $m^{-2}s^{-1}$ for one month. Vernalization of the plants at this stage is more convenient for handling and saving space.

Transfer of plants to soil
After vernalization, plantlets are placed in a growth chamber at 16°C, 12 h photoperiod, with 200 µE $m^{-2}s^{-1}$ provided by fluorescent-incandescent lamps, and 90-100% RH. One week later, plants are potted (5 cm diameter) in the soil mixture mentioned above, and covered with a plastic glass during the first week; afterwards the glass is removed (Fig. 2.4-1f). Plants are kept in this chamber for another 15 to 21 days more.

Harvest of doubled haploids seeds
Plants are transplanted to pots (30 cm diameter) with the same soil mixture, and cultivated in the greenhouse for seed production at 24±4°C under a light regime of 16/8 h light/dark (Fig. 2.4-1g). A high rate of spontaneous chromosome doubling is obtained with this protocol (around 80-90%). Haploid plants, characterized by a high number of tillers, narrow leaves and no visible hair, are discarded. However, in genotypes with a low rate of spontaneous diploidization, application of colchicine can be performed following the protocol described by Jensen (1977).

Normally seeds from 5-6 spikes are harvested from every DH line. Seeds produced throughout the year are stored at 4°C. Regular field trials and agronomic evaluation and selection of the doubled haploid lines are conducted in the breeding programs (Fig. 2.4-1h)

Efficiency and Applications
Using the protocol outlined above, doubled haploid lines were produced from 69 F_1 crosses between cultivars with great agronomic value in the Mediterranean countries, including crosses of two- x two-rowed, two- x six-rowed and six- x six-rowed, spring x spring-type, spring x winter-type and winter x winter-type.

The production of DH lines by anther culture is a long process that involves several phases, each of them with different efficiencies. The latest results obtained in our laboratory in 2001 from two F_1 crosses and their reciprocals between advanced inbreed lines provide an example. Three parental lines, CT-218, L-102 and L-225, are average in response for anther culture while one, CT-186, is a low responder. The average of the four F_1 crosses was 26.6 green plants transferred to soil from 100 anthers. From these plants, 43% did not produce seeds. As an average 10% of the plants died in the phase of acclimatization and growth, and 10 to 20% of the surviving plants were haploids. Other critical aspect is the control of photoperiod and temperature, especially in areas with continental climate, since 15 to 25% of the plants were sterile due to difficulties in controlling the growing conditions during hot season.

With the protocol described above a high number of DH lines have been produced for barley breeding programs and for research projects on the identification of QTL linked to agronomic or androgenesis traits. Both projects involve collaboration with Spanish public institutions and private companies. A concrete collaboration was also established with the Washington State University (U.S.A.). In addition, the new barley varieties 'Belén' and 'Lola', both of them registered by the "Instituto Técnico Agrícola Provincial de Albacete (ITAP S.A.- Spain), were selected from DH lines produced in our laboratory.

References

Castillo, A. M., M. P. Valles and L. Cistue, 2000. Comparison of anther and isolated microspore cultures in barley. Effects of culture density and regeneration medium. Euphytica **113**: 1-8.

Cistue, L., A. Ramos and A.M. Castillo, 1999. Influence of anther pretreatment and culture medium composition on the production of barley doubled haploids from model and low responding cultivars. Plant Cell Tiss.Org.Cult. **55**: 159-166.

Cistué, L., A. Ramos, A.M. Castillo and I. Romagosa, 1994. Production of large number of doubled haploid plants from barley anthers pretreated with high concentrations of mannitol. Plant Cell Rep. **13**: 709-712.

Devaux, P., M. Zivy, A. Kilian and A. Kleinhofs, 1996. Doubled haploids in barley. In: Proc. V International Oat Conference and VII International Barley Genetics Symposium. Scoles, G. and B. Rossnagel (Eds.) Univ. of Saskatchewen, Saskatoon, pp. 213-222.

Hunter, C.P., 1988. Plant regeneration from microspores of barley, *Hordeum vulgare* L. Ph.D. Thesis. Wye College, Univ. of London, London.

Jensen, C.J., 1977. Monoploid production by chromosome elimination. In: Applied and Fundamental aspects of Plant Cell, Tissue and Organ Culture. Reinert, J. and Y.P.S. Bajaj (Eds.) Springer-Verlag, Berlin, pp. 299-330.

Murashige, T. and F. Skoog, 1962. A revised medium for rapid growth and bioassays with tobacco tissue cultures. Physiol.Plant. **15**: 473-497.

2.5
Anther culture for doubled haploid production in barley (*Hordeum vulgare* L.)

I. Szarejko
Department of Genetics, University of Silesia, Jagiellonska 28, 40-032 Katowice, Poland

Introduction

In barley, haploids can be produced both from female and male gametophytes, through a number of techniques, including: chromosome elimination following inter-specific crosses with *Hordeum bulbosum*, androgenesis using *in vitro* culture of anthers or isolated microspores, and gynogenesis using ovule culture *in vitro*. Among these systems, only two have been developed for large-scale production in barley: androgenesis including anther and microspore culture, and wide hybridisation followed by chromosome elimination. The isolated microspore culture system, often referred to as microspore or pollen embryogenesis, provides the most efficient and uniform route to mass scale production of haploid plants. It is, however, technically demanding and requires controlled environment for donor plant growth. Using anther culture and the protocol outlined below, we were able to produce green regenerants from donor plants grown in a greenhouse or in field conditions, for a wide range of barley genotypes. Among many factors, composition of the induction medium has been found to be important or even critical for androgenic response. In barley anther culture, major improvements have been achieved by replacement of sucrose by maltose as a carbon source, the balance of organic and inorganic nitrogen compounds and the replacement of agar by Ficoll or agarose in the induction medium. In the presented protocol, liquid induction medium BAC3 (Szarejko and Kasha, 1991; Cai *et al.*, 1992) supplemented with Ficoll 400 (a high molecular weight sucrose polymer which increases medium density) gave a higher androgenic response than other media tested, due to better embryo formation. We have routinely used BAC3 medium (Szarejko and Kasha, 1991; Cai *et al.*, 1992) for anther culture of barley varieties, hybrids, mutants and other breeding materials.

Protocol

Donor plant growth

Donor plants should be grown in controlled environments. It is very important to ensure that plants are grown under optimal conditions, as factors which introduce stress upon the donor plants, have a profound effect on androgenic response. The vigour of donor plants is influenced by several parameters, such as temperature, light intensity, photoperiod, nutrition, water relations and application of pesticides.

Seedlings of winter genotypes, 2 weeks after sowing into flats are vernalized at 4°C, 8 h photoperiod, light intensity 100 µE m^{-2} s^{-1} for 8 weeks. After this period they are transferred to 15 cm diameter pots with a mixture of soil, peat and perlite (3:1:1). Seeds of spring genotypes are sown directly to pots with the same soil mixture.

The highest efficiency of barley anther culture is achieved when donor plants are grown in a controlled growth room, at relatively low temperature, 15°C during the day and 12°C during the night, or alternatively, at 12°C constant, 16 h photoperiod, light intensity at a pot rim level about 350-450 µE m^{-2} s^{-1}, humidity 60-80%. We also grow donor plants during winter/spring season in a semi-controlled greenhouse, at a temperature 10-15°C during the day and 8-12°C during the night, 16 h photoperiod. Additional light is supplied by a mixture of high pressure sodium and mercury lamps (HQI-T 400W Planta and HQL 400W DE LUXE, Osram), to obtain the final intensity of 350-450 µE m^{-2} s^{-1}. During the whole vegetation period, plants should be properly watered and fertilized weekly to maintain a vigorous growth. Fungicides and insecticides may be applied, if necessary, during plant vegetation but such treatments should be avoided 2 weeks before spike collection. It is possible to use field grown material as donor plants but lower efficiency should be expected. The more proper and uniform the growth conditions of donor plants are, the more uniform is the development of microspores, and more green plants are produced in the culture.

Spike selection and microspore staging

The stage of microspore development is the other critical factor for successful induction of androgenesis. In barley, the development of microspores within florets and florets within the spike is generally synchronous, with apical and basal florets being slightly delayed. In 6-row genotypes, anthers in the side florets are also lagging behind in their development as compared to the main florets. Tillers containing spikes at the desired developmental stage can be pre-selected on the basis of their morphology. To confirm the initial staging of microspores, anthers from the central part of the spike are excised and squashed on a microscope slide in a drop of 4% acetocarmine. The stained microspores are examined under the light microscope. The stage of microspore development can be determined on the basis of the size and the position of nucleus and vacuole in the cell (Fig. 2.5-1a-h). The highest androgenic response is achieved when the collected spikes contain microspores at mid to late uninucleate stage of development (Fig. 2.5-1d-e).

In barley, the distance between the flag leaf and the penultimate leaf indicates the proper stage of microspore development. For most barley genotypes, tillers should be collected when the distance between the flag and the penultimate leaf is 3-6 cm (Fig. 2.5-2a). For some genotypes, the flag leaf may be emerged 0-1 cm above the penultimate leaf and for others, the tips of awns should be visible above the flag leaf. The plant morphology indicating the proper developmental stage of microspores depends both on plant genotype and growth conditions. Only those spikes with anthers at the mid and mid-late uninucleate stage should be used for pretreatment (Fig. 2.5-2b, c).

Cold pretreatment

Many barley genotypes require a 3-4 week cold pretreatment before culture. The collected tillers are surface sterilized with an aerosol of 75% ethanol. Spikes containing microspores at the mid uninucleate stage are removed from leaf sheaths and placed into 2-compartment Petri dishes, with a few drops of sterile water in one compartment. Alternatively, a small plate with water can be placed inside a bigger Petri dish, containing spikes (Fig. 2.5-2d). Plates are sealed with Parafilm, wrapped in foil and stored in a refrigerator or a cold room at temperature 4-5°C for 21-28 days.

Culture media
Media composition
In the presented protocol, liquid induction medium BAC3 (Table 2.5-1), is proposed for barley anther culture (Szarejko and Kasha, 1991; Cai *et al.*, 1992). The BAC3 is based on BAC1 medium developed by (Marsolais and Kasha, 1985) but with several modifications, such as the changed ratio of nitrate to ammonium form of nitrogen, the changed source of iron, the addition of some inorganic compounds ($KHCO_3$ and $AgNO_3$), the use of maltose as a sole carbohydrate and the changed composition of growth regulators. The BAC3 induction medium contains 30% Ficoll 400 (Amersham, previously Pharmacia) as a buoyancy increasing factor and is supplemented with 6% maltose, 2 mg/L NAA and 1 mg/L BAP.

The BAC3 regeneration medium is used as a solid, with Ficoll replaced by 0.3% Gelrite™. Additionally, in the regeneration medium maltose is replaced by 3% sucrose and myo-inositol is dropped down to 100 mg/L. Different growth regulators (0.5 mg/L IAA and 0.5 mg/L kinetin) supplement the medium.

Media preparation
For most media components (macro and micro elements, vitamins and growth regulators), stock solutions 10-100x concentrated can be prepared and stored in a refrigerator or a freezer for a long period of time. Concentration and storage temperature of stock solutions is given in Table 2.5-2. To avoid precipitation, macro elements can be divided into 2-3 separate stocks. Other media components, such as: sugars, myo-Inositol, Ficoll, and agar (Gelrite) are added directly to the medium. Except for sugars and agar, the other organic components should be stored in a refrigerator.

To facilitate the preparation of hormone stock solutions, auxins should be dissolved first in a few drops of absolute ethanol or 1 N NaOH, cytokinines in 1 N HCl. For example, to prepare 100 ml of 1 mg/ml IAA stock solution:
– weigh out 100 mg IAA
– dissolve auxin in 1 ml of 1 N NaOH or absolute ethanol
– bring up to 100 ml volume with deionized H_2O.

pH of the medium is adjusted with 1 N and/or 0.1 N NaOH and HCl. All induction medium components except Ficoll are filter sterilized through a 0.22 μm pore size filter. Ficoll is dissolved separately, using ½ of the final volume of the medium and autoclaved. Then Ficoll and the other components of the medium are mixed up in sterile conditions of the flow bench.

Examples of medium preparation
BAC3 induction medium - 100 ml of final volume
– Dissolve 30 g of Ficoll in 30 ml of deionized water by heating and stirring (you can use a microwave). After dissolving, when Ficoll goes into a clear solution, bring up to a volume of 50 ml (½ of the final volume). Autoclave it.
– Mix up 3 x 10 ml of 10x concentrated macro element stock solutions: A, B and C in a 50 ml beaker.
– Add 1 ml of 100x concentrated iron stock solution.
– Add 1 ml of 100x concentrated micro element stock.
– Add 1 ml of 100x concentrated $KHCO_3$ stock solution.

- Add 1 ml of 100x concentrated $AgNO_3$ stock solution.
- Add 1 ml of 100x concentrated vitamin stock solution.
- Add 1 ml of 100x concentrated organic acid stock solution.
- Add 200 mg of myo-Inositol.
- Add 30 mg of casein hydrolysate.
- Add 6 g of maltose.
- Add growth regulators. When you have 1mg/1ml stock solutions use: 0.2 ml NAA and 0.1 ml BAP per 100 ml of induction medium.
- Bring up to the final volume of 50 ml with deionized H_2O.
- Adjust pH to 6.2 with 1 N and 0.1 N NaOH.
- Filter sterilize using a 0.22 µm pore size filter.
- Under sterile conditions mix up 50 ml of autoclaved Ficoll with 50 ml solution of other media components which have been filter sterilized.
- Wrap in foil to protect the medium from light. Keep at room temperature for about 2 days and inspect for contamination before using. The medium should be made fresh at 2-week intervals.
- Before anther inoculation, pour medium into 60x15 mm sterile Petri dishes. Pour 3 ml of medium per dish. Use 10 ml sterile pipettes for medium transfer.

BAC3 regeneration medium - 1000 ml of final volume
- Mix up 3 x 100 ml of 10x concentrated macro element stock solutions: A, B and C in a 1000 ml flask.
- Add 10 ml of 100x concentrated iron stock solution.
- Add 10 ml of 100x concentrated micro element stock.
- Add 10 ml of 100x concentrated $KHCO_3$ stock solution.
- Add 10 ml of 100x concentrated $AgNO_3$ stock solution.
- Add 10 ml of 100x concentrated vitamin stock solution.
- Add 10 ml of 100x concentrated organic acid stock solution.
- Add 100 mg of myo-Inositol.
- Add 300 g of casein hydrolysate.
- Add 30 g of sucrose.
- Bring up to the volume of 980 ml with deionized H_2O.
- Adjust pH to 5.8 with 1 N and 0.1 N NaOH.
- Add 3 g of Gelrite.
- Autoclave the medium and cool it.
- Add filter sterilized growth regulators. When you have 1 mg/1ml stock solutions, use: 0.5 ml IAA and 0.5 ml kinetin per 1000 ml of the regeneration medium. Make up 20 ml of growth regulator solution and filter sterilize using a 0.22 µm pore size filter.
- Mix up the autoclaved medium with filter sterilized growth regulators. Pour the medium into 100x20 mm sterile Petri dishes.

Anther inoculation

Anthers from the middle six to eight florets in each row of the spike are used for culture. If spikes were cold pretreated before culture, the viability of microspores should be checked before culture by acetocarmine staining. Spikes containing anthers with more than 20% non-viable microspores should be discarded. The anthers are removed from florets under the stereo-microscope using fine tipped forceps, taking care not to rupture the anthers. The planting density should be around 10-15 anthers per 1 ml of medium, i.e. 30-45 anthers per one Petri dish (60x15 mm) containing 3 ml of medium. The anthers should be plated quickly taking care to prevent spikes from drying and the medium from evaporating. The culture dishes are sealed with Parafilm and wrapped with aluminium foil.

Table 2.5-1. Composition of BAC3 induction and regeneration media

Media components	BAC3 induction medium (mg/L)	BAC3 regeneration medium (mg/L)
Macro elements		
KNO_3	2,600	2,600
NH_4NO_3	200	200
$(NH_4)_2SO_4$	400	400
KH_2PO_4	170	170
$NaH_2PO_4 \times H_2O$	150	150
$CaCl_2 \times 2H_2O$	600	600
$MgSO_4 \times 7H_2O$	300	300
Iron source		
$FeNa_2EDTA$	40	40
Micro salts		
HBO_3	5	5
$MnSO_4 \times 4H_2O$	5	5
$ZnSo_4 \times 7H_2O$	2	2
KI	0.8	0.8
$Na_2MoO_4 \times 2H_2O$	0.25	0.25
$CuSO_4 \cdot 5H_2O$	0.025	0.025
$CoCl_2 \times 6H_2O$	0.025	0.025
Other inorganic components		
$KHCO_3$	50	50
$AgNO_3$	10	10
Vitamins		
myo-Inositol	2,000	100
Pyridoxine HCl	0.5	0.5
Thiamine HCl	1	1
Nicotinic acid	0.5	0.5
Ascorbic acid	1	1
Organic acids		
Citric acid	10	10
Pyruvic acid	10	10
Carbohydrates		
Maltose	60,000	-
Sucrose	-	30,000
Growth regulators		
NAA	2	-
BAP	1	-
IAA	-	0.5
Kinetin	-	0.5
Other organic components		
Casein hydrolysate	300	300
Ficoll	300,000	-
Gelrite™	-	3,000
pH	6.2	6.2

Table 2.5-2. Preparation and handling of stock solutions

Component	Concentration of stock solution	Storage temperature ($^{\circ}$C)
Macro elements*	10x	+4
Micro elements	100x	-20
Iron	100x	-20
Other inorganics	100x	-20
Vitamins**	100x	-20
Organic acids	100x	-20
Growth regulators	1-2 mg/ml	+4

* To avoid precipitation, macro elements are divided into 3 separate stock solutions: A - [KNO_3, NH_4NO_3]; B - [$(NH_4)_2SO_4$, KH_2PO_4, NaH_2PO_4 x H_2O]; C - [$CaCl_2$ x $2H_2O$, $MgSO_4$ x $7H_2O$]
**myo-Inositol is not included in stock solution

Culture incubation
Barley anthers are incubated at 26-28°C in darkness. When anthers are cultured in a liquid medium, the medium can be replenished after 2 and 4 weeks of incubation. One ml of the induction medium is drawn off using the Eppendorf pipette (with a sterile tip) and 1 ml of a fresh medium is slowly added. This procedure is always applied for highly responsive genotypes. Care should be taken not to withdraw anthers and embryoids, and not to sink them in the medium.

At about the 25-30th day of culture calli and/or embryoids should be apparent in the induction Petri dish (Fig. 2.5-2e-h). At that time, the number of responding anthers and number of calli/embryoids per 100 anthers cultured can be estimated.

Regeneration
Beginning from the 30th day of culture, calli or embryoids about 2 mm in diameter are transferred onto a regeneration medium, at a density of about 20 structures per 100x20 mm Petri dish. Culture dishes with developing embryoids should be inspected for the presence of structures ready for transfer every 5-7 days. Regeneration plates are incubated at 22±2°C under low light conditions (50 µE m^{-2} s^{-1}, provided by the fluorescent lamps Fluora L 36W/77, Osram). After 2-3 weeks from the transfer of calli onto the regeneration medium, the number of developing plantlets (Fig. 2.5-2i,j) can be estimated. For better root formation green plantlets can be transferred into jars or vials containing regeneration medium without growth regulators, supplemented with 4 g/L charcoal (Fig. 2.5-2k).

Transfer to soil
When roots have developed, plants are potted in small pots containing a mixture of soil and peat moss and transferred into a growth room or a greenhouse at 22/17±2° C day/night temperatures, 16 h photoperiod, light intensity 350-450 µE m^{-2} s^{-1} (Fig. 2.5-2l). To prevent plants from water stress they are covered with plastic or glass caps for about 1 week. When plants establish a vigorous growth and roots overgrow the soil, they should be transferred into bigger pots with the same soil mixture as used for donor plants growth. Root tips can be collected during re-potting to evaluate the chromosome number of microspore derived plants.

Colchicine treatment of haploid plants
Whereas about 70-90% of anther culture derived barley plants obtained with this protocol are spontaneously doubled and completely fertile (Fig. 2.5-2m), the colchicine treatment is not required. The remaining haploid plants are usually discarded. If there is a need to rescue all regenerated plants, the ploidy level of regenerants should be checked before colchicine treatment.

Haploid plants to be colchicine treated should be vigorous and have at least 2-3 tillers with several leaves. Plants are removed from pots and roots are rinsed with water. Roots can be trimmed off 3 cm from the base of the stem before treatment. Plants are placed in beakers containing colchicine solution so that the crown is completely immersed in the solution. Plants are treated for 5 hours at 22°C under light. After treatment, they are carefully rinsed in running water and re-potted into bigger pots.

For preparation of 1000 ml 0.1% colchicine solution use:
- Colchicine   1 g
- Tween 20 (wetting agent)   3 ml (about 10 drops)
- Dimethyl sulfoxide (DMSO)   20 ml.

Efficiency and Applications

We have routinely used BAC3 medium for barley anther culture as it gave us better response than other media tested. Using the BAC3 induction medium and the protocol outlined below, it was possible to produce green plants from a wide range of barley genotypes, including those which have not given any androgenic response on the FHG induction medium (Savaskan et al., 1999). Among the genotypes tested were spring barley varieties and breeding lines originating from different regions (Europe, Canada, USA, Peru, Middle East), F_1 hybrids, mutants and M_1 plants derived from mutagenic treatments with physical and chemical mutagens. The average efficiency of green plant production for 10 spring barley genotypes analysed in one experiment was about 12.2 green plants per 100 anthers plated. This frequency varied from 2-3 green plants for the less responsive genotypes to about 50-60 green regenerants for the most responsive varieties (Ximing et al., 1994; Szarejko et al., 1995). North American variety 'Bruce' and Peruvian 'UNA-LaMolina 96' were among the best responding genotypes tested. Winter variety 'Igri' (the model genotype in barley androgenesis) has not been examined with the presented protocol. While embryo/calli production was very high (on average 75% responding anthers with 285 transferable calli/100 anther plated), high frequency of albino regenerants remains the main problem to be solved. It should be noted, that the method does not require the step of chromosome doubling, as 70-90% of regenerated plants are spontaneously doubled (Szarejko et al., 1997). Using the presented protocol, it was possible to obtain DH lines for breeding and molecular mapping from Polish and exotic germplasm.

References

Cai, Q., I. Szarejko, K. Polok and M. Maluszynski, 1992. The effect of sugars and growth regulators on embryoid formation and plant regeneration from barley anther culture. Plant Breed. **109**: 218-226.

Marsolais, A. A. and K. J. Kasha, 1985. Callus induction from barley microspores. The role of sucrose and auxin in a barley anther culture medium. Can.J.Bot. **63**: 2209-2212.

Savaskan C., I. Szarejko and M.C. Toker, 1999. Callus production and plant regeneration from anther culture of some Turkish barley cultivars. Turkish J.Plant Botany, **23**: 359-365

Szarejko, I., D. E. Falk, A. Janusz and D. Nabialkowska, 1997. Cytological and genetic evaluation of anther culture-derived doubled haploids in barley. J.Appl.Genet. **38**(4): 437-452.

Szarejko, I., J. Guzy, J. Jimenez Davalos, A. Roland Chavez and M. Maluszynski, 1995. Production of mutants using barley DH systems. In: Induced Mutations and Molecular Techniques for Crop Improvement. IAEA, Vienna. pp.517-530.

Szarejko, I. and K. J. Kasha, 1991. Induction of anther culture derived doubled haploids in barley. Cereal Res.Commun. **19**(1-2): 219-237.

Ximing, L., C. Savaskan, K. Polok, A. Bielawska, I. Szarejko and M. Maluszynski, 1994. The effect of media and culture conditions on androgenic response in barley. In: Reports Bot. Garden, Polish Acad. Sci. Vol. 5/6. Rybczynski, J.J., I. Szarejko, J. Puchalski and M. Maluszynski (Eds.), Warsaw. pp.487-495.

2.6
Barley isolated microspore culture protocol

K.J. Kasha, E. Simion, R. Oro and Y.S. Shim
Department of Plant Agriculture, Biotechnology Division, University of Guelph, Guelph, Ontario, Canada N1G 2W1

Introduction

Among the cereals, barley (*Hordeum vulgare* L.) has been most successful in haploid production with many new cultivars produced. Most of these have come from wide hybridisation with *H. bulbosum* pollination. However, the potential for larger numbers from the culture of microspores is greater and successful procedures for isolated microspore culture of barley are now available and being used. Not only are the plant numbers per spike greatly increased but also the induced chromosome doubling in microspore nuclei leads to 70-80% of the plants recovered being completely fertile doubled haploids. Thus, colchicine or other chromosome doubling procedures is not needed. Our objective was to develop a highly efficient isolated microspore culture system using a defined induction media. This has been achieved with time saving features and the following protocol for barley isolated microspore culture was recently published in detail (Kasha *et al.*, 2001) and will be outlined here. The use of maltose as the sugar source in media may infringe upon patent rights for commercial haploid production.

Protocol

Donor plants and growth conditions

Frequencies of plants produced are usually higher and production more consistent when microspore donor plants are grown stress free. At Guelph, donor plants are grown in a growth room at 60-70% humidity and at cool temperature of 15°C for 16 h lighted period (350-400 $\mu E\ m^{-2}\ s^{-1}$) and 12 °C for 8 h dark period/day for winter barley and at 18°C and 15°C respectively for spring barley. Care should be taken to keep the plants disease and insect free and avoid drought stress. This is best achieved by growing the donor plants in a room isolated from other barley plants. If needed, spraying with an appropriate pesticide can be done at the early stages of development. Spraying plants for pests during the final 2 to 3 weeks prior to collecting microspores appears to be deleterious to response in culture. Fertilizing donor plants once a week with a water-soluble 20:20:20 fertilizer (N-P-K) suitable for cereals is recommended. Spacing of donor plants in pots to allow good tillering is also recommended. For breeding programs, the F_1 plants are most often used as donor plants, but selected F_2 plants may be desired for some breeding objectives.

Spike collection and pretreatment

Spikes from donor plants are collected at the mid- to late-uninucleate stage (Kasha et al., 2001) when checked on a central floret near the middle of the spike. Anthers are squashed in aceto-carmine stain for staging. Subsequently, tillers at a similar stage of morphological development are cut and placed in water at 4°C until (not longer than 4 days) sufficient numbers are obtained for blender isolation.

Spikes are removed from the sheath after spraying cut tillers with 70% ethanol. Awns are pulled off spikes and spikes arranged in large Petri dishes. However, if spikes have emerged from the sheath, they may be further sterilized with 10-15% Javex liquid bleach (i.e. diluted with water to give a final concentration of 0.5 to 1.0% chlorine content) for up to 10 min and then thoroughly rinsed in sterile distilled water before removing the awns. Many labs find that contamination is higher with isolated microspore culture than using anther culture and this step of bleach on spikes is the critical step in controlling contamination in wheat.

Various pretreatments are possible and we have used three that are equally effective:
1. Spikes in the Petri dishes can be partially immersed in ice-cold 0.3 M mannitol and kept in a dark refrigerator for 4 days at 4°C.
2. Spikes may be kept at room temperature for 4 days, partially immersed in 0.3 M mannitol.
3. Place spikes in a Petri dish with 0.5 ml of sterile water at 4°C for 3 to 4 weeks. Do not submerge the spikes in water.

Of these methods, we prefer the cold plus mannitol in that it is a short time frame and the microspore stage is maintained.

Microspore isolation

Under sterile conditions (such as flow bench) and with sterile glassware and instruments (autoclaved), pick up a spike with forceps and use scissors to clip off the small florets at both ends of the spike. Then cut the remaining spike into 2-3 cm segments into a chilled (refrigerated) Warring blender container, 125 or 300 ml size, depending upon the number of spikes. Add sufficient ice-cold 0.3 M mannitol to half-fill the container or to cover the spike segments. Blend at low speed for 4-7 seconds. Blending too fast or too long can damage the microspores and reduce their viability.

Filter the slurry from the blender through four layers of cheesecloth or a coarse nylon membrane. Quickly rinse the mash with cold 0.3 M mannitol and then filter the collected filtrate through a 100 μm mesh into graduated centrifuge tubes and rinse the mesh with cold 0.3 M mannitol. Spin the tubes in a centrifuge at 50 g (900 rpm) for 5 minutes. About 0.15 ml of microspores number about 500,000, so the approximate number of plates to be used can be estimated. Decant and suspend the microspore pellet in 0.3 M cold mannitol for washing and repeat centrifugation. This washing is done 2 or 3 times until the supernatant does not show a green tinge.

Decant the mannitol and suspend the microspores in 2 ml of FHG induction medium (Table 2.6-1). Check microspore viability under the microscope and if low, place the 2 ml of suspension on top of a 20% maltose solution and centrifuge it. Dead microspores fall to the bottom so carefully pipette off the viable microspores from the top of the maltose, add cold 0.3 M mannitol and centrifuge them for collection. Suspend the microspores in FHG induction medium (Table 2.6-1).

Microspore culture media
We have routinely used the FHG media of Hunter (1987), modified by using PAA (Phenyl acetic acid) as the auxin (Table 2.6-1). 2,4-D could also be used as the auxin but we have seen better embryo structure formation and regeneration using PAA. Key features of these media are low inorganic nitrogen (NH_4NO_3), high organic nitrogen (glutamine) and maltose as the sugar source. The pH is adjusted to 5.8 and 3 g/L of Phytagel are added to solidify the medium.

Microspore plating and culture
In a vacuum filter flask connected to a vacuum source, place two sterilized No. 2 Whatman filter papers. Rinse the filter papers either with liquid medium, 0.3 M mannitol or sterile H_2O. Under low vacuum, pipette drops of FHG medium with microspores onto the filter paper so they form a compact spot of 3 or 4 layers of microspores. Transfer the top filter with microspores to a plate with solidified FHG medium and seal the plates with Parafilm or similar material. More than 2 layers of Parafilm may hinder gas exchange and result in poor growth of microspores.

We usually put a high density of 500,000 microspores in one dish, as density is important. The average 2-rowed barley spike provides about 40-50,000 good microspores while a 6-rowed spike provides about 120,000. This can vary considerably with the technical skill in blending and isolation speed. Place the culture plates in the dark at 25-28°C. Monitor the plates weekly and add a drop of liquid media on top of the microspores if they start to dry out. Microspores could also be cultured in liquid medium or as a drop directly on top of solidified media but in our experience, embryo development is slower than on filter paper.

After 3-4 weeks in culture in the dark, some embryos will reach the size of 1-2 mm in length. Good strong embryos could go directly to MS regeneration medium (Table 2.6-1) but most often we will move them to FHG differentiation medium. FHG medium is modified with 30 g of maltose, 250 mg inositol, 1,000 mg of casein, 690 mg proline per litre along with normal FHG levels of thiamine, pyridoxine, nicotinic acid and BAP. The auxin is omitted. They are left on this differentiation medium for 1-2 weeks until root and shoot initiation is well developed and then transferred to MS regeneration medium. Smaller structures are left on the FHG induction medium but the medium is replenished with drops of induction medium on filter paper or solidified medium, or by the addition of liquid medium to cultures on liquid induction medium.

Table 2.6-1. Culture media used for induction and regeneration of isolated barley microspores (reproduced with permission from Kasha et al., 2001)

Media components	FHG induction media		FHG differentiation medium (mg/L)	MS regeneration medium (mg/L)
	original (mg/L)	modified (mg/L)		
Macro salts				
KNO_3	1,900	1,900	1,900	1,900
NH_4NO_3	165	165	165	1,650
KH_2PO_4	170	170	170	170
$MgSO_4 \times 7H_2O$	370	370	370	370
$CaCl_2 \times 2H_2O$	440	440	440	440
Micro salts				
$FeNa_2 \times EDTA$	40	40	40	37.3
$MnSO_4 \times 5H_2O$	22.3	22.3	22.3	22.3
H_3BO_3	6.2	6.2	6.2	6.2
$ZnSO_4 \times 7H_2O$	8.6	8.6	8.6	8.6
$CoCl_2 \times 6H_2O$	0.025	0.025	0.025	0.025
$CuSO_4 \times 5H_2O$	0.025	0.025	0.025	0.025
$Na_2MoO_4 \times 2H_2O$	0.25	0.25	0.25	0.25
KI	-	-	-	0.83
Other components				
Glutamine	730	750	-	146
Proline	-	-	690	-
Casein hydrolysate	-	-	1,000	-
myo-Inositol	100	100	250	100
Thiamine HCl	0.4	0.4	0.1	0.4
Nicotinic acid	-	-	0.5	0.5
Pyroxidine HCl	-	-	0.5	0.5
Sucrose	-	-	-	30,000
Maltose	62,000	62,000	30,000	-
PAA	-	10	-	-
IAA	-	-	-	1
BAP	1	1	0.4-1	-
Kinetin	-	-	-	1
SeaPlaque agarose	8,000	-	-	-
Phytagel	-	3,000	3,000	3,000
pH	5.8	5.8	5.8	5.8

Regeneration
Embryos placed on MS regeneration medium (Table 2.6-1) are left in the dark for 3-4 days and then moved into low light for 8 h/day at 22-25°C. If they are in Petri plates, transfer plantlets to larger containers with solidified regeneration medium without hormones. Once plants reach the 3 to 4 leaf stage they can be transferred into your usual potting mixture in pots. These can be stored in small pots under cool conditions until it is time for field planting or grown out in large pots in a greenhouse.

Efficiency and Applications

We have tested the above procedure on more than 30 different barley genotypes including 2-row and 6-row, European and Canadian lines, malting and non-malting genotypes as well as spring and winter growth habit types. The numbers of embryos per plate have ranged from 5.5 to 15 thousand with most being over 10,000. Regeneration frequency checked on the first 500 embryos has usually been between 80 and 90% (Kasha *et al.*, 2001). The occasional genotype will produce a high percentage of albino plantlets but embryo numbers and regeneration percentage are high enough that albinism is not a problem for obtaining green plants from such genotypes.

Anni Jensen (personal communication) in Denmark has developed and used a microspore culture system similar to ours to handle a complete barley breeding program for Pajbjergfonden. It involves about 250 crosses (F_1's) per year of spring barley and 200 crosses per year of winter barley, with the goal being 400 plants per cross but 200 is still acceptable. These levels are achieved in 70% of winter barley crosses and 40% of spring barley crosses over a number of years. Two skilled persons have been able to produce well over 100,000 plants per year and plants are transplanted directly into the field in the fall or spring. At present, two cultivars, 'Jacinta' and 'Helium' have been released from this program and many good lines are in final stages of evaluation.

The high frequency of completely fertile doubled haploids means that additional chromosome doubling procedures are not required. Additional details of procedures and discussion can be found in Kasha *et al.* (2001).

Acknowledgements

The barley haploid research has been supported by grants from NSERC and AAFC and by facilities and funding by OMAFRA and is gratefully acknowledged. Kind permission from Kluwer Academic Publishers of the journal Euphytica to use excerpts from our recent publication vol. 120:370-385 by Kasha *et al.*, is gratefully acknowledged. Our thanks are also extended to numerous researchers for their valued inputs and particularly the unpublished information from Anni Jensen, Pajbjergfonden.

References

Hunter, C. P., 1988. Plant regeneration from microspores of barley, *Hordeum vulgare* L. Ph.D. Thesis. Wye College, Univ. of London, London.

Kasha, K. J., E. Simion, R. Oro, Q. A. Yao, T. C. Hu and A. R. Carlson, 2001. An improved *in vitro* technique for isolated microspore culture of barley. Euphytica **120**: 379-385.

2.7
Barley isolated microspore culture (IMC) method

P.A. Davies
South Australian Research and Development Institute, GPO Box 397, Adelaide, 5001. Australia

Introduction

The protocol described here is for the culture of immature barley microspores, which are removed from the anther prior to culture. It is referred to as isolated microspore culture (IMC). Critical factors for success of the technique include: growing conditions for the donor plants; specific developmental stage at which microspores are collected; appropriate pretreatment of microspores; and appropriate culture media and culture conditions. This technique is applicable to all barley genotypes although some have a greater capacity for microspore growth and/or plant regeneration than others. For varieties such as the highly responsive 'Igri', we have observed regeneration levels up to 2,000 green regenerants per 100 anthers cultured, which is approximately equivalent to 1,000 green regenerants per spike. For our breeding and mapping spring barley populations, it is not uncommon to observe frequencies of 200-400 greens per spike although for some genotypes it is as low as 10-20 greens per spike.

Protocol

Donor plants and growth conditions

Donor plant growth conditions should be similar to the natural optimal growth conditions for the genotype being grown. Most of the material we grow for breeding purposes is spring barley and donors are grown in controlled environment growth rooms at 17/14°C day/night with a 14 h photoperiod under sodium vapour lamps (500 µmol $m^{-2} s^{-1}$ at canopy level). Seeds are sown directly into University of California (UC) potting mix with additional commercially available slow release fertilized at twice the recommended rate. The UC mix is prepared by mixing 1,200 litres of sterilized sand with 750 litres of peatmoss with the addition of calcium hydroxide (hydrated lime, 1 kg) calcium carbonate (agricultural lime, 1.8 kg) and NPK fertilizer (Nitrophoska 12-5-14, 2 kg). The pH of this soil mix is 6.8. Other soil mixes using composted materials have produced plants, which are suitable for IMC but the UC mix provides consistent and repeatable results.

When the donor plants are 4 weeks old a commercially available liquid fertilizer is applied weekly. It is important to keep donor plants pest and disease free since many chemical control methods reduce the efficiency of microspore culture. We have found that in general sulphur sprays and synthetic pyrethroids successfully control many of the common fungal or insect problems, respectively, without significantly affecting microspore culture. However we use them only when pest or disease problems arise and avoid them if at all possible. The sulphur product is called Microsul DF (TM - registered trade mark). It is supplied as 800 g per kg sulphur as wetable sulphur in water dispersal granule form. It is made up at the rate of 3 g/L and applied as a spray. The only foliar disease we have problems with in our controlled environment rooms is powdery mildew. This spray works well for this disease. The pyrethrin is manufactured by a company called Agchem in Australia. The chemical is supplied as a liquid with 4 g/L pyrethrins and 16 g/L piperonyl butoxide. It is made up at the rate of 12.5 ml/L of water and applied as a spray. We use this spray for the conrol of aphids, which are our main insect problem in the controlled environment growth rooms.

Pretreatments

Spikes should be collected when microspores are at the early to mid-uninucleate stage of development. At this stage the spike is yet to emerge from the leaf sheath. The stem is cut at least 10 cm below the spike and the cut end placed in a beaker of water to transport the spike to the laboratory. The spike remains in the sheath until it is removed under aseptic conditions following surface sterilisation of the sheath surface with 70% ethanol. Spikes are stored in Petri dishes in the dark at 4°C for 2 to 4 weeks before microspore isolation. A small piece of sterile dish-washing sponge saturated with water is placed in each Petri dish to maintain humidity.

Microspore isolation

Anthers are removed from 7 to 10 spikes under aseptic conditions and placed in a Petri dish containing 5 to 10 ml KFWC medium with 60 g/L maltose (Sigma Grade I, M-5885). KFWC medium is the "medium C" described by Kuhlmann and Foroughi-Wehr (1989) containing 1.0 mg/L indole-3-acetic acid (IAA) and 1.0 mg/L 6-benzylaminopurine (BAP) and modified to contain either 60 or 90 g/L maltose with the omission of Ficoll 400 and barley starch (Table 2.7-1). The anthers and solution are then placed in a microblender chamber (Model 8580, Eberbach, Mich., USA; see Fig. 2.7-1a) and the volume is made up to 20 ml with additional KFWC medium before blending at high speed for 20 seconds. The blended mixture is then filtered through a 100 µm sieve and centrifuged at 100 g for 10 minutes. The sieves can be easily made with a plastic autoclavable screw-cap jar and a piece of 100 µm sieve cloth secured by a lid with a large hole cut in it (Fig. 2.7-1b). After removing the supernatant, the microspores are resuspended in a solution containing 580 mM maltose plus 5 mM of the buffer MES (2-[N-morpholino]ethanesulfonic acid, Sigma M-8250), pH 5.8, to a final volume of approximately 11 ml and poured into a sterile 10 ml volumetric flask. Using a sterile Pasteur pipette, 1 ml of a solution containing 300 mM mannitol plus 5 mM MES (pH 5.8) is gently layered onto the maltose solution containing the microspores. The volumetric flask is capped with sterile aluminium foil and centrifuged at 85 g for 8 minutes. Following centrifugation viable microspores settle at the interface between the maltose and mannitol (Fig. 2.7-1c). These are removed with a Pasteur pipette in a total volume of approximately 500 µl, made up to 5 ml with KFWC medium containing 90 g/L maltose and the total number of microspores estimated by observing a 4 µl sample on a haemocytometer. The microspore solution is then centrifuged at 100 g for 10 minutes, the supernatant is removed, and a volume of KFWC (90 g/L maltose) added to produce a final microspore concentration of 1×10^5 microspores per ml. The microspores are plated either in 9 cm diameter Petri dishes (6 ml per dish) or 5 cm Petri dishes (2 ml per dish) with the addition of the antibiotic cefotaxime (Sigma, C-7039) at a concentration of 200 µg/ml and incubated in the dark at 22-25°C for 21-28 days. Cefotaxime is very effective in controlling bacterial contamination and has no adverse effect on the barley isolated microspore cultures.

Subculture of cell colonies
Using an inverted microscope, the total colonies per plate are estimated. For optimal regeneration, colonies are plated on KFWC (60 g/L maltose) medium solidified with 0.3% Phytagel (Sigma P-8169) at a final density of between 12 to 25 colonies per cm^{-2}. The cultures are incubated at 22-25°C under a 16 h photoperiod at 50 µmol m^{-2} s^{-1} using either metal halide lamps or a combination of cool white and grolux fluoresent lamps. On this medium shoots generally grow within two to four weeks. Fig. 2.7-2a illustrates microspore derived colonies immediately before subculture to solidified medium and Fig. 2.7-2b illustrates the subsequent growth of regenerants.

Root induction
When shoots are a minimum of 1 cm in length, they are transferred to growth regulator free Murashige and Skoog medium with 20 g/L sucrose solidified with 7g/L Difco BiTek agar. Fig. 2.7-2c shows the growth of rooted shoots on this medium after 4 weeks growth.

Transfer to soil
Rooted shoots are transferred to soil in a controlled environment growth room as described above for donor plant growth. High humidity is maintained for the first 3-5 days after transplanting by covering the plants with transparent plastic sheeting. Plants are grown in UC soil in 50 mm diameter plastic tubes (Fig. 2.7-2d). After 3-4 weeks the tubes are placed on a bed of UC soil of approximately 80 mm depth in plastic boxes, which have drainage holes to prevent the soil from becoming water logged. Twenty plants can be grown per box, each of which can produce several spikes. This is an efficient method of growing a large number of doubled haploids in a relatively small area (Fig. 2.7-2e).

Chromosome doubling
On average, at least 70% of plants transferred to soil have spontaneously doubled and are fully fertile so chromosome doubling is not necessary. This varies between genotypes and the lowest frequency of doubling observed in our laboratory is 50%. Tetraploids may exist among the fertile spontaneously doubled plants but these occur at frequencies which can be ignored for the purposes of plant breeding because tetraploid barley generally performs poorly in the field and is eliminated early in the selection process. We have found that the proportion of doubled plants in populations, which have a higher regenerative capacity is usually much higher than 70%. This is possibly because we have more shoots to choose from

on solid induction medium and in choosing the largest, most vigorous shoots we are biasing the selection in favour of doubled plants.

Table 2.7-1. KFWC medium (modified medium C, Kuhlmann and Foroughi-Wehr, 1989)

Medium components	Concentration (mg/L)
Macro salts	
$(NH_4)NO_3$	165
KNO_3	1,900
KH_2PO_4	170
$MgSO_4 \times 7H_2O$	370
$CaCl_2 \times 2H_2O$	440
Micro salts	
$MnSO_4 \times 4H_2O$	22.3
$ZnSO_4 \times 7H_2O$	8.6
H_3BO_3	6.2
KI	0.82
$CuSO_4 \times 5H_2O$	0.025
$Na_2MoO_4 \times 2H_2O$	0.25
$FeNa_2EDTA \times 2H_2O$	40
Other components	
Thiamine.HCl	0.4
Pyridoxine.HCl	2.0
Nicotinic acid	2.0
Ca-pantothenate	2.0
Biotin	0.02
Na-pyruvate	10
Citric acid	10
Casein hydrolysate	300
myo-Inositol	200
Glutamine	600
IAA	1
BAP	1
Cefotaxime	200
Maltose	60,000 or 90,000
pH	5.8

References and other related papers

Davies, P.A. and S. Morton, 1998. A comparison of barley isolated microspore and anther culture and the influence of cell culture density. Plant Cell Rep. **17** (206): 210.
Hoekstra, S., M.H. van Zijderveld, F. Heidekamp and F. van der Mark, 1993. Microspore culture of *Hordeum vulgare* L.: the influence of density and osmolality. Plant Cell Rep. **12**: 661-665.
Kuhlmann, U. and B. Foroughi-Wehr, 1989. Production of doubled haploid lines in frequencies sufficient for barley breeding programs. Plant Cell Rep. **8**: 78-81.
Mordhorst, A.P. and H. Lörz, 1993. Embryogenesis and development of isolated barley (*Hordeum vulgare* L.) microspores are influenced by amount and composition of nitrogen sources in culture media. J.Plant Physiol. **142**: 485-492.
Ziauddin, A., E. Simion and K.J. Kasha, 1990. Improved plant regeneration from shed microspore culture in barley (*Hordeum vulgare* L.) cv. Igri. Plant Cell Rep. **9**: 69-72.

2.8
Doubled haploid production in wheat through wide hybridization

M.N. Inagaki
Japan International Research Center for Agricultural Sciences (JIRCAS), Tsukuba, Ibaraki, 305-8686 Japan

Introduction

Practical breeding programs for self-pollinating crops, such as wheat (*Triticum aestivum* L., $2n=6x=42$), must include a process of genetic fixation for uniformity of agronomic traits after genetic recombination to increase variation. Repeated selection of heterozygous materials can increase uniformity, but many generation cycles are required to reach homozygosity in loci associated with agronomic traits. Of great interest to plant breeders, haploid production followed by chromosome doubling offers the quickest method for developing homozygous breeding lines. Doubled haploids derived from heterozygous materials show complete uniformity when used as recombinant inbred lines in selection procedures. An efficient technique for producing doubled haploids thus complements conventional breeding programs. Two major methods for producing wheat haploids (polyhaploids, $2n=3x=21$), one using microspores (pollen) and one using megaspores (egg cells), have been examined over last two decades. The method for producing haploids from cultured pollen has developed as an anther or pollen culture technique, and is still constrained by the different responses of wheat genotypes. On the other hand, the method using megaspores in ultra-wide crosses with Panicoides subfamilial species followed by embryo rescue has recently been developed. This paper reports on progress achieved in developing a method for wheat haploid production through ultra-wide crosses since the previous publication on producing wheat haploids through the bulbosum technique (Inagaki, 1990). The recent achievements can be attributed to the collaborative research projects of JIRCAS with the International Center for Agricultural Research in the Dry Areas (ICARDA), Syria, and with the International Maize and Wheat Improvement Center (CIMMYT), Mexico.

Protocol

Donor plants and growth conditions
Plant materials
Wheat and pollen donor plants [maize (*Zea mays* L.) and pearl millet (*Pennisetum glaucum* (L.) R.Br.)] grown in natural field or greenhouse (15-25°C, 12-14 h day length) can be used for the crossing. Wheat genotypes of winter growth habit require vernalization treatment for flowering. Sowing times of seed materials are determined in order to synchronize flowering times of wheat and pollen donors. A continuous supply of pollen from pollen parents sown at one or two-week intervals should be ensured. From a practical point of view, mixed pollen from several genotypes can be used in the crossing for decreasing the genotypic variation in wheat haploid production.

Crossing
At ear emergence, wheat spikes are emasculated following removal of the central florets of each spikelet and the apical and basal spikelets (Fig. 2.8-1a). Emasculated spikes with 25-30 florets are enclosed in glassine bags. Polyethylene bags may be used for maintaining humidity after emasculation. One day before predicted wheat anthesis, emasculated spikes are pollinated with freshly collected donor pollen (Fig. 2.8-1b). On two consecutive days after pollination, the uppermost internodes of wheat culms with pollinated spikes are needle-injected with a solution of 100 mg/L 2,4-D (Fig. 2.8-1c). Most pollinated florets produce plump seeds, but the seeds do not always contain embryos. Four embryos per wheat spike on average can be obtained.

Pollen storage
Stored pollen can be used for crossing on wheat when fresh pollen is not available. Pollen is collected from donors at anthesis and screened through a sieve to remove anthers. Ten grams of pollen are spread on a paper tray and dried with gentle ventilation at 35°C under 35-40% relative humidity. Pollen water content is determined from a 0.5 g pollen sample dried at 95°C for five hours. Optimum water content for stored pollen is 10-12% for maize and 5-7% for pearl millet (Fig. 2.8-1d). The dried pollen is distributed among cryopreservation tubes. The sealed tubes are stored in liquid nitrogen (-196°C) until use. After thawing sample tubes containing pollen in a water bath at 38°C for five minutes, pollen is checked for its germination viability and used for crossing. The drying and freezing process reduces pollen viability and may affect embryo formation frequency in wide crosses.

Detached-tiller culture
In the case of wheat plants growing in distant sites, detaching tillers with spikes is feasible for wide crosses. At ear emergence wheat tillers are cut off at the base and cultured in a flask containing tap water. Wheat spikes are emasculated as described above. After pollination, tillers are cultured in a solution containing 40 g/L sucrose, 8 ml/L sulfurous acid solution (H_2SO_3 containing 6% SO_2, Wako Pure Chemical Industries LTD., cat. No.196-11005) and 100 mg/L 2,4-dichlorophenoexyacetic acid (2,4-D) for two days. They are then transferred to a solution containing only sucrose and sulfurous acid, and cultured until embryo rescue. This process takes approximately 12 days (in maize crosses) and 14 days (in pearl millet crosses) after pollination (Fig. 2.8-1e). The culture conditions are 22.5°C, 12 h day length, 60-70% relative humidity and 5,000-10,000 lux light intensity in a growth chamber.

Embryo culture
Seeds set in wheat crossed with pollen donors are somewhat smaller than selfed seed, and seeds are filled with an aqueous solution instead of solid endosperm as found in selfed seeds (Fig. 2.8-1f). After 14 (in maize crosses) or 18 days (in pearl millet crosses) of spike growth on

wheat plants, wheat seeds are removed from spikes and sterilized in a dilute solution containing 1-2% sodium hypochloride of CLOROX and few drops/L Tween-20 for 10 minutes (Fig. 2.8-1g). After rinsing with sterilized water, embryos (approximately 1.0 mm in diameter) are asceptically excised (Fig. 2.8-1h) and transferred onto half strength Murashige and Skoog medium supplemented with 20 g/L sucrose and 6 g/L agarose (Type I, Sigma, cat. No. A-6013) in test tubes or plastic dishes (Fig. 2.8-1i). Embryos are incubated at 25°C, using 16 h day length and ca. 5000 lux light intensity (Fig. 2.8-1j). Plant regeneration frequencies from embryos are 50-70%.

Plant regeneration
Cultured plants with fully developed roots and leaves are transplanted to potted soil and grown further (Fig. 2.8-1k). The chromosome number can be examined in squashed preparations of root-tips stained with acetocarmine or acetoorcein (Fig. 2.8-1l).

Chromosome doubling
At tillering, roots are cut back to about 2 cm below the crown. Plants are then immersed in colchicine solution (0.05- 0.1% colchicine, 2% dimethyl sulfoxide (DMSO) and 15 drops/L Tween-20) in a flask for five hours at room temperature (Fig. 2.8-1m). Treated plants are potted following thorough washing with tap water (Fig. 2.8-1n). Potted plants are grown in favorable, tillering-enhancing conditions. Wheat spikes from colchicine treated plants are enclosed in glassine bags to prevent outcrossing. Wheat grains are harvested at a 60-70% success rate (Fig. 2.8-1o).

After multiplying harvested grains for one generation cycle (Fig. 2.8-1p), doubled haploid lines are planted in the field for evaluating genetic variation (Fig. 2.8-1q).

Efficiency and Application
A technique for producing wheat haploids using ultra-wide crosses followed by embryo rescue has been developed over the last two decades. Significant technical advances have been achieved by using pollen selected from different subfamilial species and applying plant growth regulators. Efficient crossing procedures were developed using stored pollen and detached tiller culture. At present, doubled haploids derived from hybrid progenies can be used as recombinant inbred lines with favorable uniformity in breeding selection. This technique could thus complement conventional breeding programs and accelerate the release of new varieties in developed countries, as well as in developing countries where rapid varietal development is critical for sustainable wheat production systems.

Pollen source
Wide crosses of wheat with bulbous wild barley (*Hordeum bulbosum* L.) pollen result in the production of immature haploid embryos of wheat after preferential elimination of *H. bulbosum* chromosomes from hybrid zygotes (Barclay, 1975). However, the crossability of wheat with *H. bulbosum* is genetically controlled by the genes *Kr1* and *Kr2* located on chromosomes 5B and 5A, respectively. According to the pedigrees of wheat varieties, crossable genotypes can be traced to the variety Chinese Spring or to materials of Asian origin. Both Japanese and Chinese wheat varieties, and in particular, local varieties, are highly crossable with *H. bulbosum*. Wheat genotypes carrying the dominant *Kr* gene(s) are not crossable with *H. bulbosum* and cannot produce haploid embryos. Non-hybridization of wheat genotypes with *H. bulbosum* is due to the failure of the pollen tube to penetrate the embryo sac. Application of plant growth regulator 2,4-D increases seed setting and embryo formation when crossable genotypes are used for crosses, but cannot break the barrier of cross-incompatibility. Since the efficiency of wheat haploid production is greatly influenced by the crossability of *H. bulbosum* onto wheat, the method using *H. bulbosum* crosses is restricted to crossable wheat genotypes.

Ultra-wide crosses of wheat with members of the Panicoides subfamily have been attempted in alien genetic transfer. Maize pollen can successfully hybridize with wheat egg cells and produce hybrid zygotes (Laurie and Bennett, 1986), irrespective of the presence of *Kr* gene(s). The maize chromosomes are rapidly eliminated from hybrid zygotes, requiring artificial rescue of proembryos at early developmental stage. 2,4-D treatment after pollination is critical to enhance embryo development in wheat x maize crosses (Suenaga and Nakajima, 1989). Maize pollination and subsequent 2,4-D treatment onto wheat results in production of immature wheat embryos capable of regenerating haploid plants, even for wheat varieties that are cross-incompatible with *H. bulbosum*. Genotypic differences in embryo formation frequency may not be significant for maize genotypes. Wheat haploid production through maize crosses has been confirmed using diverse wheat varieties. Some species related to maize, such as teosinte (*Z. mays* L. ssp. *mexicana*) and eastern gamagrass (*Tripsacum dactyloides* (L.) L.), are alternative pollen donors for wheat haploid production.

Cytological evidence indicates successful fertilization and elimination of paternal chromosomes from hybrid zygotes in sorghum (*Sorghum bicolor* (L.) Moench) and pearl millet crosses, which suggests that sorghum and pearl millet are potential pollen sources for wheat haploid production. Wheat haploids were obtained at high frequencies from sorghum and pearl millet crosses followed by 2,4-D treatment after pollination. However, sorghum crosses expressed a strong genotypic barrier of wheat to embryo formation. Therefore, haploid production through crosses with maize and pearl millet appears more stable than other methods because there is less genotypic effect on haploid embryo formation.

Pollen storage and detached tiller culture
Methodologies using ultra-wide crosses require viable pollen at the time of crossing, that is, flowering of wheat and pollen donors must be synchronized. This method can only be applied in seasons and places where both wheat and pollen donors grow. In addition, the environmental conditions where wheat plants will develop after crossing are critical to efficiency of wheat haploid production. Adequate techniques for long-term pollen storage and for culturing wheat tillers with spikes under controlled conditions are helpful for doing ultra-wide crosses without having to synchronize flowering times of both parents (Inagaki *et al.*, 1997).

Maize and pearl millet pollen can be successfully preserved at ultra-low temperatures for long periods. The process of storing pollen involves both drying and freezing. Understanding the effects of drying and freezing on pollen viability is essential for achieving successful long-term pollen storage. Pearl millet pollen is relatively tolerant to drying and freezing, in contrast with maize pollen, which has a narrow water content range for maintaining viability during drying and freezing (Fig. 2.8-2). As a result, pollen storage at ultra-low temperatures does not affect embryo formation frequency in wheat x pearl millet crosses, but greatly reduces frequency in wheat x maize crosses. Stored pearl millet pollen can be used as an alternative medium for producing wheat haploids when fresh pollen is not available.

A technique for artificially culturing detached wheat tillers has been developed through physiological research on immature seed vernalization. Major components of the culture solution are sucrose as a nutrient and sulfurous acid for preventing fungal contamination. Supplementing the culture solution with 2,4-D is essential for developing haploid embryos from wheat x maize crosses. The effectiveness of detached-tiller culture in combination with pollen storage in wheat haploid production has also been demonstrated. With detached tiller culture, it is possible to collect spikes from wheat plants growing in distant sites and handle them in a laboratory.

These techniques also avoid having to synchronize flowering times of both parents and result in considerable savings in terms of labor and space required for growing parent plants. They also provide greater flexibility in when and where wheat haploid production through ultra-wide crosses can be performed.

Haploid production

Wheat haploid production through ultra-wide crosses consists of two consecutive steps: formation of immature embryos by crossing and regeneration of haploid plants from rescued embryos. Factors such as the development stage of the wheat florets used for crosses and of the embryos rescued for plant regeneration are critical to haploid production efficiency. In general, crossing at an early development stage of wheat florets results in higher frequencies of embryo formation. Artificial rescue at the appropriate embryo development stage is required to attain higher frequencies of plant regeneration. Cytological examination of regenerated plants indicates that most of them are euhaploids with 21 chromosomes. Retention of pollen donor chromosomes was detected at the proembryo stage in crosses with maize and sorghum, and at the tillering stage of plants in crosses with pearl millet and maize. This does not eliminate the possibility of translocating chromosome segments of these pollen donors to the wheat genome background through somatic associations. However, variations in agronomic traits of doubled haploids were negligible in *H. bulbosum* and maize crosses. No significant distortion of segregation ratios was found in doubled haploids produced from hybrid progenies through *H. bulbosum* or maize crosses.

Efficiency of wheat haploid production over a three-year period at CIMMYT is shown in Table 2.8-1. The production scheme and efficiency is summarized in Table 2.8-2. Production of wheat haploids is efficient enough to obtain two haploid plants per wheat spike (1.5 doubled haploids per wheat spike after chromosome doubling). The process takes approximately nine months, from sowing of plant materials to harvesting of doubled haploid grains. However, it must be noted that production efficiency varies with the availability of environment-controlled facilities.

Table 2.8-1. Efficiency of wheat haploid production using maize crosses

Year	Wheat materials	No. spikes pollinated	No. embryos obtained	No. plants regenerated
1993	F_1 plants	284	1,449	810
1994	F_3 plants	456	1,219	898
1995	F_1 plants	501	1,894	1,267
Total		1,241	4,562	2,975

Table 2.8-2. Scheme of doubled haploid production in wheat through ultra-wide crosses and its efficiency

Month	Procedure	Efficiency
0	Sow wheat materials	
2.5	Cross wheat with pollen donors	10 spikes
3.0	Embryo rescue	40 embryos
4.0	Transfer haploid plants to soil	25 plants
5.0	Treat plants with colchicine	25 plants
7.5	Cover ears with glassine bags	25 plants
9.0	Harvest doubled haploid grains	15 plants

In case of haploid production of durum wheat (*T. turgidum* L. var. *durum*) through wide crosses with maize, frequency variation among durum wheat genotypes has been reported. However, significant difference in fertilization frequency was not found. These facts are an indication that the absence of the D genome in durum wheat causes differences in development of haploid embryos from hybrid zygotes, resulting in frequency variation of haploid production.

Acknowledgments
The author would like to thank Dr. M. Tahir, ICARDA, and Dr. A. Mujeeb-Kazi, CIMMYT, for research collaboration on wheat haploid production, and Ms. Alma L. McNab, wheat program editor, CIMMYT, for her editorial review of the manuscript. This paper was republished with minor changes from the institutional journal article (JIRCAS Journal No.4: 51-62,1997) under the permission of JIRCAS.

References

Barclay, I.R., 1975. High frequencies of haploid production in wheat (*Triticum aestivum*) by chromosome elimination. Nature **256**: 410-411.

Inagaki, M.N., 1990. Wheat haploids through the bulbosum technique. In: Biotechnology in Agriculture and Forestry. Bajaj, Y.P.S. (Ed.) Springer-Verlag, Berlin, pp. 448-459.

Inagaki, M.N., T. Nagamine and A. Mujeeb-Kazi, 1997. Use of pollen storage and detached-tiller culture in wheat polyhaploid production through wide crosses. Cereal Res.Commun. **25**(1): 7-13.

Laurie, D.A. and M.D. Bennett, 1986. Wheat x maize hybridization. Can.J.Genet.Cytol. **28**: 313-316.

Suenaga, K. and K. Nakajima, 1989. Efficient production of haploid wheat (*Triticum aestivum*) through crosses between Japanese wheat and maize (*Zea mays*). Plant Cell Rep. **8**: 263-266.

2.9
Protocol for wheat (*Triticum aestivum* L.) anther culture

J. Pauk, R. Mihály, and M. Puolimatka[1]
Cereal Research Non-Profit Co., Wheat Genetics and Breeding Department, Szeged, P.O.Box 391, H-6701, Hungary; [1]Plant Production Inspection Centre, Tampereentie 51, P.O.Box 111, FIN-32201 Loimaa, Finland

Introduction

After many years of different experiments around the world on haploid production different methods have become available to the breeders. In Szeged, *in vitro* androgenesis *via* anther culture is one of the most efficient systems of homozygous line production. It can be widely applied in wheat and could make a significant contribution to the breeding of new varieties by saving time and increasing selection efficiency via doubled haploid (DH) lines. During the development of the doubled haploid production system into a routine technique, many problems have been solved (induction medium, genotype dependence, carbohydrate source, breeding applications, etc.). Induction medium is one of the most important factors for the induction of androgenesis and the subsequent development of embryoids and plantlets. The potato (P-4) medium published by Ouyang *et al.* (1983), seems to be the most effective haploid induction medium, after our minor modifications (P-4mf, Table 2.9-1). The amino acids glutamine and asparagine have been shown to enhance androgenesis and they can partially replace the potato extract in potato medium. Serine has reputedly a positive effect on the anther wall in *Nicotiana* anther culture. Marsolais *et al.* (1986) and others, have described an efficient method of green plant production by anther culture of barley (*Hordeum vulgare* L.) and wheat (*Triticum aestivum* L.) using Ficoll in the media. Ficoll prevents anthers from sinking into the medium and avoids the presence of inhibitors from agar leading to improved androgenic response. Its application has opened up a new chapter in wheat anther culture. Among the conditions and variables used in anther culture system, incubation temperature and stage of microspore development are important factors. Cold pretreatment breaks down normal microspore development (gametophytic pathway) and increases the frequency of symmetrical mitoses, which causes a further increase in the appearance of spontaneous doubled haploid formation. Short heat shock treatments at the initiation of culture are also effective. The replacement of sucrose by other sugars, notably by maltose, enhances the success of anther culture. Experiments on cereals have generally shown an increase in embryo induction or an improvement in embryo development and green plant regeneration by the use of maltose instead of sucrose (Orshinsky *et al.*, 1990).

Protocol

Donor plants and growth conditions

In our breeding programme, we combine the conventional and *in vitro* doubled haploid breeding steps. The process starts with crossing, where the breeding goal is to increase the genetic variability of basic material. The microspores of the F_1 generation have already been subjected to segregation so that anther culture can be theoretically started from F_1 plant material. But, due to the lack of possibility of selection in the F_1 population, we recommend initiating the culture of the donor spikes collected from selected F_2 plants. This selection step can be very valuable, because the plants with clear negative variants (tall, disease sensitive etc.) are removed from the programme. Thus, we recommend culturing from the F_2 or from later generations (F_3-F_4) after a very strict selection at all breeding steps. The best time to collect the donor tillers is early morning, about 2-3 hours after sunrise.

For anther culture the donor plants are grown both in nursery and in glasshouse. In Hungary, for the nursery, the winter wheat is sown in October and the donor tillers are collected for haploid production during the first 10 days of May. For glasshouse conditions, seeds are germinated at 25°C and vernalized in the dark for six weeks at 4-6°C. Plantlets are transplanted to 17 cm plastic pots. Each pot contains three to four plants. In the glasshouse, plants are grown under natural light supplemented with high-pressure sodium lights (Philips 400 W) to provide light intensity 350-400 µE m^{-2} s^{-1} at the top of the canopy. The temperature is controlled between 17°C to 25°C, depending on the phenophase. We have not found significant differences in the response of nursery or glasshouse-derived donor tillers.

In the nursery and glasshouse the avoidance of chemical pesticides on the breeding material is impossible. Our results are produced under standard glasshouse and nursery technology including herbicide, fungicide and insecticide treatments.

Pretreatments

The donor tillers containing microspores at the late uninucleated stage are cut between the 2nd and 3rd nodes and the selected donor tillers are kept in Erlenmayer flasks with tap water. Then they are covered by PVC (polyvinyl ethylene) bags to keep high humidity, and put in the dark for cold pretreatment at 4°C for about two weeks. Before isolation, the microspore developmental stage is checked under a microscope by squashing the anthers in a drop of water and examining without a cover slip. Spikes containing early and mid uninucleated microspores are surface-sterilised with 2% NaOCl containing 2 drops of Tween-80 for 20 minutes on shaker and then rinsed 3 times with sterile distilled water.

Isolation, culture of anthers and plant regeneration

The use of fine-tipped forceps is recommended for the removal of awns. One of the best forceps is the 'Aesculap BD321' (Germany). The catalogue of the World Precision Instrument also includes many good quality stainless steel fine tweezers (type: 14101).

For the isolation and culture of anthers, the following steps are needed:

1. About 2-week cold pretreatment of donor spikes is used with mid uninucleated microspores (Fig. 2.9-1a). The relatively long treatment has a positive effect on spontaneous doubled haploid production.

2. Isolation: the Chinese original technique (tik-tak) works most effectively. The donor spike is held with one hand keeping the top of head between the second and third fingers and the bottom of head between the first and fourth fingers. The awns/spiklete are removed by one move using the fine-tipped forceps in the other hand. Removing of awns from the spikelet exposes the anthers that are then picked up and placed into a Perti dish containing

a thin layer of liquid medium P-4mf (Table 2.9-1). Two hundred fifty anthers are plated per 10 ml of medium in a 90 mm dish. Avoid damage to the anthers. Incubate the dishes in the dark with high (80%) humidity for 3 days at 32°C (heat treatment).

3. After 3 days, transfer the Petri dishes to the 28°C cabinet for 5-6 weeks, in the dark. Keep high humidity in the incubator.

4. Depending on the genotype, 4-6 weeks after isolation, embryo-like structures (ELS) are visible as white beads in the liquid medium (Fig. 2.9-1b and c).

5. When the size of ELS is about 1.5 mm and the bipolar embryo morphology is easily detected, transfer the ELS carefully to solidified regeneration medium (190-2Cu, Table 2.9-1) in Petri dishes (90 mm).

6. Incubate the regeneration cultures for 3-4 weeks in culture room (16/8 h light/dark) at 24-26°C, light intensity 20 $\mu E\ m^{-2}\ s^{-1}$ provided by cool-white fluorescent tubes. After 1 week of incubation, green and albino plantlets emerge.

7. Transfer green plantlets to individual culture tubes for tillering and rooting (Fig. 2.9-1d), keeping the plantlets in a same culture room.

8. After about 4-5 weeks of regeneration, the plantlets with good roots and tillers are mature enough for transplanting into the glasshouse. Use normal soil with 20% peat for transplanting and cover the plantlets with PVC bags individually, or keep under high humidity for 1 week. After the acclimatisation of plantlets, the PVC bags are removed and the well-growing plantlets undergo ploidy level determinations.

Potato extract preparation
After testing of different media, the modified P-4 medium was found to be the most effective. Two modifications are made to the components, namely 9% maltose is used instead of sucrose and 10% Ficoll replaces the agar (P-4mf medium, Table 2.9-1). In the P-4mf medium, the potato extract is a key component. For the preparation of potato extract the following steps are needed:

1. For 100 ml potato extract clean 100 g potato tubers (a red skin potato is recommended) under tap water by brushing. Do not peel tubers as only the physical cleaning is important.

2. Cut the potato tuber into 10 mm cubes into a 300 ml Erlenmeyer flask and add 100 ml distilled water. Cover the flask with aluminium foil.

3. Boil very slowly for 30 minutes so that only a few bubbles appear during boiling.

4. Sieve through 2-3 layers of cheese-cloth to provide about 70 ml of filtrate.

5. Add 80 ml distilled water to the potato cubes and boil in the same way as in point 3.

6. Filter again as in point 4, giving about 120-125 ml potato extract. Let it stay for about 1 h and then use 100 ml/L of medium.

Modified 190-2 medium
A modified 190-2 medium has proven to be better than any other media for plant regeneration. We find a significant effect of Cu^{++} ions on rooting of plantlets. For

regeneration of ELS, 190-2Cu medium, made by increasing the $CuSO_4 \times 7\ H_2O$ by twenty times (dose of MS) to 0.5mg/L (Table 2.9-1) was used.

Table 2.9-1. Composition of the used media in wheat anther culture

Media components	P-4mf medium (mg/L)	190-2Cu medium (mg/L)
Macro salts		
KNO_3	1,150	1,000
KCl	35	40
$(NH_4)_2SO_4$	100	200
KH_2PO_4	200	300
$Ca(NO_3)_2 \times 4H_2O$	100	100
$MgSO_4 \times 7H_2O$	125	200
Iron source		
Na_2EDTA	37.3	37.3
$FeSO_4 \times 7H_2O$	27.8	27.8
Micro salts		
$MnSO_4 \times 4H_2O$	-	8
$ZnSO_4 \times 7H_2O$	-	3
H_3BO_3	-	3
KI	-	0.5
$CuSO_4 \times 5H_2O$	-	0.5
Vitamins		
myo-Inositol	-	100
Thiamine HCl	1	1
Pyridoxine HCl	-	0.5
Nicotinic acid	-	0.5
Other components		
Glycine	-	2
Sucrose	-	30,000
Maltose	80,000	-
2,4-D	1.5	-
Kinetin	0.5	0.5
NAA	-	0.5
Gelrite	-	2,800
Ficoll 400	100,000	-
Potato extract	10%	-
pH	5.8	5.8

Ploidy level determination

There are three well-known methods for ploidy level determination: the classical root tip cytology on slides under a microscope (Fig. 2.9-2a); the measurement of stomata guard cell length; the flow cytometric analysis. The simplest is the comparison of stomata guard cell length.

To determine the ploidy level of the plantlets the length of the 10 mm distal leaf segment's stomata guard cells was checked. Chlorophyll is extracted in 70% alcohol and the leaf segments are mounted in a drop of water on a glass slide with a cover slip. The length of stomata guard cells is measured by an ocular micrometer. The length of the stomata guard cells of haploids is 40-50% shorter than that of control hexaploid plants. The regenerated plants are separated into two groups of ploidy (haploids and spontaneous doubled haploids). The haploids are treated with colchicine.

Colchicine treatment
Doubling of the number of chromosomes is one of the most critical steps in the doubled haploid breeding process. With the following technology, we can obtain about 50% chromosome doubling efficiency:

1. Prepare a 0.2% colchicine with 2% DMSO solution in tap water.

2. Clean the haploid plantlets from the soil under tap water. Trim the root system back to about 2 cm. Divide the plantlets if they have many tillers and store the plantlets till next morning in tap water.

3. Treat the plantlets in the above prepared colchicine solution for 5 h in glass vials (Fig. 2.9-2b).

4. Colchicine treatment is followed by an overnight washing under tap water.

5. The shoots are then trimmed back to about 10-12 cm (depending on individuals) before transplanting into pots. Grow the treated plantlets (Fig. 2.9-2c) under 16 h daylight and at 18/15°C day/night temperature under high humidity.

6. In the case of winter wheat, the well regenerated plantlets after the colchicine treatment are vernalized for 42 days in a cool chamber at 4°C, in dim light.

7. After 42 days of vernalization the plantlets are transferred to glasshouse under standard conditions for growth (Fig. 2.9-2d). Seeds of DH plants are harvested and sown using the ear to row system in a nursery.

Efficiency and Applications

Shortening the breeding time and increasing the efficiency of the breeding programmes are very important requirements in crop plant improvement. In the last decades of the 20th century, several new methods have been developed based on different plant tissue culture methods. The excellent basic and applied research results of wheat anther culture have opened new approaches in the efficient breeding of wheat.

Our experiments have confirmed a differential response among genotypes in anther culture, with superior induction coming from segregating breeding materials (F_2) compared to the parental lines or varieties. These results support the use of anther culture for breeding. Breeding programmes are generally based on large germplasm collections. These collections should be tested for response to induction and regeneration and the best responsive varieties integrated into crossings. The *in vitro* breeding method provides another selection criteria for the breeders to use in their breeding strategy. The first DH cultivar, 'GK Délibáb', (Fig. 2.9-3) came from an F_2 selection program (Pauk *et al.*, 1995). We have also used selected F_4 to F_6 plants for anther culture, particularly when selecting for disease resistance. While this procedure does not save time in the breeding program, we have been able to establish stable lines with both quality and disease resistance that should maintain its traits for a long time. The cultivars 'GK Szindbád' and 'GK Tündér' were produced from this process. The *in vitro* haploid induction and colchicine-induced doubled haploid production is routinely used for winter wheat breeding. During the last 16-17 years three patented DH varieties were released

(GK Délibáb, GK Szindbád, GK Tündér), confirming the applicability of the *in vitro* induced doubled haploids in breeding.

Regarding the concern on gametoclonal variation that may or may not be undesirable in breeding, we produced about 100-150 DH progenies from five elite varieties through anther culture. Among DH lines derived from these five varieties, we could not measure morphologically any essential (advantageous or disadvantageous) gametoclonal variation resulting from the tissue culture.

Most of the wheat breeders use sexual crossing (A x B) to increase the genetic variability of basic material. In our experience with using anther culture we have found that selection of F_2 plants has allowed us to reduce the frequency of negative plants obtained from anther culture. Benefits of this breeding approach are: perfect homogeneity, shorter breeding time, easy and simple variety maintenance and the disadvantages are: laboratory process, relatively high number of unusable DH lines because of lack of selection for agronomic characters.

Practical aspects of doubled haploid breeding have been demonstrated by the production of valuable breeding lines and new cultivars. The first wheat cultivar of anther culture-origin 'Jinghua No. 1' was released in China in 1986 (Hu *et al.*, 1986). In Europe, 'Florin' was the first androgenic doubled haploid wheat variety released in France (De-Buyser *et al.*, 1987). In Hungary, GK Délibáb was the first DH wheat variety registered and patented in 1992 (Pauk *et al.*, 1995). Since then, in Hungary four new DH wheat varieties have been released: GK Szindbád (1996), GK Tündér (2001) from the laboratory and breeding program of the Cereal Research Non-profit Co., Szeged, and varieties 'Mv Szigma' and 'Mv Madrigál' from Agricultural Research Institute of the Hungarian Academy of Sciences at Matonvásár'(Barnabas *et al.*, 2000).

Besides these achievements many problems still exist, such as genotype dependence (to day it is less true than 20 years ago), plant regeneration, albinism etc., and must be solved to make doubled haploid production a more routine technique.

References

Barnabas, B., E. Szakacs, I. Karsai and Z. Bedo, 2000. *In vitro* androgenesis of wheat from fundamentals to practical application. In: Wheat in a Global Environment. Bedo, Z. and L. Lang (Eds.) Kluwer Acad.Publishers, Dordrecht, pp. 517-525.

De-Buyser, J., Y. Henry, P. Lonnet, R. Hertzog and A. Hespel, 1987. 'Florin': A doubled haploid wheat variety developed by the anther culture method. Plant Breed. **98**: 53-56.

Hu, D., Z. Yuan, Y. Tang and J. Liu, 1986. Jinghua No. 1 - A winter wheat variety derived from pollen sporophyte. Scientia Sinica Series B, **XXIX**: 733-745.

Marsolais, A.A., W.G. Wheatley and K.J. Kasha, 1986. Progress in wheat and barley haploid induction using anther culture. Spec.Pub.No.5.Agron.Soc.N.Z.: 340-343.

Orshinsky, B.R., L.J. McGregor, G.I.E. Johnson, P. Hucl and K.K. Kartha, 1990. Improved embryoid induction and green shoot regeneration from wheat anthers cultured in medium with maltose. Plant Cell Rep. **9**: 365-369.

Ouyang, J.W., S.M. Zhou and S.E. Jia, 1983. The response of anther culture to culture temperature in *Triticum aestivum*. Theor.Appl.Genet. **66**: 101-109.

Pauk, J., Z. Kertesz, B. Beke, L. Bona, M. Csosz and J. Matuz, 1995. New winter wheat variety: 'GK Delibab' developed via combining conventional breeding and *in vitro* androgenesis. Cereal Res.Commun. **23**(3): 251-256.

2.10
Protocol for producing doubled haploid plants from anther culture of wheat (*Triticum aestivum* L.)

B. Barnabás
Agricultural Research Institute of the Hungarian Academy of Sciences, Martonvásár, Hungary

Introduction

The use of anther culture to produce homozygous doubled haploid lines for wheat cultivar development has been widely advocated, and during the last few years has become an increasingly important technique in many breeding programs (Henry and De Buyser 1990). The anther culture procedure, however, can only become an efficient, economical part of breeding practice if it ensures the production of a sufficient number of cytologically stable doubled haploid plants from a wide range of genotypes. Since haploid induction ability has strong genotype dependence, wheat anther culture has been used more efficiently (Holme *et al.*, 1999) in certain geographical regions (China, Central and East Europe) than in others (North and West Europe). In the following protocol an efficient, reproducible anther culture methodology is presented for producing homozygous doubled haploid plants of common wheat (*Triticum aestivum* L.).

Protocol

Collection of spikes
In the case of field-grown plants the developmental stage of the spikes can be estimated by manual touching, because development is more or less synchronized in the population. Plants in the boot stage are suitable for cytological analysis. For plants grown in a greenhouse or growth chamber it is necessary to carry out individual checks.

Observation of the right developmental stage of the microspores
Samples must be randomly chosen from field-grown plants for the examination. Plants grown under artificial climates should be examined individually. Anthers from the apical, middle and basal parts of the ear are used for making squashes stained by acetocarmine. Microspores released from the anthers can be observed under a light microscope. The optimal developmental stage for *in vitro* androgenesis is the late uninuclear stage (Fig. 2.10-1a), but the mid uninuclear stage can also be used.

Sterilization of spikes
It is necessary to use pre-sterilized dishes and forceps for this procedure. Sterilization should be done in laminar air flow cabinets. The spikes are sterilized by immersion in 20% sodium hypochlorite solution for 30 minutes, then wash in two changes of sterile distilled water. Finally, spikes are transfered to sterile Petri dishes.

Inoculation of the anthers onto the induction medium
1. Preparing of the synthetic medium W14 (Ouyang *et al.*, 1984) (Table 2.10-1).

2. Autoclaving the medium for 20 min at 120°C, 90 kPa. After autoclaving, filter sterilized colchicine (Sigma C 3915) is added to the induction medium at a concentration of 40 mg/L. Colchicine is sterilized by filtration through a MILLE X-GS 0.22 µm filter (MILLIPORE).

3. Distribution of the medium into Petri dishes (90 mm in diameter, glass or plastic) 25 ml each. The procedure must be done in laminar cabinets.

4. Inoculation of the anthers. Under sterile conditions, using a pair of forceps, place no more than 100 anthers onto the medium and close the lid. Seal with Parafilm or plastic foil.

5. Early genome redoubling. Incubate the anthers on the colchicine containing medium for 3 days at 29°C in the dark to complete the first microspore mitosis (Fig. 2.10-1b). After the treatment transfer the anthers to a colchicine free induciton medium for further androgenic development.

Incubation of the anthers
Anthers are carried out at a constant $29\pm1°C$ temperature, 80% relative humidity (RH), in the dark for 1 month. As a consequence of the colchicine treatment a significant increase in the frequency of symmetrical nuclear divisions, and a considerable increase in the yield of embryos may occur in a number of genotypes (Fig. 2.10-2 a-c). During the 1-month incubation embryos and calli of microspore origin will develope from the anther cultures (Fig. 2.10-3).

Table 2.10-1. Composition of media used for wheat anther culture

Media components	W14 induction medium (mg/L)	190-2 regeneration medium (mg/L)
Macro elements		
KNO_3	2,000	1,000
$NH_4H_2PO_4$	380	-
$MgSO_4 \times 7H_2O$	200	200
$CaCl_2 \times 2H_2O$	140	-
K_2SO_4	700	-
KH_2PO_4	-	300
$(NH_4)_2SO_4$	-	200
$Ca(NO_3)_2 \times 4H_2O$	-	100
KCl	-	40
Micro elements		
$ZnSO_4 \times 7H_2O$	3	3
KI	0.5	0.5
$MnSO_4 \times 4H_2O$	8	8
H_3BO_3	3	3
$Na_2MoO_4 \times 2H_2O$	0.005	-
$CuSO_4 \times 5H_2O$	0.025	-
$CoCl_2 \times 6H_2O$	0.025	-
Iron source		
Na_2EDTA	37.3	37.3
$FeSO_4 \times 7H_2O$	27.8	27.8
Vitamins		
Nicotinic acid	0.05	0.5
Pyridoxine HCl	0.05	0.5
Thiamine HCl	2	1
Myo-Inositol	-	100
Hormones		
2,4-D	2	-
Kinetin	0.5	0.5
NAA	-	0.5
Other components		
Glycine	2	2
Sucrose	100,000	30,000
Agar	6,000	6,000
pH	5.8	6.0

Plant regeneration from microspore-derived embryos or calli

1. Transfering embryogenic structures to regeneration medium. The 190-2 (He and Ouyang, 1984) medium is used for plant regeneration (Table 2.10-1).

2. Sterilization of the medium for 20 min at 120°C, 90 kPa.

3. Distribution the medium into sterile flasks (200 ml) 20-25 ml each.

4. Inoculation of the embryogenic structures: Transfer the individual embryos or calli to the regeneration medium using a needle.

5. Incubation of the embryogenic structures at 26°C, 50% RH, 16 h daylength, 50 µmol m^{-2} s^{-1} light intensity for four weeks.

6. Transplantation of the green plantlets onto fresh medium. The medium consists of the same ingredients as the regeneration medium with the exception of plant hormones. Sucrose is reduced to 1.5%.

7. Vernalization. After one week put the dishes into a cold room (2°C, 8 h daylength with 62.5 µmol m^{-2} s^{-1} light intensity) for a 6-week period.

Planting of the green plants into pots
Free the plants from agar by gently washing with tap water and then transfer them to pots containing potting compost. Cover the plantlets within plastic foil to prevent desiccation. After 2 weeks remove the foil and transfer to larger pots where the plants will develop further.

Growth of doubled haploid regenerants
DH are grown in a greenhouse (Sept.-Nov. or Feb.-April) at 17°C constant temperature, 80% RH, 16 h day length, or in phytotron climate chambers (E15 CONVIRON) using the T3 climatic program till maturity (Table 2.10-2, Fig.2.10-4.b)

Table 2.10-2. T3 climatic programme for growing cereals in growth chambers

	Weeks					
	1	2	3	4	5	6[*]
	Temperature (°C)					
Day	16.0	16.0	17.0	17.0	18.0	18.0
Nigth	15.0	15.0	15.0	15.0	15.0	15.0
	Relative humidity (%)					
Day	65.0	65.0	65.0	65.0	65.0	65.0
Night	75.0	75.0	75.0	75.0	75.0	75.0
Lights	Hours					
1/3 On	5.15	5.15	4.45	4.45	4.14	4.15
2/3 On	7.30	7.30	7.00	7.00	6.45	6.45
3/3 On	9.30	9.30	9.30	9.30	9.15	9.15
1/3 Off	14.00	14.00	14.00	14.00	14.15	14.15
2/3 Off	16.15	16.15	16.30	16.30	16.45	16.45
3/3 Off	18.30	18.30	18.45	18.45	19.15	19.15
	Illuminance (klx)					
3/3 On	25	25	25	25	25	25
	Photosynthetic Photon Flux Density (µmol/m^{2}s)					
3/3 On	320	320	320	320	320	320

[*] For cereals, this programme is repeated until the end of flowering

Efficiency and Applications

Tests on the anther culture response of more than 30 winter and spring wheat genotypes demonstrated that a considerable quantity (10-85%) of structures of microspore origin can be induced from anthers cultured on a wholly synthetic medium W14 (Ouyang et al., 1989) and a suitable number of green doubled haploid plantlets (1-40 plants/spike) can be regenerated from the calli or microspore embryos. The advantages of the present technique are that: no pretreatments are necessary before culturing the anthers; reproducible synthetic medium is used instead of a medium containing natural plant extracts (Chuang et al., 1978); chromosome doubling in the microspores is carried out at the onset of culturing, ensuring the regeneration of truly homozygous doubled haploid plants.

The anther culture protocol described above is recommended for use mainly with Central and Eastern European wheat genotypes in which spontoneous chromosome doubling is rare. It provides satisfactory microspore induction for a wide range of genotypes and a suitable number of doubled haploid green plants. Across 32 genotypes, anther response ranged form 6.8 to 82.3%, with an average response of 26.3%, based on 2000 anthers per genotype. The percentage of green plants per 100 anthers ranged from 1.8 to 43.8%, with an average of 11.4%.

The early microspore genome reduplication method seems to be effective: the low concentracion of colchicine does not reduce the induction parameters to any great extent and in fact leads to a drastic decline in the number of albino regenerants in a number of the genotypes tested. The large number of microspore embryos developing in colchicine-treated anther cultures are capable of regenerating viable green plants with good fertility parameters (Table 2.10-3) leading to genetically stable DH progeny generations for further breeding.

Table 2.10-3. Fertility of DH1 plants regenerated from anther cultures

Genotype	Treatment	Fertile plants (%)	Seed production per fertile plant (No.)
Mv Szigma	A	25	29.5
	B	100.0**	84.8**
Mv Pálma	A	61.1	39.9
	B	92.0*	120.0**
Mv 16	A	68	19.6
	B	93.3*	63.4*
Diószegi 200	A	32.3	33.0
	B	56.6^{ns}	270.5**

A/ colchicine free induction medium (control); B/ colchicine added to the induction medium at a concentration of 0.04%. Data calculated on the average of 20 plants. *Significantly different from the control at P=0.05; **Significantly different from the control at P=0.01

Androgenetic doubled haploid development based on wheat anther culture enables breeders to dramatically reduce the time interval required from crossing to the development of a uniform line.

References

Chuang, C.C., J.W. Ouyang, H. Chia, S.M. Chou and C.K. Ching, 1978. A set of potato media for wheat anther culture. In: Proc. Symp. Plant Tissue Culture. Science Press, Beijing, pp. 51-56.

He, D.G. and J.W. Ouyang, 1984. Callus and plantlet formation from cultured wheat anthers at different developmental stages. Plant Sci.Lett. **33**: 71-79.

Henry, Y. and J. De Buyser, 1990. Wheat anther culture: agronomic performance of doubled haploid lines and release of a new variety "Florin". In: Biotechnology in Agriculture and Forestry. Vol. 13. Bajaj, Y.P.S. (Ed.) Springer-Verlag, Berlin, pp. 285-352.

Holme, I.B., A. Olesen, N.J.P. Hansen and S.B. Andersen, 1999. Anther and isolated microspore culture response of wheat lines from northwestern and eastern Europe. Plant Breed. **118**: 111-117.

Ouyang, J.W., S.E. Jia, C. Zhang, X. Chen and G. Fen, 1989. A new synthetic medium (W14) for wheat anther culture. Ann.Rep.Inst.Genet.Sin. pp. 91-92.

2.11
Wheat anther culture

S. Tuvesson, R. von Post and A. Ljungberg
SW Laboratory, Svalöf Weibull AB, SE-268 81 Svalöv, Sweden

Introduction

The following procedure for large-scale production of doubled haploids (DH) in wheat at Svalöf Weibull AB (SW) is based on protocols developed in Sven Bode Andersen's laboratory in Copenhagen (Andersen *et al.*, 1987; Tuvesson *et al.*, 1989). Since then the tissue culture programmes were rationalised to a few steps in order to speed up DH production and make the procedure cost effective. The described DH method is based on the knowledge that the response of the parental lines to anther culture influences the response of progeny combination (Tuvesson *et al.*, 1989; Zhou and Konzak, 1992). At the same time it allows the practical use of anther culture technique in current breeding programmes. The goal is to produce DH from as many breeding combinations as possible rather than reaching high values for a few selected crosses. With a simple screening procedure of parental lines non-responding types are avoided. Circa 4,000 green wheat haploids are regenerated per year in both the screening experiments and the production. The method has been used successfully in French and Swedish winter wheat breeding programmes.

Protocol

Donor plants and growth conditions
Screening for regeneration of green haploids takes place throughout the year, from mainly glasshouse grown material. Donor plants raised in the glass-house are germinated and vernalized for two months at 3°C before planting. The plants receive natural light and during the darkest months (November - February) artificial light is supplied. The donor plants are grown in 12 cm plastic pots, at 10-15°C night and 15-20°C day temperatures. DH production for the breeding programmes takes place from April to August and the donor plants are grown in the glasshouse or preferable in the field. The anther culture response from glasshouse grown material is approximately a half of the response compared to field grown donor plants.

Anther culture
The microspores used are in the early and middle uninucleate stage of development (He and Ouyang, 1984). Development of the microspores is checked under microscopy a few times during the season. Eight spikes are cut from a minimum of four plants in order to reduce the effect of plant and spike in the screening experiments (Dunwell et al., 1987). To produce DH for breeding programmes, approximately 20 spikes are used and the number of spikes may be adjusted based upon knowledge about the response of the parents. The cut spikes are placed at 7°C for a maximum of 14 days before handling. Spikes from the glass-house are sterilized with 70% ethanol using moistened filter paper before plating of the anthers and field-grown material are sterilized with 0.1% mercury chloride for 8 minutes (Ouyang et al., 1983) followed by washing with three changes of sterile water. *Warning: mercury chloride is a toxic compound and should be handled according to regulations.* Forceps are used to dissect out the anthers, which are placed in 9 cm plastic Petri dishes with substrate.

The substrate used for anther culture is 190-2 (Wang and Hu, 1984), supplemented with 9% maltose, 1.5 mg/L 2,4-D, 0.5 mg/L kinetin, and without NAA. Substrates are solidified with 0.35% Gelrite (Kelco) (Table 2.11-1). After plating the anthers, Petri dishes are placed in the dark at 28°C.

Plant regeneration
Embryos are transferred for regeneration after about 30 days of anther culture. Regeneration is achieved with 190-2 substrate, but without hormones (Wang and Hu, 1984), (Table 2.11-1). Regeneration takes place at 24°C in cool white fluorescent light (OSRAM tubes L36W/20) under a 16/8 h photoperiod. Alternatively, Petri dishes are placed in room temperature and receive normal day light. Wheat haploids in tissue culture are vigorous and have no special demands (Fig 2.11-1).

Chromosome doubling
Regenerated plants are left in Petri dishes until a good rooting system has developed after which transfer to soil takes place. It is important to keep plantlets in a humid environment until they adjust to glasshouse conditions. Colchicine treatment of plants is achieved after washing the roots and carefully wiping with filter paper. The plants are placed with roots in 0.1% colchicine + 2% DMSO in the glasshouse under a lamp for 4 hours. After washing in running water for 0.5-1 h, the plants are left over night in water and then planted in soil.

Table 2.11-1. Modified 190-2 substrate from Wang and Hu (1984) for wheat anther culture

Media components	190-2 induction medium (mg/L)	190-2 regeneration medium (mg/L)
Macro salts		
KNO_3	1,000	1,000
$(NH4)_2SO_4$	200	200
$MgSO_4 \times 7H_2O$	200	200
KH_2PO_4	300	300
$Ca(NO_3)_2 \times 4H_2O$	100	100
KCl	40	40
Micro salts		
$MnSO_4 \times H_2O$	8	8
$ZnSO_4 \times 7H_2O$	3	3
H_3BO_3	3	3
KI	0.5	0.5
Iron source		
$Na_2EDTA \times 2H_2O$	37.3	37.3
$Fe_2SO_4 \times 7H_2O$	27.8	27.8
Other components		
myo-Inositol	100	100
Thiamine HCl	1	1
Nicotinic acid	0.5	0.5
Pyridoxine HCl	0.5	0.5
Glycine	2	2
2,4-D	1.5	-
Kinetin	0.5	-
Maltose	90,000	-
Sucrose	-	20,000
Gelrite	3,500	3,500
pH	6.0	6.0

Pre-screening of parental lines for anther culture response
DH production from F_1 and F_2s is based on results from a pre-screening of parental lines using the following procedure:
1. Eight spikes are harvested from a minimum of four plants.

2. Anthers from the eight spikes are cultured according to the described procedure, except that plants are not saved for chromosome doubling.

3. The numbers of green haploids per spike are calculated.

4. The genotype producing more than one green plant per spike is accepted as a parental line.

5. No information is needed about the other parent.

Figure 2.11-1. Vigorous haploid from 'Ciano' in tissue culture. Ciano is a spring wheat variety used by SW as reference genotype. It responds well to anther culture, giving up to 18 green haploids per spike under good growth conditions.

Large-scale DH production
The following rationalisations and 'short cuts' make the protocol well suited for large-scale production of DH at a low cost.
1. The described method allows anther culture to be performed with a minimum of resources: sterile cabinet, autoclave, 28°C dark room or incubator and a glass-house.

2. Field grown donor plants for DH production give the best result in anther culture and glass-house place can be saved.

3. Screening of pure lines for DH response may take place the whole year, which allows flexibility in laboratory and greenhouse activities.

4. A single stock medium 190-2 is used for anther culture and regeneration of plantlets.

5. The development stage of microspores is not tested on a routine basis.

6. The number of anthers is not counted, which allows approximately 18 spikes to be plated in one hour. The average number of anthers per spike may be used if number of embryos or haploids need to be related to 100 anthers.

7. Embryos are neither counted nor evaluated for size or quality, but simply transferred for regeneration.

8. The screening of a genotype based on 8 spikes take about one hour of laboratory work, thus 40-50 genotypes can easily be screened in a few weeks.

Efficiency and Applications

In a wheat screening study with advanced breeding lines from Germany, France, Sweden, UK and other countries, the average response was 2.1 green haploids per spike (3.3 green haploid per 100 anthers). Forty eight percent of the genotypes produced more than one green haploid per spike (1.6 haploids per 100 anthers) and breeding combinations with these as one parent can be used for anther culture. No information is needed about the other parent.

The average response 4.7 green plants per spike (5.6 green plants per 100 anthers) was obtained from F_1 breeding combinations of French and Swedish breeding programmes. Significant correlation was observed between the number of green haploids produced by the hybrids and that produced by the parental lines that had been positively characterized in pre-screening by anther culture. There was no significant difference between the average number of green plants per spike in the F_1s and the average response of the actual parents previously characterized in anther culture. All but three combinations gave green plants in anther culture.

In the screening study and the five years of testing, the land of origin seem not to be important factors in determining the degree of response and the results are therefore of general value to the wheat breeder. The data related to how 'in-house advanced breeding lines' respond to anther culture are added to databases which include published information about commercial cultivars. Such data can be helpful in the selection of crosses for DH production for wheat breeding. Löschenberger and Heberle-Bors (1992) published information about 31 winter wheat cultivars. Other examples of such screening include 75 genotypes of winter and spring wheat cultivars (Foroughi-Wehr and Zeller, 1990), 60 genotypes from Orlov et al. (1993) and 91 winter wheat cultivars from Tuvesson et al. (2000).

We chose as a lower limit for screening one green haploid per spike for one parent. Fifty two percent of the analysed wheat genotypes responded with less than one green plant per spike. Combinations of such low responding genotypes in anther culture of F_1 and F_2 appear not to be useful. Before this screening programme was initiated, we observed that approximately 25% of the breeding combinations failed to respond with green plants. Such combinations are now avoided in principle. Only three wheat F_1's did not respond with green plants. The three non-responsive wheat combinations had a parent that responded well. For example a F_1 breeding combination between the cultivar 'Champetre', which responded with 3.0 green plants per spike and a parent that was not previously characterized in anther culture did not respond with green plants. A reciprocal effect may be a possible explanation as to these three F_1's failed to respond in anther culture.

Our first DH cultivar, 'SW Agaton' was released in 2001. Agaton derives from an anther plated in 1993 from the cross: 'Sv92865'x'Riband'. Riband was tested in anther culture early in 1993 and the donor-plants were also used as crossing parents. Riband responded with 1.2 green plants per spike and no information was taken for the other parent.

References

Andersen, S.B., I.K. Due and A. Olesen, 1987. The response of anther culture in a genetically wide material of winter wheat (*Triticum aestivum* L.). Plant Breed. **99**: 181-186.
Dunwell, J.M., R.J. Francis and W. Powell, 1987. Anther culture of *Hordeum vulgare* L.: a genetic study of microspore callus production and differentiation. Theor.Appl.Genet. **74**: 60-64.
Foroughi-Wehr, B. and F.J. Zeller, 1990. *In vitro* microspore reaction of different German wheat cultivars. Theor.Appl.Genet. **79**: 77-80.
He, D.G. and J.W. Ouyang, 1984. Callus and plantlet formation from cultured wheat anthers at different developmental stages. Plant Sci.Lett. **33**: 71-79.
Löschenberger, F. and E. Heberle-Bors, 1992. Anther culture responsiveness of Austrian winter wheat (*Triticum aestivum* L.) cultivars. Die Bodenkultur. **2**: 115-122.

Orlov, P.A., E.B. Mavrishcheva and A.N. Palilova, 1993. Estimation of the response to anther culturing in 60 geneotypes of different wheat species. Plant Breed. **111**: 339-342.
Ouyang, J.W., S.M. Zhou and S.E. Jia, 1983. The response of anther culture to culture temperature in *Triticum aestivum*. Theor.Appl.Genet. **66**: 101-109.
Tuvesson, I.K.D., S. Pedersen and S.B. Andersen, 1989. Nuclear genes affecting albinism in wheat (*Triticum aestivum* L.) anther culture. Theor.Appl.Genet. **78**: 879-883.
Tuvesson, S., A. Ljungberg, N Johansson, K.-E. Karlsson, L.W. Suijs and J.-P. Josset, 2000. Large-scale production of wheat and triticale doubled haploids through the use of a single-anther culture method. Plant Breed. **119**: 455-459.
Wang, X.Z. and H. Hu, 1984. The effect of potato II medium for triticale anther culture. Plant Sci.Lett. **36**: 237-239.
Zhou, H. and C.F. Konzak, 1992. Genetic control of green plant regeneration from anther culture of wheat. Genome **35**: 957-961.

2.12
Haploid wheat isolated microspore culture protocol

K.J. Kasha, E. Simion, M. Miner, J. Letarte and T.C. Hu[1]
Department of Plant Agriculture, Biotechnology Division, University of Guelph, Guelph, Ontario Canada N1G 2W1; [1]/*Current address: Monsanto Co., 700 Chesterfield Pkwy. N., St. Louis, MO 63198 USA*

Introduction

Haploid production in hexaploid wheats (*Triticum aestivum* L.) has been achieved mainly by wide hybridization. The first wide hybridization system using *Hordeum bulbusom* pollen was very genotype specific. However, the subsequent use of maize, sorghum or millet pollinators has largely overcome this genotype effect and the system is being used in many programmes for research and breeding. Anther culture in hexaploid wheat has given poor frequencies and high genotype specificity. The development of a highly responsive isolated microspore culture system for barley across genotypes and using a defined induction media encouraged us to also try to improve isolated microspore cultures of wheat. The following protocol has been developed primarily on 'Chris', 'Pavon 79' and 'Bob White' genotypes where response is very good and it is currently being more thoroughly tested across a range of genotypes. The protocol requires co-culture of microspores with wheat ovaries and thus it is not a completely defined media. A patent application covering this protocol has been filed (Kasha and Simion, 2001).

Protocol

The following protocols have been modified from Hu and Kasha (1999).

Donor plants and growth conditions

As in barley, wheat donor plants should be grown under stress free conditions for best results. At Guelph, a growth room is used with 16-17 h light (350-400 µE $^{-2}$ s^{-1}) at ca 20°C and 7-8 h dark period at ca 16°C. Humidity is kept at 60-70% and plants are checked daily for water requirement and are fertilized once a week. Plants should be kept free from pests and water or drought stress. Spraying with pesticides during the 2-3 weeks prior to spike collection should be avoided.

For most applications, the donor plants are F_1 hybrids that permit the production of pure breeding recombinants in the shortest time possible. In some instances, the use of selected F_2 plants as donors is desired. Purification of advanced generation lines for cultivar production may also be advantageous.

Spike collection and pretreatments

Tillers from donor plants are collected when the oldest microspores in the spike are at the mid-to late-uninucleate stage. The staining of microspores from the oldest anthers on a single spike in aceto-carmine, Alexanders stain or DAPI, permits staging the microspores based on the location of the nucleus relative to the microspore pore (Kasha *et al.*, 2001).

Subsequently, additional tillers are collected when they reach a similar stage of morphological development. This stage can vary from genotype to genotype. The tillers are collected in water and stored in a refrigerator at 4°C up to 4 days, until sufficient spikes (5-10) are collected for blender isolation of microspores.

The tillers are then sprayed with 70% ethanol and the spikes removed. With many genotypes, the top of the spike will have emerged from the leaf sheath when tillers are collected and additional sterilization with bleach is required. If spikes are awned, the awns should be cut off prior to bleach sterilization. If awns are clipped too short, the top of the floret will be exposed and bleach may enter the florets and kill the microspores. Spikes with awns trimmed are placed in jars, the jars are filled with 15% Javex liquid bleach (bleach diluted with sterile water to provide a final concentration 0.5 to 1% chlorine) and lids placed on the jars and left for 15 minutes upside down. The bleach is poured off and the spikes rinsed 3 times with sterilized water and left to stand so residual surface moisture will evaporate. Many labs have problems with contamination in using isolated microspore culture and this step is critical in wheat, as the spikes have partially emerged from the sheath.

The spikes are placed in a layer in large Petri dishes (15x150 mm) for pretreatment. Three different pretreatments have been used with good success: the spikes are partially covered with 0.4 M mannitol and placed in the dark for 5 days at room temperature; the spikes in mannitol are placed in a refrigerator at 4°C for 5 to 7 days; 0.5 ml of sterile water is added to the Petri dish with spikes and placed in the cold at 4°C for 2 to 4 weeks.

Microspore isolation

Use sterilized glassware and utensils within a flow bench to avoid contamination. Pick up the spikes with forceps, remove the small florets at the ends of the spike with scissors and then cut the spikes into 2-3 cm sections into a chilled Warring blender vessel (125 or 300 ml). Add ice-cold 0.4 M mannitol to the vessel to just cover the cut segments and blend. The speed is normally set at low and run from 3-10 seconds. Spikes of some genotypes are much softer than spikes of other genotypes and blending too long may result in high microspore mortality. If you have used mannitol pretreatment at room temperature, the anthers may be removed from the spikes and placed in a Vortex to isolate the microspores. While anther removal is much more laborious than blending, viability of the microspores tends to be higher. Since ovaries are required for co-culture with microspores, anthers from spikes used to obtain ovaries may be isolated at the same time, pretreated and isolated.

After blending, quickly filter the slurry through 4 layers of cheesecloth or a nylon mesh with a large pore size. Then put the filtrate through a 100 or 150 µm mesh into graduated centrifuge tubes and centrifuge at 50 g (900 rpm) for 5 minutes. Decant and wash the microspores with cold 0.4 M mannitol 2 to 3 times and centrifuge each wash. Keep tubes on ice as much as possible. Resuspend the microspore pellet in 2 ml of culture induction medium. Check the viability of the microspores. If it is low - below 50-60% - it is best to place the microspores on a 20% maltose solution and centrifuge it. The dead microspores will fall to the bottom and viable microspores are rescued from the top of the maltose with a pipette. Add chilled 0.4 M mannitol and centrifuge to collect the microspores.

Culture medium

Culture media are important and require refinements for each species. For wheat microspores, our best induction response has been obtained with a modified MS medium that is labelled as MMS4 (Table 2.12-1). Differences to MS include lower inorganic nitrogen (NH_4NO_3) and higher organic nitrogen (glutamine), high inositol, maltose in place of sucrose, PAA instead of IAA, Phytagel in place of agarose and the inclusion of Larcol™, i.e. arabinogalactan protein (AGP) from larch wood (Sigma L 0650). AGP greatly improves microspore viability and the frequency of embryos produced. Embryos can develop on culture medium with AGP and no ovary co-culture but response is much faster with ovary co-culture. The pH of MMS4 is adjusted to 5.8 and may be solidified with 3 g/L of Phytagel™ or Gelrite™.

Microspore plating and culture

Microspores may be cultured in liquid medium, as drops on solidified medium or placed on filter paper with the aid of vacuum. Embryo development is faster on filter paper but may require a higher density than in liquid. In liquid we plate 50,000-100,000 microspores and on filter paper about 200,000 to 400,000/plate.

Two Whatman No. 2 filter paper discs are placed on the vacuum platform of a vacuum flask. The filters are washed with medium, 0.4 M mannitol or sterilized water before filtration begins. Drops of microspores in liquid medium are placed on the filter under low vacuum to produce a thin layer (3-4 layers) of microspores. The top filter paper is transferred directly onto solidified MMS4 medium and 4-6 ovaries are placed among the microspores. The Petri dishes are sealed with Parafilm. The cultures are placed in the dark at 28°C and checked weekly. If the paper begins to dry out, drops of fresh medium are placed on the microspores. If ovaries start to show dark spots, they should be removed and replaced. Large numbers of embryos also develop in liquid MMS4 medium but they are slower in reaching a transferable size than on filter paper.

Regeneration

Once the embryos reach a size of 1-2 mm, they are transferred to a differentiation medium for 3-4 days in the dark at 25°C and then into lighted cabinets for 1-2 weeks until shoots and roots are well developed. The differentiation medium (MMS5) is similar to the induction medium but has only 30 g/L maltose. The auxin PAA (phenylacetic acid) is reduced to 0.2 mg/L or eliminated, and kinetin (0.5 mg/L) and 0.5 mg/L GA_3 are added to the medium. Also, 10 µM $CuSO_4$ (2.5 mg/L) and 355 mg/L of U2.5 amino acids (Comeau et al., 1992) are added and it is solidified. U2.5 amino acids are not readily available and casein (1 g/L) and proline (690 mg/L) may be used instead (Table 2.12-1). Embryos are placed directly on the differentiation medium.

Table 2.12-1. MS and modified MS media compositions used for embryo regeneration, wheat microspore induction (MMS4) and for embryo differentiation (MMS5)

Media components	MS original medium (mg/L)	Modified MS media	
		MMS4 induction (mg/L)	MMS5 differentiation (mg/L)
Macro salts			
KNO_3	1,900	1,400	1,400
NH_4NO_3	1,650	300	300
KH_2PO_4	170	170	170
$MgSO_4 \times 7H_2O$	370	370	370
$CaCl_2 \times 2H_2O$	440	440	440
Iron source			
$FeSO_4 \times 7H_2O$	27.8	27.8	27.8
Na_2EDTA	37.3	37.3	37.5
Micro salts			
$MnSO_4 \times 5H_2O$	22.3	22.3	22.3
H_3BO_3	6.2	6.2	6.2
$ZnSO_4 \times 7H_2O$	8.6	8.6	8.6
$CoCl_2 \times 6H_2O$	0.025	0.025	0.025
$CuSO_4 \times 5H_2O$	0.025	0.025	2.5
$Na_2MoO_4 \times 2H_2O$	0.25	0.25	0.25
KI	0.83	0.83	0.83
Other components			
Glutamine	146	975	975
Myo-Inositol	100	300	300
Thiamine HCl	0.4	0.4	0.4
Nicotinic acid	0.5	0.5	0.5
Pyroxidine HCl	0.5	0.5	0.5
Sucrose	30,000	-	-
Maltose	-	90,000	30,000
PAA	-	2	0.2
IAA	1*	-	-
BAP	-	-	-
Kinetin	1*	0.5	0.5
Larcol™	-	10	-
Agarose	8,000*	-	-
Phytagel	-	3,000	3,000
GA_3	-	-	0.5
Amino acids, U2.5	-	355	355
pH	5.8	5.8	5.8

*For regeneration on MS, these hormones are omitted and 3.0 g/L Phytagel replaces agarose.

Once shoots and roots are developing well, the structures are transferred to MS medium with 30 g/L sucrose but modified to omit hormones and using 3 g/L Phytagel or Gelrite in place of the more expensive 8 g/L of SeaPlaque agarose (Table 2.12-1), for regeneration in vials or flasks. The structures are kept in the full cabinet light at 22°C. Once 3-4 leaves develop, the plantlets are transferred to potting media in pots in a growth room.

Efficiency and Applications

From limited testing it would appear that this protocol using MMS4 induction medium may be genotype independent relative to the production of large numbers of structures, similar to our protocol for barley microspore culture. This needs further evaluation. However, the frequency of albinism appears to vary with the genotype.

A major factor in improved response is the presence of AGP (Larcol™) in the medium. In a study with Pavon 79, microspore viability was estimated on samples after 10 days in culture. In plates with no AGP and no ovaries, viability was 18.8%. In plates with ovaries and no AGP, viability was 35.5%. However, in plates with 10 mg/L AGP, viability without ovaries was 77.8% and with ovaries, 78.3%. The number of embryos developed in these same cultures was estimated after 30 days of culture. In plates without ovaries or AGP, there were some multicellular structures but no embryos. In plates with only ovaries, the average was about 1650 embryos produced. In the plates with 10 mg/L AGP, there were about 50 embryos in the absence of ovaries and 4250 in plates with ovaries and AGP. Frequencies of embryos cultured in liquid MMS4 and on filter paper were similar in the presence of both AGP and ovaries. We have tried culturing microspores in media with a range of AGP concentrations up to 1000 mg/L. Concentrations above 10 mg/L show no significant difference in microspore viability and in numbers of embryos produced.

In various studies comparing 0 and 10 mg/L of AGP in wheat microspore cultures, the frequency of embryos produced ranges between 2 and 3 fold higher in 10 mg/L. The embryos also develop faster on AGP induction medium and thus regeneration is usually better. Co-culture of wheat microspores with 4-6 ovaries per plate is recommended. However, AGP does not provide any advantage when present in regeneration medium. Regeneration of embryos transferred to regeneration medium has ranged between 40-50%. The frequency of completely fertile doubled haploids among the regenerating plants is around 70%, thus eliminating the need to apply chromosomal doubling agents to seedlings.

With our blending procedures for microspore isolation we obtain about 50,000 viable microspores per spike. Thus, embryo production with this protocol ranges between 2,000 to 4,000 embryos per spike. We have not attempted regeneration from more than 500 embryos per plate.

Acknowledgements

Funding for this research has been provided by Monsanto Co. and NSERC and is gratefully acknowledged. Rapid approval from Monsanto to send in this protocol is also gratefully appreciated, particularly the effort of Joyce Fry. Facilities and funding have also been provided by OMAFRA. Many researchers have contributed improvements to microspore culture of cereals and we have incorporated many of these into our protocol but we are unable to site them in this publication. Such contributions are gratefully acknowledged.

References

Comeau, A., P. Nadeau, A. Plourde, R. Simard, O. Maës, S. Kelly, L. Harper, J. Lettre, B. Landry and C.-A. St-Pierre, 1992. Media for the *in ovulo* culture of proembryos of wheat and wheat-derived interspecific hybrids or haploids. Plant Sci. **81**: 117-125.

Hu, T. and K.J. Kasha, 1999. A cytological study of pretreatments used to improve isolated microspore cultures of wheat (*Triticum aestivum* L.) cv. Chris. Genome 42: 432-441.

Kasha, K.J. and E. Simion, 2001. Embryogenesis and plant regeneration from microspores. International Patent Application No. PCT/CAOO/01436 (Published electronically June 14, 2001).

Kasha, K.J., E. Simion, R. Oro, Q. A. Yao, T. C. Hu and A. R. Carlson, 2001. An improved *in vitro* technique for isolated microspore culture of barley. Euphytica **120** (3): 379-385.

2.13
Production of doubled haploids in wheat (*Triticum aestivum* L.) through microspore embryogenesis triggered by inducer chemicals

M.Y. Zheng[1], W. Liu, Y. Weng[2], E. Polle and C.F. Konzak
Northwest Plant Breeding Co., 2001 Country Club Rd., Pullman, WA 99163, USA
[1] *Department of Biology, Gordon College, 255 Grapevine Rd., Wenham, MA 01985, USA*
[2] *USDA-ARS, 215 Johnson Hall, Washington State University, Pullman, WA 99164, USA*

Introduction

Production of doubled haploids (DH) from microspores by androgenesis is a proven method to obtain homozygous individuals in a single step, thus the method is very useful in plant breeding. Since homozygosity is achieved in one generation, the breeder can avoid the numerous cycles of inbreeding required by conventional breeding systems and at the same time substantially reduce the population sizes required for effective selection of superior trait combinations. An efficient doubled haploid production system is also a preferred technique to produce homozygous transgenic plants. Dominant or recessive target gene(s) to be incorporated into gametic cells by genetic transformation can be fixed in homozygous form in a single generation. Doubled haploids are also important tools in plant genome mapping and studies of embryogenesis. In this chapter, there are described two unique and efficient systems - "flask" and "fresh microspore" systems for the production of doubled haploids, effective for a wide spectrum of genotypes in common wheat (*Triticum aestivum* L.). These systems rely upon three key steps. First, the microspores are switched from their naturally programmed pathway for gametophytic development (Fig. 2.13-1a) to sporophytic development (Fig. 2.13-1b) by an inducer chemical treatment under proper physical conditions. Second, optimal culture conditions are provided, which include adequate nutrition and a favorable physical environment to help microspores elaborate their altered developmental programme leading to embryoid production, once a large population of reprogrammed microspores is obtained. Third, embryoids are germinated on solid, hormone-free medium to form doubled haploid plants. Cell divisions in treated microspores follow a well-organized and predictable pattern that leads to the formation of true "embryoids" rather than calli, ensuring success in plant regeneration. Cell divisions within the first 7 days in culture lead to the formation of a clearly defined multi-cellular structure, whose size is only 10-15 μm (diameter) larger than a microspore (Fig. 2.13-5a).

Protocol

As mentioned earlier, two systems are in routine use in the Northwest Plant Breeding Co. (NPB) research laboratory, the "flask" system and the "fresh microspore" system. Both systems were established based upon the concept of inducing embryogenesis by inducer chemicals under proper physical and physiological conditions. Once the embryogenic potential is induced, whether *via* "flask" or "fresh" microspore system, culture media and procedures for embryoid development and production of plants are identical, thus are described only once hereunder.

The "Flask" System for Doubled Haploid Production from Microspores (Liu *et al.*, 2002b)

Donor plants and growth conditions
One to three seeds are sown in each pot (20 by 25 cm) filled with premixed soil. Plants are grown in temperature-controlled greenhouses at 27±2°C (day) and 17±2°C (night) under a 17/7 h (day/night) photoperiod. Fertilizers (N, P, K) were premixed with soil at the time the seeds were sown, with subsequent applications through daily watering with water containing liquid forms of nitrogen, phosphorus and potassium (20:20:20). In general, any standard conditions for growing wheat in a greenhouse are acceptable provided that quality donor plants can be harvested. With regard to winter wheats, in order to ensure normal plant development and hence optimum culture responses, it is essential that the vernalization be complete (6-10 weeks at 6°C, depending on genotype).

Collecting tillers
Fresh tillers with spikes containing microspores at an appropriate developmental stage (mid- to late uninucleate stage) are cut at the second node from the top of the tiller, and the base of each tiller is immersed immediately in a 250 ml flask containing 30-40 ml distilled water. All of the leaves are removed by cutting leaf blades at their bases. The time between the collection of tillers and the application of pretreatment is preferably minimized to reduce the possibility of contamination by microorganisms and/or fluctuation in subsequent culture response. Microspores enclosed within the anthers in the central section of a spike should preferably be in the mid- to late uninucleate stage of development. Morphological features of tillers containing microspores at these stages can easily be established for each genotype via microscopic examination in acetocarmine stain or distilled water (Konzak *et al.*, 1999). Fresh tillers so collected are then ready for the pretreatment.

Androgenesis induction treatment
1. After removing the stem section of each tiller that was immersed in (hence in contact with) distilled water, the fresh base of each tiller is immersed immediately in an autoclaved 250 ml flask containing 50 ml inducer chemical formulations.

2. The frequently used inducer chemical formulation contains: distilled water; 0.01% (w/v) 2-hydroxynicotinic acid (2-HNA); 10^{-6} M benzylamino purine (BAP); 10^{-6} to 10^{-5} M 2,4-dichlorophenoxyacetic acid (2,4-D). Other inducer chemical formulations may be effective, but are yet to be fully evaluated (Konzak *et al.*, 1999).

3. The open end of a plastic bag (thin-walled, grocery store vegetable/fruit bag) is then wrapped around the neck of the flask and sealed around the neck with masking tape to prevent microbial contamination and excessive loss of water by evaporation (Fig. 2.13-2a).

4. The flask is then placed in an incubator set at the desired stress temperature, preferably at 33°C, but the temperature employed for certain genotypes may be higher or lower, with optimal ranges of 4 to 10°C, or 30 to 33°C.

5. The optimum period of incubation for pretreatment varies with the genotype, but is usually from about 48 h to about 72 h at 33°C, but may involve more time at lower temperature.

6. The flask containing tillers collected from greenhouse also may be stored in a refrigerator at 4°C for up to one week before subjecting them to the pretreatment with the inducer chemical formulation and temperature/nutrient stress. It is important to note that after the treatment with the inducer chemical formulation and temperature/nutrient stress, the tillers should not be stored because the microspore viability decreases rapidly with the storage, irrespective of the temperature (4-33°C tested) employed for storage. Thus, immediate processing following the pretreatment is advised.

7. Immediately after the pretreatment, the isolated embryogenic microspores typically have eight or more small vacuoles concentrically lined against the cytoplasmic membrane. These vacuoles surround the condensed cytoplasm in the center, forming a fibrillar structure (Figure 2.13-1b). In addition, the embryogenic microspores are usually, but not always, of a larger size (about 50 µm) than the average non-treated microspores (25-30 µm).

Microspore isolation
1. After the tillers have been pretreated in accordance with the methods described above, they are removed from the treatment flask in a laminar flow hood.

2. All foliage beneath the first tiller node is removed, keeping only the boot encasing the spike. Boots are placed on a paper towel and sprayed with 75% ethanol to saturation. The boots are then wrapped in the towel and left in the hood for approximately 45 minutes, or until the ethanol has fully evaporated.

3. Alternatively, boots are disinfected through immersion in 1.5% sodium hypochlorite solution (or 20% of commercial bleach solution) in a 100 ml graduate cylinder for 20 minutes.

4. The spike is aseptically removed from each disinfected boot and placed on top of a sterile (autoclaved) blender cup (Waring MC II) (Fig. 2.13-2b). For mass production, several blender cups are needed to facilitate processing in clean sterile cups.

5. Florets are cut from their bases with awns (if present) removed and allowed to drop into the open blender cup (Fig. 2.13-3a).

6. Florets obtained from one to six spikes may be used for each run of blending in 50 ml of an autoclaved 0.3 M mannitol solution.

7. After a sterilized lid is placed and secured on top of the blender cup, the cup is assembled to the MC II Waring blender (Fig. 2.13-3b). The florets are then blended at a low speed (2,200 rpm) for 20 seconds.

8. To eliminate the larger debris, we pass the resulting slurry through an autoclaved 100 µm stainless steel mesh filter (Fig. 2.13-4a). The blender cup is rinsed three times using 5 ml

of an autoclaved 0.3 M mannitol solution each time, which is also poured through the filter. Debris and residues trapped on top of the mesh filter are discarded.

9. The filtrate is then poured onto a 38 μm mesh filter (Fig. 2.13-4b). This essentially traps all of the viable microspores as well as some of the debris and dead microspores.

10. All mesh filters are designed and manufactured at NPB using our own facilities. The stainless steel mesh and the autoclavable plastic cups, can be conveniently ordered from vendors in US. The procedures to manufacture mesh filters of various sizes are described below.

11. The microspores are then rinsed off of the 38 μm mesh filter using 2 ml of 0.3 M mannitol solution and layered over 5 ml of an autoclaved 0.58 M maltose solution (Fig. 2.13-4c) and centrifuged at 100 x g for 3 minutes. Three ml of the upper band (containing embryogenic microspores) is collected (Fig. 2.13-4d) and allowed to pass through a 38 μm mesh filter.

12. The microspores retained on the mesh filter are rinsed three times using 5 ml of medium NPB-99 (Table 2.13-1) for each rinse and washed off of the filter into a Petri dish using 15 ml of NPB-99 liquid medium.

13. The lower band (junk pellet) is also resuspended using 15 ml of 0.3 M mannitol solution in a 15 ml conical tube.

14. The total number of microspores in each band is estimated under an inverted (or light) microscope using a haemacytometer.

15. The debris and junk microspores in the conical tube (pellet) are discarded after the counting. The total number of microspores is the sum of the microspores from both the upper band and the pellet.

16. Only the microspores from the upper band (now in a Petri dish) are used for subsequent culture. The junk (pellet) microspores appear to be largely those damaged during blending or those filled with starch, which typically have passed the late uninucleate stage.

17. The overlay of 0.3 M mannitol on 0.58 M maltose followed by density-gradient centrifugation is a critical step, leading to the separation of embryogenic from non-embryogenic microspores. The viable microspores having multiple vacuoles, characteristic of embryogenic microspores (Fig. 2.13-1b), are lower in density, and thus float and form a compact band at the interface between maltose and mannitol. The debris, damaged and non-embryogenic microspores, having a higher density, form a pellet at the bottom of the conical tube.

18. The embryogenic microspores in the Petri dish are then distributed into 15x60 mm Petri dishes with a final density of approximately 7×10^3 microspores/ml. An alternative procedure, which allows the collection of nearly 100% embryogenic microspores is described under the sub-title "Method for producing 100% embryogenic microspores."

Table 2.13-1. Media used in wheat microspore culture protocol

Media components	NPB-A' pretretment medium (mg/L)	NPB-99 induction medium (mg/L)	190-2 regeneration medium (mg/L)
Macro salts			
$(NH_4)_2SO_4$	-	232	200
KNO_3	-	1,415	1,000
$CaCl_2 \times 2H_2O$	148	83	-
$Ca(NO_3)_2 \times 4H_2O$	-	-	100
KH_2PO_4	136	200	300
$MgSO_4 \times 7H_2O$	246	93	200
KCl	1,492	-	40
Iron source			
Na_2EDTA	37.3	37.3	37.3
$FeSO_4 \times 7H_2O$	56	27.8	27.8
Micro salts			
H_3BO_3	3	5	3
$CoCl_2 \times 6H_2O$	-	0.0125	-
$CuSO_4\text{-}5H_2O$	-	0.0125	-
KI	0.5	0.4	0.5
$MnSO_4 \times 4H_2O$	8	5	8
$Na_2MoO_4 \times 2H_2O$	-	0.0125	-
$ZnSO_4 \times 7H_2O$	0.3	5	3
Other components			
myo-Inositol	-	50	100
Glycine	-	-	2
Nicotinic Acid	-	0.5	0.5
Pyridoxine HCl	-	0.5	0.5
Thiamine HCl	-	5	-
2-HNA	50	-	-
Sucrose	-	-	30,000
Maltose	90,000	90,000	-
Glutamine	-	500	-
Kinetin	0.5	0.2	-
2,4-D	-	0.2	-
PAA	-	1	-
Gelrite	-	-	3,000
PH	6.5	7.0	6.5

Gelrite is added after the adjustment of pH. Medium NPB-A' and NPB-99 should be filter-sterilized with 0.22 µm filter. Medium 190-2 is autoclaved at 115°C for 30 minutes.

Culture of isolated microspores and production of embryoids

1. Isolated microspores are cultured in filter-sterilized liquid NPB-99 medium in a form of suspension culture.

2. An aliquot of 2 ml media per 35x10 mm Petri dish, or 5 ml media per 60x15 mm Petri dish, at a density of approximately 7×10^3 microspores per ml is effective.

3. Microspores are co-cultured with either live ovaries or a combination of live ovaries with medium pre-conditioned by live ovaries (OVCM). The first is a co-culture of microspores with the inclusion of only live ovaries at culture initiation. The second involves the preparation of OVCM prior to the isolation of microspores.

4. Co-culture of microspores with OVCM in both flask and fresh microspore procedures is a two-step process (Zheng *et al.*, 2002). The first step deals with the production of OVCM by which live ovaries are excised and grown to condition the medium NPB-99. Live ovaries are picked aseptically from disinfected spikes with a pair of fine forceps and placed into Petri dishes (60x15 mm), each supplemented with 4 ml of NPB-99 medium. The ovary densities effective for conditioning range from 10 to 20 for each ml of medium. These Petri dishes are then sealed and incubated at 27°C for conditioning. The optimal conditioning period is 7 to 10 days. The next step (once the conditioning is completed) is the dilution of OVCM with regular NPB-99 for use in microspore cultures. Each of 1.0 ml OVCM is drawn from conditioned plates and mixed with regular NPB-99 into a final volume of 4 ml for microspore culture of one Petri dish (60x15 mm).

5. Live ovaries at 2 per 1 ml of regular medium are picked and placed into Petri dishes to co-culture with microspores.

6. Ideally, a combination of live ovaries and OVCM are included in microspore cultures. Ovaries of all genotypes have proved to be effective for co-cultures, but the ovaries from variety 'Chris' have been preferred for co-culture of all wheat microspores tested so far due to its awnless feature and ease of obtaining ovaries.

7. After the inclusion of fresh ovaries or the fresh ovaries combined with OVCM, all Petri dishes are sealed with Parafilm and incubated in the dark at 27°C for embryoid development.

8. Embryogenic microspores begin their first cell division after approximately 12 h in culture. Thus, the responsiveness to the inducer chemical formulation can be determined within 1 day following the pretreatment. Multi-cellular structures, still enclosed within the microspore wall or exine, are formed in approximately one week (Fig. 2.13-5a).

9. In approximately one more week, the exine wall ruptures and the pro-embryoids emerge, to grow into immature (Fig. 2.13-5b) and then, mature embryoids (Fig. 2.13-5c).

10. Obtaining microspores with a fibrillar cytoplasmic structure (Fig. 2.13-1b) is a pre-requisite for embryogenesis, but only those that proceed to proembryoid development will eventually develop into mature embryoids, which are then able to germinate to produce plants (Fig. 2.13-5d).

Germination of embryoids to produce plantlets

1. When embryoids reach the size of 1 to 2 mm, they are transferred aseptically to autoclaved solid 190-2 medium (Table 2.13-1) at a density of 25-30 embryoids in each 100x15 mm Petri dish.

2. The Petri dishes are sealed with Parafilm and are placed under continuous fluorescent light at room temperature (22°C).

3. In approximately two weeks, green plants develop and are ready for transfer to soil.

4. Green plants are raised in a greenhouse, much like those plants grown from seeds. Due to a large number of green plants obtained and fair levels of doubling frequency for many genotypes, use of colchicine or caffeine may not always be necessary.

5. Caffeine or colchicine treatments are needed only for genotypes exhibiting extremely low doubling frequencies. Seeds produced on any plants are instantly homozygous, and so can immediately be used for rapid evaluation and selection in breeding, for analyses in genetic research, or for selection and evaluation of transgenic plants in biotechnology. We have found that spontaneous doubling occurs with many genotypes, and that chimera plants are rare without the chromosome doubling treatment.

Method for transplanting plantlets and doubled haploid production

1. When plantlets have 2 or more leaves and the shoots are 4 cm or longer, they can be transplanted into small pots and raised in the greenhouse.

2. The plantlets in the Petri dish are washed with running water so the Gelrite and medium are washed away.

3. In preparation for transplanting, small (5x5x5 cm) pots are filled with soil and watered thoroughly. Soils are premixed with fertilizers (N, P, K).

4. Plantlets are planted in small pots in such a way that roots are placed in a hole and soil is gently pressed by fingers to enable good root-soil contact.

5. The pots are placed in a plastic tray (Fig. 2.13-6a). A transparent plastic cover tray can be placed over the pots in the tray to maintain a high relative humidity around the plants while light still penetrates the cover (Fig. 2.13-6b). It is essential to maintain a high relative humidity for the plantlets during the first week after transplanting.

6. Plants are watered every 2 days or whenever the soil appears to be dry. The plastic cover tray can be lifted gradually so that plants will become acclimated to the greenhouse conditions.

7. In about 2 to 3 weeks, plants grow vigorously. The plastic cover can be removed, and the plants can be transplanted to larger, 20x25 cm pots for doubled haploid production.

8. Green plants are raised in the greenhouse, much like plants grown from seeds. If plants are identified as haploid (as summarized below), colchicine or caffeine can be applied to induce chromosome doubling. Seeds produced on any plants are instantly homozygous.

9. Winter wheat may be vernalized either as germinating plantlets in the Petri dishes or as plantlets transferred to greenhouse soil in pots.

Method for identifying doubled-haploid plants

Doubled haploid plants can be identified in a variety of ways. The flow cytometry method is based on the fact that the genomic DNA content of somatic cells in haploid plants is only half of that in doubled haploid plants. However, the method requires expensive instruments to isolate and measure the DNA content of samples. The method is useful when a large number of samples are to be examined.

Stomata size on haploid plant leaves is generally smaller than those on doubled haploid plant leaves. Estimation of stomata size is a simple method to identify doubled haploid plants at the seedling stage. Haploid plants can be cut and treated with a doubling agent such as colchicine or caffeine. However, the physiological condition and developmental stage of plants may affect the accuracy of its practical utilization for determining if a plant is a doubled haploid or a haploid. This is troublesome because of the fact that the quality of microspore-derived embryoids varies, which affects the physiological condition of plantlets. A good correlation between stomata size and ploidy level of cells should be established for genotypes of interest before this method is employed routinely. The stomata were measured when the plants were from 15-30 cm in height, and with four to seven leaves. The tip of the oldest fully expanded leaf was removed and viewed under a 10x lens (100 magnification). With genotypes tested in our laboratory, the majority of plants having stomata less than 50 µm were found to be non-fertile (haploid), whereas those plants having stomata greater than 59 µm tended to be fertile (doubled haploid). This technique can be used to eliminate haploid plantlets at early stages when a large number of green plants are recovered from embryoids, hence reducing the workload.

The most reliable method for identifying doubled haploid plants is to count the chromosome number of cells and to examine the fertility of plants. However, counting chromosome numbers is time-consuming and labor intensive. A fertility check is possible only at a later plant developmental stage.

When large numbers of plantlets are obtained and if more than 50% of plantlets are often spontaneously doubled haploids, no additional work may be necessary because an appreciable number of doubled haploid plants are produced. Haploid plants, when identified, can be eliminated from the pots to lessen the workload. Spontaneously chimera plants of haploid/doubled haploid are few but may be induced by the chromosome doubling treatment.

The Fresh Microspore Culture System for Doubled Haploid Production (Zheng et al., 2001)

Donor plants and growth conditions
Conditions for growing donor plants are the same as described in the "flask" system.

Collecting tillers
Fresh tillers that contained microspores at the mid- to late uninucleate stage were cut at the second node from the top of the tiller, and the base of each tiller is immersed immediately in a 250 ml flask with 30-40 ml of distilled water. All leaves were removed by severing leaf blades at their bases. Morphological features of tillers containing microspores at the mid- to late uninucleate stage can easily be established for each genotype *via* microscopic examination of microspores in aceto-carmine stain or distilled water (Konzak et al., 1999).

Isolation of microspores
1. All foliage beneath the first node is removed, keeping only the boot encasing the spike.

2. Boots are then disinfected through immersion in 1.5% sodium hypochlorite solution in a 100 ml graduated cylinder for 20 min, followed by 3 rinses with sterile distilled water over 3 min.

3. The spikes are aseptically removed from each disinfected boot and placed on top of a sterile blender cup (Waring MC II) that has been autoclaved.

4. Florets are cut from their bases with awns (if present) removed, and allowed to drop into an open blender cup.

5. Florets obtained from three to six spikes are mixed for each run of blending in 50 ml of autoclaved 0.3 M mannitol solution.

6. After a sterile lid is placed and secured on top, the blender cup is assembled to the Waring MCII blender.

7. The blending is executed at 14,000 rpm for 20 s. Note a higher blending speed (rpm) is needed for spikes without being subjected to pretreatment. This is due to a much tougher tissues with freshly sampled spikes.

8. To eliminate the larger debris, the resulted slurry is passed through a 100 μm stainless steel mesh filter. The blender cup is rinsed three times with 5 ml of 0.3 M mannitol each, which is also poured through the filter.

9. The filtrate is then poured onto a 38 μm mesh filter, which essentially traps all viable microspores and some debris and dead microspores.

10. The microspores are then rinsed three times on the filter with 5 ml each of medium NPB-A' (Table 2.13-1), then washed off the filter into a Petri dish (100x15 mm) using 15 ml of NPB-A'.

Treatment with inducer chemical formulation

1. Using 60x15 mm Petri dishes, the concentration of 2-HNA is adjusted to 0.18 mM in NPB-A'. The microspore density is held constant at approximately 10,000 per ml as determined with a haemocytometer, with a total volume of 4 ml per dish. Other inducer chemicals may also be dissolved in NPB-A' prior to adding to the pretreatment cultures.

2. Once microspores and 2-HNA are gently mixed together, the Petri dishes are sealed with parafilm and incubated in the dark for 2 d at 22-24°C. At this room temperature range, the treatment time may vary from 38 to 52 h according to the genotype.

3. Following the incubation, the microspores are recovered on a 38 μm stainless steel mesh filter, and then rinsed once with 5 ml of 0.3 M mannitol.

4. Microspores trapped on the 38 μm mesh are rinsed off the filter with 2 ml of 0.3 M mannitol onto 5 ml of 21% (0.58 M) maltose in a 15 ml conical tube.

5. The conical tube is capped and then centrifuged at 150 g for 3 min.

6. Using a pipette the embryogenic microspores are collected from the band at the interface between 21% maltose and 0.3 M mannitol, and transferred to another 38 μm mesh filter.

7. Microspores retained on the mesh are rinsed three times with 3 ml of medium NPB-99 each (Table 2.13-1).

8. The mesh filter is then inverted and microspores are gently rinsed off of the mesh into a sterile 60x20 mm Petri dish.

9. Microspores are then equally distributed to various Petri dishes of 60x20 mm with a minimal density of 4×10^3/ml as determined with a haemocytometer, and with a total volume of 4-5 ml of NPB-99.

Induction culture
After adjusting the density with NPB-99, 8-10 fresh wheat ovaries are transferred aseptically to each Petri dish and culture continues as described previously.

Germination of embryoids to produce plantlets
From this point onward, all procedures are as described in the "flask" system.

Method for making mesh-filters
Mesh-filters are an integral part of the isolated microspore culture systems. These mesh filters have different pore sizes so that microspores of different size can be separated. These mesh filters are autoclavable. They are made of thermoplastic polypropylene, which does not melt at autoclave temperature of 115°C. The stainless steel meshes are heat-sealed to plastic cylinders, tubes, and beakers. No chemical glues are used, to avoid possible toxic effects.

Materials
- Stainless steel mesh cloths of various pore sizes
- Autoclavable (115°C) 15 ml polypropylene centrifuge tubes
- Autoclavable 50 ml polypropylene beakers
- Autoclavable 1 and 5 ml polypropylene pipette tips
- Autoclavable plastic cylinders of various sizes.

Tools
Hot plate, aluminum foil, scissors, fine sandpaper.

Mounting method
Tubes are cut into cylinders 1.2 cm in height. One cm of the sharp end of pipette tips is removed. Beakers are cut in half. Bottoms of cylinders are cut off and discarded; the upper cylinder parts are used for making the filters. A hot plate is set at high to provide the heat for sealing the mesh into the plastic. A piece of aluminum foil is set onto the top surface of the hot plate. A piece of mesh cloth is set on the aluminum foil. Using one hand to hold one end of the cylinder, set the other end onto the mesh cloth. After a few seconds, the end of the plastic cylinder melts by and becomes sealed to the mesh cloth. At this point, the aluminum foil paper, cylinder and mesh complex is removed from the plate, leaving the mesh and cylinder complex intact. After about 2 minutes, the mesh-cylinder complex cools down, and the mesh-filter can be removed from the surface of the foil. The excess mesh cloth is cut off from around the edge of cylinder. Rough edges of the mesh can be smoothened by rubbing the mesh edges with fine sandpaper. After washing, the filters can be autoclaved at 115°C for 30 minutes.

Polypropylene plastic beaker filters made with stainless steel meshes of pore sizes 38, 43, 50 and 100 μm are used in the current invention (Fig.2.13-7 and 8). The small tube filters fit into 30x10 mm or larger Petri dishes. The beaker filters can be set into other 50 ml beakers of same type to achieve sequential filtration. The pipette tip filters are used for medium replenishment, since the filter tips prevent the uptake of developing microspores.

Method for Producing 100% Embryogenic Microspores
Obtaining a purified population of androgenic microspores is important during the process of embryoid induction. It was observed that up to 50% of the initially androgenic microspores stopped cell division within 7 to 10 days in the induction medium NPB-99. The dead

microspores may plasmolyze and release cytoplasmic substances. The released substances may negatively affect growth of the normally developing androgenic microspores, by changing the chemical composition and osmolality of the induction medium. It also may be essential to maintain a highly purified androgenic microspore population for certain biological applications, i.e. genetic transformation of microspores.

A population of nearly 100% dividing microspores (Fig. 2.13-9a) can be obtained, simply by the use of a mesh filter with a pore size of 50 μm to filter out the dead microspores. This separation seems best done about 7 to 10 days after culture of the microspores in NPB-99 induction medium has been initiated. The medium with microspores is pipetted into an autoclaved 50 μm mesh filter. The 2 to 16 celled multi-cellular dividing microspores are retained on the surface of the filter, because their diameters are larger than 50 μm, while the dead microspores pass through the filter. The dividing microspores in the filter are rinsed 3 times with NPB-99 medium, and collected by washing the filter from the reverse side of the filter twice with NPB-99 medium, 5 ml each time. These dividing microspores are re-suspended in NPB-99 medium, and plated. Nearly all of these microspores are androgenic, and will continue their development into embryoids and/or calli (Fig.2.13-9b).

Efficiency and Applications

With these two systems, 20–50% of microspores in anthers are converted from gametophytic into sporophytic development, triggered by inducer chemicals under proper conditions. Under adequate culture conditions, embryogenic microspores isolated from a single spike can produce up to 5,000 or more green plants. There are two highly practical and efficient systems for the delivery of inducer chemical formulations to microspores. The "flask" system provides for contact with the inducer chemical through the vascular system of the plant *via* the base of sampled tillers. Transpiration facilitates movement of the chemical into the anthers and microspores. The reprogrammed microspores typically have eight or more small vacuoles. These vacuoles surround the condensed cytoplasm in the center, forming a fibrillar structure (Fig. 2.13-1b). The "fresh microspore" system provides direct contact between the inducer chemical formulation and the microspores. In this case, microspores are first isolated from tillers sampled afresh, and then mixed with a solution containing the inducer chemical and osmoticum. The reprogrammed microspores have the same morphology as those isolated from the other delivery system.

Even though reprogramming is the pre-requisite for DH production, the extent and repeatability of success also depends upon the appropriate culture conditions following the reprogramming step. These conditions include a proper balance of nutrients in a defined medium, a favorable physical environment and addition of live ovaries and medium pre-conditioned by live ovaries to the culture (Zheng *et al.*, 2001; Zheng *et al.*, 2002; Liu *et al.*, 2002a). Absence of wheat ovaries and ovary-conditioned medium in this phase typically leads to aborted or abnormal embryogenic development by many induced and dividing microspores. This may explain the sporadic, yet often unrepeatable and limited success in microspore cultures by many laboratories.

It should be further noted that microspores being collected for culture represent only a fraction of reprogrammed embryogenic microspores. At least 50% of embryogenic microspores are usually lost in the isolation process. Many microspores are either trapped in residues or still enclosed within the wall of broken anthers. In addition, a large portion of microspores collected through isolation may be damaged by the blending force hence lose their embryogenic potential although triggered by the inducer chemical.

These systems should permit the wide use of DH technology, providing powerful tools for crop improvement and genetics research. We have tested also a wide range of genotypes using this technology. Although genetic differences in response to culture do exist and may not have been tested yet, we continue to make improvement to the technology although the inducer chemical is able to induce high yields of embryoids, even in commonly known recalcitrant genotypes, i.e. 'Waldron' and 'WPB 926' (Table 2.13-2). For some genotypes considered responsive, treatment with the inducer chemicals routinely yield more than a 50-fold increase in embryoid and green plant numbers compared to the non-treated (Table 2.13-2).

In a survey of 31 genotypes carried out from 1998 to 2001, including both spring and winter wheat, there was quite a range in response but only three were classified as low, having fewer than 100 embryos/calli and fewer than 50 green plants per spike. Ten were classified as high, having over 500 green plants per spike and 18 were moderate, having 50 to 500 green plants per spike. The data was based on results estimated from the transfer of 20 or more calli or embryoids. The percentage of regeneration ranged from 36 to 100% and that of green plants ranged from 8 to 100%.

Table 2.13-2. Fifty fold plus increase in embryoids responding to signal chemical treatment by recalcitrant and responsive genotypes per spike

Genotype	Untreated	2-HNA treatment*
Chris	126	6,294
Pavon76	100	4,965
WED202-16-2	86	4,305
WPB 926	0	250
Waldron	2	348

*Embryoid production from embryogenic microspores as triggered by 2-HNA at a concentration of 100 mg/L, dissolved in aqueous solution in the flask in contact with the basal section of tillers.

Preliminary studies indicate that the signal technology is also effective in other species, i.e. barley and maize. Research is well under way to adapt similarly efficient systems for DH production in these and other species. As a result of our research on doubled haploid production, four patent applications had been filed. More patents likely will be filed in 2002.

References

Konzak, C.F., E.A. Polle, W. Liu and M.Y. Zheng, 2002. Methods for generating doubled haploid plants. U.S. patent 6, 362, 393.
Liu, W., M.Y. Zheng and C.F. Konzak, 2002a. Improving green plant production via isolated microspore culture in bread wheat (*Triticum aestivum* L.). Plant Cell Rep. **20**: 821-824.
Liu, W., M.Y. Zheng, E.A. Polle and C.F. Konzak, 2002b. Highly efficient doubled-haploid production in wheat (*Triticum aestivum* L.) via induced microspore embryogenesis. Crop Sci. **42**: 686-692.
Zheng, M.Y., Y. Weng, W. Liu and C.F. Konzak, 2002. The effect of ovary-conditioned medium on microspore embryogenesis in common wheat (*Triticum aestivum* L.). Plant Cell Rep. **20**: 802-807.
Zheng, M.Y., W. Liu, Y. Weng, E. Polle and C.F. Konzak, 2001. Culture of freshly isolated wheat (*Triticum aestivum* L.) microspores treated with inducer chemicals. Plant Cell Rep. **20**: 685-690.

2.14
Isolated microspore culture in maize (*Zea mays* L.), production of doubled-haploids *via* induced androgenesis

M.Y. Zheng[1], Y. Weng[2], R. Sahibzada and C.F. Konzak
Northwest Plant Breeding Co., 2001 Country Club Rd., Pullman, WA 99163, USA
[1] *Department of Biology, Gordon College, 255 Grapevine Rd., Wenham, MA 01985, USA*
[2] *USDA-ARS, 215 Johnson Hall, Washington State University, Pullman, WA 99164, USA*

Introduction

Whenever we mention the use of doubled haploids (DH), the idea of achieving instant homozygosity immediately comes to mind. The potential uses of DHs, however, reach far beyond practical plant breeding. DH systems are widely useful in plant genome mapping, and in many areas of basic research such as investigations of *in vitro* embryogenesis and developmental biology. DH lines also may be used to create unique germplasm, for *in vitro* screening of plants possessing traits such as pathogen and pest resistances, improved quality, and stress tolerance produced either *via* conventional breeding or transgenic means. Although various approaches including *in-situ* gynogenesis, *in vitro* gynogenesis, and *in vitro* androgenesis have been used to produce maize DH plants, the only potentially efficient approach is *in vitro* androgenesis, production of haploid/DH through anther/microspore cultures. The procedures and conditions for anther culture have been reviewed extensively elsewhere (Büter, 1997). In isolated microspore cultures, microspores are usually freed from anthers and other tissues by mechanical methods, including macerating the anthers against a stainless steel sieve with a glass rod, chopping anthers with a razor blade, blending anthers in an electric blender or pulverizing anthers with a dispersing tool. Among various isolation techniques, blending seems to be less stressful to the microspores. No conclusive evidence has been available concerning the optimal explant type as well as optimal blending regime. Nevertheless, it is generally agreed that blending should be done in a medium that provides adequate osmolarity and at a relatively low temperature. In addition to anthers, spikelets or tassel segments may be used as explants for microspore isolation. Following the isolation, microspores are usually suspended in induction medium similar to those used in anther culture (Büter, 1997). Embryo-like structures (ELS) may appear 14-21 days after culture initiation, somewhat earlier than after anther culture initiation. Factors affecting isolated microspore cultures include growth conditions for donor plants, pretreatment conditions prior to the isolation, isolation procedures, microspore plating density, concentrations and nature of ingredients used in induction and plant regeneration culture, and physical conditions during the culture process (Zheng *et al.*, 2002). Generally, low yields of embryoids (especially for elite germplasm), difficulty in plant regeneration and low frequencies of chromosome doubling are major limiting factors that have hindered the use of maize microspore culture technology for practical breeding (Büter, 1997). Shipping tassels even overnight has not proved satisfactory. In this Chapter, we describe an efficient system for production of maize DH plants from isolated microspores although further improvements can yet be made. From the onset of the project, our efforts were directed step by step to addressing the major problems in previous systems, i.e. improving the frequency of cells induced into androgenesis, increasing ELS yields, establishing a sound plant regeneration system to produce high frequencies of green doubled haploids. In our research, we combined an inducer chemical formulation (Zheng *et al.*, 2001) with physical stress as developed for wheat microspore culture (Konzak *et al.*, 1999), which acts to switch microspores from their gametophytic to a sporophytic development, to achieve high frequencies of embryogenic microspores. By then optimizing the induction culture medium and conditions, we were able to produce high yields of calli/embryoids and of DH plants. Over 98% of regenerated plants are green, 60-65% of healthy green plants have produced seeds upon sib-pollinations. This system was first established using a sweet corn germplasm 'Seneca 77' and has since been found applicable to other types of maize. As a result of our success, a utility patent was applied for, and is pending with United States Patent and Trademark Office.

Protocol

Donor plants and growth conditions
One seed is sown into pre-mixed soil in each 25 x 30.5 cm pot. Plants may be grown in either a greenhouse or a growth chamber with a day/night temperature regime at 30/20°C, and photoperiod of 16/8 h. In preparation for planting, pots filled with pre-mixed soil are first watered to saturation. One seed is dropped in a 1.5 cm (depth) hole in the middle of the pot and then fully covered by the soil. Further application of fertilizer is achieved by daily watering, which contains 200 to 250 ppm of water-soluble fertilizer of nitrogen (N), phosphorus (P_2O_5) and potassium (K_2O), 20% each. In general, any standard conditions for growing maize in a greenhouse or growth chamber are acceptable provided that quality donor plants can be harvested.

Sampling
1. For every genotype or hybrid, a correlation between plant morphology and developmental stage of microspores can be established by microscopic examination.

2. Before the tassel enclosed in the boot is about to emerge, 2-3 top florets of a tassel are picked out by a pair of long forceps with one hand firmly holding the base of the boot and the plant.

3. Sampled florets are brought back to the laboratory where anthers are taken out of florets and crushed with a glass rod in a droplet of acetocarmine or 0.3 M mannitol solution on a slide.

4. The developmental stage of the microspores on the slide is then scored under an inverted or a light microscope.

5. If most microspores are at late uninucleate to early binucleate stages, the whole tassel is ready to be sampled. Once established for each genotype/hybrid, the relative location between the boot and the enclosed tassel can be used as a convenient criterion for sampling, assuming relatively stable growing conditions in a greenhouse or growth chamber.

6. Plants are cut at 1-2 nodes below the base to which the tassels attached.

7. All foliage, except 3-4 leaves immediately outside the tassel, is removed.

8. These 3-4 leaves are trimmed to just 2.5 cm longer than the tassel itself and the tassel is placed in a flask (or other container) with the base in contact with distilled water in the flask.

9. The flask is then brought back to the laboratory for further processing.

10. Tassels so harvested can be disinfected immediately when the schedule permits, or stored in a refrigerator at 4°C.

11. For storage, the flask with a tassel is wrapped in a plastic grocery vegetable bag, which is then sealed by masking tape. The flask is then placed in a refrigerator at 4°C.

12. Tassels may be stored this way for 1-3 days with no or negligible detrimental effects on microspore viability. However, storage beyond 3 days without further processing may be detrimental, and hence is strongly discouraged.

Pretreatment
1. In a laminar flow hood, the remaining foliage encasing the tassel is removed and the tassel is then taken out of the boot for disinfection (Fig. 2.14-1a).

2. The tassel may be disinfected as a whole or first separated into florets before being subjected to disinfection treatment.

3. The whole tassel is submerged into a 20% commercial bleach (or 1.2% sodium hypochlorite) solution in a 100 ml graduate cylinder for 20 min covered by Parafilm or plastic film, during which periodic shaking is applied.

4. The bleach solution is then poured out, and the tassel is washed three rinses with autoclaved dH_2O.

5. The florets are then removed from the tassel/branch with two pairs of forceps (flamed and sterile) and placed into a 100 mm (diameter) Petri dish for androgenesis induction pretreatment.

6. In case florets are separated first and then disinfected (Fig. 2.14-1c), they are transferred directly to a 100 mm (diameter) Petri dish for pretreatment after the disinfection step, which is identical to the handling of a whole tassel.

7. In each 100 mm Petri dish, 150-200 florets are floated over 10-15 ml of MMA (Table 2.14-1) medium consisting of primarily an inert sugar plus inducer chemical(s), usually 2-HNA (2-hydroxynicotinic acid), which switch(es) microspores from gametophytic to sporophytic development.

8. Petri dishes are sealed with Parafilm and placed in incubators with temperatures ranging from 5±1°C to 10±1°C in the dark for 8-14 days.

Table 2.14-1. MMA medium for pretreatment of tassels or florets

Medium components	Concentration (mg/L)
KCl	1,492
$MgSO_4 \times 7H_2O$	246
$CaCl_2 \times 2H_2O$	148
KH_2PO_4	136
H_3BO_3	2
KI	0.5
$MnSO_4 \times H_2O$	8
$ZnSO_4 \times 7H_2O$	0.3
FeNaEDTA	56
Mannitol	54,7
Ascorbic acid	50
2-HNA	100
pH	5.7-6.0

Isolation of microspores
1. After the androgenesis induction pretreatment, 100 mm Petri dishes with florets are brought to a laminar flow hood and florets in the Petri dishes are transferred into an MC-II Waring blender cup.

2. 50–60 ml of isolation medium (Table 2.14-2) is added to the blender cup. As an alternative, 50-60 ml of 0.3 M mannitol or 6% maltose + 50 mg/L ascorbic acid can be used as isolation medium.

3. Place the lid onto the blender cup.

4. Assemble the MC-II blender cup to Waring blender (Fig. 2.14-1d). Blend at 16,000 rpm for 10 seconds, and 14,000 rpm for 10 seconds.

Table 2.14-2. Isolation medium

Medium components	Concentration (mg/L)
Ascorbic acid	50
Biotin	0.1
Nicotinic acid	10
Proline	100
Mannitol	54,700
pH	5.7-6.0

5. Transfer blender cup back to the laminar flow hood and filter the slurry first through a 100 µm mesh filter (Fig. 2.14-2a). Then, the filtrate is collected and filtered through a 50 µm mesh filter (Fig. 2.14-2b).

6. Microspores retained on top of the 50 µm mesh filter are rinsed three times with 2 ml of 0.3 M mannitol solution each, then washed off the filter into a 60 mm (diameter) Petri dish with 2 ml of 0.3 M mannitol solution.

7. The mixture of microspores and 0.3 M mannitol solution is then layered over 5 ml of 21% maltose in a sterile conical centrifuge tube (15 ml).

8. The tube is balanced off and centrifuged at 750 rpm for 2 min. Viable microspores form one or two band(s) at the interphase of the mannitol and maltose solutions, while debris and damaged microspores form a pellet at the bottom of the tube (Fig. 2.14-2c).

9. Pipette and transfer the band(s) with 3ml solution into another 15 ml conical centrifuge tube. Centrifuge at 1,400 rpm for 1.5 to 2 min. The induced microspores form a band on top of the aqueous phase (Fig. 2.14-2d).

10. Add 1 ml of isolation medium on top of the band slowly then collect microspores in the band carefully and transfer them with a pipette to another 50 µm mesh filter. Allow the solution to pass through while microspores are retained on top of the mesh filter.

11. Rinse microspores trapped on the mesh filter three times with 2 ml each of induction medium IND (Table 2.14-3). Microspores are then rinsed off the mesh filter and into a 20x60 mm Petri dish with 2 ml of induction medium IND (Fig. 2.14-3a).

Table 2.14-3. Media recipes for embryo induction and plant regeneration

Media components	IND medium* (mg/L)	Reg-I medium** (mg/L)	Reg-II medium (mg/L)	Reg-III medium (mg/L)
Macro salts				
$(NH_4)_2SO_4$	-	463	-	-
KNO_3	2,500	2,830	2,500	2,500
$CaCl_2 \times 2H_2O$	176	166	176	176
NH_4NO_3	165	-	165	165
KH_2PO_4	510	400	510	510
$MgSO_4 \times 7H_2O$	370	185	370	370
Iron source				
Na_2EDTA	37.3	8.2	37.3	37.3
$FeSO_4 \times 7H_2O$	27.8	5.6	27.8	27.8
Micro salts				
H_3BO_3	1.6	10	1.6	1.6
KI	0.8	0.4	0.8	0.8
$MnSO_4 \times 4 H_2O$	4.4	5	4.4	4.4
$ZnSO_4 \times 7 H_2O$	1.5	5	1.5	1.5
$CoCl_2 \times 6 H_2O$	-	0.0125	-	-
$CuSO_4 \times 5 H_2O$	-	0.0125	-	-
$Na_2MoO_4 \times 2 H_2O$	-	0.0125	-	-
Other components				
myo-Inositol	-	100	-	-
Glycine	2	-	2	2
Nicotinic acid	0.5	1	0.5	0.5
Pyridoxine	-	1	-	-
L-Proline	400	2,880	-	-
Thiamine HCl	0.5	10	1	1
Asparagine	15	-	-	-
Sucrose	50,000	30,000	30,000	30,000
Maltose	70,000	-	-	-
Casein hydrolysate	-	100	-	-
Glutamine	125	-	146	146
BAP	-	-	2	-
Kinetin	0.4	-	2	-
2,4-D	1.2	2	-	-
PAA	1	-	-	-
NAA	-	-	1	-
Gelrite	-	2,700	4,000	4,000
pH	5.7-6.0	5.7-6.0	5.7-6.0	5.7-6.0

* IND is a liquid medium that is filter sterilized through 0.2 μm filter
** Reg-I (Nageli *et al.*, 1999), Reg-II, Reg-III are all semi-solid media that are sterilized by autoclave (20 psi at 120°C, 25 min.)

12. Microspore density is assessed through a haemocytometer under an inverted microscope and the microspores are evenly divided into Petri dishes for induction culture. In all cultures, the density of microspores is made approximately $2\text{-}7 \times 10^4$ ml^{-1}.

Induction culture

1. 4-6 wheat ovaries (preferably from variety. 'Chris,' although ovaries from other wheat cultivars work as well) are added into each Petri dish with the isolated maize microspores. All Petri dishes are sealed with Parafilm and placed in an incubator in the dark with a preset temperature of 27-28°C.

2. The induction medium is refreshed one week after the culture initiation, and wheat ovaries are replaced at four-week interval. The medium is refreshed by removing half (1.5-2.0 ml) of the old medium with a pipette and adding the same amount of fresh medium. Refreshment serves to remove toxic substances that are released by dead or degenerated microspores and/or prevent excess change in medium osmolarity, normally resulting from the breakdown of sucrose by enzymes of dividing microspores. During this culture period, close monitoring of the culture should be practiced.

3. The first cell divisions typically start after 3 days in culture (Fig. 2.14-3b). Multi-cellular structures typically are clearly defined after one week in culture (Fig. 2.14-3c). Pro-embryoids emerge out of the exine in about 11-14 days following the culture initiation (Fig. 2.14-3d).

4. The first group of embryoids/calli becomes visible to the eye approximately 21 days from the culture initiation. Once embryoids/calli reach the size of 2-3 mm in diameter (Fig. 2.14-4a), they are transferred to regeneration medium to allow direct or indirect plant regeneration.

Regeneration culture

1. Embryoids or regenerable calli, which are yellowish and compact, are transferred directly to regeneration/germination medium Reg-II (Table 2.14-3) for plant regeneration/ development.

2. The emerged shoots (Fig. 2.14-4b) are transferred to Reg-III for rooting (Fig. 2.14-4c, d).

3. Calli of poor quality (appearing loose, watery and white in color) can be first transferred to medium Reg-I (Table 2.14-3) to induce embryogenic competency.

4. Following such transfer, calli are kept in the dark at 28°C for 1-2 weeks prior to their transfer to regeneration medium Reg-II for plant development.

5. Petri dishes with embryoids or regenerable calli on Reg-II are kept under light for two weeks.

6. Once shoots grow to approximately 2-3 cm in height, they are transferred to tissue culture tubes containing regeneration medium Reg-III (Table 2.14-3), for root initiation (Fig.2.14-4c).

7. About 7-10 days following the transfer to regeneration medium Reg-III, regenerated plantlets (Fig. 2.14-4d) are ready for transfer to a greenhouse or growth chamber for further growth to facilitate the examination of chromosome doubling or seed production.

8. It is important to note that one regenerable callus or embryoid often gives rise to multiple green plants, which allow subsequent sib-pollinations for seed production.

Raising doubled haploids in a greenhouse

1. Once the root system is well established and the shoot reaches 5-6 cm or higher, plantlets can be transplanted directly to pots of 25x30.5 cm.

2. A tool (preferably a small spatula) is first employed to loosen the contact between the agar and the tube wall.

3. The plantlet is then gently removed from the tube and the agar is carefully rubbed off the roots. Care should be taken to avoid any unnecessary damage to the root system.

4. In preparation for transplanting, pots are filled with soil and watered to saturation. Soils are premixed with fertilizers (N, P, K).

5. The plantlets are then planted in pots in such a way that roots are placed in a hole and soil is gently pressed around the plant by fingers to enable good root-soil contact.

6. Pots are watered once a day or whenever the soil appears to have dried somewhat. It is beneficial to keep a high humidity for the first few days following transplanting.

7. Four to seven days after transplanting, plantlets are established in the pots and begin to grow rapidly (Fig. 2.14-5a,b).

8. In a few weeks (varies with the maturity date of genotypes), cobs and tassels emerge from plants (Fig. 2.14-5c), doubled haploid plants show normal fertility with full seed-set following sib-pollination.

9. Alternatively, smaller plantlets (with shoot ≤ 4 cm) may be first transplanted to plastic flats that have transparent plastic covers in a tray to maintain a high relative humidity while light still penetrates the cover. After a week or so, acclimated plantlets are retransferred to pots of normal size as described above.

References

Büter, B., 1997. *In vitro* haploid production in maize. In: *In Vitro* Haploid Production in Higher Plants. Jain, S.M., S.K. Sopory, and R.E. Veilleux (Eds.) Kluwer Academic Publishers, Dordrecht, pp. 37-71.

Konzak, C.F., E.A. Polle, W. Liu and M.Y. Zheng, 2002. Methods for generating doubled haploid plants. U.S. patent 6,362,393.

Nageli, M., J.E. Schmid, P. Stamp and B. Büter, 1999. Improved formation of regenerable callus in isolated microspore culture of maize: impact of carbohydrates, planting density and time of transfer. Plant Cell Rep. **19**: 177-184.

Zheng, Y., C.F. Konzak, Y. Weng and R. Sahibzada, 2002. Methods for generating doubled haploid maize plants. U.S. patent application.

Zheng, Y., W. Liu, Y. Weng, E. Polle and C.F. Konzak, 2001. Culture of freshly isolated wheat (*Triticum aestivum* L.) microspores treated with inducer chemicals. Plant Cell Rep. **20**: 685-690.

2.15
Anther culture of maize (*Zea mays* L.)

B. Barnabás

Agricultural Research Institute of the Hungarian Academy of Sciences, Martonvásár, Hungary

Introduction

The potential of doubled haploids (DH) in maize breeding has long been recognised. By the use of DH-producing techniques various gene combinations can be fixed in homozygous form in a short time. Moreover, DH lines possessing high fertility and favourable agronomic characters can be utilized directly in heterosis breeding to produce new hybrids which satisfy market demands. Anther culture could be a useful tool for creating maize inbred lines. However, the successful application of anther culture techniques in maize breeding is largely dependent on the androgenic responses of the genotypes and on the frequency of induced or spontaneous genome doubling in plants of microspore origin. Since the pioneer research of Chinese scientists (Kuo *et al.*, 1978) intensive studies have been carried out to improve the culture conditions, leading to greater androgenic response and the construction of highly responsive maize stocks (Genovesi and Collins, 1982; Dieu and Beckert, 1986; Petolino and Jones, 1986). As most of the genotypes responsive to anther culture have been found in non-commercial maize germplasm, culturability must be transferred from responsive, exotic germplasm (mainly of Chinese origin) into elite types. During the last few years an efficient, reproducible anther culture methodology has been elaborated in the Agricultural Research Institute of the Hungarian Academy of Sciences, Martonvásár, Hungary. DH lines of exotic origin selected for high androgenic capacity, good adaptability to local climatic conditions and satisfactory agronomic characters were chosen as crossing partners to introduce microspore induction ability into the elite lines used in breeding. Single cross hybrids in which the exotic parent represents 50% of genotype, or hybrids backcrossed to the elite lines (BC_1, BC_2; 25 and 12.5%, respectively) have been used as anther donors to develop new DH lines *via* anther culture. This technique makes it possible to produce a large number of microspore-derived calli or embryos capable of regenerating viable, fertile DH plants from a wide range of genotypes which contain a minimum of 12.5% exotic (Chinese) germplasm.

Protocol

Donor plants and growth conditions
Both field-grown and phytotron-grown plants can be used as anther donors.

Growth conditions for anther donor plants raised in phytotron chambers
Germinate the seeds on pre-sterilized filter paper moistened with sterile distilled water and put in covered Petri dishes (10 per dish). Transfer the dishes to a growth chamber (Conviron TCL) at 26°C, using a 8/16 h day/night photoperiod with 100 µE m^{-2} s^{-1} light intensity. After about 4-5 days transfer the seedlings to 5 L pots filled with potting compost and grow the plants in a phytotron chamber (Conviron GB-48) at 18/15°C day and night temperature using a 16/8 h photoperiod at 200 µE m^{-2} s^{-1} light intensity for 2 weeks. Then change the temperature to 20/17°C for development of anthers in two weeks, then to 22/20°C for the further development of the plants. The relative humidity in the chamber should be over 80%.

Cold pretreatment of anthers
Harvest the tassels from the donor plants just before their emergence from the whorl, when the microspores should be in the mid uninucleate stage of development (Fig. 2.15-1a). After collection cover the excised tassels with aluminium foil and store them at 7°C for one week.

Examination of the developmental stage of the cold pretreated microspores
1. Use anthers from the main branch and from the lateral branches of the tassel for making squashes stained with acetocarmine. Microspores released from the anthers can be observed under a light microscope.

2. Microspores in the mid uninucleate stage will continue the development during cold pretreatment and they reach the late uninucleate stage (Fig. 2.15-1b) by the end of cold storage.

3. The optimal developmental stages for *in vitro* androgenesis are the late uninucleate and premitotic stages.

Surface sterilization of the tassels
It is necessary to use pre-sterilized dishes and forceps for this procedure. The whole procedure should be carried out in a laminar air flow cabinet.
   Parts of the tassels which contain late uninucleate or premitotic microspores should be sterilized with 20% sodium hypochlorite for 30 minutes and then washed 3 times with sterile distilled water. Isolate the anthers under aseptic conditions and inoculate them onto the surface of induction medium.

Preparation of the induction medium.
1. For the induction of microspore-derived structures (calli and embryos) it is best to use a modified YP medium (Genovesi and Collins, 1982), (Table 2.15-1).

2. Autoclave for 20 minutes at 120°C, 90 kPa.

3. Distribute the medium into Petri dishes (90 mm in diameter, glass or plastic) 25 ml each. The procedure must be done in laminar air flow boxes.

4. Inoculate anthers under sterile conditions, using a pair of forceps. Place no more than 100 anthers onto the medium and close the lid. Seal with Parafilm or plastic foil.

Table 2.15-1. Composition of media used for maize anther culture

Media components	YP induction medium (mg/L)	Modified N6 regeneration medium (mg/L)
Macro salts		
KNO_3	2,500	2,830
NH_4NO_3	165	-
$CaCl_2 \times 2H_2O$	176	166
KH_2PO_4	510	400
$MgSO_4 \times 7H_2O$	370	185
$(NH_4)_2SO_4$	-	463
Micro salts		
$ZnSO_4 \times 7H_2O$	8.6	1.5
KI	0.83	-
$MnSO_4 \times 4H_2O$	22.3	4.4
H_3BO_3	6.2	1.6
$Na_2MoO_4 \times 2H_2O$	0.25	-
$CuSO_4 \times 5H_2O$	0.025	-
$CoCl_2 \times 6H_2O$	0.025	-
Iron source		
Na_2EDTA	37.3	37.3
$FeSO_4 \times 7H_2O$	27.8	27.8
Vitamins		
Thiamine HCl	0.5	-
myo-Inositol	100	-
Growth regulators		
2,3,5-TIBA	0.1	-
Kinetin	-	1.0
NAA	-	0.5
Other componentss		
L-Asparagin	150	-
Casein hydrolysate	500	-
Sucrose	120,000	40,000
Activated charcoal	5,000	-
Gerlite	2,500	-
Agar	-	6,000
pH	5.4	5.8

Early genome doubling
Early genome doubling is necessary for the genotypes in which spontaneous doubling is rare. After autoclaving for 20 min at 120°C, 90 kPa, the colchicine (Sigma C 3915) is added to the induction medium at a concentration of 30 mg/L after filtration through a MILLEX-GS 0.22 μm filter (MILLIPORE). The anthers should be inoculated on the colchicine-containing medium and left until the microspores complete their first mitotic division (approx. 3-4 days). After the transfer of anthers to a colchicine-free induction medium, culture them for further androgenic development.

Anther incubation

Culture the anthers at a constant temperature of 29±1°C at 80% of relative humidity in the dark for 1 month. The asymmetric type of the first mitotic microspore division seems to be characteristic to the development of callus (Barnabas *et al.*, 1999) (Fig. 2.15-2a,b). However, as a consequence of using colchicine for treatment of induced anthers, a moderate increase in embryo yield can be observed (Fig. 2.15-2c). The colchicine treatment does not basically modify the anther response of the given genotypes (Fig. 2.15-3). During the one month incubation period, a considerable number of calli and embryo-like structures of microspore origin will be developed from the anther cultures.

Plant differentiation from microspore-derived structures

Remove calli and embryo-like structures from anthers cultured for one month and transfer them directly to N6 regeneration medium (Table 2.15-1).

1. Sterilize the medium for 20 minutes at 120°C, 90 kPa.

2. Distribute the medium into Petri dishes and sterile flasks (250-500 ml) 25 ml each.

3. Inoculate the microspore-derived structures: Put the individual calli or embryoids on the surface of the regeneration medium using a pre-sterilized needle.

4. Incubate the Petri dishes at a constant temperature of 26°C under 16 h illumination (50 μmol s^{-1} m^{-2}) until some of the regenerating plantlets (Fig. 2.15-4a) reach a length of 1.0–1.5 cm, after which the plantlets should be placed in larger glass containers for further growth (2 weeks) on the same growth medium as before, but without hormones and with reduced (2%) sucrose content (Fig. 2.15-4b).

Planting the green plants into soil

Transplant the well-differentiated, healthy green plantlets into peat pellets (5 cm diameter) from AS Jiffy, Products Ltd., and cover them with plastic foil for around 5-6 days to prevent desiccation.

Growth of the plants under artificial climate

Transfer the plants into 20 L plastic pots filled with maize soil and grow to maturity in a phytotron climate chamber, as described above.

Efficiency and Applications

The protocol described above made possible to select and develop number of DH lines of partly or fully exotic origin (Fig.2.15-5,6) (see examples in Table 2.15-2), which can be used efficiently to transmit *in vitro* androgenic ability into non-responsive elite lines of maize. These lines (DH 105, DH 109) have high haploid induction capacity (50-90%) and good green plant regeneration ability (20-30%); moreover, they are adaptable to the Central Europe climatic conditions and both have satisfactory agronomic features.

The above mentioned lines can be used to introduce *in vitro* androgenic ability into elite lines, since not only the F$_1$ hybrids produced from them, but even the second backcrossed generation still have satisfactory responsiveness in anther culture, allowing new DH lines to be developed from these crosses (Table 2.15-3).

Table 2.15-2. Characterization of DH lines suitable for transmitting *in vitro* androgenic ability into non-responsive elite lines

Genotype	Origin	Plant height (cm)	Ear height (cm)	Ear length (cm)	No. of kernel rows	Time to 50 % tasselling (days)
DH 105	SR88 × Chi 592	167	65	13	14	85-88
DH 109	Chi 592	155	68	16	11	80-85

Table 2.15-3. *In vitro* androgenic capacity of various SC and BC generations of maize

Crosses (female x male)	Induction frequency of microspore-derived structures[1] (%)	Green plants[2] (%)
A632 × DH109	46.9	22.9
DH105 × HMv5405	96.8	18.4
DH109 × OH43	123.9	8.2
Mo17 × DH109	68.2	8.7
BC_1	4.8	5.0
SR88 × DH109	21.2	13.2
BC_1	7.6	11.6
CM7/1 × DH109	25.0	12.8
BC_1	18.2	15.7
DH109 × CM7/1	23.6	11.9
BC_1	17.2	14.3
HMv5405 × DH109	51.2	4.1
BC_1	22.5	3.6
BC_2	13.2	6.1
DH109 × HMv5405	92.0	8.0
BC_1	23.9	3.9
BC_2	28.9	3.7

[1] Per 100 inoculated anthers
[2] As a % of the microspore-derived structures cultured

Since not all genotypes used for this purpose have a high tendency to spontaneous chromosome doubling, treating the induced anthers with colchicine may help to overcome the problems of diploidization. The data in Table 2.15-4 indicate that in justified cases the use of colchicine to treat anther cultures may provide a large number of fertile DH plants, thus helping to widen the genetic basis for further doubled haploid breeding.

Table 2.15-4. Effect of colchicine treatment on plant regeneration capacity and on the fertility of the regenerants

Genotype	Treatment	No. of embryos cultured	Plant regeneration (%)	Fertile regenerants (%)
A632×DH109	Control	63	22.9	22.5
A632×DH109	0.03% colchicine added to the induction medium	85	15.3	80.1
	LSD$_{5\%}$		17.5	45.9

References

Barnabas, B., B. Obert and G. Kovacs, 1999. Colchicine, an efficient genome-doubling agent for maize (*Zea mays* L.) microspores cultured *in anthero*. Plant Cell Rep. **18**: 858-862.

Dieu, E. and M. Beckert, 1986. Further studies of androgenetic embryo production and plant regeneration from *in vitro* cultured anthers in maize (*Zea mays* L.). Maydica **31**: 245-259.

Genovesi, A.D. and G.B. Collins, 1982. *In vitro* production of haploid plants of corn via anther culture. Crop Sci. **22**: 1137-1144.

Kuo, C.S., A.C. Sun, Y.Y. Wang, Y.L. Gui, S.R. Gu and S.H. Miao, 1978. Studies on induction of pollen plants and androgenesis in maize. Acta Bot.Sinica, **20**: 204-209.

Petolino, J.P. and A.M. Jones, 1986. Anther culture of elite genotypes of maize. Crop Sci. **26**: 1072-1074.

2.16
Laboratory protocol for anther culture technique in rice

F.J. Zapata-Arias
Plant Breeding Unit, FAO/IAEA Agriculture and Biotechnology Laboratory, Seibersdorf, Austria

Introduction

Rice provides 23% of global human energy and 16% of protein per capita. As the world's population is expanding rapidly, it is necessary to increase the rice production from today's 320 to 420 million tons in the next 20 years. Although conventional breeding techniques have considerably increased the productivity of modern crops, the application of *in vitro* techniques could speed up further crop improvement by shortening the breeding process, and overcome some of the substantial agronomic and environmental problems that have not been solved using conventional techniques. The doubled haploid technique has been extensively shown to be a valuable tool in the improvement of various crops including rice. The technique offers the quickest method of making heterozygous breeding lines homozygous, and permits greater selection discrimination between genotypes within any generation. It enables the immediate evaluation of quantitative characters, since all genes are fixed in homozygous stage and desirable alleles cannot be lost through segregation in future generations. Presently more than 20 rice varieties have been obtained using this technique.

Protocol

Donor plants and growth conditions

1. Use only one source of rice seed and accession number to ascertain repeatability of the experiments. Note the source and accession number of seeds used for panicle collection.

2. Separate partial and completely filled seeds, grade them according to their specific gravity (Fig. 2.16-1a). The optimum specific gravity should be between 1.1 and 1.2. Solutions of different gravities will be prepared by dissolving various amounts of sodium chloride in water to give specific gravity range of 1.00 to 1.20 with 0.05 intervals. Specific gravity (SG) should be measured using a hydrometer. A random sample of 200 grains should be used per replication. Five replications per variety should be made. The grains should be sequentially immersed and stirred into increasingly more concentrated salt solutions. Grains that rise should be isolated and graded as having SG lower than that of the solution in which they were immersed while grains which sank exceeded the SG of the solution. Grains should be air-dried for 30 minutes and those that sank should be transferred to the next more concentrated solution. Behavior of grains in each salt solution should be recorded. Percent grain distribution based on specific gravity should be determined.

3. Break the dormancy of clean seeds of F_1 hybrids or varieties in the oven at 50°C for 3-5 days. Place seeds in a Petri dish lined with moistened filter paper for 12-24 h.

4. Sow the seeds in seed boxes and water sufficiently to wet the soil.

5. At twenty-one days after seeding, transplant the seedling to around 15 cm diameter pots containing soil, fertilized at the rate of 2.5 g $(NH_4)_2 SO_4$, 1.25 g P_2O_5 and 0.75 g K_2O per pot.

6. Keep the plants in the screen house (Fig. 2.16-1b) and water them in the morning and afternoon. Apply pesticides as required for your area and types of insects.

Panicle collection and pre-culture treatments

1. Collect disease-free panicles (also called boots) from the primary tillers between 7-9 am or 4-6 pm. It is at this time that the developmental stage of the microspores would be just before the nuclear mitosis stage. Literature reveals that microspores at this stage respond well to anther culture. The distance of the flag leaf auricle to that of the next leaf should be between 5 to 9 cm (Fig. 2.16-1c). This distance would approximate the mid uninucleate to early binucleate stages of microspore development.

2. Wash the panicles thoroughly with tap water to remove surface dirt, remove the leaf blades, spray the panicles with 70% ethanol then wrap in paper towel moistened with distilled water (Fig. 2.16-1d).

3. Place the panicles, or the plated anthers, in the incubator at 8-10°C for 8 days. It has been observed that chilling or cold treatment prior to culture enhances androgenesis in rice. Reports indicate that low temperature shocks delay senescence, retain the microspore viability longer and prevent their abortion, thus increasing the number of viable microspores, which are destined to form embryos.

Sterilization of panicles (boots)
1. Select the florets, which contain anthers at the proper stage (Fig. 2.16-1e). The anthers should not exceed half of the floret length. Lemma and palea should be slightly green. This approximately corresponds to the late uninucleate - early binucleate microspore stage which are the most responsive stages for callus induction in rice.

2. Surface-sterilize florets with 20% Chlorox [commercial bleach containing 5.2% (w/v) NaOCl solution] for 20 min. In severe cases of contamination, wipe panicles with a cotton swab saturated with 70% alcohol or rinse in 70% alcohol with a drop of Teepol (liquid detergent) for 2-3 min before disinfecting with Chlorox solution.

Determination of microspore stage
1. Fix spikelet samples in a 3:1 ethanol-acetic acid fixation solution with 1.9% $FeCl_3$ added as a mordant.
   Preparation of fixation solution
   - Add 1 part of glacial acetic acid in 3 parts ethyl alcohol (v/v). Preparation of fixation solution should be done inside the fumehood.

   Preparation of acetocarmine stain
   - Dissolve 2 g of carmine or basic fuchsin in 100 ml of 45% acetic acid in a 500 ml flask.
   - Reflux slowly for 8 h.
   - Let it cool overnight and filter with a fluted paper the next day.
   - Pour the filtrate into an amber bottle.
   - The stain is better if aged for at least 2 weeks.

2. Keep fixed spikeletes in the incubator at 5-10°C at least overnight or transfer to 70% ethanol for permanent preparation.

3. Pick out anthers with fine forceps and place on a glass slide.

4. Crush anthers in 2% aceto-carmine stain using a flattened needle.

5. Cover with a glass slip and observe microspore stage under a light microscope (Fig. 2.16-1f-i).

Anther plating and callus induction
Perform all the steps inside a laminar flow hood (Fig. 2.16-2a-b)
1. Cut the base of the floret just below the anthers with sharp surgical scissors.

2. Using forceps, pick the floret at the apex and tap on the rim of the Petri dish so that the anthers fall down on the callus induction medium (liquid or semisolid form of N6 (Chu *et al.*, 1975) or MS (Murashige and Skoog, 1962) containing 0.5-2 mg/L 2,4-D. Composition of induction media is given in Table 2.16-1 and 2.

3. Inoculate 60 anthers per Petri plate (60x15 mm) containing 6 ml of induction medium.

4. Keep the plated anthers in the dark at 24±2°C.

5. Observe callus formation four weeks after anther planting (Fig. 2.16-2c,d)
   - Callus production efficiency can be determined in anthers plated on semisolid medium by taking the percentage of the number of anthers responding over the number of anthers plated. In liquid medium, the percentage is taken by dividing the

total number of calli produced by the number of anthers planted. Callus induction is best in liquid medium; however, the determination of callus induction efficiency is obtained more accurately in a semisolid medium. But, if semisolid medium is used, especially in the anther culture of F_1 hybrids, there is the danger of mixing calli originated from segregating microspores, particularly if the calli are allowed to grow big. Plants regenerated from mixed calli may be misinterpreted as segregation when plants are grown in the field.
- Although N6 medium was developed mostly for *japonica* varieties, it has also proved to induce callus in *indicas*. The reason for this may be that N6 medium is low in NH_4 concentration (total NH_4 is equal to 7 meq/L*, and in MS medium is equal to 20.61 meq/L)
- In general, *japonica* rice varieties are more susceptible to higher NH_4 concentrations than *indica* varieties.
- The synthetic auxin 2,4-D (0.5-2 mg/L) is the most commonly used auxin. When the concentration of 2,4-D is higher than 2 mg/L callus is induced but plant regeneration is reduced).

*Ionic concentrations should be given in milliequivalents per liter (meq/L). To obtain mM concentrations, divide the meq/L figure by the valence of the ion.

Media preparation
Stock solutions
- Formulate stock solutions to facilitate media preparation.
- Dissolve nutrients, one at a time in warm (not boiling) distilled water.
- Group the chemicals in such a way that precipitation will not occur.
- Dissolve water-insoluble compounds with 2 to 3 drops of 1N NaOH (for acidic compounds) or 1N HCl (for alkaline compounds).
- Bring the solutions to the desired volume.
- Store the different stock solutions in labeled amber bottles at 8-10°C. Use them within a month after preparation.

Culture media preparations
- Mix the media components according to the final concentration of each chemical.
- For a semi-solid medium preparation, add agar and melt it by heating before adjusting to the final volume.
- Adjust the pH using 1N NaOH or 1N HCl when the medium is relatively cold (about 40°C). The usual pH range requirement is at pH 5.6 to 5.8. For semisolid medium with a gelling agent, which solidifies easily, even at temperature higher than 40°C like gelrite, adjustment of pH is done prior to the addition of the gelling agent.

Media sterilization
- The media should be autoclaved for 15 min at 15 psi.
- For specific experimental requirements, use filter sterilization with a vacuum pump for large volumes, and syringes for microfiltration.
- Filter-sterilize thermolabile chemical solutions, such as some hormones and amino acids, and then add them to the autoclaved medium.
- Store medium preferable at 8-10°C

Calli plating and plant regeneration
1. Pick up calli of 2 to 3 mm in diameter with a sterile flattened needle or a microspatula.

2. Place the calli into test tubes (Fig. 2.16-2e) containing a semisolid MS medium (Table 2.16-1).

3. As a growth regulator use kinetin (0.5-4 mg/L), sucrose as a source of carbon. Calli will then develop shoots and roots (Fig. 2.16-2f,g).

4. However, when roots are slow growing, transfer the plantlets to semisolid MS medium without growth regulators.

5. The cultures should be incubated under cool, white, fluorescent light (1,000 lux) in light/dark conditions of 9/15 h, respectively, at $25\pm2°C$.

Planting in culture solution
1. Select plantlets with vigorous rooting systems, wash them thoroughly with tap water to remove the agar and callus adhering to the roots.

2. Record the number of plants per test tube.

3. Loosely wrap the base of the plantlets originating from one callus with foam.

4. Insert the plantlets into one hole of a 60-hole Styrofoam board. Place the Styrofoam over a plastic tray filled with Yoshida's culture solution. (Table 2.16-3)

Growing the plants in culture solution
1. Grow the plants in the screen-house (Fig. 2.16-2h) and whenever possible the temperature should be around $21/29°C$ night/day for two weeks under normal day length 9-10 h day/14-15 h night). The culture solution should be changed every five days.

2. Individualize the plants in a batch. Replant individual plants on a Styrofoam and grow in the culture solution in the screen house for another week.

3. Transfer each plant to a 15 cm pot containing fertilized soil. Determine the total number of regenerated major lines, plants coming from one test tube are consider as one line, and should be nominated as follows AC1-1, AC1-2…AC1-5.

Haploids vs. diploids
Anther culture derived green plants could not be distinguished as to ploidy level at the time of transplanting from culture flasks into pots. However, as plants started to grow, they could be identified readily. Haploid plants are sterile, shorter, with narrower and shorter leaves, numerous tillers, smaller panicles with fewer secondary branches and smaller spikelets.

Diploidization by colchicine treatment
1. Cut the root and upper portions of the plants at around 4 cm and 15 cm from the plant base, respectively (Fig. 2.16-2i).

2. Immerse the plants in a 0.1% colchicine solution for 4-5 hours. Use enough solution to immerse 5 cm of the stem. Grow the plants in the screen house until maturity.
   Preparation of colchicine
   - Dissolve 1.0 g of colchicine in 20 ml dimethysulfoxide (DMSO).
   - Make up the volume with water to 1 liter.
   - Add 3 drops of Tween 20 to the solution.

Warning: Colchicine is a carcinogenic compound. Avoid inhaling the dust and observe all precautionary measures such as wearing gloves and mask when handling the chemical.

Table 2.16-1. Preparation of modified Murashige and Skoog medium

Stock	Components	Final concentration (mg/L)	Stock solutions (g/L)	Amount of stock per liter of medium (ml)
MS-1	NH_4NO_3	1,650	82.5	
	KNO_3	1,900	95	
				20
MS-2	$MgSO_4 \times 7H_2O$	370	37	
	$MnSO_4 \times 4H_2O$	22.3	2.23	
	$ZnSO_4 \times 7H_2O$	10.58	1.058	
	$CuSO_4 \times 5H_2O$	0.025	0.0025	
				10
MS-3	$CaCl_2 \times 2H_2O$	440	44	
	KI	0.83	0.083	
	$CoCl_2 \times 6H_2O$	0.025	0.0025	
				10
MS-4	KH_2PO_4	170	17	
	H_3BO_3	6.2	0.62	
	$Na_2MoO_4 \times 2H_2O$	0.25	0.025	
				10
MS-5	$FeSO4 \times 7H_2O$	27.85	2.785	
	$Na_2EDTA \times 2H_2O$	37.25	3.725	
				10
Vitamins			(mg/L)	
	Nicotinic acid	0.5	100	0.5
	Pyridoxine HCl	0.5	100	0.5
	Thiamine HCl	0.1	100	0.1
	Glycine	2	100	2.0
Others	myo-Inositol	100	-	-
	Sucrose	30,000	-	-
	Agar	8,000	-	-
	pH	5.8	-	-

Screen-house operations
Bagging
Bag each panicle of the anther culture derived plants before the onset of anthesis to prevent cross-pollination with neighboring rice plants. This is done to maintain the attained homozygosity. Place panicle inside the glassine bag and secure with a paper clip or rubber band at the panicle base.

Harvesting
1. Harvest panicles in which 85% of grains have already turned straw gold in color.

2. Put harvested panicle in envelopes or paper bags, label properly and sun-dry for 3 days, or place in a dryer at $70^{\circ}C$ for a day to attain a 12-14% moisture content. Hand-thresh the panicles and determine the fertility percentage.

3. Store grains at 8-10°C, plant seeds for increasing seed number in preparation for higher seed requirement in field evaluation. Take extra care in handling these processes to prevent mixture or contamination of seeds from different plants.

Field screening of anther culture-derived plants
The field screening process will provide a wide array of valuable information on important physiological and agronomic characteristics like grain yield (yield components and harvest index), days to flowering, plant height, tiller number and grain quality characteristics as well as reactions to major insect pests and diseases.

Table 2.16-2. Preparation of N6 medium

Stock	Components	Final concentration (mg/L)	Stock solutions (g/L)	Amount of stock per liter of medium (ml)
G-1	KNO_3	2,500	250	10
G-2	$(NH_4)_2SO_4$	134	13.4	
	$MgSO_4.7H_2O$	150	15	
				10
G-3	$NaH_2PO_4.H_2O$	150	15	10
G-4	$MnSO_4.H_2O$	10	1.0	
	H_3BO_3	3	0.3	
	$ZnSO_4.7H_2O$	1.5	0.15	
				10
G-5	KI	0.75	0.075	10
G-6	$Na_2MoO_4.2H_2O$	0.25	0.025	
	$CuSO_4.5H_2O$	0.025	0.0025	
	$CoCl_2.6H_2O$	0.025	0.0025	
				10
Iron salt and Ca	Na_2Edta	37.25	3.725	
	$FeSO_4.7H_2O$	27.85	2.785	
	$CaCl_2.2H_2O$	150	15	
				10
myo-Inositol		160	-	-
Vitamins			(mg/100 ml)	
	Nicotinic acid	1	100	1
	Pyridoxine HCl	1	100	1
	Thiamine HCl	10	100	10
Hormones	2, 4-D	1	10	10
	BAP	0.5	10	5
	IAA	0.5	10	5
Others	Sucrose	20,000	-	-
	Agar	8,000	-	-
	pH	5.6-5.8	-	-

Table 2.16-3. Preparation of Yoshida's culture solution (Yoshida et al., 1976)

Stock	Components	Stock solution (g/L)	Amount of stock per 4 liters of solution (ml)
A	NH_4NO_3	914	5
B	$NaH_2PO_4 \cdot 2H_2O$	403	5
C	K_2SO_4	714	5
D	$CaCl_2$	886	5
E	$MgSO_4 \times 7H_2O$	3,240	5
F	Trace elements*		
	$MnCl_2 \times 4H_2O$	15	5
	$(NH_4)_6Mo_7O_{24} \times 4H_2O$	0.74	5
	H_3BO_3	9.34	5
	$ZnSO_4 \times 7H_2O$	0.74	5
	$CuSO_4 \times 5H_2O$	0.31	5
	$FeCl_3 \times 6H_2O$	77	5
	Citric acid (monohydrate)	119	5
	pH	5.0	

*Trace elements. Dissolve separately, then combine with 500 ml concentrated H_2SO_4. Make up to 10 liters volume with distilled water.

Efficiency and Applications

Two hundred and nine anther culture derived plants were regenerated from 18 varieties and 16 F_1 crosses and analyzed for their ploidy level. Of them, 88 were haploids (42.8%), 118 diploids (56.0%), 1 triploid and 2 tetraploids. From this study we also concluded that haploid production is genotype dependent. From 36 plants obtained from variety 'Taipei 309,' 24 were haploid and 11 had a diploid chromosome number. Of 17 plants from Taipei 309/'Tatsumimochi,' only 1 was haploid and 16 were diploid and presumed to be doubled haploids (Mercy and Zapata, 1986).

References

Chu, C.C., C.C. Wang, C.S. Sun, C. Hsu, K.C. Yin, C.Y. Chu and F.Y. Pi, 1975. Establishment of an efficient medium for anther culture of rice through comparative experiments on the nitrogen sources. Sci.Sin. **18**: 659-668.

Mercy, S.T. and F.J. Zapata, 1986. Chromosomal behavior of anther culture derived plants of rice. Plant Cell Rep. **5**: 215-218.

Murashige, T. and F. Skoog, 1962. A revised medium for rapid growth and bioassays with tobacco tissue cultures. Physiol.Plant. **15**: 473-497.

Yoshida, S., D. A. Forno, J. H. Cock and K. A. Gomez. (1976) Laboratory manual for physiological studies of rice. International Rice Research Institute, Los Banos, The Philippines.

2.17
Triticale anther culture

S. Tuvesson, R. von Post and A. Ljungberg
SW Laboratory, Svalöf Weibull AB, SE-268 81 Svalöv, Sweden

Introduction

The present long-term tissue culture programme aims at directing double haploid (DH) production of triticale towards responsive genotypes and at the same time allowing the practical use of anther culture in current breeding programmes. It is hoped that this can be achieved without limiting plant breeders to a few selected crosses. The goal of this investigation was to apply wheat methods to triticale, thus suggesting a strategy for large-scale DH production based on a screening procedure of parental lines and directing DH production of triticale towards predicted responsive F_1 and F_2s. Furthermore, the tissue culture programmes were rationalized in order to make this procedure cost effective.

Note: The presented protocol is based on the text prepared for wheat anther culture, Chapter 2.11, except of significant changes and data related to triticale DH production.

Protocol

Donor plants and growth conditions

The criterion used to select triticale breeding combinations for DH production was that one of the parents should give at least one green haploid per spike in anther culture. No information was needed concerning the other parent.

Screening for regeneration of green haploids took place throughout the year, from mainly glasshouse grown material. Donor plants were sown in peat in a mini glasshouse ('Minigro', 60x22 cm from VEFI, N-3255 Larvik) and vernalized in a cold-room for two months at 3°C before planting. Vernalization took place under warm white light (LUMA tubes 58W/L-83) using a 15/9 h photoperiod. The plants were raised in the glasshouse in 12 cm plastic pots, 10 cm between the pots with 10-15°C night and 15-20°C day temperature. The plants received natural light and during the darkest months (November-February) artificial light was supplied. DH production for the breeding programmes took place from April to August and the donor plants were grown in the glasshouse or, preferably in the field.

Anther culture

The microspores used were in the early and middle uninucleate stage of development (He and Ouyang, 1984). Development of the microspores was checked under microscopy a few times during the season. Eight spikes were cut from a minimum of four plants in order to reduce the effect of plant and spike in the screening experiments (Dunwell *et al.*, 1987). To produce DH for breeding programmes, approximately 20 spikes were used and the number of spikes may be adjusted based upon knowledge about the response of the parents. The cut spikes were placed at 7°C in the dark with the cut ends in water for a maximum of 14 days before handling. Spikes from the glasshouse were sterilized with 70% ethanol using moistened filter paper before plating of the anthers and field-grown material was sterilized with 0.1% mercury chloride for 8 minutes (Ouyang *et al.*, 1983) followed by washing with three changes of sterile water. Forceps were used to dissect out the anthers, which were placed in 9 cm plastic Petri dishes with medium. Anthers from the spike were placed in one Petri dish and dishes were sealed with plastic film. The medium used for anther culture was 190-2 (Wang and Hu, 1984), supplemented with 9% maltose, 1.5 mg/L 2,4-D, 0.5 mg/L kinetin, and without NAA. Medium was solidified with 0.35% Gelrite (Kelco) (Table 2.17-1). After plating the anthers, Petri dishes were placed in the dark at 28°C.

Regeneration

Embryos were transferred for regeneration after ca 35 days of anther culture.

Regeneration was achieved in 9 cm plastic Petri dishes with 190-2 substrate, but without hormones (Table 2.17-1). The dishes were sealed with plastic film. Regeneration took place at 24°C in cool white fluorescent light (OSRAM tubes L36W/20) under a 16/8 h photoperiod.

Chromosome doubling

Regenerated plants were left in Petri dishes until a good rooting system had developed after which transfer to soil took place. It is important to keep plantlets in a humid environment until they adjust to glasshouse conditions. In our experiments we used mini glasshouses with lids ('Minigro', 60x22 cm from VEFI, N-3255 Larvik) to secure a humid environment. Colchicine treatment of plants was achieved after washing the roots and carefully wiping with filter paper. The plants were placed with roots in 0.1% colchicine + 2% DMSO solution in the greenhouse under a lamp for 4 h. After washing in running water for 0.5-1 h, the plants were left overnight in water and then planted in soil. The presented chromosome doubling method resulted in 41% of DH. Arzani and Darvey (2001) reported a higher success rate (82.3% doubling) using a hydroponic recovery system.

Table 2.17-1. Modified 190-2 media for triticale anther culture

Media components	190-2 induction medium (mg/L)	190-2 regeneration medium (mg/L)
Macro salts		
KNO_3	1,000	1,000
$(NH_4)_2SO_4$	200	200
$MgSO_4 \times 7H_2O$	200	200
KH_2PO_4	300	300
$Ca(NO_3)_2 \times 4 H_2O$	100	100
KCl	40	40
Micro salts		
$MnSO_4 \times H_2O$	8	8
$ZnSO_4 \times 7 H_2O$	3	3
H_3BO_3	3	3
KI	0.5	0.5
Iron source		
$Na_2EDTA \times 2H_2O$	37.3	37.3
$Fe_2SO_4 \times 7H_2O$	27.8	27.8
Vitamins		
Myo-Inositol	100	100
Thiamine HCl	1.0	1
Nicotinic acid	0.5	0.5
Pyridoxine HCl	0.5	0.5
Other componenets		
Glycine	2.0	2
2,4-D	1.5	-
Kinetin	0.5	-
Maltose	90,000	-
Sucrose	-	20,000
Gelrite	3,500	3,500
pH	6.0	6.0

Large-scale DH production
The following rationalizations and 'short cuts' make the protocol well suited for large-scale production of DH at a low cost.

– The described method allows anther culture to be performed with a minimum of resources: sterile cabinet, autoclave, 28°C dark room or incubator, 24°C light growth room and a glasshouse.

– A single stock medium 190-2 is used for anther culture and regeneration of plantlets.

– The development stage of microspores is not tested on a routine basis.

– Embryos are neither counted nor evaluated for size or quality, but simply transferred for regeneration with a small spoon.

Efficiency and Applications

The screening programme consisted of 38 genotypes (8 cultivars and 30 advanced breeding lines) and gave a total of 1,624 green haploids. The average yield was 5.3 green haploids per spike (4.9 green plants per 100 anthers), ranging from 0.0 to 31.3 green haploids per spike. In the parental screening, eight commercial triticale cultivars were included (Tuvesson et al., 2000). Examples of well responsive commercial varieties are 'Bogo' with 4.4 green haploids per spike, 'Fidelio' with 3.3 green haploids per spike and 'Dagro' with 3.0 green haploids per spike. An example of a non- responsive variety is 'Moreno' (0.0 green haploids per spike). Of the 38 genotypes tested in the triticale screening experiment, 29 responded by giving more than one green haploid per spike. The actual parental lines used for crosses for DH production responded with 5.0 green plants per spike, ranging from 1.0 to 16.2 green plants per spike.

DH production from 21 Swedish and Dutch triticale F_1 and F_2 breeding combinations gave 1,211 green plants. On average 6.2 green plants were obtained per spike (4.7 green plants per 100 anthers). The response ranged from 0.8 to 13.6 green haploids per spike, with all of the combinations in the experiment responding with green haploids. There was no significant difference between the average response of the F_1s and F_2s and the average response of the actual parents that were previously characterized in anther culture. A significant correlation ($r = 0.299$) occurred between the number of green haploids produced by F_1s and F_2s and that produced by the actual parental lines characterized by anther culture.

Information on triticale anther culture is limited, compared to the information available for other cereals such as wheat and barley. In general, triticale appears to respond poorly to anther culture (Karsay et al., 1994; Ryöppy, 1997) and a major limitation of the method is the genotype effect. We observed that modern triticale lines responded well to DH production through anther culture using the presented method. Therefore, this methodology will be of practical value in various breeding programmes. In the screening study, 13% of the triticale genotypes responded by giving more than 10 green haploids per spike. The protocol presented here made the anther culture method for DH production possible in 94% of triticale crosses when based on the screening results.

A major problem is that many albinos are produced. In the present work the number of albinos was estimated to be 2.6 times higher than the number of green plants. The employment of isolated microspore technique might lead to a better frequency of regenerated green plants.

References

Arzani, A. and N.L. Darvey, 2001. The effect of colchicine on triticale anther-derived plants: microspore pre-treatment and haploid-plant treatment using a hydroponic recovery system. Euphytica **122**(2): 235-241.

Dunwell, J.M., R.J. Francis and W. Powell, 1987. Anther culture of *Hordeum vulgare* L.: a genetic study of microspore callus production and differentiation. Theor.Appl.Genet. **74**: 60-64.

He, D.G. and J.W. Ouyang, 1984. Callus and plantlet formation from cultured wheat anthers at different developmental stages. Plant Sci.Lett. **33**: 71-79.

Karsay, I., Z. Bedö and P.M. Hayes, 1994. Effect of induction medium pH and maltose concentration on *in vitro* androgenesis of hexaploid winter triticale and wheat. Plant Cell Tiss.Org.Cult. **39**: 49-53.

Ouyang, J.W., S.M. Zhou and S.E. Jia, 1983. The response of anther culture to culture temperature in *Triticum aestivum*. Theor.Appl.Genet. **66**: 101-109.

Ryöppy, P.H., 1997. Haploidy in triticale. In: *In Vitro* Haploid Production in Higher Plants. Jain, S.M., S.K. Sopory and R.E. Veilleux (Eds.) Kluwer Accademic Publishers, Dordrecht, pp. 117-131.

Tuvesson, S., A. Ljungberg, N Johansson, K.-E. Karlsson, L.W. Suijs and J.-P. Josset, 2000. Large-scale production of wheat and triticale doubled haploids through the use of a single-anther culture method. Plant Breed. **119**: 455-459.

Wang, X.Z. and H. Hu, 1984. The effect of potato II medium for triticale anther culture. Plant Sci.Lett. **36**: 237-239.

2.18
Protocol for anther culture in hexaploid triticale (x *Triticosecale* Wittm.)

M. Wędzony
Department of Plant Physiology, Polish Academy of Sciences, Kraków, Poland

Introduction

Various aspects of triticale anther culture have been widely studied. Two recent positions drew the attention of the author since they show interesting alternative protocols (Immonen and Robinson, 2000; Arzani and Darvey, 2001). Immonen and Robinson (2000) used modified W14 medium (Ouyang *et al.*, 1989) and tested various pretreatments. It appeared that two weeks of cooling at 4° C followed by three days at 32°C at the beginning of the induction phase (heat shock) raised green plant regeneration in comparison to cooling alone. Arzani and Darvey (2001) used hydroponics system to reverse sterile haploid plants back to the vegetative stage and to efficiently double their chromosome numbers to obtain DH lines. This option maximizes the efficiency of this method. The method of triticale anther culture described here is based on C17 induction medium (Wang and Chen, 1983) and 190-2 regeneration medium (Zhuang and Xu, 1983). Modifications and technical details have been elaborated in the Department of Plant Physiology, Polish Academy of Sciences in Krakow, Poland by the author and I. Marcinska. In our laboratory, several other media were tested with the same plant material and showed neither better nor more reliable results. Some initial results of a collaborative work were published (Marciniak *et al.*, 1998; Ponitka *et al.*, 1999).

Protocol

Donor plants and growth conditions
Spring triticale are sown directly into pots (3 parts garden soil, 2 parts Sphagnum substrate, 1 part sand). Winter triticale kernels are sown in perlite pre-soaked with Hoagland's salt solution (Table 2.18-1). Two-three day old seedlings are put to vernalization at 4-6°C, with a 12/12 h day/night photoperiod for 9 weeks. Later, they are potted in soil and grown in a glasshouse with supplementary illumination of halogen bulbs with average radiation density 1000 µmol (photons)·m^{-2}·s^{-1}. The additional illumination is applied to prolong the daytime and as a supplementary light during unfavourable weather conditions. Plants are grown under 16/8 h day/night photoperiod at 21/17°C, respectively. The same glasshouse and conditions are used to grow potted haploid plants, and doubled haploids to their maturity. Starting with shooting period, plants are fertilized with 1 L Hoagland's solution (Table 2.18-1) per 10 L pot twice a week.

Pretreatment
Tillers are collected when microspores in the central part of spikes are at the middle to late uninucleate stage. Stems are cut about 10 cm beneath the highest node and dipped in Hoagland's solution in glass jars. Each tiller has to be labelled with a genotype name and date of collection. Flag leaves are not removed during the pre-treatment period. The collected material is kept in a fridge in darkness at 4°C for 14 days.

Sterilization of material and isolation of anthers
Halved spikes, devoid of leaves and stems, are put into glass 150 ml Erlenmeyer flasks, 5-6 spikes per flask. They are immersed in 70% ethanol for 1 min, then in a 10% solution of commercial bleach containing also detergent (Domestos) for 15 minutes. Flasks are gently shaken by hand several times during sterilisation. Later spikes are washed 5 times in sterile water. The initial two washes are for 3 minutes, and following three washes for 10 minutes, all with intensive agitation of the flasks.

Under sterile laminar flow cabinets, anthers are dissected from spikes with fine forceps and then randomly and evenly distributed on the surface of the medium. It is essential not to squeeze anthers during isolation. Petri dishes with a diameter of 6 cm are used for culturing 100-130 anthers. Additionally, ten ovaries are usually placed randomly in every dish for conditioning of the medium. Petri dishes are tightly sealed with strips of thin clinging film (polyethylene food-foil). This better protects the culture from drying out in comparison to Parafilm (Sigma).

Induction phase
Induction mC17 medium with modifications is given in Table 2.18-1. All ingredients are autoclaved with the exception of vitamins, amino-acids and growth regulators. The latter are added to the medium after filter sterilization (pore diameter 20 µm) when it has cooled down to about 40°C.

Petri dishes with culture are placed in a cabinet at 29±1°C for 30 days. In three weeks, anthers usually turn dark yellow with brown spots. Starting with the fourth week, some anthers break longitudinally and white, compact structures appear (Fig. 2.18-1a). They often look like globular embryos and sometime continue to develop embryonic organs like scutellum (Fig. 2.18-1b,c) or coleoptile (Fig. 2.18-1d), therefore we call them embryo-like structures. Many androgenic structures develop into amorphic masses, later showing somatic embryogenesis on their surface. In that case, we call them androgenic calli. Both types of structures and all kinds of intermediate ones co-exist in the same anther culture (Fig. 2.18-2 to 4). When the androgenic structures (AS) i. e. embryo-like structures and calli reach a diameter of 1 mm or higher, they can be transferred to the regeneration medium.

Table 2.18-1. Composition of media used for Triticale anther cullture

Media components	Hoagland salts (mg/L)	mC17 induction medium (mg/L)	m190-2 regeneration medium (mg/L)
Macro nutrients			
KNO_3	660	1,400	1,000
$(NH_4)_2SO_4$	-	-	200
$Ca(NO_3)_2 \times 4H_2O$	940	-	100
$MgSO_4 \times 7H_2O$	520	150	200
$NH_4H_2PO_4$	120	-	-
KH_2PO_4	-	200	300
NH_4NO_3	-	300	-
$CaCl_2 \times 2H_2O$	-	150	-
Micro nutrients			
$MnSO_4 \cdot H_2O$	3.4	-	-
$MnSO_4 \cdot 4H_2O$	-	11.2	8
$ZnSO_4 \times 7H_2O$	0.2	8.6	3
H_3BO_3	2.8	6.2	3
KJ	-	0.86	0.5
$CuSO_4 \times 5H_2O$	0.1	0.025	-
$Na_2MoO_4 \times 2H_2O$	0.025	-	-
$CoCl_2 \times 6H_2O$	-	0.025	-
Iron source			
$Na_2EDTA \times 2H_2O$	37.8	37.8	37.8
$FeSO_4 \times 7H_2O$	27.8	27.8	27.8
Vitamins			
Nicotinic acid	-	0.5	0.5
Pyridoxine HCl	-	0.5	0.5
Thiamine HCl	-	1	1
Biotin	-	1.5	-
Folic acid	-	0.5	-
myo-Inositol[1]	-	100	100
Other components			
Glycine	-	2	2
Glutamine	-	-	100
Sucrose	-	-	30,000
Maltose	-	90,000	-
2,4-D	-	2.0	-
Kinetin	-	0.5	0.5
NAA	-	-	0.5
Agarose	-	6,000	-
Agar	-	-	6,000
(or Agargel[TM])*			(5,000)
pH (before autoclaving)	6.7	5.8	6.0

*A blend of agar and agar substitute; SIGMA A 3301

Regeneration phase

For regeneration, we use the modified m190-2 medium (Table 2.18-1). The medium is solidified with agar or agar-gel. We have not noticed any influence of the solidifying agent on the regeneration rate. For regeneration, 90 mm Petri' dishes are used during the first phase of regeneration (Fig. 2.18-1e) and later Magenta boxes (SIGMA V8380) for rooting (Fig. 2.18-1f).

AS not smaller than 1 mm were collected twice, at 30-33 and 37-40 days of the induction culture. Later, it is still possible to find new AS in the induction plates, however, their regeneration ability is usually dramatically lower. Therefore, in large-scale experiments we do not collect AS after day 40 of the induction period. AS are transferred to 90 mm Petri dishes (approx. 30 AS per dish) (Fig. 2.18-1e). Initially, they are kept in a fridge at 10°C and darkness for 7 days then moved to 20-23°C and illumination 16/8 h day/night photoperiod. Resulting green plantlets (GP) are subsequently transferred to fresh medium in Magenta boxes for better rooting (Fig. 2.18-1f), and albino plants are immediately removed from dishes when they appear. Cultures were let to regenerate plants up to 6 weeks after the passage.

Colchicine treatment

Colchicine solution (0.1% colchicine, 4% DMSO, 0.3% Tween 20, 0.025% GA_3) was applied in two ways in order to double chromosome number in haploid plants. With the first colchicine protocol, we worked with 4-6 week old green plants growing *in vitro*, with the second one we worked with plants rooted in soil. Colchicine is a poisonous and cancerogenic substance, therefore strict safety precautions have to be used to protect people and the environment.

Colchicine protocol 1

The treatment is most effective when applied to green plants two weeks after their transfer to Magenta boxes for rooting (Fig. 2.18-1f). At that time, their root system usually consists of 5-7 young roots and they have begun to produce new shoots. If the light period starts at 6 am, the treatment should start at 9-10 am, because the rate of cell divisions is the highest at that time.

1. The colchicine solution is poured over rooted plantlets growing in the medium, sufficient to make a 1 cm layer above the medium surface. Plants are treated like that for 6 h at 25°C under light intensity of 1,000 $\mu mol \cdot m^{-2} \cdot s^{-1}$.

2. The remnants of solution and agar are removed, and the plants are washed in several changes of water. Old roots that had developed *in vitro* are cut off, leaving fragments of 5 cm in length.

3. Plantlets are placed in the aerated Hoagland's solution (Table 2.18-1) for 10 days and kept at 10°C under a light intensity of 250 $\mu mol \cdot m^{-2} \cdot s^{-1}$ with a photoperiod 12/12 h. The solution is replaced with a fresh one, on the first, second, and fifth day after the colchicine treatment. We use a laboratory pump to aerate the solution, however, regular aquarium equipment could be utilised for the same purpose.

4. Subsequently, viable green plants are potted in perlite soaked with Hoaglands solution or in soil (Fig. 2.18-1f) and moved to a greenhouse for 10-14 days before initiating vernalization.

5. Vernalization at 4-6°C can be shortened to 7.5 weeks in this case, since treatment in 10°C serves also as a vernalization period.

6. Later, plants are re-potted into bigger vessels and grown in a greenhouse until maturity. Each plant should be individually isolated with paper or cloth bag during flowering, since triticale, especially under greenhouse conditions, can easily cross-pollinate.

The rate of plant survival with this method is 98%, the rate of doubled haploids among survivors is 75%. The additional advantage of the method is that doubled haploids produce well-filled spikes and average seed set was over 100 kernels per plant. This method is effective but expensive since a lot of colchicine is used.

Colchicine protocol 2
This protocol is applied to rooted green plants cultured in soil at the tillering stage.
1. Soil is washed off plants and their roots cut back to a length of 5-6 cm. Stronger plants can be divided into 2-3 parts, thus vegetative cloning is achieved. Be sure to label plants correctly at that stage.

2. The plants are dipped in the colchicine solution to cover 5-7 cm of shoot bases, for 7.5 h under light intensity of 1,000 µmol·m^{-2}·s^{-1} and temperature about 25°C. The solution should be aerated during this period.

3. After the treatment, plants are thoroughly washed with running water and placed in aerated Hoagland's solution for ten days at 10°C as described in the colchicine protocol 1.

4. Then plants are re-potted in soil, allowed to recover for 10-14 days in glasshouse conditions and then put into vernalization conditions for 7.5 weeks.

Later stages are the same as after the colchicine protocol 1. The rate of plant survival with this method is 85%, the rate of plants with chromosome doubling among survivors is 80%. Kernels usually appear in fertile sectors of otherwise sterile spikes and only 50% of plants produced more than 100 kernels per plant. Colchicine use is lower in comparison to the colchicine protocol 1 since the colchicine solution can be reused efficiently three times.

Remarks
The protocol described above is effective also with some modifications.
- It was successfully used to induce DH in material collected from field conditions
- In the induction medium, 2 mg/L 2,4-D can be replaced with the same amount of dicamba, or a mixture of 2,4-D and dicamba, or 2,4-D and picloram, 1 mg/L each. In some experiments, mixtures were superior to 2,4-D alone.
- If the cooling of AS at the beginning of regeneration phase is omitted, efficiency drops by a few percent.
- Both plastic and glass Petri dishes are used with similar success. The Magenta boxes can be replaced with various types of glass jars without affecting protocol efficiency.

Detrimental effect of some factors was noticed:
- High temperature (over 25°C) and treatment with herbicides, pesticides or fungicides prior to collection of material negatively affects the efficiency of the protocol.
- On the other hand, infectious diseases and pests on the donor plants make it very difficult to establish a sterile culture. For this reason, having healthy donor material became the key to success of this method.

- Temperatures over 25°C during the regeneration phase negatively affects the rate of plant regeneration and proportion of green to albino plants.

Efficiency and Applications

The presented method was used to obtain doubled haploid lines of winter and spring hexaploid ($2n = 6x = 42$) triticale (×*Triticosecale* Wittm.). The average frequencies of green plant regeneration were 6.4 green plants per 100 anthers based on 15 winter varieties. Each genotype was used in at least two experiments with a minimum three replications each. Across 22 hybrids, the average yield was 5.2 green plants per 100 anthers cultured. Variation between replications of the same experiment usually did not exceed 10% of the average value of given genotype, while differences between experiments reached up to 20% in some cases suggesting that uncontrolled factor(s) influenced the results. One of them could be the season of a year that affects biological clock and quality of daylight. The best results were usually obtained from mid January to May, while the worst were from September to mid December. The principal factor determining the results in every experiment was donor plant genotype.

The rate of spontaneous duplication of chromosome complements usually varied from 25 to 35%, only exceptionally reaching 55% in the described method. Therefore, doubling of chromosomes appears necessary. Several methods of colchicine application were tested with variable success, so the two most efficient methods were presented here.

Acknowledgements
The protocol presented in this paper was developed in collaboration with Dr. A. Ponitka, Dr. A. Ślusarkiewicz-Jarzina and MSc. J. Woźna from the Institute of Plant Genetics Polish Academy of Sciences (Poznań, Poland) and with Dr. K. Marciniak from DANKO Plant Breeding Ltd. (Choryń, Poland).

References

Arzani, A. and N.L. Darvey, 2001. The effect of colchicine on triticale anther-derived plants: microspore pre-treatment and haploid-plant treatment using a hydroponic recovery system. Euphytica **122**(2): 235-241.
Immonen, S. and J. Robinson, 2000. Stress treatments and ficoll for improving green plant regeneration in triticale anther culture. Plant Sci. **150**: 77-84.
Marciniak, K., Z. Banaszak and M. Wedzony, 1998. Effect of genotype, medium and sugar on triticale (x*Triticosecale* Wittm.) anther culture response. Cereal Res.Commun. **26**(2): 145-151.
Ouyang, J.W., S.E. Jia, C. Zhang, X. Chen and G. Fen, 1989. A new synthetic medium (W14) for wheat anther culture. Ann.Rep.Inst.Genet.Sin. 91-92.
Ponitka, A., A. Slusarkiewicz-Jarzina, M. Wedzony, I. Marcinska and J. Wozna, 1999. The influence of various *in vitro* culture conditions on androgenic embryo induction and plant regeneration from hexaploid triticale (x*Triticosecale* Wittm.) anther culture. J.Appl.Genet. **40**(3): 165-170.
Wang, P. and Y. Chen, 1983. Preliminary study on prediction of height of pollen H2 generation in winter wheat grown in the field. Acta Agron.Sinica. **9**: 283-284.
Zhuang, J.J. and J. Xu, 1983. Increasing differentiation frequencies in wheat pollen callus. In: Cell and Tissue Culture Techniques for Cereal Crop Improvement. Hu, H. and M.R. Vega (Eds.) Science Press, Beijing, pp. 431-432.

2.19
Protocol of triticale (x *Triticosecale* Wittmack) microspore culture

J. Pauk, R. Mihály, T. Monostori and M. Puolimatka[1]
Cereal Research Non-Profit Co., Wheat Genetics and Breeding Department, Szeged, P.O.Box 391, H-6701, Hungary; [1]*Plant Production Inspection Centre, Tampereentie 51, P.O.Box 111, FIN-32201 Loimaa, Finland*

Introduction

Triticale (x *Triticosecale* Wittmack) is a spontaneous synthetic amphiploid cereal that has been considerably improved through breeding. Triticale is currently grown on about 2 million hectares worldwide. After the initial period, intensive breeding research on triticale was started in the early 1950s by Árpád Kiss in Hungary. Nowadays, the centre of European triticale breeding is in Poland. Polish breeders released excellent varieties that have had a good performance, especially in Europe. Since the induction of the first triticale anther culture derived haploid plantlets by Wang *et al.* (1973), the method of anther culture has been essentially modified and improved. Theoretically, two *in vitro* cell and tissue culture methods are used to induce androgenesis: anther and microspore culture. The anther culture method seems to be more laborious. The blender isolation protocol provides a sufficient number of microspores to avoid the necessity for anther isolation. In cereals, microblending isolation was earlier established for barley (Olsen, 1991) and wheat (Ziauddin *et al.*, 1992), confirming its potential. The present protocol describes an *in vitro* process for blender-isolated triticale microspores leading to doubled haploid lines from microspore derived embryo-like structures (ELS). Embryogenesis was observed in isolated triticale microspores. This phenomenon also occurs in microspore culture of other cereals. From the ELS green plants can be regenerated, but albinism is still a typical problem of triticale microspore culture. Depending on the genotype, more than half of the regenerants may be albino.

Protocol

Donor plants and growth conditions

Seeds of winter triticale donor genotypes are sown in a nursery in autumn. In our experiment, each donor triticale genotype is a complete hexaploid ($2n=6x=42$, AABBRR). The donor varieties used in the crosses were Polish 'Tewo', 'Moniko' and 'Presto', all registered varieties in Hungary. Donor tillers are cut between the 2nd and 3rd node with scissors when the primary anthers of the most mature florets contain mid uninucleate microspores. The developmental stage of the microspores is determined under light microscope after squashing the anthers in a drop of water. Except for the flag leaf, the leaves are cut and the tillers are put into Erlenmeyer flasks containing fresh tap water. Tillers are covered by a PVC bag to maintain high humidity and transferred for cold pretreatment.

Pretreatment

Microspore donor tillers are cold pretreated under a dim fluorescent light at 4°C for about 2 weeks. A vernalization room (chamber) is used for cold pretreatment of donor tillers. During the two weeks of pretreatment, the donor tillers are kept in Erlenmeyer flasks filled with tap water and covered by PVC bags. During pretreatment it is very important that the level of tap water in the flasks is maintained. During the two weeks of cold pretreatment, the micropores develope a bit. We only use the late uninucleate (80%) and early binucleate (20%) micropores for isolation. At the end of pretreatment, the microspore developmental stage is checked again. If the developmental stage of microspores is earlier or later than the above suggested stage, these donor tillers are not used for DH production.

In vitro steps for microspore isolation and culture

The most characteristic steps of the triticale microspore culture from haploid cells to DH lines are demonstrated in Figure 2.19-1a-g.

1. After cold treatment, about ten to fifteen spikes containing late uninucleate (80%) to early binucleate (20%) microspores are removed from the leaf sheath and sterilized in 2% sodium hypochlorite (plus some drops of Tween 80/L) for 20 minutes on a rotory shaker (120 rpm), and then rinsed three times in sterile distilled water using 150-200 ml dH_2O per rinse.

2. The sterilized heads are cut into 1 cm sections using sterilized scissors and put into 100 ml Waring Micro Blender container (Eberbach Corporation, Ann Arbor, Michigan, USA) filled with 60 ml of 0.3 M autoclaved mannitol solution (5.47 g mannitol/100 ml distilled water).

3. Microspores in mannitol (0.3 M) are isolated by blending twice for 5 s at low speed, each time monitoring the maceration visually through plastic cap of vessel. If some spike fragments have not been cut in suspension, do an extra 5 s blending at low speed. After 2-3 minutes rest, start the separation of debris from the living microspores.

4. The crude microspore suspension from the container is filtered through a 160 μm sterile nylon sieve, then the microspore suspension is filtered again through a 80 μm sieve. Use about 20 ml 0.3 M mannitol to wash the microspores from the vessel through the nylon sieve.

5. Transfer the whole extract, by pipette, into 10 ml sterile centrifuge tubes (with cups). The number of centrifuge tubes depends on the final quantity of filtrate (about 100 ml). After

balancing the centrifuge baskets, centrifuge at 800 rpm for 5 min at laboratory temperature (22°C).

6. Remove the supernatant with a transfer pipette. The pellet is suspended in 2 ml 0.3 M mannitol. Check the microspores viability under an inverted microscope.

7. Pipette 4 ml 0.58 M autoclaved maltose into sterile centrifuge tubes. Carefully suspend the microspore suspension by a pipette and layer to the top of 0.58 M maltose.

8. Tubes are centrifuged at 600 rpm, for 10 minutes. The fraction of viable microspores is located in a fine band at the mannitol/maltose interphase. Collect the band carefully with a 2 ml pipette and transfer the collected microspores to sterile centrifuge tube. Transfer one band-derived microspores (1-2 ml) into a 10 ml sterile centrifuge tube and add 8 ml 0.3 M mannitol. Carefully suspend the microspores. Leave the tubes for about 10 minutes.

9. Centrifuge the microspore suspension as before, remove the supernatant, be careful with the last drop (a thin layer of mannitol has to remain over the pellet).

10. Add 1 ml filter sterilized liquid culture medium (Table 2.19-1) to the microspore pellet and suspend very carefully. Check the freshly isolated microspores under inverted microscope as previously. Determine the total number of microspores with a hemocytometer, and fit the final density to $60\text{-}80 \times 10^3$ microspores/ml of culture medium.

11. Culture 2 ml aliquots in 35 mm plastic sterile Petri dishe. Keep cultures in dark at 28°C and after 2 weeks add 500 µl of fresh culture medium (Table 2.19-1). Parafilm-closed cultures are kept in darkness at 80% humidity until embryoids appear on the top of the culture medium.

Microspore culture
After the final centrifuging (point 9), the pelleted microspores are suspended in exact quantities of culture medium. The quantity of microspores is estimated using a Burker (or other type) chamber. The viability of microspores is identified by staining with fluorescein diacetate (Widholm, 1972). This data can help in the evaluation of microspore culture response. Subsequently, the culture density is adjusted to $60\text{-}80 \times 10^3$ microspores/ml by adding culture medium. Two ml of microspore suspension are put into a plastic Petri dish (CORNING 35x10 mm disposable sterile suspension culture dishes are recommended).

For microspore culture, basic 190-2 medium (Zhuang and Xu, 1983) is supplemented with 3 mM L-glutamine and 175 mM maltose (Table 2.19-1). The medium can include the following growth regulator combinations: 1.5 mg/L 2,4-D and 0.5 mg/L kinetin (190-D/K) or 10 mg/L PAA (190-PAA), but the medium without hormones (190-0) gives the best results. The pH of medium is adjusted to 5.8 with 1 M KOH. Osmotic pressure is checked and adjusted to the value of 235 mOSkg^{-1} H$_2$O using Osmomat 030-D osmometer. The medium is filter sterilized (Millex GP syringe driven filter unit 0.22 µm) and stored at laboratory temperature. Medium older than 4 weeks is not used.

By the 5-6th week of subculture, well developed ELS (about 1 mm size) are transferred onto Gelrite solidified 190-2Cu medium (Table 2.19-1) and exposed to 16 h photoperiod provided by cool white fluorescent tubes with 20 µmol m^{-2} s^{-1} photon density, in a growth culture chamber at 28°C. Gelrite solidified 190-2Cu medium is used with sucrose replacing maltose. For the regeneration of ELS collected from the plated cultures, 190-2Cu medium is applied (Table 2.19-1). For induction of regeneration CORNING 100x20 mm disposable sterile suspension culture dishes, and for the individual growth of regenerants 25x90 mm sterilized glass tubes are recommended.

Regeneration and transfer of green plantlets
The bipolar embryos start to germinate on regeneration medium and on the apical part of ELS green or albino leaf primordia appear. The individual plantlets (about 1 cm size) are transferred to the single culture tube including the same regeneration medium, 190-2Cu. Copper ion significantly helps to obtain well rooted triticale plantlets.

Table 2.19-1. Composition of media used in triticale microspore culture

Component	190-0 culture medium (mg/L)	190-2Cu regeneration medium (mg/L)
Macro salts		
KNO_3	1,000	1,000
$(NH4)_2SO_4$	200	200
$MgSO_4 \times 7H_2O$	200	200
KH_2PO_4	300	300
$Ca(NO_3)_2 \times 4 H_2O$	100	100
KCl	40	40
Micro salts		
$MnSO_4 \times 4H_2O$	8	8
$ZnSO_4 \times 7 H_2O$	3	3
H_3BO_3	3	3
KI	0.5	0.5
$CuSO_4 \times 5H_2O$	-	0.5
Iron source		
$Na_2EDTA \times 2H_2O$	37.3	37.3
$Fe_2SO_4 \times 7H_2O$	27.8	27.8
Vitamins		
myo-Inositol	100	100
Thiamine HCl	1	1
Nicotinic acid	0.5	0.5
Pyridoxine HCl	0.5	0.5
Others		
Glycine	2	2
L-Glutamine	438	-
NAA	-	0.5
Kinetin	-	0.5
Maltose	63,000	-
Sucrose	-	30,000
Gelrite	-	2,800
pH	5.8	5.8

Requirements for growing H/DH plants

The well-tillered and rooted plantlets (about 4 weeks in regeneration conditions) are transplanted into non-sterilized 1:1 ratio peat and sandy soil mixture. During the following 2 weeks, the plantlets are acclimatized in a growth cabinet (Conviron) or in greenhouse at 80% relative humidity. The light intensity is 200 µmol m^{-2} s^{-1} with a 16 h light and 8 h dark photoperiod. Subsequently, the plants are grown under standard greenhouse conditions. Haploid plantlets are colchicine treated before vernalization. Vernalization is carried out in a cold chamber for 6 weeks at 2-4°C under continuous fluorescent light at 40 µmol m^{-2} s^{-1} photon density.

Ploidy level determination

Two different ploidy level determination methods are suggested: root tip cytology and estimation of ploidy level. Chromosome numbers are determined from root tip preparations. Donor plant for root tips is transported to a cool chamber at 4°C for 24 h. The pretreated tips are collected and treated in saturated oxiquinolin suspension for 5 h and fixed in 3:1 ethanol and glacial acetic acid solution. Root tips are hydrolyzed in 1 M HCl at 60°C for 10 min before staining in acetocarmine. Squash preparations are made in 45% acetic acid and chromosomes are counted in three well spread cells per root tip.

The length of the stomatal guard cell is determined using an ocular micrometer from a 10 mm distal leaf segment taken from microspore derived plants. Chlorophyll is extracted in 70% alcohol and the leaf segments are mounted in a drop of water on a glass slide with a coverslip. The length of stomatal guard cells is measured (length in lengthwise of stomatal guard cell) using a micrometer ocular. The length of stomatol guard cell of haploids is shorter by 40-50% when compared to a control hexaploid plant.

Colchicine treatment and DH production

The chromosome number of haploids should be doubled to get homozygous fertile plants. It is one of the most critical steps of triticale DH production (Fig. 2.19-1g). If the protocol is rigorously followed, about 50-70% doubling efficiency is obtained.

1. Prepare 0.2% colchicine, 2% DMSO in tap water.

2. Clean the haploid plantlets from the soil under tap water. Trim the root system back to 2 cm. Divide the plantlets for side tillers in the case of well-tillered plantlets and store the plantlets till next morning in tap water.

3. Treat the plantlets in the above prepared colchicine solution for 5 h in a glass vial.

4. Colchicine treatment is followed by an overnight washing under tap water.

5. The shoots are then trimmed to about 10-12 cm before transplanting into pots. Grow the treated plantlets under 16 h daylight and at 18/15°C day/night temperature at high humidity.

6. In the case of winter wheat, the well-regenerated triticale plantlets after colchicine treatment are vernalized for 42 days in a cool chamber at 4°C, in dim light.

7. After 42 days of vernalization the plantlets are transferred to greenhouse under standard growth conditions. The seeds of the DH plants are harvested and sown with an ear to row system in a nursery.

Efficiency and Applications

In our experiments, the viable mechanically isolated microspores have been successfully cultured in three different induction media, but a medium without hormones produced the best results. Each genotype, (Moniko, Presto, Tewo and their hybrids) responded in microspore culture and regeneration experiments. These genotypes were tested in a previous anther culture experiment and were found to be responsive. For laboratories beginning work with triticale microspore culture, we suggest using, at first, anther culture-responsive triticale genotypes. The blender-isolation technique claims a perfect sieving and separation of viable microspores from debris. After the microblending we use two different hole-size mechanical sieving through nylon sieve. The separation of viable microspores from dead cells is made by density gradient centrifugation. This technical solution was first suggested for viable protoplast preparation. After the successful protoplast "cleaning", the technique was successfully adapted in barley microspore isolation to obtain nice, cytoplasmatically dense, viable microspores (Olsen 1991).

Efficiency of the method is well represented in our published results (Pauk *et al.* 2000). In microspore culture of F_1 Tewo x Moniko, 9.5×10^4 ml^{-1} of micropores were cultured and 87 embryoids were obtained by the 6th week of culture. This efficiency is significantly lower than the top result of barley microspore culture using model genotype, 'Igri'. While we have not enough practice with different genotypes or varieties, the triticale microspore culture looks relatively easy for direct microspore embryogenesis.

The determination and comparison of stomatal guard cell length is a very effective routine method for separation of haploids from the spontaneous diploids. Generally the ploidy level determinations have confirmed that the microspore culture-derived plants are mainly haploids (90%). The haploids treated with a suitable colchicine solution will produce fertile seeds for further tests.

References

Olsen, F.L., 1991. Isolation and cultivation of embryogenic microspores from barley (*Hordeum vulgare* L.). Hereditas **115**: 255-266.
Pauk, J., M. Poulimatka, K.L. Toth and T. Monostori, 2000. *In vitro* androgenesis of triticale in isolated microspore culture. Plant Cell Tiss. Org. Cult. **61**: 221-229.
Wang, Y.Y., C.S. Sun, C.C. Wang and W.I. Chien, 1973. The induction of the pollen plantlets of Triticale and Capsicum annuum from anther culture. Scientia Sinica **16**: 147-151.
Widholm, J.M., 1972. The use of fluoresceine diacetate and phenosafranine for determining viability of cultured plant cells. Stain Technology **47**: 189-194.
Zhuang, J.J. and J. Xu, 1983. Increasing differentiation frequencies in wheat pollen callus. In: Cell and Tissue Culture Techniques for Cereal Crop Improvement. Hu, H. and M.R. Vega (Eds.) Science Press, Beijing, pp. 431-432.
Ziauddin, A., A. Marsolais, E. Simion and K.J. Kasha, 1992. Improved plant regeneration from wheat anther and barley microspore culture using phenylacetic acid (PAA). Plant Cell Rep. **11**: 489-498.

2.20
Protocol for doubled haploid production in hexaploid triticale (x *Triticosecale* Wittm.) by crosses with maize

M. Wędzony
Department of Plant Physiology, Polish Academy of Sciences, Krakow, Poland

Introduction

The method of triticale doubled haploid (DH) production by crosses with maize, as described here, was successfully used to obtain DH lines of winter and spring hexaploid ($2n = 6x = 42$) triticale (×*Triticosecale* Wittm.). The method is based on the protocols described for hexaploid wheat (Suenaga, 1994; Matzk and Mahn, 1994). Modifications and technical details have been elaborated in the Department of Plant Physiology, Polish Academy of Sciences in Krakow, Poland by the author and I. Marcinska (Wedzony et al., 1998a; Wedzony et al., 1998b). We have been working with this protocol since 1994, and its efficiency was compared with anther culture of triticale (Wedzony et al., 2000). The comparison shows that the effectiveness of both methods is related to the frequency of embryo induction. However, with maize crosses we were able to obtain doubled haploid triticale lines in materials recalcitrant to androgenic induction and in genotypes yielding exclusively albino plants in anther culture system. We have not found in the scientific literature any other successful protocols that use crosses with maize to produce doubled haploids in triticale. Methods to obtain DH triticale lines using anther or isolated microspore culture can be found in the other chapters of this Manual.

Protocol

Donor plants and growth conditions

Seeds of spring triticale are sown directly to pots (3 parts garden soil, 2 parts Sphagnum substrate, 1 part sand). Winter triticale genotypes are sown in perlite pre-soaked with Hoagland salt solution (Table 2.20-1). Two to three day-old seedlings are put to vernalization at 4-6°C, with a 12/12 h day/night photoperiod for 9 weeks. Later, they were potted in soil and grown in a glasshouse with supplementary illumination of halogen bulbs with average radiation density 1000 μmol (photons)$\cdot m^{-2} \cdot s^{-1}$. Additional illumination is applied to prolong the light period as a supplementary light during unfavourable weather conditions. Plants are grown under 16/8 h day/night photoperiod at 21/17°C, respectively. The same glasshouse conditions are used to grow maize, potted haploid plants, and doubled haploids to their maturity.

For two days, maize kernels are germinated in Petri dishes on filter paper soaked with Hoagland solution. Then they are put in soil at a depth of 5-6 cm and pots are placed in the same greenhouse that is used for triticale. To correlate flowering of maize and triticale, it is necessary to start germination of the first maize plants three weeks before the end of vernalization of triticale, or before sowing of spring triticale. The germination of maize can be finished two weeks after the last vernalized triticale plants are put into greenhouse, or last spring cultivars are sown. We germinate 4 maize plants twice a week when we expect about 100 triticale spikes to be pollinated per week. Hybrid sweet maize genotypes are the best pollinators. After testing several varieties, we have chosen the Polish cultivar 'Gama' as the most suitable for this purpose. We have never tested 'Seneca 60', the best wheat pollinator, since seeds were not available to us.

Starting at the time of triticale tillering, when maize plants have reached the stage of 5th leaf, plants are supplied with 1 L of the Hoagland solution per 10 L of soil twice a week. Any treatment with herbicides, pesticides or fungicides are strictly avoided starting two weeks prior to triticale flowering, since it was found that they can negatively influence the results of experiments.

Pollination with maize and treatment with dicamba

Three days before anthesis, the upper and basal spikelets and the central florets of spikelets are removed from triticale spikes. The remaining florets are emasculated by hand and spikes are isolated until one week after pollination, except for moments of treatments. When the flowers at the central part of triticale spikes are at the stage of flowering (glumae partially opened and fresh stigma with a brushy appearance visible), pollination is performed with freshly collected maize pollen. It is important to remember that the maize pollen looses its ability to germinate on triticale stigma 20-30 minutes after it leaves the maize anther. We use metal foil to collect maize pollen by shaking panicles over it. Then we brush collected pollen into a glass Petri dish and apply it with a fine brush to the triticale stigmas. At the end of each day, the old maize anthers are gently removed by hand from panicles.

Dicamba (3,6-dichloro-o-anisic acid) stock solution is obtained by dissolving 1 g of the substance in 10 ml 50% ethylene alcohol. The final solution of 100 mg/L is obtained by dilution in distilled water and the pH is adjusted to 5.5. An acidic pH is important, since dicamba in that condition is more efficiently transported into cells. The dicamba is applied to stigmas between 24 and 48 h after pollination. Drops of the solution are placed with a syringe among the glumes of the emasculated florets. Distinct enlargement of ovaries can be noticed as soon as 3 days after the treatment. Five days after treatment, the isolation bags are removed and kernel-like structures are let to grow *in planta* until 18-21 day after pollination. Proper watering and nutrition of plants during this period is essential.

Table 2.20-1. Composition of media

Ingredients	Hoagland salts (mg/L)	M190-2 (mg/L)	MS2 (mg/L)
Macro nutrients			
KNO_3	660	1,000	1,900
$(NH_4)_2SO_4$	-	200	-
$Ca(NO_3)_2 \times 4H_2O$	940	100	-
$MgSO_4 \times 7H_2O$	520	200	370
$NH_4H_2PO_4$	120	-	-
KH_2PO_4	-	300	170
NH_4NO_3	-	-	1,650
$CaCl_2 \times 2H_2O$	-	-	440
Micronutrients			
$MnSO_4 \times H_2O$	3.4	-	-
$MnSO_4 \times 4H_2O$	-	8	22.3
$ZnSO_4 \times 7H_2O$	0.2	3	8.6
H_3BO_3	2.8	3	6.2
KJ	-	0.5	0.83
$CuSO_4 \times 5H_2O$	0.1	-	0.025
$Na_2MoO_4 \times 2H_2O$	0.025	-	0.25
$CoCl_2 \times 6H_2O$	-	-	0.025
Iron source			
$Na_2EDTA \times 2H_2O$	37.8	37.8	37.8
$FeSO_4 \times 7H_2O$	27.8	27.8	27.8
Vitamins			
Nicotinic acid	-	0.5	1
Pyridoxine HCl	-	0.5	0.5
Thiamine HCl	-	1	0.5
Biotin	-	1	-
Folic acid	-	1	-
myo-Inositol	-	100	100
Others			
Gycine	-	2	2
Casein hydrolysate[1]	-	500	-
Sucrose[2]	-	50,000	30,000
2,4-D	-	-	2
Agar	-	6,000	6,000
pH (before autoclaving)	6.7	5.5	6.0

[1]Do not add to rooting medium; [2]In rooting medium lower sugar level to 30,000 mg/L

Sterilization of material and isolation of embryos

Enlarged ovaries (kernel-like structures) are isolated from spikes 18-21 days after pollination (Fig. 2.20-1a). The optimal period of isolation is genotype dependent and should be experimentally adjusted. The ovaries are immersed in 70% ethanol for 1 min, then in a solution of 10% commercial bleach containing also detergent (Domestos) for 15 minutes. Vessels are gently shaken by hand several times during sterilization. Later ovaries are washed

5 times in sterile water. The initial two washes are for 3 minutes and the following three washes for 10 minutes, all with extensive agitation of the flasks.

Under sterile laminar flow cabinet and with help of binocular stereo-microscope, embryos are dissected with needles from kernel-like structures and placed on the surface of the germination medium with scutellum (if distinguishable) in touch with medium. We put 3 embryos in 6 cm diameter dishes, or use 3 cm dishes for single embryos. Petri dishes are tightly sealed with strips of thin clinging film (polyethylene food-foil).

Embryos in this type of inter-specific cross develop in the absence of endosperm. They are attached with a spiky end (suspensor) to the micropylar part of ovule located in narrow end of kernel-like structure. The remaining part of the ovary cavity is usually filled with liquid. Exceptionally, traces of endosperm tissue can be found. Sometimes, before opening the kernel-like structure, the embryo is already detached from its proper position and floats in the liquid. It is visible as a compact white or slightly yellowish structure. Shapes differ from regular zygotic embryos (Fig. 2.20-1.b-d). Embryos can be at the globular stage (Fig. 2.20-1b) or at more advanced stages (Fig. 2.20-1c,d). Scutellum is generally the most irregular structure, while the embryo-axis organs are usually better formed. Embryo size and the stage of development achieved *in planta* depend on variety. Globular embryos are not able to germinate *in vitro*, however callus tissue can be induced from them and plants can be obtained this way.

Direct germination of embryos

For germination of embryos with embryo axis organs formed at time of excision (Fig. 2.20-1c,d), modified M190-2 regeneration medium (Zhuang and Xu, 1983) is used (Table 2.20-1). All ingredients are autoclaved with exception of vitamins, amino acids and growth regulators. The latter are added to the medium after filter sterilization (pore diameter 20 µm) when it has cooled down to about 40°C.

The Petri dishes with the embryos are kept in darkness at 18°C for about 4 weeks. Then they are brought to a growth chamber at 20-25°C and a day/night regime 16/8 h. During the dark period dishes should be checked periodically since some embryos can germinate earlier. If green plants appear, they should be transferred to the light conditions (Fig. 2.20-1e,f). Even embryos irregular in shape often produce regular plantlets. Some embryos develop double axis (twin embryos). When shoots of the plantlets have 2-4 cm they are moved to Magenta boxes (SIGMA V 8380) with the same medium (M190-2) but without casein hydrolysate and with sugar content lowered to 30,000 mg/L. This gives plants more space to grow and promote rooting. The rooted plantlets are either subjected to the colchicine treatment, according to the colchicine protocol 1, or they are potted in soil and treated with colchicine later, according to the colchicine protocol 2 (see methods described in Protocol 2.18).

Callus induction and plants regeneration

This option has been used in our laboratory to obtain haploid plants from small, undifferentiated, globular embryos (Fig. 2.20-1b). Callus can also be induced from bigger embryos (Fig. 2.20-1c,d). Embryos are put on MS2 medium (modified Murashige and Skoog medium, Table 2.20-1) in darkness at 25-28°C for 4 weeks. Induced calli are then transferred to the MS medium devoid of growth regulators and placed in dim light at 25°C until they start to regenerate green shoots. When shoots reach a few centimetres in length they are separated from calli and placed on half strength MS medium for better rooting. At that stage, plants were either subjected to the colchicine protocol 1, or potted in soil and treated with colchicine later according to the colchicine protocol 2 (see Protocol 2.18).

Remarks

Maize is a demanding plant and the growth conditions described above are not optimal for maize plants. If a warmer chamber is available, it should be used with night temperatures not dropping below 20°C. It is necessary to remember that maize should be germinated and grown under temperatures higher than 13°C. Even temporary cooling of maize plants at fast growth stage can result in pollen sterility. If the method fails to produce embryos, maize pollen viability should be checked as one of probable causes of failure. The second, most common mistake is pollination of too old flowers. Young flowers, a day prior to flowering and on the day of flowering, are most receptive. Triticale and wheat stigmas may loose their receptivity to maize pollen while still being receptive to the pollen of the same species.

Efficiency and Applications

With the use of maize pollination, from 1.5 to 19 green doubled haploid lines per 100 pollinated flowers were obtained by direct germination of haploid embryos. The average number of haploid plants using 14 varieties was 8.1 plants per 100 pollinated florets and from 17 hybrids it was 9 green plants. While the influence of genotype is clearly visible, we have not found so far any genotype completely recalcitrant to the method. The best responding winter triticale were varieties 'Bogo' and 'Salvo'. In comparison to hexaploid wheat, hexaploid triticale genotypes more frequently yield exclusively small embryos. Our studies have shown that the frequency of haploid embryo formation is controlled independently from the ability of the embryos to reach advanced stages of differentiation, enabling germination into plants (data not published). In genotypes yielding great numbers of small haploid embryos, plants can be obtained by regeneration from haploid calli. This alternative way increases the number of triticale genotypes that can give satisfactory results with maize crosses. Similar to anther culture, we found seasonal variation in success with the use of maize crosses. The best results were usually obtained from mid January to May, while the frequency of embryo formation drops by as much as 30% in autumn.

Spontaneous doubling of chromosome complements has never been observed in plants obtained *via* crosses with maize. Therefore, chromosome doubling is always necessary. We use the same methods of chromosome doubling as described for haploid plants obtained by anther culture (Protocol 2.18).

Acknowledgements

The protocol presented in this paper was developed in collaboration with Dr. K. Marciniak from DANKO Plant Breeding Ltd. (Choryń, Poland) and with Dr. A. Ponitka, and Dr. A. Ślusarkiewicz-Jarzina from the Institute of Plant Genetics, Polish Academy of Sciences (Poznań, Poland).

References

Matzk, F. and A. Mahn, 1994. Improved techniques for haploid production in wheat using chromosome elimination. Plant Breed. **113**: 125-129.

Suenaga, K., 1994. Doubled haploid system using the intergeneric crosses between wheat (*Triticum aestivum*) and maize (*Zea mays*). Bull.Natl.Inst.Agrobiol.Resour. **9**: 83-139.

Wedzony, M., H. Goral and E. Golemiec, 2000. Prospects for breaking genetic barriers in triticale doubled haploid production. Vortr.Pflanzenzüchtg. **49**: 125-135.

Wedzony, M., I. Marcinska, A. Ponitka, A. Slusarkiewicz-Jarzina and J. Wozna, 1998a. Factors influencing triticale doubled haploid production by means of crosses with maize. In: Proc. 4th Int.Triticale Symp. Red Deer, Canada. Vol. 1. Juskiw, P. (Ed.) International Triticale Association, Alberta, Canada, pp. 45-52.

Wedzony, M., I. Marcinska, A. Ponitka, A. Slusarkiewicz-Jarzina and J. Wozna, 1998b. Production of doubled haploids in triticale (*xTriticosecale* Wittm.) by means of crosses with maize (*Zea maize* L.) using Picloram and Dicamba. Plant Breed. **117**: 211-215.

Zhuang, J.J. and J. Xu, 1983. Increasing differentiation frequencies in wheat pollen callus. In: Cell and Tissue Culture Techniques for Cereal Crop Improvement. Hu, H. and M.R. Vega (Eds.) Science Press, Beijing, pp. 431-432.

2.21
Protocol for rye anther culture

S. Immonen[1] and T. Tenhola-Roininen

MTT Agrifood Research Finland, Plant Production Research, Crops and Biotechnology, Myllytie 10, FIN-31600 Jokioinen, Finland; [1]Current address: FAO, Viale delle Terme di Caracalla, 00100 Rome, Italy

Introduction

Rye (*Secale cereale* L.) is a cross-pollinated cereal, which has a strong self-incompatibility system. Population varieties of rye are highly heterozygous. Genes for self-fertility can, however, be introduced from wild relatives for purposes of hybrid rye breeding. In population breeding, selfing is detrimental, as rye suffers from severe inbreeding depression. The protocol for rye anther culture was developed for cross-pollinating rye for purposes of population breeding and genetic marker research. Rye has been generally considered to be recalcitrant to anther culture. The source of self-compatibility, a weedy wild relative *S. vavilovii* (L), has shown better response (Immonen, 1999). However, the use of a self-fertile species in population breeding would be problematic and therefore a protocol was needed for cultivated rye. The motivation for developing a broadly applicable rye anther culture protocol for rye is three-fold: in hybrid rye production homozygous doubled haploids could facilitate the process because, although the time needed for testing the combining ability cannot be reduced, the differentiation of quantitative traits is more effective with DH lines; in marker studies DH materials make development of mapping populations faster and mapping easier than is otherwise the case with obligatory cross-breeding crops; in population breeding doubled haploids could be a potential means for fixing a trait or increasing its prevalence in the population, providing that the trait can be selected for either visually or through marker-assisted selection at plantlet stage, or self-incompatibility could be circumvented. The latter could be reached through a recurrent selection scheme provided that the DH method to be used in subsequent cycles is economical and efficient, and the restrictions of self-incompatibility are overcome.

Protocol

Donor plants and growth conditions

In breeding programmes, tillers may be collected from donor plants grown in the field, but in our experiments glasshouse grown donor plants give better response. Glasshouse conditions should be optimised to give a constant day/night regimes and temperature. Seeds are sown in 14 cm diameter pots, three seeds in each, in peat soil mix, which is fertilized with 1% fertilizer solution (Kekkilä´s Superex Taimi: 19% N, 4% P, 20% K). At the 3 leaf stage, the plants are given additional micronutrient fertilization (0.5% Kemira Hivenliuos-I: 9.2% N, 9.2% urea-N, 0.2% B, 0.01% Co-EDTA-chelate, 0.24% Cu-EDTA-chelate, 0.51% Mn, 0.02% Mo, 0.46% Zn). The plants are grown under 20/15°C day/night temperature regime, with at least 16 h photoperiod in natural light, supplemented with fluorescent lamps. The total light intensity is approximately 280-350 μmol m^{-2} s^{-1}. Spring materials are sown in March or in August, 5-6 weeks prior to the planned collection of tillers and the plants are ready for anther excision approximately in 8 weeks which includes the cold pretreatment of spikes. Winter ryes require vernalization, making a total growth period of up to 21 weeks before collecting tillers (2-3 weeks initial growth, 12 weeks vernalization and 6 weeks growth to heading). Long-term experience with several cereals has shown that mid-winter and mid-summer periods are not favourable for initiating anther cultures.

Vernalization

Three methods are used for vernalizing winter rye materials:

- The seeds are sown in pots in an unheated glasshouse in October. They reach approximately the 5-leaf stage before winter frost. The temperature is below 0°C for 2-3 months. During vernalization the plants are grown under natural illumination. The temperature is increased in March and stimulates growth. A 16 h photoperiod is provided as for spring ryes. The donor plants begin heading in April-May in our conditions (Finland).

- The seeds are sown in pots and when the plantlets have reached the 5-leaf stage they are moved to cold rooms where they are kept in dim light for 3 months at 2°C with 25 μmol m^{-2} s^{-1} illumination.

- The seeds are sterilized in 1.2 % sodium hypochlorite for 15 min. after which they are rinsed once in 70 % ethanol and 4-5 times in sterile water. They are sown in Petri dishes on a thin layer of germination agar (100 mg/L CaSO$_4$ x 2H$_2$O and 8 g/L agar). The dishes are kept at 4°C for three weeks to induce germination and at 2°C in darkness for 6 weeks for vernalization, after which the seeds are sown in pots and grown following the regular process. There may be genotypic differences regarding the vernalization time, and some materials may require more than ten weeks.

The second option requires the same time as glasshouse vernalization but is not dependent on season. With the third vernalization option, growth from sowing to collection of tillers takes 15 weeks, although there may be varietal differences regarding the optimal time on agar.

Microspore developmental stage

The developmental stage of the microspores is determined for collecting the rye spikes at an optimal stage. At the initiation of culture, the majority of the microspores should be at the late uninucleate stage or just starting the first pollen grain mitosis (PGM). At the late uninucleate stage the vacuole is well developed and visible and the nucleus is located in the periphery, opposite the germ pore. During the first PGM, the vacuole is still visible, but then gradually disappears (Figure 2.21-1a,b). In order to determine the microspore stage, an anther is excised from the middle of the spike and placed onto a microscope slide in a drop of water, which provides an adequate image and is easier and faster than preparation with commonly used acetocarmine. The microspores are squeezed out by squashing the anther with a scalpel. A cover glass is placed on top and the microspores are observed under a dissecting microscope. It is practical to establish an anther length/developmental stage profile by studying anthers from a few spikes. With this profile, an adequate approximation of the microspore developmental stage is possible on the basis of the anther length. However, the relation between anther length and microspore developmental stage varies and the profile must be established for each batch of spikes separately. Spikes are usually ready for harvest when they are about to emerge from the flag leaf. The microspores develop slightly during pretreatment and their correct stage must be checked just prior anther excision.

In rye, microspore developmental stage seems to affect induction and regeneration differently: the induction rate from microspores at early uninucleate stage is poor and reaches a peak at early binucleate stage. However, the regeneration ability of the embryo-like structures (ELS) diminishes as they derive from more mature microspores. The first PGM seems to be a crucial stage. For maximal yield of green plants, microspores at the late uninucleate stage or first PGM are optimal (Immonen and Anttila, 1998).

Cold pretreatment of spikes
The spikes are collected in the beginning of the flowering period and the tillers are cut when the microspores are slightly premature for culturing - that is at the mid to late uninucleate stage. All leaves except the flag leaf are removed. The tillers are then placed with their cut ends in tap water in a container, which is covered with a plastic bag. The tillers are stored in darkness at 4°C for three weeks. In our experiments, the most responsive rye materials do not require such a long cold pretreatment, but with several lines it was necessary for triggering induction. With very recalcitrant genotypes even longer pretreatment periods of up to 28 days may be beneficial. A long cold pretreatment also seems to enhance spontaneous chromosome doubling and is therefore routinely used (Immonen and Anttila, 1999).

Surface sterilization
For anther excision, the spikes must be surface sterilized. They are removed from the leaf sheath and the awns are cut. The florets at the base and at the tip of the spike are also removed. The spikes are placed in jars, which contain 15% solution of commercial bleach (Oy FF-Chemicals Ab) (1.2% sodium hypochlorite solution) and a couple of drops (0.5%) of polyoxyethylenesorbitan monooleate (Tween 80), as a wetting agent. The spikes are agitated in the jars for 20 minutes after which they are rinsed several times with sterile water. From this stage on, the spikes and anthers are handled aseptically in a laminar flow cabin.

Anther excision and plating
Randomisation of anthers from different spikes between Petri dishes and treatments is a standard practise in experiments. Also in practical application it may be beneficial to mix anthers from different spikes and thereby obtain a more even response between Petri dishes. We have observed that the embryo-like structures appear more often in adjacent anthers than in anthers, which have no response near them (Fig. 2.21-1c). This may indicate that an initial response enhances response in neighbours. In research experiments, 800-1,000 anthers of each rye line have been plated for each treatment. In practical tests, if for instance DH-plants are needed from a particular F_1, the number of anthers may be much higher.

In comparing two Petri-dish sizes (3.4 cm and 5.4 cm diameter), ELS and calli deriving from the larger dishes showed significantly better regeneration ability than those deriving from the small dishes. Anther densities of 30–50/cm^3 give the highest regeneration rates (Immonen and Anttila, 2000).

Culture medium for induction

Basal induction medium
In our experiments, culture medium has appeared less critical for rye anther culture response than, for instance, the microspore developmental stage and the cold pretreatment. Induction and green plants have been obtained using several media that differ in their mineral composition, with total nitrogen ranging from 4.3 mM to 60 mM and the relation of ammonium to nitrate nitrogen from 34:66 to 0:100, and even from medium containing solely organic nitrogen (Immonen and Anttila, 2000). Medium x genotype interactions were not observed regarding development of embryo like structures in the comparative testing of various basal media. However, two media, 190-2 (Wang and Hu, 1984) and W14 (Ouyang *et al.*, 1989), both of which have lower total nitrogen content than the commonly used MS medium, repeatedly supported superior development of ELS. The ELS from the latter medium had relatively better regeneration efficiency. For rye, W14 medium has been chosen for routine use. The medium is supplemented with vitamins and iron, as described for MS medium and with L-glutamine at 3.4 mM (= 500 mg/L). The medium is adjusted to pH 5.8 prior to autoclave sterilization in glass bottles (15 min, 121°C, 20 psi). The L-glutamine component is filter sterilized and added to the medium after autoclave sterilization. Ten ml of medium is measured in each Petri dish using a sterile pump fitted into the bottle.

Solidification of basal medium
The medium is solidified with 0.3% (w/v) of Phytagel (gellum gum from Sigma Chemical Co., also known as gelrite), which according to Deimling and Flehinghaus-Roux (1996) is a suitable solidifying agent for rye anther culture – equal to agarose and superior to agar. With some rye genotypes liquid medium supplemented with 10% Ficoll 400 (Pharmacia Biotech, to increase buoyancy) may promote induction and regeneration efficiency. As solid medium is less costly and easier to use, it has been chosen for routine use in anther culture.

Carbohydrate
Deimling and Flehinghaus-Roux (1996) discussed the favourable effect of maltose over sucrose as carbohydrate source. Six percent would seem to be most cost-effective concentration of maltose for anther culture (Immonen and Anttila, 2000).

Growth regulators
Growth regulators, auxin in particularly, are used to trigger induction and enhance the production of ELS. Following experimentation of different combinations of auxins and kinetin, we have opted for 2.3 µM kinetin (0.5 mg/L) with 9 µM auxin (2 mg/L) (Immonen and Anttila, 2000). The auxin 2,4-D is used in the routine procedure, as it doesn't require filter sterilization.

Preparation of medium
Induction medium based on W14 salts (1 L)

W14 macro nutrients (10x)	100 ml
W14 micro nutrients (100x)	10 ml
MS iron (200x)	5 ml
MS vitamins (100x)	10 ml
2,4-D (0,2 mg/ml)	10 ml
Kinetin (1 mg/ml)	0.5 ml
Maltose	60 g
Phytagel	3 g
L-glutamine (1g/20 ml, filter sterilized)	10 ml
pH	5.8

Induction medium based on 190-2 salts (1 L)
190-2 macro nutrients (10x)	100 ml
190-2 micro nutrients (100x)	10 ml
MS iron (200x)	5 ml
MS vitamins (100x)	10 ml
2,4-D (0,2 mg/ml)	10 ml
Kinetin (1 mg/ml)	0.5 ml
Maltose	60 g
Phytagel	3 g
L-glutamine (1g/20 ml, filter sterilized)	10 ml
pH	5.8

Regeneration medium based on MS medium with half strength salts (1 L)
MS macro nutrients (10x)	50 ml
MS micro nutrients (100x)	5 ml
MS iron (200x)	5 ml
MS vitamins (100x)	10 ml
IAA (1 mg/ml)	0.5 ml
Kinetin (1 mg/ml)	1 ml
Sucrose	30 g
Phytagel	3 g
pH	5.8

Preparation of stock solutions

W14 macro nutrient stock (10x) (g/L)
KNO_3	20
$NH_4H_2PO_4$	3.8
$MgSO_4 \times 7H_2O$	2.0
$CaCl_2 \times 2H_2O$	1.4
K_2SO_4	7.0

W14 micro nutrient stock (100x) (g/L)
$MnSO_4 \times 4H_2O$	0.8
$ZnSO_4 \times 7H_2O$	0.3
H_3BO_3	0.3
KI	0.05
$CuSO_4 \times 5H_2O$	0.005
$CoCl_2 \times 6H_2O$	0.005
$Na_2MoO_4 \times 2H_2O$	0.001

190-2 macro nutrient stock (10x) (g/L)
KNO_3	10.0
$(NH_4)_2SO_4$	2.0
$MgSO_4 \times 7H_2O$	2.0
KCl	0.4
KH_2PO_4	3.0
$Ca(NO_3)_2 \times 4H_2O$	1.0

190-2 micro nutrient stock (100x)	(g/L)
MnSO$_4$ x 4H$_2$O	0.8
ZnSO$_4$ x 7H$_2$O	0.3
H$_3$BO$_3$	0.3
KI	0.05

MS macro nutrient stock (10x)	(g/L)
KNO$_3$	19.0
NH$_4$NO$_3$	16.5
MgSO$_4$ x 7H$_2$0	3.7
KH$_2$PO$_4$	1.7
CaCl$_2$ x 2H$_2$0	4.4

MS micro nutrient stock (100x)	(g/L)
MnSO$_4$ x 4H$_2$O	2.23
ZnSO$_4$ x 7H$_2$O	1.06
H$_3$BO$_3$	0.62
KI	0.083
CuSO$_4$ x 5H$_2$O	0.0025
CoCl$_2$ x 6H$_2$O	0.0025
Na$_2$MoO$_4$ x 2H$_2$O	0.025

MS iron stock (200x)	(g/L)
FeSO$_4$ x 7H$_2$0	5.57
NaEDTA x 2H$_2$0	7.45

MS-vitamin stock (100x)	(g/L)
Glycine	0.2
myo-Inositol	10.0
Nicotinic acid	0.05
Pyridoxine HCl	0.05
Thiamine HCl	0.01

The macro and micro nutrient and iron stock solutions are stored in refrigerator. The vitamin and growth regulator solutions are stored in suitable aliquots in plastic tubes in deep freeze.

Incubation conditions
Petri dishes are sealed with parafilm (Parafilm 'M'®, Laboratory film, American National Can) and incubated at 25°C in darkness for seven to ten weeks. The cultures can be incubated in a phytothrone or in a room, or container, where the temperature can be kept constant. Development of calli and ELS (Fig. 2.21-1d,e) is first observed after six weeks of culture, but there are genotypic differences in the speed ELS start forming and developing.

Subculture of calli and embryos
ELS and compact calli of approximately 2 mm in diameter are transferred to regeneration medium for plant regeneration through direct embryogenesis (Fig. 2.21-1f,g). The medium may have a relatively low concentration of mineral salts, and in our laboratory 190-2 and half strength MS medium have been used. The regeneration medium is supplemented with MS vitamins and iron, 2.85 µM IAA (0.5 mg/L) and 4.65 µM kinetin (1 mg/L), 3% sucrose and 0.3% Phytagel. The pH is adjusted to 5.8 prior to autoclave sterilization as described before. Baby food jars and Magenta vessels are used, containing 30 ml or 50 ml of medium, respectively. The regeneration vessels are kept in a phytothrone at 25°C with a light intensity

of about 25 µmol m^{-2} s^{-1}. They may also be placed on illuminated shelves (both shelves and phytothrone have Philips Fluorescent lamps PH TLD 58W/33).

Growth of plantlets
After approximately 5 weeks of incubation the regenerated green plants are transferred to 5.5 cm diameter perforated pots containing a soil peat mixture. It needs to be noted that plants derived from calli can be genetically identical. The plantlets are sprayed with water regularly and kept covered by a plastic lid or sheet for one week to maintain high moisture. They are then grown on in similar glasshouse conditions as described above for donor plants (Fig. 2.21-1h).

Determination of ploidy
DNA ploidy is most reliably determined by flow cytometry. The protocol for determining DNA ploidy with flow cytometer involves protoplast isolation and is described by Immonen et al. (1999). However, we also use chromosome counting from root tip cells, where root tips are collected early in the morning when the cells are actively dividing. Arrested metaphases are stained with orcein.

Protocol for chromosome counting in rye using orceine staining
Solutions

Colchicine	100 mg (0.05% w/v)
8-OH-quinoline	50 mg
Dimethyl sulfoxide	100 drops
H$_2$O	total volume 200 ml

Mix for 3 h
Prepare fresh every time.

Orceine	10 g
Glacial acetic acid	225 ml
H$_2$O	total volume 500 ml

Boil for 3 hours mixing, cool down and store in refrigerator.

Collecting root tips
1. Pot plantlets in perforated pots or jiffy pots, which are placed on a thin layer of vermiculite or sand.

2. Once thick roots penetrate through the pot, collect root tips (ca 5 mm) in the morning for colchicine treatment and gently remove pieces of vermiculite or sand. You can also take root tips from plants still on medium.

Colchicine treatment
1. Place a filter paper on a large Petri-dish. With pencil draw sections on the paper and number them for different genotypes.

2. Using Pasteur pipette add colchicine solution so that the paper is wet through, but the root tips will not float.

3. Place 3-4 root tips of each genotype on numered places. Cover the dish and store in room temperature for 3-4 h.

4. Transfer the root tips into small numbered test tubes or bottles containing orcein solution. Close the vessels and store in refrigerator. Staining takes approximately a week.

Preparing the slide
1. Move the root tips from orcein solution into another set of small test tubes into 45% acetic acid.

2. Heat the test tube above flaim for 5-10 seconds.

3. Place one root tip on a microscope slide and remove excess acetic acid with filter paper.

4. Using a microscalpel or blade squeeze the inner cells carefully from the tip of the root trying not to include the epidermis tissue.

5. Add acetic acid and place the coverslip on top.

6. Tap lightly with a glass rod to separate the cells into a homogenous solution.

7. Heat the sample above the flame for 5 seconds to remove air bubbles and press with thumb through filter paper.

8. Study the preparation under the microscope. If the cells are in clusters, add a drop of acetic acid under the corner of the cover glass, heat quickly and press again with thumb.

Colchicine treatment for chromosome doubling
Colchicine doubling is not normally done as usually 50% of plants are doubled spontaneously. If the number of DH plants is low then more can be produced by colchicine treatment. The following procedure can be used to double haploid plants if required. The plants are treated with colchicine when root tips have emerged through the perforated pots. The roots are rinsed under tap water and the ends are cut so that the remaining roots are approximately 2 cm long. The plants are placed in Duran test tubes, which are filled with colchicine solution (0.05% colchicine, 2% DMSO in distilled water, care should be taken as colchicine is a carcinogen) so that the roots are covered. The plants are kept in the colchicine solution for 5 h after which they are rinsed with tap water for five times. They are then left in the last rinsing water over night. The treated plants are repotted in peat soil mixture in 14 cm diameter pots.

Efficiency and Applications

The prerequisites for using DH technology in breeding are: a) the production method can be applied to a large number of genotypes, b) large numbers of green DH regenerants can be produced at reasonable cost, and c) the green regenerants are of high quality, displaying the genetic variability of the cross. Even with limited success rates, haploids and doubled haploids can be used in research and genetic studies and the technology may be tested for improving particular traits of value. The potential applications and real benefits from rye haploid production in breeding or research are yet to be demonstrated. Due to genotypic limitations to rye androgenesis, there is no published information of its practical or research application to-date. In our laboratory, the methodology is mainly applied for establishing markers for important traits such as reduced straw length and pre-harvest sprouting resistance. The method is also being tested for reduced straw length in breeding populations by passing segregating material through anther culture and selecting for short straw for the subsequent population.

The protocol was developed using spring ryes in which some responsiveness was observed. The spring rye line 'Jo 02', (that has a Finnish landrace, but also a Central Europe rye in its pedigree) proved to be a relatively responsive line. The promising results were subsequently tested and methods further developed with winter ryes and breeder's materials. The protocol described here has been sufficient for producing adequate numbers of doubled haploids from previously untested parents and F_1 progenies in developing mapping populations and for testing practical applications. Over 50 lines have been tested with only 5 giving no green plants, the best line remains 'Jo 02' (30 green plants/100 cultured anthers). However, the results vary from time to time and the price of single DH-plants can thus be quite high. Flow cytometry of a range of material has shown that over 50% of the green regenerants are spontaneous doubled haploids (the lowest is 'Voima' at 27%, the highest is 'EM-1' at 83%). Cold pretreatment seems to increase the proportion of spontaneous DH plants. Spontaneous DH plants have reduced vigour due to inbreeding depression but they produce viable pollen and seed when pollinated within or between populations. Colchicine treatment has been used only to a limited extent and can produce a high level of sterility. This may have been due to poor doubling rate or an adverse effect of the treatment itself. Earlier results from rye anther culture studies are summarised in a review by Deimling and Flehinghaus-Roux (1996), which forms the basis of the present protocol.

References

Deimling, S. and T. Flehinghaus-Roux, 1996. Haploidy in rye. In: *In vitro* Haploid Production in Higher Plants. Jain, S.M., S.K. Sopory and R.E. Veilleux (Eds.) Kluwer Academic Publishers, Dordrecht, pp. 181-204.

Immonen, S., 1999. Androgenic green plants from winter rye, *Secale cereale* L., of diverse origin. Plant Breed. **118**: 319-322.

Immonen, S. and H. Anttila, 1998. Impact of microspore developmental stage on induction and plant regeneration in rye anther culture. Plant Sci. **139**: 213-222.

Immonen, S. and H. Anttila, 1999. Cold pretreatment to enhance green plant regeneration from rye anther culture. Plant Cell Tiss.Org.Cult. **57**: 121-127.

Immonen, S. and H. Anttila, 2000. Media composition and anther plating for production of androgenetic green plants from cultivated rye (*Secale cereale* L.). J.Plant Physiol. **156**: 204-210.

Immonen, S., A. Tauriainen and O. Manninen, 1999. Assessment of green regenerants from rye and triticale anther culture. In: Anther and Pollen: from Biology to Biotechnology. Clement, C., E. Pacini and J.-C. Audran (Eds.), Springer-Verlag, Berlin, pp. 237-245.

Ouyang, J.W., S.E. Jia, C. Zhang, X. Chen and G. Feng, 1989. A new synthetic medium (W14) for wheat anther culture. Ann.Rep.Inst.Genet.Sin. pp. 91-92.

Wang, X.Z. and H. Hu, 1984. The effect of potato II medium for triticale anther culture. Plant Sci.Lett. **36**: 237-239.

2.22
Microspore culture of rye

S. Pulli and Y.-D. Guo
Laboratory of Plant Physiology and Molecular Biology, Department of Biology, University of Turku, 20014 Turku, Finland

Introduction

Rye (*Secale cereale* L.) is an important cereal in Europe, with approximately 90% of global rye production concentrated in this area. In subtropical low latitude areas, rye is grown primarily as a late-fall, early-spring forage and, secondarily, for grain. Production of inbred lines in a heterozygous crop such as rye is relatively difficult and time consuming. Doubled haploid breeding of cross-pollinated crops such as rye, will be a great help when selecting and fixing the desirable recessive genes which are agronomically attractive, and which can not be obtained by conventional breeding method. Doubled haploids are also very useful in test crosses and particularly in hybrid breeding. As the regenerated plants are homozygous for all loci, several generations of inbreeding for establishing pure lines are avoided and selection for desired traits can be performed on the regenerants or *in vitro* on the developing plantlets. Production of haploids and the subsequent development of doubled haploid lines have proven valuable in plant breeding and biotechnology. Pioneering work on rye anther culture was done in Wenzel and Thomas's laboratory in 1970's. Calli, embryos and green plants have been produced from anther culture of cultivated rye (*S. cereale* L.) and semi-wild rye (*S. vavilovii* Grossh) (Wenzel and Thomas, 1974). Rye is relatively recalcitrant to anther culture because of the poor induction of embryo/callus and the difficulties in regenerating green plants. Deimling and Flehinghaus-Roux (1996) reported successful regeneration of plants from isolated microspore culture of semi-wild rye (*S. vavilovii* Grossh) SC35. Culture of isolated microspores eliminates the risk of plants arising from diploid tissue (septum, anther wall and tapetum) and, the efficiency of isolated microspore culture is high enough for breeding purposes. The isolated microspore culture and green plant regeneration of rye (*S. cereale* L.) were first reported by Guo and Pulli (2000).

Protocol

Donor plants and growth conditions
Varieties of spring and winter rye can be used. Seeds of donor plants are sown and germinated into a peat soil mix in a glasshouse under controlled conditions employing 25/18°C, and a 16 h photoperiod. The required light intensity is about 100 µmol m^{-2} s^{-1} supplemented by fluorescent lamps. The plantlets are watered with tap water supplemented with N:P:K (16:9:22) composite fertilizer every 2-3 days. For winter rye, vernalization is needed at 4°C, 8 h photoperiod (30 µmol m^{-2} s^{-1}) for 10-12 weeks.

Microspore developmental stage
The microspore development stage is assessed when 20-30 mm of the culms have emerged from the leaf sheath. Microspores are examined microscopically by aceto-carmine staining. Spikes with microspores at the mid to late uninucleate stage are harvested and temporarily stored (not more than one week) with stalks in water at 4°C in darkness. Developmental stage varies with genotype and growing condition.

Anther pretreatment
Spikes are surface sterilized in 15% sodium hypochlorite (10% available chlorine) with one drop of Tween-20 for 15 min followed by three repetitive washes with sterile, distilled water. Anthers are excised aseptically from florets and pretreated by culturing on 0.3 M mannitol at 10-12°C for 4-7 days in Petri dishes.

Microspore isolation
Anthers obtained from mannitol pretreatment are macerated in a glass tube (100x15 mm) or by a pestle with small amount of 0.2 M cold (4°C) mannitol (0.21 Osm kg^{-1} H_2O of osmotic pressure) and filtered through a 100 µm nylon mesh. The filtrate is rinsed twice in 0.2 M mannitol followed by centrifugation at 28 g for 2 min, repeated 2 to 3 times. The microspore pellet obtained from centrifugation procedure is resuspended in the required amount of PG-96M medium supplemented with 6% maltose and hormones (2,4-D, 1.5 mg/L, kinetin, 0.5 mg/L). The osmotic pressure is adjusted to somewhere in the range 0.22 to 0.28 Osm kg^{-1} H_2O. Liquid culture medium is sterilized by filtration to avoid the degradation of some components. The final density of microspores is checked with a haemocytometer and adjusted to $5x10^4$ per milliliter. Seven milliliters of microspore suspension are dispensed into a 100x15 mm sterile Petri dish. The Petri dishes are wrapped with double layers of parafilm and the microspores incubated at 27°C in darkness.

PG-96M induction medium
The PG-96M induction medium has been modified from PG-96 (Guo et al., 1999) and is recommended for rye microspore culture (Table 2.22-1).

Plant regeneration
After 3-5 weeks of microspore culture, embryos are counted and transferred to the solid 190-2 medium (Wang and Hu, 1984) containing 2% sucrose, 0.3% (w/v) phytagel (Sigma), plant growth regulators (2,4-D, 0.5 mg/L; BA, 0.5 mg/L) and cultured at 25°C, 16/8 h photoperiod with the light intensity of approximately 100 µmol $m^{-2}s^{-1}$. The plantlets obtained are labelled individually, then potted and placed in a glasshouse under plastic covers in order to maintain a high humidity.

Colchicine treatment
Colchicine treatment may not be necessary as up to 90% of plants exhibit spontaneous doubling. However, haploid plants can be treated as follows. Prior to transfer of the plantlets

to soil, roots are immersed in a 0.1% colchicine solution containing DMSO (0.1%) for 3 h followed by several 5 min washes in water.

Table 2.22-1. Composition of PG-96M basal medium for rye microspore culture

Components	mg/L
Macro nutrients	
KNO_3	1,500
$(NH_4)_2SO_4$	150
KH_2PO_4	125
$MgSO_4 \times 7H_2O$	200
$Ca(NO_3)_2\ 4H_2O$	200
KCl	50
Iron source	
$FeSO_4 \times 7H_2O$	27.8
$Na_2EDTA \times 2H_2O$	37.3
Micro nutrients	
$MnSO_4 \times 4H_2O$	10
$ZnSO_4 \times 7H_2O$	3
H_3BO_3	5
KI	0.83
$Na_2MoO_4 \times 2H_2O$	0.25
$CuSO_4 \times 2H_2O$	0.025
$CoCl_2 \times 6H_2O$	0.025
Others	
Ascorbic acid	5
Proline	10
Aspartic acid	5
Citric acid	5
Biotin	0.05
Glycine	1
myo-Inositol	100
Thiamine HCl	10
Pyridoxine HCl	1
L-Glutamine	250
Nicotinic acid	0.25
Glutathione	5
L-Serine	10
Casein hydrolysate	100
L-Alanine	5
2,4-D	1.5
Kinetin	0.5
Maltose	60,000
pH	5.7

Efficiency and Applications

In our study (Guo and Pulli, 2000), twenty-one spring and winter rye varieties were used as donor materials, eleven of which produced embryos/calli and green plants, the best being 'Jussi', 'Amilo', 'Florida 401', 'Anna', 'Talovskaja' and 'Voima'. The results indicated that the androgenic embryogenesis capacity of rye was not genetically related to winter or spring habit. The spring genotype Florida 401 showed high capacity of androgenic embryogenesis whereas 'Bonel' and 'Florida Dwarf' showed quite low capacity. Androgenic embryogenesis capacity of winter-type genotypes Jussi and Voima also showed apparent differences. The major genetic control was expressed in the efficiency of androgenic embryogenesis and green plant regeneration ability from embryos/calli.

There was no universal embryos/calli induction medium for all of the genotypes we tested. The PG-96M induction medium (Table 2.22-1) was modified from PG-96 (Guo *et al.*, 1999) for use in rye microspore culture. For most genotypes, the PG-96M induction medium gave better results, especially in Florida 401 and Jussi. For rye microspore culture, maltose proved to be the best carbohydrate source, especially at the 6% level.

Solutions of different concentrations partially affect growth and morphogenesis through nutritional value, and partly *via* their varying osmotic potential value. It is crucial for the microspores to survive from the beginning of culture induction. In rye microspore culture, this is one of the key stages for subsequent culture. Microspores are very sensitive to environmental osmotic pressure. Unsuitable environmental osmotic pressure will cause microspore cells to shrink or break, leading to death. According to the results of this study, suitable osmotic pressure for rye microspore culture ranged from 0.22 to 0.28 Osm kg^{-1} H_2O, with a survival percentage of 70%-74% after 2 days of culture (Guo and Pulli, 2000).

In rye microspore culture, mannitol pretreatment proved beneficial to microspore vitality at the beginning of culture and during subsequent cell division and embryogenesis. After mannitol pretreatment, vitality percentage of microspores increased strongly. Beneficial effect of mannitol pretreatment has been shown to be due to improved sugar uptake.

Solid 190-2 medium (Wang and Hu, 1984) containing 2% sucrose and 0.3% (w/v) phytagel (Sigma) was used as the basic regeneration medium, 190-2 regeneration medium with 0.5 mg/L kinetin and 0.5 mg/L NAA produced the highest green plant regeneration frequency. The relatively small, compact embryos/calli produced more green plants. The highest green plant regeneration frequency was obtained from relatively younger (4-8 weeks culture) and smaller (1-2 mm in size) embryos/calli.

Spontaneous doubling of haploids to doubled haploids had been observed in our study. Twenty-two (90.9%) doubled haploids (2n=14) were found in 24 random samples without colchicine treatment of genotypes Amilo, 'Talovskaja', Jussi, Florida 401 and 'Wheeler'. Morphologically, haploid plants are less vigorous and with narrow leaves. The phenotype of DH plants varies with donor plants. In general DH plants have short-stalks with short internodes and more tillers compared to haploid plants. The average height of DH regenerants Amilo was 49.8 cm whereas the average height of donor plants variety Amilo was 78.2 cm; the average tiller number of DH regenerants was 17.7 whereas the average tiller number was 9.2 for Amilo donor plant. These DH plants were vigorous, the microspore/pollen development and the seed set capability of the studied DH plants were normal.

References

Deimling, S. and T. Flehinghaus-Roux, 1996. Haploidy in rye. In: *In vitro* Haploid Production in Higher Plants. Jain, S.M., S.K. Sopory and R.E. Veilleux (Eds.) Kluwer Academic Publishers, Dordrecht, pp. 181-204.

Guo, Y.-D. and S. Pulli, 2000. Isolated microspore culture and plant regeneration in rye (*Secale cereale* L.). Plant Cell Rep. **19**: 875-880.

Guo, Y.-D., P. Sewon and S. Pulli, 1999. Improved embryogenesis from anther culture and plant regeneration in timothy. Plant Cell Tiss.Org.Cult. **57**: 85-93.

Wang, X.Z. and H. Hu, 1984. The effect of potato II medium for triticale anther culture. Plant Sci.Lett. **36**: 237-239.

Wenzel, G. and E. Thomas, 1974. Observations of the growth in culture of anthers of *Secale cereale* L. Z.Pflanzenzüchtg. **72**: 89-94.

2.23
Oat haploids from wide hybridization

H.W. Rines
Plant Sciences Research, U.S. Department of Agriculture, Agricultural Research Service, and Department of Agronomy and Plant Genetics, University of Minnesota, 411 Borlaug Hall, 1991 Upper Buford Circle, St. Paul, MN 55108, USA

Introduction

Haploids of cultivated oat (*Avena sativa* L., $2n=6x=42$) have been produced from wide hybridizations with Panicoidae species, particularly maize (*Zea mays* L.), by procedures similar to ones described for the production of wheat (*Triticum* sp.) haploids from wheat x maize hybridizations (Laurie and Bennett, 1988; Rines *et al.*, 1997). As in wheat, oat haploid production by the maize wide cross method appears to be much less genotype restricted than haploid production by anther culture. However, in oat the plant recovery frequencies reported to date of 1-2% of florets maize-pollinated is much lower than that attained with wheat and, like for oat haploid production in oat anther culture (Kiviharju *et al.*, 2000), not yet adequate for routine use in breeding. Also, there are important differences in the types and in the reproductive behavior of plants recovered in oat x maize compared to wheat x maize hybridizations. These include on occasion maize chromosome retention in the recovered oat plants and partial self-fertility in oat haploid plants (Riera-Lizarazu *et al.*, 1996). These differences in products can be detrimental in trying to produce doubled haploids routinely for breeding, but on the plus side they have led to the recovery of valuable novel materials for genetic studies. The following protocols described for haploid and doubled haploid oat plant recovery and characterization are ones in our current use, but modifications in factors such as oat and maize genotype effects, post pollination treatment, embryo rescue medium hormone components, and plant growth conditions are being investigated in ours and other laboratories.

Protocol

Donor plants and growth conditions

Donor plants can be grown in the field, glasshouse, or growth cabinet; however, because flowering times of the oat and maize need to coincide, it is often necessary to use glasshouse or growth cabinets for one or the other species and to have multiple planting dates of one or both parents. For oat plants grown in controlled environments, growth conditions of 20°C day/15°C night with short days of 12 h light period for 6-8 weeks and a shift to long days of 16 h light to promote reproductive development have produced vigorous plants with large panicles. Maize plants in controlled environments are grown in long days of 14 h or more of light and warmer temperatures; e.g., 25°C day, 20°C night. Providing good light levels of 300-400 $\mu E\ m^{-2}\ s^{-1}$ at canopy top in growth cabinets or from supplemental light in autumn and winter glasshouses, as well as maintaining good plant nutrition with additions of fertilizer to soil or watering solutions, are important for optimal donor plants of both oat and maize.

Emasculation, pollination and post-pollination treatment

Emasculation of oat florets in preparation for application of maize pollen may be done either through removal of the oat anthers or by clipping the florets with scissors. The latter method may either cut the developing anthers to halt their development or allow maize pollination before the oat anthers dehisce. Removal of the oat anthers usually results in higher frequencies of embryos recovered but is more tedious and labor demanding than clipping the florets. Oat florets are produced in spikelets of a panicle with each spikelet composed of an outer glume overlapping an inner glume, with both surrounding a basal floret, a smaller secondary floret, and sometimes a tertiary floret. Because spikelets in an oat panicle develop over several days in an apical to basal fashion, emasculation by either method is done when the primary floret in the apical (most mature) spikelet is at anthesis. This timing tends to maximize the number of receptive florets for maize pollination two days later. Usually 20 to 30 primary florets are emasculated with those in the uppermost spikelets beyond and the lowermost ones before optimal stage at pollination. In the anther removal technique, the inner glume is pulled back using forceps, the secondary (and tertiary, if present) floret is removed, the palea is also pulled back, and the exposed anthers removed and discarded being careful not to injure the developing stigma. The palea and glumes are then restored to normal position protecting the primary floret stigma. In the cut emasculation technique, the spikelets are simply clipped with scissors at a point corresponding to the base of the secondary floret, thus removing the secondary florets and clipping the tips of the developing anthers of the primary florets. In both techniques, after emasculation, the panicles are covered with glassine bags to keep out stray oat pollen and to retain moist air around the emasculated florets.

Two days later the bags are removed. Stigmas of the florets in the anther removal technique are exposed for maize pollen application by clipping the florets above the stigmas. With the cut floret emasculation technique the stigmas are already exposed among the anther remnants. Freshly collected maize pollen is distributed lightly over emasculated florets with the aid of an artist's brush while using one's forefinger and thumb to gently position the individual florets upward. The glassine bags are then replaced over the panicles. Two days later the bags are removed, the florets sprayed to dripping with a solution of 100 mg/L 2,4-dichlorophenoxyacetic acetic acid (2,4-D) and 50 mg/L gibberellic acid (GA_3) to promote caryopsis development, and the bags replaced.

Embryo rescue and plant recovery

Caryopses are harvested from the oat spikelets at about 14 days post maize pollination. Caryopses developing from oat self-fertilization, which occasionally occur with the cut floret emasculation technique, are readily distinguished by their plumpness and are discarded. Caryopses from an individual plant are placed in a small plastic Petri dish (60x15 mm) and surface-sterilized in a sterile-air transfer hood by a 5 min treatment (with occasional agitation of the plate) in a solution of 2.5% sodium hypochlorite (one-half strength commercial bleach) containing 2 drops per 100 ml of a wetting agent, such as Triton-X-100 (Fisher Scientific). The sterilizing solution is decanted away, and the process is repeated 3 times with sterile water to rinse away the sterilizing solution. After decanting the final rinse solution, embryos are dissected from the caryopses using flame sterilized instruments and a dissecting microscope in a transfer hood. Rescued embryos are placed on half-strength Murashige and Skoog (MS) medium containing 2% sucrose and 0.2% Phytagel (Sigma) in 100x25 mm Petri plates or other appropriate containers. The embryos are placed in a 20°C dark incubator until embryo germination is observed, at which time they are transferred to a 20°C cabinet with 12 h per day of ~100 µE m^{-2} s^{-1} light. Plantlets attaining 8-10 cm shoot growth are transferred to pots containing potting mix or soil, watered with half-strength MS major elements solution, and placed in a growth cabinet under conditions described earlier for oat donor plant growth. The young transplants are covered with an inverted glass beaker for the first several days to help prevent desiccation until they become established.

Doubled haploid plant recovery
Haploid oat plants, if left to grow and mature with a shift to long-day light periods of 14-16 h when they have tillered and attained a height of 30-40 cm, will often set some seed. This seed set results from formation of unreduced gametes. Alternatively, colchicine treatments may be applied to the haploid oat plants to induce somatic doubling with seed set *via* regular gamete formation. For colchicine treatment, early tillering plants are removed from pots, the soil or potting mix washed from the roots, the roots trimmed to about 2 cm length, and the shoots trimmed to about 4 cm length. The trimmed plants are placed in a solution of 0.05% (w/v) colchicine and 2% (v/v) dimethyl sulfoxide (DMSO) sufficient to cover the crown portions of the plants in a container with an aerator. The aerator could be such as one used in fish aquariums, the chemicals used are harmful and care is needed. After 4 h treatment at 20°C, the plants are removed from the colchicine solution, rinsed for several minutes in running tap water, and placed back into pots with soil. The plants are protected, e.g., covered with an inverted beaker, for the first few days to aid their acclimation to normal growth. After the plants have attained a height of 30-40 cm they are transferred to long day conditions to promote reproductive development.

Characterization of recovered plants
Recovered doubled haploid plants can be checked for evidence of normal euploid oat chromosome number ($2n$=42) and absence of maize chromosomes by cytological analysis of root-tip meristem cells. Root tips can be collected either from roots of seeds germinated on filter paper or from roots exposed when a young plant and surrounding soil are carefully removed from a pot. Root tips of ~1 cm length are pretreated by placing them on filter paper saturated with a solution of 0.5 g/L colchicine, 0.025 g/L 8-hydroxyquinoline, and 1.5% (v/v) DMSO at room temperature for 3.5 to 4 h to arrest cells at metaphase prior to fixing in 3:1 ethanol: glacial acetic acid for 2 days at room temperature and storage at -20°C. Maize chromosomes, if present, can be detected by their small size relative to oat chromosomes in stained (e.g., aceto-carmine) microscope slide preparations, or more definitively by genomic *in situ* hybridization using fluorescent labeled maize genomic DNA as a probe (Riera-Lizarazu *et al.*, 1996). Abnormal chromosome numbers can also be detected in meiotic cell preparations and also by observations of frequent micronuclei in microspores of such plants (Rines *et al.*, 1997). Maize chromosome presence can also be tested in the initial haploid plants developing from rescued embryos; however, these plants can be chimeric for retained maize chromosome(s) and may not transmit them meiotically to the doubled haploid offspring (Riera-Lizarazu *et al.*, 1996). An alternative, highly sensitive test for the presence of maize chromatin in primary haploids or derived doubled haploids from oat x maize

hybridization is a molecular test for *Grande-1*, a highly repeated DNA sequence distributed throughout the maize chromosomes. Primers and polymerase chain reaction (PCR) conditions are described in detail in Kynast et al. (2000).

Efficiency and Applications

The presence of maize chromosomes in oat x maize F_1 plants or in derived doubled haploid lines and progeny can be detected either cytologically or with a molecular probe, as described in the above protocol. Lines with an individual maize chromosome added to a haploid oat complement have been recovered for all 10 maize chromosomes and fertile oat-maize disomic additions ($2n=42+2$) for all but maize chromosome 10 (Kynast et al., 2001 and unpublished results). Such lines are valuable as maize genomics tools for physical mapping and other studies because each contains only a portion of the maize genome, a single chromosome; furthermore, the maize chromosome can be fragmented by radiation to yield oat lines containing only a segment of an individual maize chromosome. Such lines also may be a source of novel oat germplasm with genes for traits useful in oat improvement introduced from maize. As a routine step, oat doubled haploids produced from oat x maize crosses should be checked either cytologically or by a molecular test for maize chromosome presence before use in breeding.

Another major difference between oat haploids derived from wide hybridization or any means and haploids of most cereals is that oat haploids may be partially self-fertile with up to 50% seed set, through the production of unreduced gametes (Rines et al., 1997). Furthermore, up to 25% of progeny plants from haploid self-fertilization may be aneuploid. These aneuploids have great value in studies to associate oat genes to chromosome but are not desired in breeding materials. Treatment of oat haploid plants with colchicine to produce somatic doubling can result in plant sectors which set seed that are larger in size than those produced from unreduced gametes in haploid florets and are of more normal $2n$ euploid constitution. Seed from colchicine-doubled sectors of haploid oat plants should be harvested separately from any seed on the remainder of the plant to reduce the chance presence of aneuploid in seed for breeding use. The relatively low frequency of embryos recovered, usually 2-10% of maize-pollinated oat florets, combined with often low frequencies of less than 20% of these embryos germinating into vigorous plants has made it difficult to conduct experiments of adequate scope for statistical comparisons of factors influencing plant recovery frequencies. Genotypic differences in response among both oat lines and maize pollen sources have been noted, but these may be confounded by how the lines respond to the plant growth conditions employed. In general, the more vigorous the plants, the higher frequency and quality of embryos is recovered.

Acknowledgments
The author gratefully acknowledges contributions of colleagues and students to the development over the years of the described protocol, including E.V. Ananiev, G. Chen, L.S. Dahleen, D.W. Davis, R.G. Kynast, S.D. Maquiereira, R.L. Phillips, O. Riera-Lizarazu, and C.D. Russell. Funding support was provided in part by The Quaker Oats Company, the USDA-NRI program under grant no. 96-35300-3775, and the National Sciences Foundation under grant no. 9872650.

References

Kiviharju, E., M. Poulimatka, M. Saastamoinen and E. Pehu, 2000. Extension of anther culture to several genotypes of cultivated oat. Plant Cell Rep. **19**: 674-679.
Kynast, R.G., O. Riera-Lizarazu, M.I. Vales, R.J. Okagaki, S.D. Maquieira, E.V. Ananiev, W.E. Odland, C.D. Russell, A.O. Stec, S.M Livingston, H.A. Zaia, H.W. Rines and R.L. Phillips, 2001. A complete set of maize individual chromosome additions to the oat genome. Plant Physiol. **125**: 1216-1227.

Laurie, D.A. and M.D. Bennett, 1988. The production of haploid wheat plants from wheat x maize crosses. Theor.Appl.Genet. **76**: 393-397.
Riera-Lizarazu, O., H.W. Rines and R.L. Phillips, 1996. Cytological and molecular characterization of oat x maize partial hybrids. Theor.Appl.Genet. **93**: 123-135.
Rines, H.W., O. Riera-Lizarazu, V.M. Nunez, D.W. Davis and R.L. Phillips, 1997. Oat haploids from anther culture and from wide hybridizations. In: *In Vitro* Production of Haploids in Higher Plants. Vol. 4. Cereals. Jain, S.M., S.K. Sopory and R.E. Veilleux (Eds.) Kluwer Accademic Publishers, Dordrecht, pp. 205-221.

2.24

Haploid and doubled haploid production in durum wheat by wide hybridization

P.P. Jauhar
United States Department of Agriculture-Agricultural Research Service, Northern Crop Science Laboratory, Fargo, North Dakota 58105, USA

Introduction

Production of durum haploid plants in large numbers would be necessary for their meaningful use in fundamental genetic research and exploitation in practical breeding. Several methods of extracting haploids have been tried in durum wheat. These include anther culture and chromosome elimination after hybridization with maize, *Zea mays* L. (Sarrafi *et al.*, 1994; Almouslem *et al.*, 1998). *Hordeum bulbosum*-mediated haploid production has been successfully achieved in bread wheat var. Chinese Spring with the crossability alleles *kr1* and *kr2* (Barclay, 1975), but this method has yielded little success with durum wheat (O'Donoughue and Bennett, 1994). Our experience has shown that maize-induced haploid production is the method of choice for both hexaploid wheat and durum wheat. The protocol for maize-mediated durum haploid production and factors influencing the efficiency of this technique are described in this article. Methods that use commercially acceptable genotypes for extracting haploids will maximize the value of these research efforts. The success of the maize technique in producing haploids can be attributed to several factors. Genotype of the maternal parent plays an important role in haploid production. Different durum genotypes yield varying frequencies of haploid embryos and subsequent haploid plantlets. Some cultivars yield a high frequency of embryos but then embryo to plantlet conversion is not efficient. The ultimate goal is to produce the maximum number of haploid plants. We found that durum cultivar 'Medora' gives the highest efficiency of conversion of embryos into plantlets. This genotypic response may be related to response to the medium used. Another most important factor seems to be treatment with $AgNO_3$ by mist spraying, which is effective in delaying abscission of young embryos. Thus, the enhanced development of embryos due to sustained growth increases their chances of conversion into haploid plantlets. Appropriate concentrations of $AgNO_3$ should be determined for a durum genotype used in the work. Certain chromosomes

---

Mention of a trademark or proprietary product does not constitute a guarantee or warranty of the product by the USDA or imply approval to the exclusion of other products that also may be suitable.

affect the ability to produce haploids. We produced haploids in the durum cultivar 'Langdon' (with the full chromosome complement) and its substitution line 5D(5B) in which chromosome 5B was replaced by 5D. Our results indicate that substitution of 5D for 5B enhances haploid production. Hybridization with maize followed by chromosome elimination has proved successful in producing haploids of durum wheat. Our experience has shown that this technique, if carefully planned, can yield a large number of haploids in commercial durum cultivars (Almouslem et al., 1998).

Protocol

Preparation of plant material

Seed of selected durum cultivars (to be used as the female parent) and seed of a maize cultivar(s) (to be used as the pollen donor) are planted weekly for 4 to 5 weeks in a glasshouse in 30 cm plastic pots (three pots per cultivar with three seeds per pot) filled with an appropriate soil mix (in our work, Fisons Sunshine Mix No. 1 from Sun Gro Horticulture, Bellevue, Washington, USA). The glasshouse is maintained at 22±4°C with light from 36 automatically timed 400 W sodium vapor lights on a 16 h day and 8 h night regime. Plants are fertilized with a suitable fertilizer. We used Sierra General Purpose Greenhouse controlled-release fertilizer, 15-10-10 N-P-K plus minor elements, supplied by Scotts-Sierra Horticultural Products, Marysville, Ohio, USA. Planting of female and male parents is staggered to ensure a continuous supply of durum spikes and maize pollen (approximately 60 days after initial planting).

Emasculations and pollinations

In wheat, the inflorescence is a spike consisting of a main axis, called rachis, which bears the spikelets. Each spikelet consists of two glumes that enclose five alternating florets borne on a short axis called rachila. Each floret is enclosed by a lemma and a palea. Generally, the two outside florets (primary and secondary) are more developed (older) and the inner three are less developed (younger).

Durum spikes at the appropriate stage (generally when fully emerged from the flag leaf) are hand emasculated, prior to anthesis while the anthers are green. A few spikelets at the base (that are not well developed) are removed, and well-developed, healthy spikelets are only used for emasculation. The top portion of each spikelet is carefully cut off, exercising caution that none of the anthers are cut. Cutting the top portion enables the removal of unwanted florets and facilitates emasculation of the two selected florets. In each spikelet, all florets but the two outer, well-developed ones are removed. Thus, generally, the three florets in the center are discarded. Anthers (3 from each floret) are then gently removed with forceps. The emasculated spikes are covered with glassine bags (4 cm by 15 cm) to prevent unwanted pollination.

Two to four (sometimes up to five) days after emasculation, the feathery, receptive stigmas are pollinated with fresh maize pollen collected by tapping tassels on a piece of cellophane paper. Using the flat tip of forceps, pollen is carefully applied to stigmas of the emasculated florets. Pollinations are generally done in the mornings around 9 to 10 o'clock and each spike is preferably pollinated twice on successive days. Pollinated spikes are then covered with glassine bags, which are appropriately labeled.

Post-pollination treatments

There are three important stages leading to the successful production of haploid plants: 1) fertilization of the durum egg with maize male gamete and formation of a healthy zygote; 2) haploid embryo formation via elimination of the maize chromosomes; and 3) development of the haploid embryos into mature, haploid plants. We found that successful fertilization of durum cultivars by maize pollen results in large numbers of embryos, which must be successfully rescued for large-scale production of haploid plantlets.

Small, under-developed embryos are relatively difficult to rescue regardless of the medium used. Therefore, optimum embryo development is essential and may be maximized by appropriate post-pollination treatments that prevent premature abscission. Ethylene induces abscission in response to auxin(s) produced by the plant itself. The application of $AgNO_3$ likely delays the abscission process, presumably by reducing ethylene levels in the plant, and keeps the embryo in the floret much longer, thereby facilitating its development. We mist-spray hormonal solutions using Tween-20 (1% by volume) as a surfactant. We found that post-pollination treatment with 3 mg/L 2,4-D is effective in causing caryopsis swelling, but a combined treatment with silver nitrate ($AgNO_3$) at 120-180 mg/L arrests early abortion of the floret, and hence of the embryo inside, long enough to enhance embryo development. The spraying with hormonal solution is started 24 h after pollination and continued until the spikes are harvested.

For haploid embryo production, we found that the treatment with 3 mg/L 2,4-D combined with 180 mg/L $AgNO_3$ give the best results (yielded the most embryos). However, mist-spraying with 3 mg/L 2,4-D plus 120 mg/L $AgNO_3$ proved consistently better in terms of haploid plant production. These responses may vary with the genotype of durum wheat.

Embryo rescue

Cold treatment of spikes
Ten days to two weeks after pollination with maize pollen, durum spikes are regularly checked for caryopses development. The spikes with well-developed caryopses, sometimes even with small caryopses, are harvested 16 to 18 days after pollination, and refrigerated at 4°C for 24 to 48 hours. This cold treatment enhances the success of embryo rescue presumably by breaking any dormancy. The duration of the cold treatment may vary with the genotype.

The culture technique
The caryopses containing embryos are surface-sterilized with 15% (v/v) commercial bleach, containing 6.15 % sodium hypochlorite, for 10 to 20 minutes, followed by treatment with 70% ethanol for 3-5 min, and then rinsed three times with sterile distilled water.

Small embryos can be seen in otherwise empty, without endosperm, caryopses (Fig. 2.24-1a). Lack of endosperm in a caryopsis serves as the initial criterion for finding haploid embryos. Under aseptic conditions in a laminar flow hood, embryos are dissected from caryopses (Fig. 2.24-1b) and cultured on hormone-free MS medium (Murashige and Skoog, 1962) supplemented with 30 g/L sucrose and 8 g/L purified agar (Sigma) in Petri dishes (100×15 mm). This modification of the MS medium improves embryo germination and growth. For smaller embryos, addition of 1 mg/L of kinetin is helpful to assist their growth.

The cultured embryos are incubated for 4-6 weeks in the dark at 25°C, and transferred to fresh medium every 3-4 weeks. Coleoptiles start developing (Fig. 2.24-1c,d) in 2-5 weeks, depending on the size of the embryo at the time of culturing, and roots generally develop later. When small primary roots are visible (Fig. 2.24-1e), the embryos are transferred to a lighted incubator for 1-4 weeks (maintained at 25°C, and 16 h light, 8 h dark regime, light provided by 40 W cool and warm fluorescent bulbs) until they develop into healthy, green plantlets (Fig. 2.24-1f). Generally within a week of transferring under light, the plantlets start developing chlorophyll. The green plantlets are planted in 5 cm Jiffy peat pellets and transferred to a lighted growth room (maintained at 21±2°C; 16 h light, 8 h dark regime) for acclimatization for 7-10 days. The green plantlets are then transferred to the greenhouse.

Endosperm nursing technique
Embryos of some genotypes fail to germinate even on the modified MS medium described above. In some such recalcitrant cases, we have successfully rescued embryos using the endosperm nursing technique. A milky to soft dough stage endosperm obtained from a normal caryopsis is used as supplemental nourishment in combination with the MS medium. A milky endosperm (from a caryopsis sterilized as described earlier) is first placed on the MS medium and an embryo is then placed in the endosperm (Almouslem *et al.*, 1998). Thus, the embryo derives nourishment from the endosperm and perhaps also from the medium in which the endosperm is embedded, increasing its chances of germination and growth.

Frequency of embryos and haploid plantlets
Using the protocol described above, we obtained a high frequency of embryos. However, the yield of haploid plants was relatively low, although some haploids were extracted from all the seven durum cultivars and three important cytogenetic stocks we used. The total number of embryos produced varied from 2.7% to 12.1% of the florets pollinated in the seven cultivars. However, the yield of haploid plants varied from 0.2 to 4.7% of the florets pollinated. We produced a total of 142 green haploid plants and no albinos. This is the largest number reported so far for durum wheat. However, the yield of haploid plants may be further improved by using appropriate genotypes and by manipulating the media formulation.

Characterization of haploid plants
Haploid plantlets of durum wheat are generally slower in growth, smaller in size, and have narrower leaves with smaller stomata on the leaf surface, compared to the parental disomic plants. Some of these are diagnostic features of haploids. However, the surest method of detecting haploids is by chromosome counts, and it is important to confirm haploid status before inducing chromosome doubling to produce doubled haploids. Plantlets in the Jiffy pellets offer an easy and ideal material and the best occasion for counting somatic chromosomes.

Somatic chromosomes are best studied from root tips, although shoot tips may also be used, if necessary. Generally, primary roots emerge from the Jiffy pellets. About 1.5 cm long, plump tips of the emerging roots from each plantlet are trimmed and placed in individual pre-chilled glass vials with distilled water at 4°C. Each vial is appropriately labeled and placed in a refrigerator for 18-24 h. This cold treatment helps condense the somatic chromosomes, facilitating their spreading in root-tip squashes. After cold treatment, the root tips are fixed in vials containing freshly prepared fixative made up of a 1:3 (v/v) mixture of glacial acetic acid and 95% ethanol. The vials are left at room temperature for 24 h and then stored in a refrigerator for a few weeks, but chromosome counts are best done from fixed root tips within a week after fixation.

The fixed root tips, with or without hydrolysis with HCl, can be squashed in various stains, e.g., acetocarmine, basic fuchsin, or toluidine blue. However, we routinely use carbol fuchsin for staining chromosomes. A root tip is removed from the fixative and transferred to a drop of 45 % acetic acid in a watch glass (or treated with 45 % acetic acid in a vial) for 2-5 min until the white color of the tip starts to clear. The root tip is then placed on a clean slide in a drop of carbol fuchsin, and the root cap is removed with a razor blade or scalpel. After removing the root cap, the tip is gently squashed out with a sharp needle to release cells into the stain. The remainder of the root tip or any other debris is then removed from the slide and a cover slip carefully placed on the stain such that no air bubbles are trapped. Spreading of somatic chromosomes is achieved by gently tapping all over the cover slip and then pressing the slide by placing it between folds of bibulous or blotting paper. The initial preparation is viewed under a 10× objective and 10× eyepiece lens (i.e., 100× magnification) to rapidly find dividing mitotic cells and look for good spreads, and chromosomes are counted under a 40× objective. In well-spread preparations, 14 chromosomes of the haploid complement can be easily counted, with one dose each of the satellited chromosomes 1B and 6B; 1B has a smaller satellite (Fig. 2.24-2a).

Using fluorescent GISH (genomic *in situ* hybridization), we identify the seven somatic chromosomes each of the A and B genomes (Fig. 2.24-2c) (Jauhar *et al.*, 1999). Meiotic chromosomes are best studied from pollen mother cells obtained from immature anthers at the appropriate stage. Generally, seven unpaired chromosomes (univalents) are observed at meiotic metaphase I in the presence of *Ph1* (Fig. 2.24-2b). Fluorescent GISH is also an excellent technique for identifying the seven meiotic chromosomes of each of the two genomes (Fig. 2.24-2d).

Development of doubled haploids (DH)
Cytologically confirmed haploid plantlets at the 2 or 3-leaf stage are treated with a 0.2% aqueous solution of colchicine at 25°C. The seedlings are immersed up to the crown level in colchicine for 4 h in a beaker that is adequately aerated. The beaker is under constant aeration using an aquarium pump with an air stone (air diffuser) to disperse the air into the liquid. The seedlings are then thoroughly washed with running water for at least 30 minutes and planted in pots filled with soil mix described above. Care must be taken when handling colchicine as it is harmful.

Spontaneous doubling
Reports of seed set on haploid plants are very rare. However, we observed viable seed set on several of our durum haploids (Jauhar *et al.*, 2000). The seed set varied with the genotype, and the cultivar Langdon with 2.75 seeds per haploid had the greatest seed set. The seeds are the result of fusion of unreduced male and female gametes, which are formed by first division restitution during meiosis. Lack of pairing because of the presence of *Ph1* (Fig. 2.24-2b,d) seems to induce meiotic restitution and is, thus, a prerequisite for chromosome doubling. The instant homozygosity derived through meiotic non-reduction is interesting (Jauhar *et al.*, 2000).

Notes
Our experience has shown that the maize-induced chromosome elimination technique is the method of choice for haploid production in wheat, in general, and in durum wheat in particular. In durum wheat, the yield of haploid embryos in all cultivars is much higher (2.7 to 13.3 % of the florets pollinated) than the yield of haploid seedlings (0.2 to 4.7 %). The success of embryo to plantlet conversion may be improved by using appropriate genotypes and more suitable media formulations. I feel that further research along these lines would be rewarding.

Acknowledgments
The protocol reported here was developed as a team effort by myself and my lab staff, visiting scholars, and graduate students: A. Baset Almouslem (Fulbright Scholar), Terrance Peterson (Technician), Münevver Doğramacı-Altuntepe (Ph.D. student). I especially would like to acknowledge the excellent technical help of Terrance throughout this study.

References

Almouslem, A.B, P.P. Jauhar, T.S. Peterson, V.R. Bommineni and M.B. Rao, 1998. Haploid durum wheat production via hybridization with maize. Crop Sci. **38**: 1080-1087.

Barclay, I.R., 1975. High frequencies of haploid production in wheat (*Triticum aestivum*) by chromosome elimination. Nature **256**: 410-411.

Jauhar, P.P., A.B Almouslem, T.S. Peterson and L.R. Joppa, 1999. Inter- and intragenomic chromosome pairing in haploids of durum wheat. J. Hered. **90** : 437-445.

Jauhar, P.P., M. Dogramaci-Altuntepe, T.S. Peterson and A.B. Almouslem, 2000. Seedset on synthetic haploids of durum wheat: cytological and molecular investigations. Crop Sci. **40**: 1742-1749.

Murashige, T. and F. Skoog, 1962. A revised medium for rapid growth and bioassays with tobacco tissue cultures. Physiol. Plant. **15**: 473-497.

O'Donoughue, L.S. and M.D. Bennett, 1994. Comparative responses of tetraploid wheats pollinated with *Zea mays* L. and *Hordeum bulbosum* L. Theor.Appl.Genet. **87**: 673-680.

Sarrafi, A., N. Amrani and G. Alibert, 1994. Haploid regeneration from tetraploid wheat using maize pollen. Genome **37**: 176-178.

2.25
Haploid and doubled haploid production in durum wheat by anther culture

P.P. Jauhar

United States Department of Agriculture-Agricultural Research Service, Northern Crop Science Laboratory, Fargo, North Dakota 58105, USA

Introduction

Haploid plants are invaluable tools in several areas of biological research. They have been successfully employed in various facets of basic research in cytogenetics, genetics, and in plant breeding. Anther culture techniques have attracted attention as supplementary tools in wheat improvement. Durum wheat, also called macaroni wheat, (*Triticum turgidum* L., $2n=4x=28$; AABB genomes) is an important cereal crop used for human consumption worldwide. It is most commonly used for making macaroni, noodles, and other pasta products. Durum wheat is widely grown in several European countries including Italy, France, Turkey, Romania, and Ukraine, and in northern Great Plains of Canada and the United States. For producing haploids, anther culture has been a favored choice in many species because of the availability of numerous microspores in each anther, each of which could potentially produce a haploid plant. Although anther culture has been successfully employed in producing haploids in numerous crop plants, including hexaploid bread wheat (*Triticum aestivum* L., $2n=6x=42$; AABBDD genomes), this technique has not proven very successful in durum wheat (Dogramci-Altuntepe *et al.*, 2001). During anther culture, microspores are induced to produce haploid plantlets after going through various developmental stages. This process of haploid production is influenced by several factors.

Mention of a trademark or proprietary product does not constitute a guarantee or warranty of the product by the USDA or imply approval to the exclusion of other products that also may be suitable.

Protocol

Donor plants and growth conditions
Growth chamber: temperature - 21°C (generally there is no fluctuation); photoperiod - 16 h; light intensity - 8 Cool white and 7 warm white fluorescent lights 160 W each, 10 incandescent lights 40 W each; soil type - Fisons Sunshine Mix 1 (source: SunGro Horticulture, Bellevue, Washington, USA); fertilizer - Sierrra General Purpose Greenhouse controlled-release fertilizer, 15-10-10 N-P-K plus minor elements supplied by Scotts-Sierra Horticultural Products, Marysville, Ohio, USA.

Greenhouse: temperature - 21-23°C; photoperiod - 16 h; light intensity - 36 automatically-timed 400 W sodium vapour lights on a 16 h day and 8 h light regime as supplementary to natural lighting; soil type - same as in growth chamber; fertilizer - same as in growth chamber.

Field: the durum material was planted in a field in Casselton in North Dakota; soil type - medium clay; fertilizer - anhydrous ammonia and mono ammonium phosphate were used giving a total of 84 kg/ha N and 45 kg/ha P, no K was applied.

Preparation of plant material
Immature spikes are collected when the awns emerge 1-2 cm from the flag leaf sheath and are placed in a beaker containing distilled water. The beaker with the spikes is covered with plastic wrap, labeled, and refrigerated at 4°C for seven days. This cold treatment improves the anther culture response.

After cold treatment, the spikes are removed from the boot (flag leaf sheath) and transferred to 50 ml tubes with caps. The spikes are then sterilized with 70% ethanol for one minute, followed by treatment with 1% sodium hypochlorite + 0.1% Tween 20 for eight minutes, and then rinsed with sterile double distilled water four to five times. After the last rinse, the spikes are left in the labeled tubes with a small amount of water and capped to prevent them from drying.

Staging microspores
Generally, spikelets from the middle of the spike are selected for anther culture. Before starting anther culture, each spike is examined for the appropriate microspore developmental stage. The central anther from the outer floret of a spikelet is removed and squashed in a drop of 1% acetocarmine on a glass slide so that microspores from the anther are released and suspended in the stain over which a cover slip is placed. The slide is then examined under a light microscope and anthers containing microspores at the mid uninucleate stage (Fig. 2.25-1a) are selected for culturing. If an anther from a floret is at the desired stage, the remaining two anthers are cultured, as are the anthers from other florets on the spike.

Anther plating
Appropriately staged excised anthers are cultured on suitable induction medium (Table 2.25-1) under aseptic conditions in a laminar flow hood. Twenty anthers are cultured in each well of a 6-well Petri plate containing 5 ml of medium. The anthers are placed on the medium in such a way that each lobe is in contact with the medium (Fig. 2.25-1b). The culture plate is then covered, sealed with parafilm, labeled, and transferred to a dark incubator at 28°C.

Stages of anther culture and development of embryoids and plantlets
Microspores inside the cultured anthers start to develop into calli (Fig. 2.25-1c), or embryoids, or both after about 2.5 to 4 weeks. The calli initiated on the BAD-1 induction medium are then transferred to the BAD-1 differentiation medium (Fig. 2.25-1d) when they

are ~ 2-3 mm or larger in size. Free-floating calli originating from microspores are also observed. It is important to have the density of the media high enough that the calli do not sink but rather float. Calli resulting from anther filaments can be identified by the lower cell density and looser cell adhesion and are eliminated prior to sub-culturing because such filament calli are likely to result in normal diploid durum plantlets.

Embryoids with shoot initials or other tissue formations (with some chlorophyll) are observed (Fig. 2.25-1e) 4-6 weeks after initial culturing and are then transferred to light (Fig. 2.25-1f). The Figure 2.25-1f shows a well-differentiated embryoid with small shoot initials. Within a week after that, primary roots can generally be seen developing into green plantlets (Fig. 2.25-1g). Albino plantlets are commonly observed (Fig. 2.25-1h). Chimeric plantlets (consisting of albino and green shoots) are also occasionally observed.

Media formulations
The androgenetic response of durum cultivars is generally low. The low frequency of embryoids is further exacerbated by their development into albino plantlets. The composition of a medium is an important factor for inducing microspores to develop into green plantlets.

The most suitable medium may vary with the species and perhaps with a genotype in a species. The source and amount of total nitrogen and the type of growth regulator are important factors. Gelling agents sometimes also make a big difference. The osmotic potential of each type of sugar or gelling agent has a significant impact on the production of callus or embryoid. It is not surprising, therefore, that numerous different formulations have been tried. Four induction media have demonstrated the ability to produce calli from microspores of durum wheat: BAC-1 (Marsolais and Kasha, 1985); BAD-1 and BAD-3 (Trottier *et al.*, 1993), and M-42 (Kao *et al.*, 1991). We modified M-42 (a barley anther culture medium) by using 2 g/L L fructose, 4 g/L glucose, 50 g/L sucrose and 130 g/L Ficoll as a buoyancy increasing agent. We recommend BAD-1 for calli induction (Table 2.25-1). After initiating enough calli (5-6 weeks after culture) we induce shoot and root initiation by transferring to BAD-1 differentiation medium (modified by substituting Ficoll with agar) and keeping in the dark (Table 2.25-1). After shoot/root initiation, small plantlets are transferred to the MS regeneration medium and placed in a lighted growth room when they start developing chlorophyll. A modified MS medium (Murashige and Skoog, 1962), using 2 mg/L kinetin, 1 mg/L indole acetic acid (IAA), and 30 g/L maltose instead of sucrose, gives better results than standard MS medium (Table 2.25-1). Generally, after 2 weeks individual plantlets are split and sub-cultured.

Media preparation
Because the quantity of some ingredients is too small to be measured accurately in one liter of medium, stock solutions are prepared for the BAD-1 and MS media. The stock solutions not only enable accurate preparation of media but also facilitate long-term storage at -20°C.

Appropriate quantities and concentration of stock solutions are prepared. For example, macronutrient stock solutions are prepared in 10× concentrations individually for all media, whereas micronutrient stocks are prepared in 100× concentrations. The solutions are appropriately aliquoted (10 ml solution into 15 ml tubes in our lab), labeled, and stored in a -20°C freezer. Similarly, the $FeSO_4$+Na-EDTA stock, vitamin stock, and amino acid stock solutions are prepared, aliquoted and frozen.

Chromosome studies
Chromosomes of durum wheat are best studied from root tips, although shoot tips may also be used, if necessary. Generally, primary roots emerge from the Jiffy pellets. About 1.5 cm long, plump tips of the emerging roots from each plantlet are trimmed and placed in individual pre-chilled glass vials with distilled water at 4°C. Each vial is appropriately labeled and placed in a refrigerator for 24 hours.

Table 2.25-1. Media composition

Components	BAD-1 induction (mg/L)	BAD-1 differentiation (mg/L)	MS regeneration (mg/L)
Macro salts			
KNO_3	1,500	1,500	1,900
NH_4NO_3	200	200	1,650
NaH_2PO_4	150	150	-
$MgSO_4 \times 7H_2O$	300	300	370
$(NH_4)_2SO_4$	400	400	-
KH_2PO_4	170	170	170
$CaCl_2 \times 2H_2O$	600	600	440
Micro salts			
KI	0.8	0.8	0.83
$MnSO_4$	5	5	16.9
$Na_2MoO_4 \times 2H_2O$	0.25	0.25	0.25
$ZnSO_4 \times 7H_2O$	2	2	8.6
$CoCl_2 \times 6H_2O$	0.025	0.025	0.025
$CuSO_4 \times 5H_2O$	0.025	0.025	0.025
H_3BO_3	5	5	6.2
Iron source			
Na_2EDTA	37.3	37.3	37.3
$FeSO_4 \times 7H_2O$	27.8	27.8	27.8
Vitamins			
Pyridoxine HCl	0.5	0.5	0.5
Ascorbic acid	1	1	-
Nicotinic acid	0.5	0.5	0.5
Thiamine HCl	1	1	0.1
myo-Inositol	1,000	1,000	100
Organic acids			
Citric acid	10	10	-
Pyruvic acid	10	10	-
Hormones			
2,4-D	8	8	-
Kinetin	-	-	2
IAA	-	-	1
Amino acids			
Casein hydrolysate	300	300	-
Asparagine	60	60	-
Arginine	30	30	-
γ-Aminobutyric acid	80	80	-
Serine	55	55	-
Alanine	30	30	-
Cysteine	10	10	-
Leucine	10	10	-
Isoleucine	10	10	-
Proline	10	10	-
Lysine	10	10	-
Phenylalanine	5	5	-
Tryptophan	5	5	-
Methionine	5	5	-
Valine	5	5	-

Components	BAD-1 induction (mg/L)	BAD-1 differentiation (mg/L)	MS regeneration (mg/L)
Glycine	2.5	2.5	2
Histidine	2.5	2.5	-
Threonine	2.5	2.5	-
Glutamine	975	975	-
Carbon sources			
Sucrose	60,000	60,000	-
Glucose	17,500	17,500	-
Maltose	-	-	30,000
Others			
Ficoll	100,00	-	-
Agar (purified)	-	8,000	8,000
Coconut water	-	-	-
pH	5.8	5.8	5.8

This cold treatment helps condense the chromosomes, facilitating their spreading in root tip squashes. After cold treatment, the root tips are fixed in vials containing freshly prepared fixative made up of a 1:3 (v/v) mixture of glacial acetic acid and 95% ethanol. The vials are left at room temperature for 24 h and then stored in a refrigerator for a few weeks, but chromosome counts are best done from fixed root tips within a week after fixation.

The fixed root tips, with or without hydrolysis with HCl, can be squashed in various stains, e.g., acetocarmine, basic fuchsin, or toluidine blue. However, we routinely use carbol fuchsin for staining chromosomes. A root tip is removed from the fixative and transferred to a drop of 45 % acetic acid in a watch glass (or treated with 45% acetic acid in a vial) for 2-5 minutes until the white color of the tip starts to clear. The root tip is then placed on a clean slide in a drop of carbol fuchsin, and the root cap is removed with a razor blade or scalpel. After removing the root cap, the tip is gently squashed out (or gently macerated for even spreading) with a sharp needle to release cells into the stain. The remainder of the root tip or any other debris is then removed from the slide and a cover slip carefully placed on the stain such that no air bubbles are trapped. Spreading of somatic chromosomes is achieved by gently tapping all over the cover slip and then pressing the slide by placing it between folds of bibulous or blotting paper. The initial preparation is viewed under a 10× objective and 10× eyepiece lens (i.e., 100× magnification) to rapidly find dividing mitotic cells and look for good spreads, and chromosomes are counted under a 40× objective. Figure 2.25-1i shows 14 somatic chromosomes of a haploid durum plantlet.

Fluorescent genomic *in situ* hybridization (GISH) is used to study details of intergenomic translocations or other chromosomal abnormalities. Among the green regenerants, several disomic ($2n = 28$), polyploid, and aneuploid plants are observed. It is possible that haploids initially induced are converted into disomic plants or higher polyploidy during culture. In addition to numerical changes, several structural alterations are induced during culture. These abnormalities include dicentric chromosomes, acentric fragments, Robertsonian translocations and sister chromatid exchanges. These aberrations may constitute the bases of gametoclonal and somaclonal variations observed in DH plants.

Efficiency and Applications

Anther culture response is influenced by several factors, including the genotype of the donor plant and the environmental conditions under which it is grown (Dogramci-Altuntepe *et al.*, 2001). Genotypic differences in anther culture response are well known in durum wheat. The medium for anther culture is very important for regeneration of haploid plants. We have observed that the genotype, growth condition, and the media have highly significant effects on anther culture response. The interactions, genotype × growth condition, genotype × medium, are also highly significant. Anthers from plants grown in the glasshouse or field respond to anther culture, although the field-grown material generally gives better anther culture response than the glasshouse material. We observed anther response on all media tested (BAC-1, BAD-1, BAD-3 and M-42), although the best response was obtained on BAD-1.

Of the ten durum genotypes studied, we observed green plantlets from only three genotypes: 'Ege-88', 'Diyarbakir-81' and 'Dicle.' Manipulation of media formulation may help improve recovery of green regenerants.

Acknowledgments

The protocol reported here was developed as a team effort by myself, lab staff and graduate students, principally Münevver Doğramacı-Altuntepe (Ph.D. student) and Terrance Peterson (technician).

References

Dogramci-Altuntepe, M., T.S. Peterson and P.P. Jauhar, 2001. Anther culture-derived regenerants of durum wheat and their cytological characterization. J.Hered. **92**: 56-64.
Kao, K.N., M. Saleem, S. Abrams, M. Pedras, D. Horn and C. Mallard, 1991. Culture conditions for induction of green plants from barley microspores by anther culture methods. Plant Cell Rep. **9**: 595-601.
Marsolais, A.A. and K.J. Kasha, 1985. Callus induction from barley microspores. The role of sucrose and auxin in a barley anther culture medium. Can.J.Bot. **63**: 2209-2212.
Murashige, T. and F. Skoog, 1962. A revised medium for rapid growth and bioassays with tobacco tissue cultures. Physiol.Plant. **15**: 473-497.
Trottier, M.-C., J. Collin and A. Comeau, 1993. Comparison of media for their aptitude in wheat anther culture. Plant Cell Tiss.Org.Cult. **35**: 59-67.

2.26
Anther culture and isolated microspore culture in timothy

S. Pulli and Y.-D. Guo
Laboratory of Plant Physiology and Molecular Biology, Department of Biology, University of Turku, 20014 Turku, Finland

Introduction

Timothy (*Phleum pratense* L.) is one of the basic forage grasses in the world and the most important grass in winter snow covered northern latitudes. It is considered to be allohexaploid in nature ($2n=6x=42$). Timothy is valued for its winter hardiness, good palatability, and moderate feeding quality. Timothy breeding is problematic due to its allohexaploid genome and strong outcrossing. Doubled haploidy is a useful tool as homozygous lines can be obtained in a single generation by androgenic cell culture techniques, thus saving many generations of backcrossing to reach homozygosity by traditional means. Niizeki and Kati (1973) first described timothy anther culture and callus formation from cultured anthers. Successful timothy anther culture and green plant regeneration procedures have been reported by Abdullah *et al.* (1994) and Guo *et al.* (1999). However, the anther culture method has been relatively inefficient in the case of timothy where the very small anthers require lengthy and technically demanding microscopic isolation. Timothy microspore culture offers an efficient method for haploid production (Guo and Pulli, 2000). An advantage of microspore embryogenesis is that regeneration from other tissues, e.g. the diploid tissues of the anther (septum, anther wall and tapetum) is avoided.

Protocol

Donor plants and growth conditions
Timothy seeds are sown into pots and grown for 4-6 weeks in a glasshouse. Plants are then transferred to a 4°C chamber and vernalized for further 10-12 weeks, after which the plants are removed to a glasshouse under the following growth conditions: 18°C/13°C, day/night temperature and a 16 h photoperiod at 100 µmol $m^{-2} s^{-1}$ supplemented by fluorescent lamps.

The physiological condition of donor plants can significantly affect the subsequent response of anthers and isolated microspores. Cold pretreatment of donor plants is used in androgenic cell culture of timothy, the optimal length of exposure to low temperature varies with genotype, but is usually 2-4 weeks at 4°C in darkness.

Anther inoculation
When the timothy heads have extended 3-4 cm out of their sheaths, the developmental stage of the microspores is examined microscopically by aceto-carmine staining. The spikes are cut, labelled, bagged and placed into a bottle containing a sufficient amount of water to cover the cut ends. They are then transferred to a cool room at 4°C in darkness for a cold pretreatment of 2-4 weeks, depending on the genotype in question. Next, the spikes are surface sterilized in 20% sodium hypochlorite (with one drop of Tween-20), this is done on a shaker for 15 min, then washed several times with sterilized distilled water. The anthers are excised and transferred into Petri dishes containing either liquid or solid induction medium (Table 2.26-1). One hundred anthers are collected per Petri dish (6 cm) with normally 3 to 5 replicates. The Petri dishes are sealed with Parafilm and placed into a dark incubation chamber at 27°C.

Microspore culture
For timothy microspore culture, two different techniques are employed; maceration by microblending and maceration with a pestle. For isolation by microblending, spikes are cut into small pieces (3-5 mm) with sterile scissors, and placed in a small high-speed blender container with 15 ml chilled (4°C) mannitol (0.2 M, 0.21 Osm kg^{-1} H_2O of osmotic pressure). The material is then blended at low-high-low speed for 10 seconds at each speed, this is repeated three times. In pestle maceration, the small spike pieces are placed in a chilled (4°C) mortar with 15 ml 0.2 M chilled (4°C) mannitol and macerated. Following the spike maceration, the suspension is filtrated through 80 µm nylon mesh, pelleted by centrifugation at 40 g for 2 min, repeated twice.

Microspore purification
After microspore isolation, the microspore pellet is resuspended in 2 ml 0.2 M mannitol, this is overlaid with 8 ml of 21% maltose monohydrate in a Pyrex® centrifuge tube (100x15 mm) and centrifuged at 40 g for 5 min. The very pure and uniform microspores are located in a band at the mannitol/maltose interphase, these are collected with a pipette and placed into a centrifuge tube, 8 ml 0.2 M mannitol are added and the microspores spun down at 28 g for 3 min.

Microspore pretreatment and culture
Microspores are resuspended in the required amount of embryo/callus induction medium (Table 2.26-1), final density of microspores is determined with a haemocytometer and adjusted to $2x10^4$ per milliliter. Two milliliters of microspore suspension are dispensed into each 50x15 mm sterile Petri dish. The cultured microspores are treated by heat shock at 31°C for 24 h, followed by incubation in the same conditions as described above for timothy anther culture.

Table 2.26-1. Composition of PG-96 basal induction medium

Components	mg/L
Macro nutrients	
KNO_3	1,500
$(NH_4)_2SO_4$	150
KH_2PO_4	125
$MgSO_4 \times 7H_2O$	200
$Ca(NO_3)_2 \times 4H_2O$	200
KCl	50
Micro nutrients	
$MnSO_4 \times 4H_2O$	10
$ZnSO_4 \times 7H_2O$	3
H_3BO_3	5
KI	0.83
$Na_2MoO_4 \times 2H_2O$	0.25
$CuSO_4 \times 5H_2O$	0.025
$CoCl_2 \times 6H_2O$	0.025
Iron source	
$FeSO_4 \times 7H_2O$	27.8
$Na_2EDTA \times 2H_2O$	37.3
Others	
Ascorbic acid	10
Proline	20
Aspartic acid	10
Citric acid	10
Biotin	0.05
Glycine	2
myo-Inositol	100
Thiamine HCl	10
Pyridoxine HCl	1
L-Glutamine	500
Nicotinic acid	0.5
Glutathione	10
L-Serine	20
Casein hydrolysate	200
L-Alanine	10
2,4-D	1.5
Kinetin	0.5
Maltose	90,000-130,000
pH	5.7

Plant regeneration

For timothy anther- and microspore culture, embryos and callus-like structures with diameters over 1 mm are counted and transferred to the solid 190-2 medium (Wang and Hu, 1984) after 5-8 weeks of culture. Regeneration medium 190-2 contains 2% sucrose and 0.3% (w/v) phytagel (Sigma) supplemented with plant growth regulators 2,4-D, 0.5 mg/L; BA, 0.5 mg/L and a relatively high amount of copper (5 µM $CuSO_4$, to promote plant regeneration).

The embryos and calli are cultured at 25°C, with a 16 h photoperiod and light intensity of 5-100 μmol m^{-2} s^{-1}. Regenerated plants are transferred to a glasshouse, vernalized (for conditions see above) and eventually planted into the field.

Chromosome doubling and chromosome counting
Chromosome doubling with colchicine is essential, as the frequency of spontaneously doubled plants is 59% and 66% for timothy anther culture and microspore culture, respectively. In our work, we treat all plants with colchicine and do not discriminate between haploids and doubled haploids. If however, you wish to treat only haploid plants, these can be selected, they are usually weaker, paler green and with very narrow leaves, confirmation of ploidy can be carried out using Fuelgen staining of root tip cells. Roots of plantlets are immersed in a 0.1% colchicine solution containing DMSO (0.1%) for 3 h followed by several water washes (care must be taken when handling colchicine as it is a carcinogen). The plants are then potted up and once re-growth is established, the ploidy level of timothy regenerants is determined by the Feulgen staining of root tip cells.

Efficiency

Anther culture and isolated microspore culture methods are established for the androgenic embryogenesis of timothy (*Phleum pratense* L). Embryos/calli have been obtained and green plants can be regenerated.

Genotype is an important factor in the induction of androgenic haploids. In our studies, embryos were obtained from 16 genotypes out of the 28 studied in anther culture, and from 6 genotypes out of 12 studied in microspore culture (Guo *et al.*, 1999; Guo and Pulli, 2000). The most responsive cultivars were: 'Adda', 'Käppa', 'Alma' and 'Grinstadt' for anther culture and Alma, Grinstadt and 'Kämpe II' for microspore culture.

For most genotypes, the PG-96 liquid medium with 6% maltose monohydrate (Table 2.26-1) produces high embryo yields and liquid medium is preferred over solid medium in timothy anther culture. In microspore culture, liquid PG-96 medium is employed.
The optimum developmental stage of microspores is between the late uninucleate and binucleate stages. Microspores have a remarkable capacity to develop into haploid plants *via* embryogenesis *in vitro*. Stress treatment triggers the induction of this sporophytic pathway. The heat shock method: 31°C for 24 h causes "swelling" of the cultured timothy microspores. Microspore diameter increases by 150% - 200% over the original size and the nucleus becomes clearly visible. The swollen microspores have specific cytological characters, comprising a large central vacuole, thin tonoplast, parietal cytoplasm and a peripheral nucleus. Many microspores between very late uninucleate and early binucleate stages will "swell" after the heat shock, whereas younger and older microspores will not. Only "swollen" microspores have the potential to continue to develop, divide and finally form embryos/calli. For microspore culture, maceration of spikes in a blender and purification of microspores between a mannitol/maltose interface gives a relatively high percentage of vital cells. Cold pretreatment applied to the donor plants (spikes) increases embryo yield in anther culture. Heat shock promotes initiation of embryogenesis in microspore culture.

Green plant regeneration is a crucial and difficult step. The occurrence of albino plants is also a serious limiting problem in androgenic cell culture of timothy. Albinism can be reduced by low light intensity conditions during the regeneration phase. In microspore culture, a relatively high level of copper (5 μM CuSO$_4$) is added to the regeneration medium to promote plant regeneration.

Among the regenerated plantlets derived from anther culture, the chromosome number was checked from 41 random samples of Adda green plantlets. The number of doubled haploids and haploids were 24 (58.5%) and 14 (41.5%), respectively. In microspore culture,

DH plants with chromosome number $6n=42$ were 65.6%, and haploid plants with chromosome number $3n=21$ were 25.0%.

References

Abdullah, A.A., S. Pedersen and S.B. Andersen, 1994. Triploid and hexaploid regenerants from hexaploid timothy (*Phleum pratense* L.) via anther culture. Plant Breed. **112**: 342-345.

Guo, Y.-D. and S. Pulli, 2000. An efficient androgenic embryogenesis and plant regeneration method through isolated microspore culture in timothy (*Phleum pratense* L.). Plant Cell Rep. **19**: 761-767.

Guo, Y.-D., P. Sewon and S. Pulli, 1999. Improved embryogenesis from anther culture and plant regeneration in timothy. Plant Cell Tiss.Org.Cult. **57**: 85-93.

Niizeki, M. and F. Kati, 1973. Studies on plant cell and tissue culture, III. *In vitro* production of callus from anther culture of forage crops. J.Fac.Agric.Hokkaido Univ. **57**: 293-300.

Wang, X.Z. and H. Hu, 1984. The effect of potato II medium for triticale anther culture. Plant Sci.Lett. **36**: 237-239.

2.27
Doubled haploid induction in ryegrass and other grasses

S.B. Andersen
The Royal Veterinary and Agricultural University, Department of Agricultural Sciences, Section Plant Breeding and Crop Science, Thorvaldsensvej 40, DK-1871 Frederiksberg C, Denmark

Introduction

Various grass species have various and diverse uses, they provide forage and pasture for cattle and sheep, grass lawns in private gardens, parks, golf course and fields sports. For these purposes grasses are bred intensively and cultivar seed marketed world-wide. Most of these grass species are heavily cross breeding in nature and therefore highly heterozygous and genetically non-uniform. Most agronomically interesting traits of grasses are affected by many genes and environmental factors, which complicates breeding and genetic studies. Formation of haploid and chromosome doubled haploid plants in such species is a new genetic tool, which may alleviate some of these problems in the future. The first green regenerants from anther culture of perennial ryegrass (*Lolium perenne* L.) were reported by Stanis and Butenko (1984) and the first large scale production of haploid and diploid green plants was reported by Olesen *et al.* (1988). The present protocols will function well with perennial ryegrass and Italian ryegrass (*Lolium multiflorum* Lam.) and also will work for orchard grass (*Dactylis glomerata* L.) (Christensen *et al.*, 1997) and hexaploid timothy (*Phleum pratense* L.) (Abdullah *et al.*, 1994). These protocols will probably also function with several other grasses such as tall fescue (*Festuca arundinacea* Schreb.), meadow fescue (*Festuca pratensis* Huds.) and various hybrids between *Festuca* and *Lolium* for which regeneration of plants from anther or panicle culture has been reported with slightly different media and methods.

Protocol

Donor plants and growth conditions
Donor plants for haploid induction by anther culture range from young seedlings to clones maintained for some time. Types with demand for vernalization for flowering will mostly need 2-3 months at low temperatures (2-5°C) with a low light intensity (20-40 W/m^2), before grown to produce flowers. This vernalization for flowering may be obtained under natural conditions in the field or in growth chambers. Vernalized plants can then be raised for flowering in 5 L pots with peat under glasshouse conditions with 10-15°C night and 15-20°C day temperature.

Spike collection and pretreatment
Spikes with microspores at the mid to late uninucleate stage of pollen development are used. For ryegrass this pollen developmental stage is approximately when half to two third of the spike has emerged from the sheath. For hexaploid timothy the correct stage of pollen development is approximately when spikes have emerged 3-4 cm above the sheath. For orchard grass, pollen development in different florets is not synchronised and a general morphological sign of correct pollen developmental stage cannot be provided. The pollen developmental stage in harvested spikes is assessed cytologically by squash preparations of anthers from one of the middle flowers. Pollen developmental stages of grasses are very similar to the ones used for anther culture of wheat and barley and can be seen clearly after staining with carmine acetic acid or even without any staining. Since there is variation in development among clones within cultivars, it is important to assess the correct stage for the material you are working with. Spikes for anther culture may be stored with their cut ends submerged in water for three to five days at 3-5°C as a pretreatment, or they can be used immediately for anther culture.

Anthers isolation and plating
Spikes are sterilized before anther isolation: the leaves are removed and the spike dipped briefly (5-8 seconds) in 70% ethanol with an added drop of Tween per litre. This is followed by treatment for 8 minutes in 3% Korsolin (Bode Chemie Gmbh & Co., Hamburg) and three rinses with sterile water. Anthers are isolated from the spikes under aseptic conditions with a pair of forceps and placed directly on R2M induction medium (Table 2.27-1) in 9 cm Petri dishes with 25-30 ml of substrate.

Anther culture and plant regeneration
Embryos and callus induction are achieved by culturing for 35 days at 26-27°C in darkness. Embryos and callus emerging from the cultured anthers are transferred to medium 190-2 (Table 2.27-1), with 20-30 structures per 9 cm Petri dish containing 20-30 ml of substrate. Cultures for plant regeneration are incubated for 30 days at 25°C with continuous white fluorescent light (15 W/m^2). Green plantlets regenerated are transferred for further development and rooting to 9 cm Petri dishes with 20-30 ml 190-2 medium without any growth hormones and further incubated in the culture regime as for plant regeneration. Green plants with well-developed roots can be established in peat in a shaded glasshouse at 15-20°C.

Chromosome doubling
Green plants from ryegrass anther culture generally have a high frequency of spontaneous chromosome doubling so that 50-70% of the plants are diploid when established in peat. A low percentage of the plants (5-8%) may double twice to reach the tetraploid ploidy level. .In anther culture of hexaploid timothy, a high proportion (70%) of the plants chromosome double from the triploid back to the hexaploid ploidy level. A comparable high level of

spontaneous chromosome doubling has been observed for tetraploid orchard grass anther culture where most green plants chromosome double to the tetraploid ploidy level spontaneously during culture. For these reasons plants derived from grass anther cultures do not generally need to be artificially chromosome doubled.

Media

The induction medium R2M is the medium described as 190-2 by Wang and Hu (1984) modified to contain 9% maltose, 1.5 mg/L 2,4-D, 0.5 mg/L kinetin and devoid of NAA. Because of the high concentration of sugar (maltose) in this medium, the maltose should be dissolved in half of the water volume and autoclaved separately from the other media ingredients to avoid serious browning reactions. When cooled to approximately 60°C after autoclaving, the two media solutions can be mixed. All media are solidified with 0.35% gelrite (Kelco) and the pH is adjusted to 6.0 before sterilization by autoclaving (Table 2.27-1).

Table 2.27-1. Composition of media used in anther culture of ryegrass and other grasses

Components	R2M induction medium (mg/L)	190-2 regeneration medium (mg/L)
Macro elements		
KNO_3	1,000	1,000
$(NH_4)_2SO_4$	200	200
KH_2PO_4	300	300
$CA(NO_3)_2 \times 4H_2O$	140	140
$MgSO_4 \times 7H_2O$	200	200
KCl	40	40
Micro elements		
$MnSO_4 \times H_2O$	4.9	4.9
KI	0.5	0.5
$ZnSO_4 \times 7H_2O$	3	3
H_3BO_3	3	3
Iron source		
Na_2EDTA	37.3	37.3
$FeSO_4 \times 7H_2O$	27.8	27.8
Vitamins		
Thiamine HCl	0.5	0.5
Nicotinic acid	0.25	0.25
Pyrodixine HCl	0.25	0.25
Glycine	1	1
Myo-Inositol	200	200
Others		
Maltose	90,000	-
Sucrose	-	30,000
2,4-D	1.5	-
NAA	-	0.25
Kinetin	0.5	0.25
Gelrite	3,500	3,500
pH	6.0	6.0

Requirements for growth of H/DH plants

Most plants regenerated from anther cultures of ryegrass show severe signs of inbreeding depression. They are weak and slow growing and although they have doubled their chromosome number to the diploid level and induced flowering normally, they generally produce little seed. Some of the plants, approximately 15-20 percent, show good vegetative growth but in most cases, the plants have reduced seed set during cross pollination with normal fertile pollinators. Because of the strong gametophytic incompatibility system in these species, self-pollination of anther culture derived plants to produce homozygous lines generally results in very few seeds. In addition, it is difficult to provide strict isolation to avoid undesired pollen contamination during such forced selfing. There are indications that the sporophytic self-incompatibility of the species may break down to promote higher self-fertility if DH plants are repeatedly selfed for several generations (Madsen et al., 1993). For higher ploidy grasses, like timothy and orchard grass, inbreeding depression among anther derived plants seems to be less severe, although general ability for seed set of the derived plants in these species is not yet known.

Efficiency and Applications

A prominent feature of anther culture systems in ryegrass is the extremely strong effect of genotype. Anther culture of most ryegrass clones from different cultivars will produce quite large number of microspore derived embryos with the present protocols. In such genetically wide plant material approximately 15 percent of the cultured anthers can be expected to produce microspore derived callus and embryos. Many responding anthers will produce more than one callus or embryoid, so that a magnitude of 50-60 microspore derived structures can be obtained per 100 cultured anthers. Therefore, even if some ryegrass clones are considerably more responsive than others with respect to embryo/callus formation, these genetic differences are not limiting because most ryegrass clones will produce enough microspore derived structures. In addition, the regeneration of microspore derived calli or embryos into plants is not a major problem as, in most cases, 20-30 percent of structures produce plants.

The major genetic effect from the donor plant is on the ability to produce green regenerants. Most ryegrass (*Lolium perenne* and *L. multiflorum*) clones will produce almost entirely albino regenerants. If different plants of ryegrass varieties are tested with the anther culture protocol, 1-2 percent of the clones are likely to have a low response capacity. Such rare clones will generally regenerate plants in the anther culture, among which 2-3 percent turn green (Olesen et al., 1988). Ryegrass clones with higher capacity to regenerate green plants are very rare among common material. High responsive clones, which will regenerate more than 50 percent green plants from anther culture can be obtained *via* recombination. If clones with low response (2-3 percent green regenerants) are hybridized, some of their offspring will have a high ability to respond, so that more than 50 percent of green plants can be regenerated in anther culture (Halberg et al., 1990). High responsive ryegrass clones (more than 50 % green plants regenerated) can be used as "inducers" to provide the genetic ability for green plant regeneration to common clones of the cultivar *via* hybridization. If the high responsive clones are crossed with random clones from the cultivar then the hybrid offspring will be able to regenerate 2-5 percent green plants in anther culture (Madsen et al., 1995). There are indications that tetraploid types of ryegrass have a higher general capacity for green plant regeneration than most diploid material. Response patterns with other grasses using these protocols are still not well clarified. Experiments with the hexaploid timothy has produced low frequencies of microspore derived embryos from the cultured anthers (0.6 embryos per 1,000 anthers) but high frequency of green plants (more than 50 %) (Abdullah et al., 1994). Considerably higher response with timothy anther culture has been reported using the PG-96 culture medium (Guo et al., 1999). Anther cultures from tetraploid orchard grass has produced a considerably higher rate of embryo formation (133 embryos per 100 anthers) and more than half of the regenerated plants turned green (Christensen *et al.*, 1997).

References

Abdullah, A.A., S. Pedersen and S.B. Andersen, 1994. Triploid and hexaploid regenerants from hexaploid timothy (*Phleum pratense* L.) via anther culture. Plant Breed. **112**: 342-345.

Christensen, J.R., E. Borrino, A. Olesen and S.B. Andersen, 1997. Diploid, tetraploid, and octoploid plants from anther culture of tetraploid orchard grass, D*actylis glomerata* L. Plant Breed. **116**: 267-270.

Guo, Y.-D., P. Sewon and S. Pulli, 1999. Improved embryogenesis from anther culture and plant regeneration in timothy. Plant Cell Tiss.Org.Cult. **57**: 85-93.

Halberg, N., A. Olesen, I.K.D. Tuvesson and S.B. Andersen, 1990. Genotypes of perennial ryegrass (*Lolium perenne* L.) with high anther-culture response through hybridization. Plant Breed. **105**: 89-94.

Madsen, S., A. Olesen and S.B. Andersen, 1993. Self-fertile doubled haploid plants of perennial ryegrass (*Lolium perenne* L.). Plant Breed. **110**: 323-327.

Madsen, S., A. Olesen, B. Dennis and S.B. Andersen, 1995. Inheritance of anther-culture response in perennial ryegrass (*Lolium perenne* L.). Plant Breed. **114**: 165-168.

Olesen, A., S.B. Andersen and I.K. Due, 1988. Anther culture response in perennial ryegrass (*Lolium perenne* L.). Plant Breed. **101**: 60-65.

Stanis, V.A. and R.G. Butenko, 1984. Developing viable haploid plants in another culture of ryegrass. Dokl.Biol.Sci.Akad.Nauk.SSSR. **275**: 249-251.

Wang, X.Z. and H. Hu, 1984. The effect of potato II medium for triticale anther culture. Plant Sci.Lett. **36**: 237-239.

2.28
Microspore culture in rapeseed (*Brassica napus* L.)

J.B.M. Custers
Plant Research International, Wageningen University and Research Centre, P.O. Box 16, 6700 AA Wageningen, The Netherlands

Introduction

In the 1980s, it was demonstrated that high embryo yields could be obtained by culturing isolated microspores from rapeseed *Brassica napus* in simple, hormone free media, without going through an intermediate callus phase (Keller *et al.*, 1987; Pechan and Keller, 1988). Efficient embryogenesis was induced by high temperature stress treatment of freshly isolated microspores. Since then, *B. napus* microspore culture has been studied extensively in many disciplines, and has evolved to become one of the best described model systems for studying the process of androgenesis. *Brassica napus* microspore cultures are used as models in studying the initiation of embryogenic development in plants. One focus is the unravelling of the molecular mechanisms underlying the induction of embryogenesis and early embryogenic development (Custers *et al.*, 1994; Custers *et al.*, 2001). A microspore culture protocol used routinely at Plant Research International is described below, it uses the homozygous *B. napus* 'Topas' line DH4079, which is a responsive genotype used in fundamental research. The protocol also works well for applied purposes, where doubled haploids are produced from various *B. napus* genotypes and accessions for breeding programmes. Both summer and winter rapeseed types are responsive, although clear genotypic dependency exists. A strong point of the protocol is that, with few exceptions, its basic component parts have not been changed since the 1980s. The protocol is based mainly on the extensive manual for *B. napus* microspore culture issued from the University of Guelph (Coventry *et al.*, 1988).

Protocol

Materials
Equipment
- conditioned growth room or phytotron
- laminar flow cabinet
- refrigerated centrifuge
- incubators or growth rooms (22° and 32°C)
- light/UV microscope, with filter combinations for DAPI and FDA
- inverted microscope
- autoclave and apparatus for medium filter-sterilization
- further common lab equipment: balance, pH metre, autoclave, fridge, freezer.

Tools for isolation of microspores
In one isolation procedure, usually microspores from three batches of flower buds are isolated, for which following tools are needed:
- 3 tea baskets (tea eggs) for disinfecting the buds
- jar (250 ml) with 2% NaOCl (2% sodium hypochlorite + 2 drops of Tween 20) at 4°C (can be re-used)
- 6 honey jars of 450 ml with sterile tap water at 4°C
- 3 plungers; we use the backside of the plunger of a disposable 35 ml syringe (Fig. 2.28-1b). The plungers are kept sterile in a closed jar with 'Bacillol' or ethanol 70%
- NLN-13 medium at 4°C (200 ml), for washing as well as culture of the microspores
- 3 sterile disposable buckets with screw lids (inside diameter should correspond with plunger size) or sterile beakers of a similar size
- sterile 10 ml pipettes with bulbs or automated dispensers
- 3 sterile filter funnels, each with 2 layers of 50 µm nylon filter + sterile Erlenmeyer flasks (wrapped in Al-foil and autoclaved in advance)
- 3 sterile 10 ml centrifuge tubes (blue cap)
- Fuchs-Rosenthal counting chambers
- Petri dishes, 6 cm diameter
- sterile yellow pipette tips.

Culture containers for successive parts of the procedure
- 3 or 6 cm Petri dishes, for culturing the microspores
- 9 cm Petri dishes, for germination of microspore-derived embryos
- honey jars with a 1 cm layer of vermiculate, for growing plantlets up to transfer into soil.

Media
- NLN-13 medium, for both isolation and washing of the microspores as well as for their eventual culture, is prepared from stock solutions (Tables 2.28-1 and 2.28-2) that are stored in the freezer at -20°C. As an alternative, the NLN medium is also commercially available as ready-for-use powder. Be cautious however as powder mixtures sometimes contain salts in a different formulation to aid stability or in order to prevent precipitation when dissolving in water. Upon preparation, the liquid NLN-13 medium is filter-sterilized and can be stored for 1-3 months at 4°C.
- B5-Gamborg medium, in its original formulation (Gamborg *et al.*, 1968), supplemented with 1% sucrose, solidified with 0.7% Oxoid purified agar L28 (Oxoid LTD, Basingstoke, UK) or with 0.6% BBL agar (Becton Dickinson and Company, Lockeysville, USA), and sterilized by autoclaving, is used for germination of the

microspore derived embryos. Liquid B5-Gamborg medium is used for their conversion into plantlets.

Chemicals
- DAPI stain to visualise the DNA of the nuclei: 4',6-diamidino-2-phenylindole (DAPI) powder is dissolved in water or in 50% ethanol at a concentration of 1.5 mg/ml, then diluted to 2.5 µg/ml in a nuclei extraction buffer consisting of 10 mM Tris-HCl (pH7.0), 10 mM spermine-tetrahydrochloride, 10 mM NaCl, and 200 mM hexyleneglycol. Triton X100 at 1% (v/v) is added as a permiabilizing agent for the microspores. This DAPI mixture is diluted 1:1 with glycerine to reduce desiccation of the slides, and then covered with Al-foil and stored in the refrigerator. Alternatively, DNA staining dyes can be purchased as ready-to-use solutions, for instance from Partec
- FDA is used to check cell vitality: fluorescein diacetate (FDA) is dissolved in acetone and stored as 0.2 mg/ml stock solution in the dark at $-20°C$.
- ABA phytohormone to mature the microspore derived embryos: abscisic acid (ABA) is dissolved in aqueous $NaHCO_3$ and is stored as 5 mg/ml stock at $-20°C$.
- colchicine is used to double the chromosome number: fresh solutions are necessary; it dissolves readily in water.

Donor plants and growth conditions
Plants of summer rapeseed, *Brassica napus* 'Topas' line DH4079 are germinated and after the 3-5 leaf stage grown in 12 cm pots with a fertile horticulture soil in a controlled environment room with a 16 h photoperiod of 150 $\mu E.m^{-2} s^{-1}$ SON-T or HPI-T (Philips) light at a constant 18°C. After 4-5 weeks, just before the inflorescences start to develop, plants are transferred to a similar controlled environment room, but with a lower constant temperature, 10°C. Application of pesticides to the plants, either as spray or as soil drench, is forbidden, as it affects vitality of the microspores. Therefore, to prevent mildew infestation of the plants, relative humidity in the climate room is kept at 55-60%. Aphids, when detected on a plant, should be removed together with the leaf or the whole plant they infect. Plants are watered with a liquid fertilizer (N:P:K = 20:10:20) at two days intervals, but generally are kept quite dry in order to prevent sudden rise in humidity in the room. After the first 1-3 flowers have opened, the inflorescences can be used as donors for microspore culture (Fig. 2.28-1a). Usually, entire inflorescences are excised from the plants and brought into the laboratory where suitable buds are collected.

Isolation of microspores and preparation of microspore suspensions
1. Switch on the refrigerated centrifuge to start pre-cooling to 4°C.

2. Switch on the laminar flow cabinet and disinfect it with ethanol.

3. Harvest 5-10 inflorescences and bring them into the laboratory. From this point on everything should be done as quickly as possible, preferably within one hour, and under 4°C conditions. To reach the latter, all tools and media used are pre-cooled in the refrigerator before use. In the preparation of large volume cultures, where large numbers of flower buds have to be collected, it is recommended that the picked buds are placed on top of a bed of ice.

4. Collect buds of the desired bud lengths (between 3.2 and 3.5 mm) with fine tweezers and sort them into three size classes by using a measuring gauge. Make each class as uniform as possible by judging for identical shape and morphology of the individual buds. From plants grown under our conditions, buds 3.2-3.3 mm in length will give mostly hundred percent

unicellular microspores, whereas 3.4-3.5 mm buds have a certain percentage microspores that have passed through first pollen mitosis. Twelve to 15 buds per class should be enough (keeping record of the numbers of buds can be useful for your own records). One bud can contain over 100,000 microspores of which 50-70% can easily be isolated. Isolation of a higher proportion by stronger squeezing is harmful to vitality of the microspores.

5. From here on, continue the work in the laminar cabinet. Hands should be cleaned with 70% ethanol or 'Sterilium'.

6. Transfer the buds into the tea eggs to be sterilized for 10 min in 2% NaOCl. Shake the tea-eggs from time to time vigorously to make sure that there are no remaining air bubbles attached to the buds. The Tween present in the 2% NaOCl decreases the surface tension of the 'waxy' layer on the bud so that the buds make good contact with the hypochlorite.

7. Rinse the tea eggs sequentially 3 times in the sterile water for about 1, 4, and 10 min each.

8. Rinse the plungers (after the buds) in the sterile water.

9. Transfer the buds to the disposable buckets and add 2 ml of sterile NLN-13 medium.

10. Squeeze the buds with the plunger by a turning pressure movement (4 half turns) to free the microspores.

11. Rinse the plunger with 2 ml NLN-13 medium, and place it back in the tap water jar. Repeat treatment 10 with the two other batches of buds.

12. Place the sterile filter funnels + two layers of nylon filter on top of the sterile centrifuge tubes standing in a solid rack so that they do not topple over (Fig. 2.28-1c).

13. Pour the suspension over filter funnel, rinse the bucket with 2 ml NLN-13 medium and pour it over the filter, rinse the filter with another 2 ml.

14. Adjust the volume of all the tubes (10 ml).

15. Centrifuge the suspensions with microspores for 3 min at 100 g; centrifuge temperature at 4°C.

16. Pour off the supernatants and resuspend the pellets with 10 ml medium.

17. Repeat steps 14 and 15, centrifugation and resuspension, twice. After the last washing step, resuspend the microspores in 1 ml NLN-13.

18. Take a sample of 20 µl to count the microspores in the Fuchs-Rosenthal counting chamber. The number of microspores counted in 5 sub-cells multiplied by 16,000 gives the density of microspores per ml.

19. Add NLN-13 medium to achieve the correct density for culture (4×10^4 microspores/ml).

20. Pour the suspension into Petri dishes for culturing. Depending on the volume obtained and the preferred experimental design, Petri dishes of 3 cm, 6 cm, or 9 cm in diameter can be filled with 1 ml, 3 ml, and 10 ml microspore suspension respectively. Tissue culture quality

Petri dishes are used to improve spreading of the small volume of microspore suspension in the dish.

21. Seal Petri dishes with Parafilm and incubate cultures at 32°C for 72 h in complete darkness, and then transfer them to 25°C again in the dark.

Table 2.28-1. Ingredients of the NLN-13 medium used for isolation, washing and culture of the *Brassica napus* microspores; composition of stock solutions and final concentrations in medium

Ingredients	Stock No. (strength)	Amount (mg or g)	Total volume (ml)	Stored aliquots (ml)[x]	Final concentration (mg/L)
Macro elements		(g)	1,000	50	
KNO_3	1 (20x)	2.5			125
$Ca(NO_3)_2 \cdot 4H_2O$		10			500
$MgSO_4 \cdot 7H_2O$		2.5			125
KH_2PO_4		2.5			125
Micro elements		(mg)	250	5	
$MnSO_4 \cdot 4H_2O$	2 (200x)	1,250			25
$ZnSO_4 \cdot 7H_2O$		500			10
H_3BO_3		500			10
$Na_2MoO_4 \cdot 2H_2O$		12.5			0.25
$CuSO_4 \cdot 5H_2O$		1.25[xx]			0.025
$CoCl_2 \cdot 6H_2O$		1.25[xx]			0.025
Iron source		(g)	250	5	
NaFe(III)EDTA	3 (200x)	2			40
Organic components I		(mg)	100	1	
Nicotinic acid	4 (1000x)	500			5
Thiamine HCl		50			0.5
Pyridoxine HCl		50			0.5
Folic acid		50			0.5
Biotin		5g			0.05
Glycine		200			2
Organic components II		(g)	500	20	
L-Serine	5 (50x)	2.5			100
L-Glutamine		20			800
Glutathione		0.75			30
Myo-inositiol		2.5			100
Carbohydrate					
Sucrose					130,000

[x] Aliquots to be combined to achieve the correct medium composition (Table 2.28-2).
[xx] For convenience, first dissolve both these micro elements at 10x higher strength

Table 2.28-2. Aliquots of stock solutions to combine for preparing 1 L of NLN13 medium [x]

Stock solution No.	Amount
Stock 1	50 ml
Stock 2	5 ml
Stock 3	5 ml
Stock 4	1 ml
Stock 5	20 ml
Sucrose	130 g

[x] pH of the medium is adjusted at 5.8 and the medium is filter sterilized

Testing composition and vitality of the microspore population

1. Use the small amounts of suspension left over in the centrifugation tubes for the characterisation of the cultures at starting date. Spin down this material in Eppendorf tubes (2-4 min, 4000 rpm), remove supernatant until approximately 15 µl is left, and resuspend the pellet in this volume.

2. For analysis of the developmental stage of the microspores, transfer 7 µl of the dense suspension into another Eppendorf tube and add a similar volume of DAPI mixture. Upon mixing both volumes, 10 µl is mounted on a microscope slide and covered with a cover slip. After about 4 hours (or overnight) the samples can be observed under the UV microscope. The developmental stage of 100 microspores is scored. For safety reasons, we strongly recommend that small volume samples are prepared, which exactly fit under the cover slip, so that no carcinogenic DAPI can spill over the microscope glass, risking contamination of the working area. All disposables containing DAPI, including the microscope slides should be discarded according to your institute's safety rules.

3. In order to assess microspore vitality, first make a working solution from the FDA stock by 1/10 dilution in NLN-13. For analysis, dilute this solution ten times with microspore suspension, incubate 10 min in darkness, place the cover slip over the sample, and then determine the percentage of fluorescent microspores under the UV microscope.

4. For both analyses above, initial samples can also be taken from the dense microspore suspension immediately after step 17.

Observation and treatments during culture

1. For observation of early development in culture, during the first 7 days, collect 50-100 µl samples from a dish, spin microspores down in an Eppendorf tube, and prepare the pellet for DAPI observation according to steps 1 and 2 of previous section (Fig. 2.28-2a-f). An inverted microscope should be used for direct observation of the microspores in culture.

2. For later observation of the cultures, when embryos are larger than 0.3 mm, use a stereomicroscope.

3. After 10 days, determine the number of well-developed embryos per ml of the original suspension. Dilute cultures 2-3 times with fresh NLN-13 medium when numbers are higher than 1,000 per ml. After 14 days, dilute the cultures further until 50-100 embryos per 1 ml, and transfer dishes approximately four days later to a rotary shaker at 50 rpm.

4. After 21-28 days, when cotyledon-stage embryos are present (Fig. 2.28-3a), ABA is added to the culture medium at a concentration of 3 mg/L to induce embryo maturation. After one week of ABA treatment the embryos are ready for germination.

Embryo germination, chromsome doubling and transfer to soil
1. Incubate embryos on solid B5 medium with 1% sucrose, 25 embryos per 9 cm dish, and place dishes in culture room with 16 h photoperiod at 22°C.

2. After 4-6 days, when germination has begun, orientate embryos in a vertical position if needed. Thereafter germination will proceed fast (Fig. 2.28-3b).

3. After 2-3 weeks, transfer young seedlings that show first leaf development into honey jars with vermiculite wetted with liquid half strength B5 medium supplemented with 0.05% colchicine. Keep the honey jars in culture room with 16 h photoperiod at 22°C.

4. Seedlings that are actively growing after 4-6 days are almost ready for transfer to soil. Then, open the jars partially or for short periods, or replace the lids by sheets of paper to get the plantlets prepared for transfer into soil. If needed, add some water to prevent wilting of the plantlets, and continue this treatment for 1 week.

5. Remove the plantlets from the jars, transfer them with the plugs of vermiculite attached in plain water to dilute surplus of nutrients, especially sucrose, for approximately 1 h and transplant them in a tray with a peat mix. The trays are covered with a polyethylene lid, placed in a glasshouse, and if needed covered with wet cheesecloth to prevent extensive temperature increase inside caused by solar radiation.

6. At the second day after transfer to soil, start to slightly lift the lids to harden off the plants. Remove the lids after one week, and transplant the plants into pots another week later.

7. Wait until plants flower and determine, based on pollen production in the flowers, whether plants are doubled haploid or haploid.

Efficiency and Applications

Other rapeseed *Brassica napus* cultivars are also responsive in this protocol, although their embryo yield will likely be lower than in the highly responsive DH4079. The early growing period until inflorescence appearance can also be performed in a regular glasshouse. Of importance is that plants have been kept for at least two weeks at low temperature conditions prior to harvesting flower buds for isolating microspores.

Other laboratories regularly recommend to grow plants during development of the inflorescences at a day/night temperature regime of 10/5°C, but under this condition, many growth cabinets can not achieve a low relative humidity, and consequently moulds and fungi can develop freely. The problem mainly occurs in the tropics with high ambient temperatures. For tropical environments we recommend that the growth cabinet is adjusted to a low temperature that it can reasonably maintain a low relative humidity. At temperatures of 15-21°C, donor plants also produce buds with potentially embryogenic microspores. The main effect of increasing donor plant temperature is that the number of suitable buds to be harvested per inflorescence decreases; the extension growth of the entire inflorescence is faster with higher temperature, resulting in bigger size differences among buds.

Donor plants can be used a second time by cutting them back quite drastically; approximately 10 cm main stem should be removed. This will allow primary auxiliary shoots to develop new inflorescences again, ready for use 2-3 weeks later.

Desired bud lengths given in this protocol are for *B. napus* 'Topas' line DH4079 under our conditions only. Since the developmental stage of microspores in buds of the same size vary between donor plants of different genotypes and physiological conditions (e.g.

volume of pot soil plays a role), one should determine for each different genotype or condition the bud size with the most suitable microspore population. Preferably, a test should be made with DAPI one day prior to preparing a culture. Finally, after observation of the cytological stages of the microspores in flower buds of different sizes, buds should be collected of the size containing a large fraction of the microspores at the late unicellular stage accompanied with 10-40% early binucleate microspore. This is the best combination to achieve a high percentage (up to 10%) of embryos.

Actually, for an optimal result of microspore isolation, the number of flower buds should not be higher than 12-15 per batch. Higher numbers of buds can not be kept together in one handling under the plunger for squeezing, so that squeezing has to be repeated too frequently, which causes higher amounts of dead microspores in suspension. If for preparing large volume cultures high numbers of flower buds are needed, for instance more than 50, these should be divided over more batches and squeezed in succession. Squeezing efficiency decreases with increasing bud numbers per batch, which results in lower yields of microspores released per bud and decreased microspore vitality.

Since we observed a higher frequency of dead microspores lacking nuclei after strong squashing or real homogenisation treatment, we prefer moderate squeezing which only tears the anthers open, whereupon the microspores can freely emerge from the anthers and spread in suspension.

In freshly isolated microspore suspension, the following developmental stages and cytological characteristics are to be observed:
- early unicellular microspore; 19-21 µm in diameter, translucent cell, thin exine wall, nucleus central in the cell,
- mid unicellular microspore; 20-22.5 µm in diameter, trilobulor shaped, thick exine wall, pale yellow, nucleus central in the cell,
- late unicellular microspre; 21-23 µm in diameter, trilobulor shaped, exine wall thick and rigid, yellow, nucleus displaced to the periphery of the cell and starts to enlarge in preparation of the first pollen mitosis,
- early bicellular pollen; 22.5-25 µm in diameter, generative and vegetative nuclei close to another, both of the same density,
- mid bicellular pollen; 24.5-26 µm in diameter, cell shape becomes more globular, generative nucleus small and dense, and vegetative nucleus large and diffuse, both have migrated from another,
- late bicellular pollen; 25-26.5 µm in diameter, globular shaped cell, also the generative nucleus has migrated to the centre of the cell, vegetative nucleus more diffuse,
- tricellular pollen; 26-28 µm in diameter, two small very dense sperm nuclei, vegetative nucleus very diffuse.

The given microspore and pollen sizes are from measurements carried out in our DAPI mixture, whereas sizes in NLN-13 culture medium are roughly 5% larger. Contrary to the above series of successive microspore and pollen stages, in the text the term microspore is mostly used to indicate the entire population of both microspores and pollen.

As specific development in an embryogenic culture, upon 32°C heat treatment, many microspores exhibit an interruption of their normal gametophytic development. They enter a few cycles of relatively random, sporophytic cell divisions (Fig.2.28-2a,d), resulting in the formation of multi-cellular clusters within the microspore exine walls. After four to five days of culture, the exine ruptures (Fig. 2.28-2b,e) and further cell divisions lead to the formation of globular embryos (Fig. 2.28-2c,f). Then, differentiation results in the organised body plan of a heart shaped embryo. Finally, successful cultures of *B. napus* 'Topas' DH4079 produce 3-10% embryos. The remaining unsuccessful microspores in cultures can be divided in roughly three groups: those with no divisions, which is generally the fate of too young, mid unicellular microspores; those in which sporophytic cell division begins, but the division pattern is irregular and an embryo is not produced; and those where gametophytic development continues as shown by early or mid bicellular pollen in the microspore population. After 2-4 days in culture, the latter group is visible as distinctly oval shaped bodies, they resemble mature pollen and are trinucleate at that moment. Notably, they suddenly disappear due to an aberrant type of pollen germination, namely the formation of large blisters, which mostly collapse.

The timing of the addition of fresh medium and culture dilution described here should not be adhered to too rigidly, as the need for fresh medium depends on the embryo density and the growing speed that is wished. With a low density culture of 10-25 embryos per ml, medium refreshment is usually not needed, whereas with densities higher than 200-400 embryos per ml a delay in dilution inhibits embryo extension growth. This can even be used as a measure to regulate the culture process, for instance to determine a date convenient for starting the germination of the embryos.

For chromosome doubling we treat the microspore derived seedlings with colchicine at the end of their tissue culture period. Colchicine is added to the liquid medium and transpiration by the plants is stimulated to improve intake of the chromosome doubling agent. We prefer this method over other well known chromosome doubling procedures, because: *in vitro* chromosome doubling by applying 50-100 µM colchicine during the first 2-3 days of microspore culture generally decreases frequency of embryogenesis; immersion of the seedlings just prior to transfer into soil overnight in 0.05% colchicine + 0.5% DMSO solution mostly affects seedling vitality badly and inhibits their recovery after planting in soil; and treatment of the mature plants by application of colchicine (0.1-0.3%) in the lower leaf axils, and thereafter removal of all existing shoots, is a very late and time comsuming treatment.

Generally, frequencies of plants with vital pollen are 50-70% upon colchicine treatment, whereas only 5% fertile plants are present in untreated controls. Flow cytometric analysis has shown that almost all fertile plants are diploid, and that variant ploidy levels, i.e. triploids and tetraploids, hardly occur.

References

Coventry, J., L. Kott and W.D. Beversdorf, 1988. Manual for microspore culture technique for *Brassica napus*. University of Guelph, Guelph.

Custers, J.B.M., J.H.G. Cordewener, M.A. Fiers, B.T.H. Maassen, M.M. Van Lookeren Campagne and C.M. Liu, 2001. Androgenesis in *Brassica*; a model system to study the initiation of plant embryogenesis. In: Current Trends in the Embryology of Angiosperms. Bhojwani, S.S. and W.Y. Soh (Eds.) Kluwer Academic Publishers, Dordrecht, pp. 451-470.

Custers, J.B.M., J.H.G. Cordewener, Y. Nöllen, H.J.M. Dons and M.M. Van Lookeren Campagne, 1994. Temperature controls both gametophytic and sporophytic development in microspore cultures of *Brassica napus*. Plant Cell Rep. **13**: 267-271.

Gamborg, O.L., R.A. Miller and K. Ojima, 1968. Nutrient requirements of suspension cultures of soybean root cells. Exp.Cell Res. **50**: 151-158.

Keller, W.A., P.G. Arnison and B.J. Cardy, 1987. Haploids from gametophytic cells - recent developments and future prospects. In: Plant Tissue and Cell Culture. Green, C.E., D.A. Somers, W.P. Hackett and D.D. Biesboer (Eds.) Allan R. Liss, Inc., New York, pp. 223-241.

Pechan, P.M. and W.A. Keller, 1988. Identification of potentially embryogenic microspores in *Brassica napus*. Physiol.Plant. **74**: 377-384.

2.29
Protocol for broccoli microspore culture

J.C. da Silva Dias
Instituto Superior de Agronomia, Technical University of Lisbon, Tapada da Ajuda, 1349-017 Lisboa, Portugal

Introduction

Broccoli (*Brassica oleracea* var. *italica*) is a vegetable cole crop of considerable interest to breeders and seed companies since its area of production has increased in recent years mainly due to improved nutritional qualities, e.g. anticancer properties. Breeding companies strive more and more to bring F_1 hybrid seeds onto the market. F_1 cultivars ensure high uniformity, offer high yields due to heterosis, allow rapid production and selection of desired genotypes and provide protection of plant breeder's rights and the markets of seed companies. The production of hybrid cultivars requires homozygous parental lines. Inbred lines in broccoli can be produced by recurrent selfing, a procedure that takes time (6 to 7 generations of selfing) and is labour intensive due to sporophitic self-incompatibility. An alternative way to obtain pure inbred lines in one generation is the production of doubled haploid (DH) lines by microspore culture. This protocol can also be applied to other cole crops such as head and tronchuda cabbages using small adaptations.

Protocol

Basic essential equipment and supplies
The list of supplies presented below is for flower bud collection, bud surface sterilization and microspore isolation and incubation.
- sterile tea baskets (many as the bud samples)
- 1,200 ml of cold sterile distilled water
- four sterile beakers of 400 ml
- sterilizing solution: sodium-dichloroisocyanurate ($C_3Cl_2N_3O_3Na$; Sigma®) sterilizer and Nonidet P-40 (Sigma®) detergent or other sterilizer and wetting agent such as 2% chlorine solution with Tween 20
- sterile fine tweezers (not serrated) and 1 pair of scissors
- one bottle (250 ml) of sterilized cold NLN-13 medium
- one bottle (250 ml) of sterilized cold ½ NLN-13 medium
- sterile disposable Pasteur pipettes
- sterile beakers of 50 ml (many as the number of samples/accessions)
- sterile smooth plungers from 50 ml syringes
- sterile sets of 50 ml or 100 ml Erlenmeyer flasks with 4.5 cm diameter funnel and double nylon mesh filter with 45 µm each
- sterile disposable centrifuge tubes of 11 or 12 ml (as many as the number of samples/accessions) in a rack
- haemocytometer
- micropipettes (P10 and P20) with respective tips
- DAPI (4', 6-diamino-2-phenylindole) fluorochrome
- sterile disposable 10 ml pipettes with a bulb
- 60 or 90 mm Greiner® plastic Petri dishes
- digital pachymeter/vernier calliper
- Parafilm (Parafilm M® or Nescofilm®)

Cold refers to fridge with storage temperature (4-5°C). The water needs to be distilled or doubled-distilled and autoclaved. The beakers, tea baskets, tweezers and scissors should be sterilized in a stove by dry heating. Remove possible condensed water before using. New sterilized instruments should be used for each operation, if these are in short supply, tweezers, scissors and others should be dipped in 80% alcohol, flamed and cooled before re-use. Sterile plastic disposable material such as pipettes, centrifuge tubes and Petri dishes, should always be opened under laminar flow with sterilized scissors and should never be re-used.

Donor plants and growth conditions
The conditions under which donor plants are grown are an important consideration for successful culture of broccoli microspores. Temperature is the most important and critical factor because sudden changes in the temperature regime, mainly after floral differentiation, could greatly diminish the embryogenic responses, because that stress alters the level of growth regulators, amino acids, carbohydrates or lipids in the plants. The plants grown at low temperatures yield the most responsive microspores. The growth temperatures of donor plants should not exceed 20°C. For example temperature regimes of 20°C day/15°C night or constant 18°C, with a 16/8 h photoperiod and a photosynthetic photon flux density of 150-200 µE m^{-2} s^{-1} given by 'warm-white' tubular fluorescent lamps are adequate. Such temperatures delay flowering and produce a wider range of bud sizes compared to higher growth temperatures.

Young inflorescences yielded more embryogenic microspores than older inflorescences and are preferred, but plants to be harvested should have initiated flowering if their open buds had not been removed. In broccoli and cauliflower, the main inflorescence is cut off to allow secondary racemes to develop, this reduces the density of buds and competition among buds. Young inflorescences yield more embryogenic structures than older ones therefore donor plants should not be kept too long. The open flowers are great sink sources and should also be removed.

The quality of bud material is a major factor for success in this protocol. Young inflorescences should be used and bud density should be reduced through removing the main flower stalk and pruning densely budded racemes. Plants should not be allowed to retain open flowers.

Do not vernalize the plants too young (with less than 12 leaves) and place them after 3-4 leafs (2 weeks after germination) in plastic pots (minimum 16 cm diameter; ideal 20-22 cm) that will allow them to grow without restrictions. Use a 1 sand: 2 commercial compost (such as Levington M2®), fertilized with 20-25 g of a slow-release fertilizer such as Blaukorn® 12N:12P:17K:2MgO or Osmocote Plus® 15N:11P:13K:3MgO. Later, after vernalization at 4°C, for 4 to 8 weeks according to genotype requirements, fertilize the plants when necessary with a liquid fertilizer, e.g. Bayfolan® 6N:4P:6K. Plants should be maintained under good nutritional status, but avoided nitrogen excess, and free from diseases and pests; such factors can all influence the responsiveness of microspores to culture. Senescent leaves should be removed promptly to avoid health problems.

Collecting of flower buds

Microspore culture should be performed only on the basis of single plants. Since the developmental stage of the microspores is the most important parameter affecting the success of microspore culture in broccoli and in other coles, only potentially embryogenic buds should be collected. Some hours before bud collection (or in the previous day) you should make a note of microspore development by microscopic observation (8x40 magnification). This is done in order to select flower buds with the maximal number of microspores at the late uninucleated stage, with about 10-30% binucleate pollen, which are the most embryogenic in broccoli. Microspore staging can also be evaluated using fluorescent microscopy and DAPI staining which allows good visualisation of the nucleus.

The length of the flower buds at the correct stage is used as a more easy reference for later collections. For each donor plant it is necessary to determine, for each time of bud collecting, the range size of the buds that will allow the optimal response.

When collecting flower buds, use fine tweezers to pick the buds off the plants leaving the pedicel attached as much as possible (to avoid its oxidation). Alternatively, you can take the branches with the flower buds but this will affect your next collection. Keep the buds cool in the fridge or in chilled Petri dishes, and work fast until plating the microspore suspension (maximum one hour).

Select the buds by size near the reference length using a digital pachymeter and make bud length groups. Please keep in mind that ideally we should plate only single buds. So try to have groups with few buds since it is better than groups with many buds. Put selected flower buds in tea baskets with the respective label.

Bud surface sterilization

In the laminar flow bench place 4 chilled beakers of 400 ml with 300 ml of sterile distilled water. The beakers with the water should be taken out of the fridge only at the moment you need them. In one of the beakers make a sterilizing solution with 5 g of the sterilizing agent [1g of sodium-dichloroisocyanurate in 60 ml of sterilized water; 1.7% (w/v)] and 4 to 5 drops of Nonidet P-40 (1 to 2 drops per 100 ml] and dissolve it using a Pasteur pipette.

Place each of the closed tea baskets with the buds in the sterilizing solution for 6 min (skake it gently to avoid air bubbles) and rinse three times in the other beakers containing the sterile water (6 min each time).

Composition and preparation of media

Table 2.29-1. Composition of the media used in microspore culture and regeneration of broccoli plants

Components	NLN-13 medium (mg/L)	Gamborg B5 medium (mg/L)	MSS medium (mg/L)
Macro nutrients			
KNO_3	125	2,500	1,900
$Ca(NO_3)_2 \times 4H_2O$	500	-	-
$MgSO_4 \times 7H_2O$	125	-	370
KH_2PO_4	125	-	-
$CaCl_2$	-	113.24	332.2
$(NH_4)_2SO_4$	-	134	170
$MgSO_4$	-	122.09	-
NaH_2PO_4	-	130.5	-
NH_4NO_3	-	-	1,650
Micro nutrients			
$MnSO_4 \times H_2O$	25	10	16.9
$ZnSO_4 \times 7H_2O$	10	2	8.6
H_3BO_3	10	3	6.2
$Na_2MoO_4 \times 2H_2O$	0.25	0.25	0.25
$CuSO_4 \times 5H_2O$	0.025	0.025	0.025
$CoCl_2 \times 6H_2O$	0.025	0.025	0.025
KI	-	0.75	0.85
Iron source			
NaFe(III)EDTA	40	-	-
$EDTANa_2 \times 2H_2O$	-	37.25	37.5
$FeSO_4 \times 7H_2O$	-	27.85	27.8
Vitamins			
Nicotinic acid	5	1	0.5
Thiamine HCl	0.5	10	0.1
Pyridoxine HCl	0.5	1	0.5
Folic acid	0.5	-	-
Biotin	0.05	-	-
Glycine	2	-	2
myo-Inositol	100	100	100
Amino acids			
L-serine	100	-	-
L-glutamine	800	-	-
Glutathione	30	-	-
Carbohydrates			
Sucrose	130,000	20,000	10,000
Glucose	-	-	10,000
Others			
Agar (bacteriological)	-	9,000	9,000
Activated charcoal	-	-	500
pH	6.1	5.8	5.8

Stock solutions used for the preparation of NLN-13 medium
Stock 1 – Macronutrient salts (for 10 L of NLN-13 culture medium). Final volume of 500 ml. Keep in the freezer (-18°C) in 10 portions of 50 ml.

KNO_3	1.25 g
$Ca(NO_3)_2 \times 4H_2O$	5 g
$MgSO_4 \times 7H_2O$	1.25 g
KH_2PO_4	1.25 g

In the ½ NLN-13 medium the concentration of major salts is reduced to 50%.

Stock 2A – Micronutrient salts. Final volume of 50 ml. Keep in the freezer (-18°C) in 8 portions of 6.25 ml.

$CuSO_4 \times 5H_2O$	10 mg
$CoCl_2 \times 6H_2O$	10 mg

Stock 2B – Micro nutrient salts (for 50 L of NLN-13 culture medium). Final volume of 250 ml. Keep in the freezer (-18°C) in 50 portions of 5 ml.

$MnSO_4 \times H_2O$	1.25 g
$ZnSO_4 \times 7H_2O$	0.5 g
H_3BO_3	0.5 g
$Na_2MoO_4 \times 2H_2O$	0.0125 g
Stock 2A	6.25 ml

Stock 3 – Iron: NaFe(III)EDTA salt (for 20 L of NLN-13 culture medium). Final volume of 100 ml. Keep in the freezer (-18°C) in 20 portions of 5 ml.

NaFe(III)EDTA	0.8 g

Stock 4 – Vitamins (for 25 L of NLN-13 culture medium). Final volume of 25 ml. Keep in the freezer (-18°C) in 25 portions of 1 ml.

Nitsch and Nitsch vitamin	2.71 g

Stock 5 – Amino acids (for 20 L of NLN-13 culture medium). Final volume of 500 ml. Keep in the freezer (-18°C) in 25 portions of 20 ml.

L-serine	1.25 g
L-glutamine	10 g
Glutathione	0.375 g

Preparation of 1 L of NLN-13 medium
For the preparation of 1L of NLN-13 medium add the following volumes of each stock solution:

Stock 1	50 ml
Stock 2B	5 ml
Stock 3	5 ml
Stock 4	1 ml
Stock 5	40 ml

In the ½ NLN-13 medium the concentration of major salts (stock 1) is reduced to 50%.

Add stocks in order to 800 ml double distilled water and allow each to dissolve completely before adding the next ingredient. Bring up to 1 L, pH 6.1 and filter sterilize (eg. with a 50 ml sterile disposable syringe with a Agrocap® filter of 0.2 µm from Gelman Sciences). The medium should be kept in the fridge at 4°C.

Activated charcoal suspension preparation
Activated charcoal suspension preparation and addition of a 0.1 ml or 0.2 ml drop to the 60 mm or 90 mm microspore culture dishes is made following Gland *et al.* (1988) and Dias (1999) procedures. Autoclave suspension of 1 g of Merck®2186 activated charcoal, 0.5 g of Sigma® type IX agarose and 100 ml of double distilled water. During autoclaving the activated charcoal appears to associate with the agarose but on cooling remains suspended. It is important to associate the charcoal with agarose because suspended charcoal without agarose adheres to the microspores and hampers embryogenesis.

Microspore isolation and incubation

1. Cool the centrifuge to 4°C, take all the materials except the water and the medium, and put them in the laminar flow cabinet. If you do not have a cooled centrifuge, you can do everything in a standard desktop centrifuge but all materials should be cold.

2. Put two tubes for each sample in the rack and mark them. Put one sterile pipette in one tube for each sample and a third pipette in the medium bottle (take this bottle out of the fridge only now). Open the 50 ml sterile beakers and pour 1.5 ml of NLN-13 medium (Table 2.29-1) into each beaker.

3. Transfer each sample of buds from the tea baskets into a 50 ml beaker with 1.5 ml of NLN medium and squeeze them gently (do not mash) with the plunger/piston from a sterile disposable 50 ml syringe. Never use a blender to do it because the blending of the buds will result in a release of starch into the medium and break apart the microspores, resulting in a decrease in embryogenesis potential. Avoid mixing up the samples.

4. Open the Erlenmeyer flasks with funnel and nylon mesh filter (45 µm) and pour the content of the beakers into the funnel. Rinse the plunger of the syringe and beaker with some millilitres of medium and pour this into the funnel too.

5. Transfer the contents of the Erlenmeyer flasks into the centrifuge tubes and add NLN-13 medium up to about 10 ml. Centrifuge the extract of the test tubes for 3 min at 900 rpm (max. 100 g) without using the brake. The precipitation of microspores should be gentle and so the centrifuge should start slowly and then increase the speed gradually to a maximum of 100 g.

6. Discard the supernatant using a pipette. Add 10 ml of NLN-13 culture medium and resuspend the microspores pellet by inverting the tubes gently and/or shaking them very gently.

7. Repeat last steps twice. Use ½NLN-13 for the final suspension. After the last centrifugation check if the microspores have precipitated properly. If not, centrifuge again until the supernatant becomes clear.

8. After the last adding of medium, with a pipette place 30 µl of suspension over a lamina and 10 µl of DAPI (2.5 µg/ml). Dispose the lamella and take the preparation to a dark place to observe later on under a fluorescence microscope. At the same time place 20 µl of the suspension of each centrifuge tube in a haemocytometer chamber (e.g. Neubauer Improved®) to determine the concentration of microspores. Before counting the microspores on the haemocytometer, make sure that the NLN mixture containing the microspores is well mixed. Count the total number of microspores in 5 squares delineated

by double lines with the microscope magnification 100x and multiply by 10^4. This gives the number of microspores per ml. The total number of microspores is equal to the number of microspores per ml x the original volume. An accurate count is essential for proper dilution to the required density.

9. Dilute to a concentration of 40,000 microspores per ml, mix well and dispense/pour into Petri dishes: 10 ml into 90 mm dishes or 5 ml into 60 mm dishes. Add a 0.1 or 0.2 ml drop of activated charcoal solution to the 60 mm or 90 mm dishes respectively. Cover with three layers of parafilm to avoid desiccation and place in a lightproof box.

10. Incubate the Petri dishes at 32.5°C in the dark in an incubator during 24 h. Then, put them in the growth chamber at 25°C in the dark (still in the lightproof box).

11. After one day of incubation, microspores should begin to swell and 3 days later the first signs of division should appear. After one week, many cells should be dividing and some early pro-embryos (3 or 4 divisions) should be present.

Embryo transfer and culture
The stage at which embryos are transferred to solidified medium is critical. Embryos transferred at the cotyledonary stage resulted in the highest frequency of plant regeneration. We typically obtain regeneration frequencies in broccoli ranging from 27% to 68%. The frequency varies widely and is a function of genotype and quality of embryos transferred to solid medium.

1. When embryos become visible, after 2.5-3 weeks of platting, transfer the box with the Petri dishes to a gyratory shaker at 60 rpm (gentle) at 24°C in the dark for one week. Then using sterile tweezers, transfer the individual embryos (with 2-5 mm long) under sterile conditions to dishes with solid Gamborg B5 medium (Table 2.29-1), with 2% sucrose, lacking growth regulators and pH 5.8, solidified with 0.9% bacteriological agar. The dishes (e.g. 90 mm Petri dishes with 11 ml of B5 medium) are sealed with Parafilm and labelled.

2. I possible, during the first week place the dishes containing embryos in a growth chamber at low temperature (10°C or less) with 8/16 h photoperiod and a photosynthetic photon flux density of 50-100 µE m^{-2} s^{-1} given by 'warm-white' tubular fluorescent lamps. Then place the dishes at 24°C day/21°C night and 16/8 h photoperiod for 3-4 weeks to allow the embryos to develop into plantlets, which can be transferred directly to soil. Transfer before, and when adequate, plantlets from plates to GA7 Magenta® boxes (using 35 ml of B5 medium) or other larger container.

3. Instead of Gamborg B5 medium, which is the most frequently used, the MSS medium [a MS medium modified by the addition of sucrose (10 g/L), glucose (10 g/L) and activated charcoal (0.5 g/L), Table 2.29-1] can be used. MSS medium works well, especially when normal embryo quality structures are not obtained (this is possibly induced by the presence of glucose in the medium, which is more available than sucrose for the regeneration of embryos).

Plantlet transfer to soil
You can expect variation in *in vitro* plantlet quality. If all the plantlets are not required, only the more promising ones should be selected for planting into compost. Although callus-like forms may develop into normal plants if a shoot apex is present, leafy plantlets with good development are best for planting.

1. Remove plantlets from the B5 medium and transfer them into individual Jiffy-7® (expanded in a solution of 1 L of bi-distilled water, 4 ml of Bayfolan® 6N:4P:6K liquid fertilizer and 0.8 g of Ridomil MZ 72® fungicide, with 64% of mancozebe and 8% of metalaxil or other similar products). With a pair of scissors, trim lightly the roots and the leaves of the plantlets before placing them into Jjffys.

2. For acclimatization, place the Jiffys in a propagator with the nutritive solution and the fungicide previously described, in a growth room with 24°C day/21°C night and 16/8 h photoperiod and a photosynthetic photon flux density of 50 µE m^{-2} s^{-1} given by 'warm-white' tubular fluorescent lamps. Do not over-wet at the beginning with the liquid fertilization and fungicide. Fertilization is only crucial when the plants become established. In some protocols fertilizer is not added until plants are well established. Do not forget also that newly transferred plantlets should be in the propagator with closed vents to retain moisture and avoid desiccation. After one week you could start opening the vents of the propagator and later, before plants start to be crowded, the propagator cover is removed.

3. Instead of Jiffies you can use cell trays. In both cases do not forget that transferred plantlets should be individually labelled with a plastic tag. This tag should identify the plant during the entire process.

Chromosome doubling
Chromosome doubling of microspore culture derived plants is an important problem in the practical application of microspore culture technology, because breeding programmes need a large number of genetically stable, homozygous doubled haploid plants with a high level of fertility.

Since spontaneous doubling occurs randomly and is extremely genotype dependent, it is important to ascertain the level of spontaneous diploids in the genotype used. In our work with broccoli we found 43% to 88% of spontaneous diploids and in other coles from 7% to 91%. For genotypes with over 60% spontaneous doubling, it is not necessary to induce doubling.

Colchicine is the most frequently used chromosome-doubling agent in plants. The procedure: select 30 day old (7-8 leaves) plants or older and group them according to identification; trim the roots and wash completely with warm tap water (don't trim tops); immerse the roots of the plants in a 0.25% working solution of colchicine (1% stock solution: 5 g of colchicine in 500 ml of distilled water) for 5 h and place the plants under intense lights for 1.5 h up to 3 h (use rubber gloves when handling colchicine and colchicine-exposed plants); pour off the solution and rinse roots thoroughly in tap water; replant the treated plants in a soil mixture; water sparingly until plants are established and do not fertilize until growth is evident. The tops of the plants will die after the colchicine treatment and the plants should be cut back to the first new buds (10-15 cm stem left). Sectors of the new growth with good development will be doubled. After vernalization, the flowers on the doubled sectors should be bagged to produce fertile selfed seeds. This procedure produces doubling rates from 53% to 71%, but has disadvantages: it may cause substantial plant mortality and reduce plant fertility (this may be overcome in the next generation); it is labour intensive and time consuming; it involves handling large quantities of colchicine, which is toxic and carcinogenic.

Instead of this plant treatment with colchicine, and when chromosome doubling is necessary, we recommend an *in vitro* method to increase the frequency of chromosome doubling of microspore culture derived plants. When used in low concentrations (0.01 – 0.02%), colchicine only inhibits cell division cycle for a short time, after which cells contain a doubled set of chromosomes that can continue into mitosis. The optimal time for *in vitro* colchicine treatment is the first 12 h after microspore isolation. After the treatment, the medium is changed by centrifugation and resuspension of the microspores in fresh medium. This direct microspore genome reduplication method proved to be extremely effective and can increase embryogenesis induction, one problem however is that tetraploids can also be produced. The use of other anti-microtubule agents such as trifluralin, oryzalin and pronamide has so far not been proven for cole crops. Doubled haploids arising either spontaneously or by chemical induction appear normal for agronomic and compositional characteristics.

Identification of the ploidy level of microspore derived regenerants
There are several ways of identifying the level of ploidy of the regenerated plants. The most used are: chromosome counting; flow cytometry analysis; phenotypic identification; and indirect methods, such as counting chloroplast number in the stomata guard cells and estimation of size of stomata guard cells.

Chromosome counting by Feulgen or aceto-carmine staining of root tip cells at mitotic metaphase in cole crops/broccoli is delicate and time consuming (15 plants/h for two cytologists) as chromosomes are very small and the number of metaphases depends on root growth. Phenotypic identification is based on the fact that the haploid plants are easily recognized by male sterile and smaller flowers, and by other morphological characteristics such as prolonged flowering, slender branching habit and narrow leaves. Flowering is the most reliable screen, but is inconvenient, as it requires plant culture over several weeks. Flow cytometry analysis is the most efficient (30-40 plants/h for two cytologists) and warrantable method of identification of ploidy level, but ploidy cytometers are very expensive. Indirect methods are useful, cheap and easy. A correlation has been established between ploidy and chloroplast numbers in the stomatal guard cells of aerial parts of plants. In cole crops, haploid plants have a chloroplast number between 6 to 9, diploids 10 to 15, and tetraploids 20 to 25. There is also a good correlation between the length of the stomata guard cell and ploidy level of the aerial parts of broccoli.

Efficiency and Applications

Successful broccoli microspore culture and an improvement in the existing cole protocols, to make this technique available for the purpose of routine breeding, was recently described by Dias (2001) for 10 different broccoli genotypes ('Arcadia', 'Green Duke', 'Green Valiant', 'Hi-Crown', 'Marathon', 'Mariner', 'Packman', 'Shogun', 'SDB3' and 'SDB9'). Embryo yields were significantly increased in almost all of the 10 broccoli genotypes by incubating microspore cultures at 32.5°C for 1 day, as compared to the standard incubation at 30°C for 2 days. Treatment of microspores for 48 h at 32.5°C produced less than optimal results suggesting that broccoli microspores are more sensitive to high temperatures than those of *B. napus*. The use of ½ NLN-13 medium yielded greater number of embryos than the standard NLN-13, Nitsch and Nitsch medium (Nitsch and Nitsch, 1967) modified by Lichter (1982) and Takahata and Keller (1991) with 13% sucrose. The magnitude of the response to the reduction of the concentration of major salts by half in the NLN medium varied with the different genotypes. High embryogenic broccoli cultivars, such as Shogun, SDB9, and Green Valiant, presented a better response to the reduction by half of the concentration of major salts in NLN-13. This reduction never produced a detrimental effect on embryo yield of the ten different broccoli genotypes and seems to have no effect in the subsequent development of embryos in plants.

Dias (1999) reported that embryo yields increased significantly in nine genotypes of coles (*B. oleracea*) after the addition of activated charcoal. This work represented a significant advance in the development of microspore technology. Successful microspore culture in broccoli has also been described by Takahata and Keller (1991) and Duijs *et al.* (1992).

References

Dias, J.S., 1999. Effect of activated charcoal on *Brassica oleracea* microspore culture embryogenesis. Euphytica **108**: 65-69.
Dias, J.S., 2001. Effect of incubation temperature regimes and culture medium on broccoli microspore culture embrogenesis. Euphytica **119**: 389-394.
Duijs, J.G., R.E. Voorrips, D.L. Visser and J.B.M. Custers, 1992. Microspore culture is successful in most crop types of *Brassica oleracea* L. Euphytica **60**: 45-55.
Gland, A., R. Lichter and H.G. Schweiger, 1988. Genetic and exogenous factors affecting embryogenesis in isolated microspore cultures of *Brassica napus* L. J.Plant Physiol. **132**: 613-617.
Lichter, R., 1982. Induction of haploid plants from isolated pollen of *Brassica napus*. Z.Pflanzenphysiol. **105**: 427-434.
Nitsch, C. and J.P. Nitsch, 1967. The induction of flowering *in vitro* in stem segments of *Plumbago indica* L. I. The production of vegetative buds. Planta **72**: 355-370.
Takahata, Y. and W.A. Keller, 1991. High frequency embryogenesis and plant regeneration in isolated microspore culture of *Brassica oleracea* L. Plant Sci. **74**: 235-242.

2.30
Microspore culture of *Brassica* species

A. Ferrie
Plant Biotechnology Institute, National Research Council of Canada, Saskatoon, SK, S7N 0W9, Canada

Introduction

The production of haploid plants from isolated microspores of *Brassica napus* (L.) was first reported by Lichter in 1982 (Lichter, 1982). The embryo yields from these initial experiments were low but have since increased through manipulation of the culture technique. Methods for achieving chromosome doubling and producing doubled haploid plants have also been developed. Microspore culture is about ten times more efficient than anther culture and is the preferred method for doubled haploidy. The microspore doubled haploid methodology is now employed routinely in many *B. napus* breeding programmes around the world. The varieties 'Q2', 'Quantum', and 'Armada' have been developed *via* microspore culture. The production of double haploids from buds to fertile plants is shown in Figure 2.30-1. The developmental sequence of a microspore to an embryo is shown in Figure 2.30-2.

Protocol

B. napus

Donor plants and growth conditions

Genotypic variation for microspore culture response does exist. The genotype 'Topas DH4079' (spring type) is the most responsive of the *B. napus* genotypes. It is very important that the donor plants to be used for microspore culture are healthy, vigorous and free of insects or disease. Plants can be grown in growth cabinets, glasshouse or field. We have found that the best conditions for growing donor plants are in the growth cabinet. Glasshouse and field conditions should only be used with highly embryogenic lines, as there is usually a decrease in embryogenesis under these conditions. Seeds are sown in 15 cm pots filled with a mix of soil, sand and peat moss (4:1:1) or a commercial mix (e.g. Redi-earth soil mix). Usually, two seeds are sown per pot and these are thinned when the plants are at the 2-3 leaf stage. Slow-release fertilizer (14-14-14 Nutricote 100) is added to each pot at time of sowing. The plants are grown at 20/15°C, 16 h photoperiod, light intensity 35,000-50,000 lux. Plants are watered three times a week with 0.35 g/L of 15:15:18 (15% N, 15% P, 18% K) fertilizer. When first buds are observed on the plants (about six weeks after seeding), the temperature of the growth cabinet is lowered to 10/5°C. Productive donor plants can be kept for up to six months. In order to maintain healthy, productive plants, dead leaves are removed as well as opened flowers to prevent pod development. Spraying for disease and pests should be avoided, affected plant parts should be removed, biological control can be used, especially in glasshouse material.

Developmental stage of the microspores:

1. The developmental stage of the microspores can be determined using 2 µg mL^{-1} DAPI (4'6-diamino-2-phenylindole dihydrochloride).

2. A drop of DAPI is applied to the microspore preparation on a slide. The slide is left for 20-30 min, then observed under a fluorescence microscope at 365 nm excitation.

3. Developmental stage can be correlated with bud size for ease of bud selection. Developmental stage of the microspores and bud size can vary depending on the genotype and donor plant growing conditions.

In vitro procedure

Once the buds have been harvested from the plant, it is important that the microspore isolation be done quickly. Embryogenic frequency decreases with delays in the culture procedure.

Selection and sterilization

1. Inflorescences are removed from the donor plants and placed on moist paper towels. These buds can be stored in the refrigerator if bud selection is delayed or takes longer than 10 minutes. Approximately 50-75 buds with microspores at the mid to late uninucleate stage are selected. This can be based on size (usually 3-4 mm). Do not select buds that are damaged or open.

2. The buds are placed in lipshaw baskets or tea eggs and surface sterilized in 6% sodium hypochlorite for 15 minutes on a shaker followed by three to five minutes washes with sterile water. Usually six baskets (50–75 buds/basket) are used per microspore isolation experiment.

Isolation of microspores
1. After sterilization, the buds are transferred with sterile forceps from the lipshaw baskets to sterile 50 ml beakers.

2. The buds are crushed in 5 ml of half strength B5-13 medium (B5 medium supplemented with 13% sucrose) (Gamborg *et al.* 1968) with a glass rod. The buds should be crushed hard enough to break the anther and release the microspores but gentle enough as to not damage the microspores.

3. This microspore suspension is then filtered through a 44 µm nylon screen cloth into a 50 ml sterile centrifuge tube.

4. The beaker and filter are rinsed three times with 5 ml of half strength B5-13 and poured through the filter (final volume is 20 ml).

5. The crude microspore suspension is centrifuged at 130-150 g for 3 minutes, the supernatant is decanted and 5 ml half-strength B5-13 is added to the pellet. Washing and centrifugation is repeated two more times for a total of three washes.

6. The number of microspores is determined using a haemocytometer. After the second centrifugation step, a drop of microspore suspension is placed on a haemocytometer and the number of microspores is counted.

7. Microspore viability can be determined using fluorescein diacetate (FDA). The FDA stock solution can be prepared by dissolving 0.01 g of FDA in 10 ml of acetone. Add 10 µl of 0.1% FDA to 0.5 ml of microspore suspension. Let this stand for 5 minutes after which time, one can determine the number of living and dead cells using a fluorescence microscope. The living cells will fluoresce green.

8. After the third centrifugation step, the supernatant is discarded and the required amount of modified Lichter medium (Lichter, 1982) is added to achieve a density of 10^5 microspores per ml. The medium used is NLN supplemented with 13% sucrose and 0.83 mg/L potassium iodide but without potato extract and adjusted to pH 5.8. Ten ml of microspore suspension is dispensed into each 100x15 mm sterile Petri dishe. The dishes are sealed with Parafilm, labeled with the date, experiment number, genotype and other pertinent information.

Culture conditions
1. The dishes are placed in an incubator set at 32°C in the dark.

2. After 72 h, the dishes are removed from the 32°C incubator and put into a 24°C incubator in the dark, for the remainder of three weeks. Embryos can be observed within ten days after culture.

Embryo culture
1. After three weeks, the embryos are counted. Notes are taken on embryo quality. Embryos should be at the cotyledonary stage. Embryos are then placed on a gyratory shaker (70 rpm) in a tissue culture room (22°C, 16 h photoperiod, light intensity 6500-8000 lux). Embryos remain on the shaker until becoming green, usually in a week.

2. Once green, the embryos are plated in 100x15 mm Petri dishes containing solid B5 medium (1% Difco agar, 1% sucrose, pH 5.8). Ten embryos are plated per Petri dish. Dishes are

sealed with Parafilm and labeled with the necessary information (experiment number, date). The dishes are placed in a tissue culture room and maintained at 22°C with a 14 h photoperiod and a light intensity of 150 µmol m^{-2} s^{-1}. In order to enhance normal regeneration, the embryos can be placed in the cold (4°C) for 2–3 weeks prior to plating on B5 medium.

Plantlet culture
Plantlets develop after three weeks. The normal plantlets (i.e. with shoot and root development similar to zygotic seedlings) are transferred to 150x25 mm Petri dishes containing solid medium (B5 with 0.8% Difco agar, 2% sucrose, pH 5.8). The dead leaves are removed and the roots are trimmed. Plates are placed in a tissue culture room, 12 h photoperiod, 22°C for three weeks. Several subcultures may be required.

Plantlet transfer to soil
When the plantlets have a well-established root and shoot system, usually 3 weeks after transfer to the large dishes, they are removed from the Petri dishes, and the agar is gently washed from the plantlets. Dead leaves are also removed. The plantlets are put into flats containing a soil-less mix or into peat pellets. The flats are maintained in a walk-in growth cabinet (20/15°C, 16 h photoperiod). These flats are covered to maintain a high humidity. Over a period of a week, the lids are lifted slightly each day to allow airflow and hardening of the young plantlets. After another week, they are transplanted into 15 cm pots and kept in the glasshouse.

Doubled haploid seed production
The doubled haploid plants in the glasshouse are well maintained so as to allow maximum seed set. The dead leaves are removed, and insect and diseases are controlled. Insects can be controlled chemically or biologically. Biological control does not harm the doubled haploids as much as chemical spray. Flowering usually takes place within 2-3 weeks and seed set within 4-6 weeks after transplanting. The plants are generally ready to harvest 9-12 weeks after transplanting. Plants and flowers will need to be agitated during flowering to ensure suitable levels of fertilization and seed set.

Chromosome doubling
Doubling the chromosome number is necessary in the *Brassica* species, as spontaneous doubling is very low. Doubled haploidy can be achieved by treating the microspores *in vitro* or treating the plantlets or plants with a chromosome doubling agent. We routinely use colchicine for *in vitro* chromosome doubling in *B. napus* experiments. Other chromosome doubling agents can also be used (i.e. trifluralin). Colchicine must be used with caution. Avoid inhalation and skin contact with the dry powder. The liquid form is not volatile, however protective clothing, such as gloves and eye protection should be worn. Aqueous solution of 0.34% colchicine is made by dissolving 3.4 g of colchicine in 1 L of water. The solution can be stored for a few days in the fridge, but should be made up fresh every week. Colchicine is disposed off appropriately. For liquid, collect the waste solution and rinse water in containers. These should be labeled with the date, chemical and concentration of the chemical. For solid waste, this should be collected, double bagged, and labeled with the appropriate information.

In vitro chromosome doubling
1. Microspores are cultured as outlined above. NLN-13 medium with colchicine (10^{-4} M) is substituted for regular NLN-13 culture initiation medium. After 72 h, medium is changed from NLN-13 with colchicine to NLN-13 without colchicine.

2. The medium and microspores are removed from the Petri dish by pipette into a sterile 50 ml centrifuge tube. These are centrifuged as before (130–150 g for 3 minutes). The supernatant

is removed and discarded appropriately. The same amount of NLN-13 medium without colchicine is added and 10 ml is dispensed into the same Petri dishes. The Petri dishes are resealed with Parafilm. There is a slight decrease (<10%) in embryogenesis when colchicine is added to the medium, however it is a more efficient system than colchicine treatment of plantlets or plants. The plantlets produced from *in vitro* colchicine treatment appear more normal and vigourous than those treated at the plantlet stage. Results show that the number of doubled plants produced by the *in vitro* protocol is greater or equal to the number produced by the plantlet treatment.

Plantlet treatment
Prior to transferring resulting plantlets to soil, the plantlets are removed from the Petri dishes, the agar is washed off, and dead leaves are removed. The roots and crown are submerged in a 0.34% solution of colchicine for 1.5 h. After treatment, both the roots and crown are rinsed in water and the plantlet is transferred to a soil-less mix and grown in the glasshouse or growth cabinet. These flats are covered as described above. Doubled haploid or haploid plants are determined by pollen production or the lack of pollen, respectively.

Mature plant treatment
Early chromosome doubling techniques involved growing the plant to the flowering stage and determining if the plant had spontaneously doubled (i.e. checking for the presence or absence of pollen). If the plant was haploid, then the inflorescences would be cut back and the soil would be washed from the roots. The roots and the crown would then be submerged in a colchicine solution (0.2% for 5–6 h) with aeration. This method is more time consuming and requires more resources (chemical, personnel) than either the plantlet treatment or the *in vitro* approach.

Other *Brassica* species

The basic *Brassica* protocol is used for the other *Brassica* species, with slight modifications depending on the species and genotype. Generally, uninucleate microspores are selected, however bud size will vary depending on the species, genotype and growing conditions. Correlation of the bud size and developmental stage of the pollen grain should be determined and optimized for each species and genotype. The basal medium used for washing the microspores is B5 with 13% sucrose and the basal medium used for culturing the microspores is NLN. *Brassica* microspores generally require a heat shock treatment (>30°C) for 24–72 h, although there are inductive treatments that eliminate this heat shock requirement. For some *Brassica* species, like *B. nigra*, routine microspore culture protocols are not published. For *B. carinata*, doubled haploids have been produced using the *B. napus* protocol, with NLN-13 (Barro and Marti, 1999) or NLN-10 (Ferrie, unpublished) media. The microspore culture protocol for *B. rapa* and *B. juncea* are outlined here. The *B. napus* protocol is used with the following modifications for the different species.

Brassica rapa

Genotype
The line CV-2 was identified as highly embryogenic, producing up to 400 embryos/bud. This line (reference number 91-6294) is available from Dr. Kevin Falk, Agriculture and Agri-Food Canada Research Centre, 107 Science Crescent, Saskatoon, SK, Canada, S7N 0X2.

In vitro conditions
Microspores at the mid-late uninucleate stage are used (bud size is about 2-3 mm). Induction medium (NLN) is modified to contain 17% sucrose, 0.83 mg/L potassium iodide, and 0.1 mg/L

benzyladenine (BA), without potato extract and without glutamine. The pH is adjusted to 5.8. Dishes are placed in a 32°C incubator. After 48 h, the microspore suspension is removed from the dishes by pipette and put into a 50 ml centrifuge tube. The tubes are centrifuged as in the microspore culture protocol for *B. napus* (3 min, 130–150 g). The supernatant is removed and the same amount of NLN-10 medium is added. The NLN-10 medium has 10% sucrose, 0.8 g/L glutamine but no benzyladenine. *In vitro* colchicine treatment is the same as the *B. napus* protocol. The dishes are resealed with Parafilm and put into a 24°C incubator for the remainder of three weeks.

Doubled haploid seed production
The *B. rapa* doubled haploids are weak compared to doubled haploids of *B. napus*. Because *B. rapa* is generally self-incompatible, seed production requires bud pollination, treatment of flowers with sodium chloride (3%) spray or carbon dioxide. The CO_2 treatment was the most productive method in terms of seed set. Plants with open flowers are selected and pollen is brushed onto the stigma with a sterile paint brush. The plants are bagged and placed in a sealed chamber and treated with 10% CO_2 for four hours. A second CO_2 treatment may be given two or three days later on subsequent flowers. Hand pollinations can also be conducted. Prior to flowering, the buds are carefully opened with forceps and the stigma is dusted with pollen from a mature anther. The bud is then closed. The inflorescence is identified with a tag and then covered with a glassine bag. After three days, the glassine pollination bags are removed to allow pod development. The plants are generally ready to harvest 6-9 weeks after transplanting.

Brassica juncea

For the genotypes that we have worked on, we use the *B. napus* protocol as previously described with the following modifications:
Buds 2-3 mm in size are used. Microspores are cultured in NLN with 17% sucrose and 0.1 mg/L BA but without glutamine. Cultures are incubated at 35°C for 48 hours. Medium is then changed to NLN with 13% sucrose without BA but with glutamine (0.8 g/L). Cultures are incubated at 24°C. Embryos develop after three weeks. Before transferring embryos to solid medium, the embryos are placed in the cold for two to three weeks. Resulting double haploid plants are placed in the glasshouse and watered lightly, as *B. juncea* does not do well under wet conditions.

Microspore culture media

Preparations for culture
- Media stock and concentrates can be made in advance and stored in a freezer. Media can be made up a few days in advance and kept in the refrigerator until use.
- B5 is autoclaved and NLN is filter sterilized.
- Keep a media log book and record what medium was made up, when, and its components.
- Add each component in order according to the list of ingredients. Use this list to check off the components as they are added.
- Make sure each component is completely dissolved before adding the next component.
- Start with less water than is required, once all the media components are added (with the exception of the agar), make up to final volume.
- After making liquid medium or after using it, the bottle is usually kept at room temperature overnight to check for contamination.
- Liquid media can be stored in the refrigerator for a few days; solid media can be stored at room temperature.

B5 wash medium (Gamborg *et al.*, 1968)
- The stock solutions are dissolved in double distilled water. We use half-strength B5 wash. Therefore, 100 ml of stock will make 2 L of medium.
- Sucrose (13%) is added and pH is adjusted to 6.0 by using 0.5 N HCl or 0.5 N KOH.
- The solution is made to volume and dispensed into 500 ml bottles.
- The medium is autoclaved at 15 psi, 121°C for 20 minutes.

B5 solid medium
- The protocol is the same as for B5 wash medium but with the addition of Difco agar and a lower concentration of sucrose, as described in the protocol (1% sucrose for embryo culture and 2% sucrose for plantlet culture).
- Agar is added to the bottle before adding the liquid medium. A 1 L bottle can be filled with 800 ml. This additional air space will prevent boiling over of the medium.
- Medium can be dispensed into magentas or Petri dishes once it has cooled to a temperature that is comfortable to work with. Swirl the bottles to allow mixing.
- Medium can be solidified and stored in the bottle and melted by microwave when required.

Stock solutions for B5 medium
Vitamins (1000x)

myo-Inositol	100 g
Nicotinic acid	1 g
Pyridoxine HCl	1 g
Thiamine HCl	10 g

Dissolve each compound before adding the next vitamin and bring up to a final volume of 1 L with double distilled water. Freeze in 10 ml aliquots.

Micro nutrients (1000x)

$MnSO_4 \times H_2O$	10 g
H_3BO_3	3 g
$ZnSO_4 \times 7H_2O$	2 g
$Na_2MoO_4 \times 2H_2O$	0.25 g
$CuSO_4 \times 5H_2O$	0.025 g
$CoCl_2 \times 6H_2O$	0.025 g

Dissolve each compound before adding the next micronutrient and make up to 1 L with double distilled water. Freeze in 10 ml aliquots.

Potassium iodide

KI	0.75 g

Bring up to 1 L with double distilled water. Store refrigerated in the dark.

B5 medium (10x concentrated)

KNO_3	30 g
$MgSO_4 \times 7H_2O$	5 g
$NaH_2PO_4 \times H_2O$	1.5 g
$CaCl_2 \times 2H_2O$	1.5 g
$(NH_4)_2SO_4$	1.5 g
Sequestrene 330 Fe	0.28 g

Add 10 ml of vitamin stock (1000x).
Add 10 ml of micro nutrient stock (1000x).
Add 10 ml of KI stock.

Dissolve each compound before adding the next and bring up to a final volume of 1 L with double distilled water. Freeze in 100 ml aliquots.

Note: 100 ml of 10x concentrate makes 1 L of media (full strength).

NLN medium (Lichter, 1982)
The stock solutions are dissolved in double distilled water. Full strength NLN is used.
- Sucrose (10, 13, or 17%) is added and the pH is adjusted to 5.8 by using 0.5 N HCl or 0.5 N KOH.
- The solution is made to the final volume. The medium is pre-filtered twice using glass prefilters and 0.8 µm milipore, then it is filter sterilized using 0.2 µm bottle-top filter units.

Stock solutions for NLN medium:
Vitamins (1000x)

Thiamine HCl	0.5 g
Nicotinic Acid	5 g
Pyridoxine HCl	0.5 g
Glycine	2 g
Biotin	0.05 g
Folic Acid	0.5 g
myo-Inositol	100 g

Dissolve each compound before adding the next vitamin and bring up to 1 L with glass-distilled water. Freeze in 10 ml aliquots.

Micro nutrients (Murashige and Skoog, 1962) (1000x)

$MnSO_4 \times 4H_2O$	22.3 g
H_3BO_3	6.2 g
$ZnSO_4 \times 7H_2O$	8.6 g
$Na_2MoO_4 \times 2H_2O$	0.25 g
$CuSO_4 \times 5H_2O$	0.025 g
$CoCl_2 \times 6H_2O$	0.025 g

Dissolve each compound before adding the next micronutrient and bring up to 1 L with double distilled water and freeze in 10 ml aliquots.

Potassium iodide
    KI   0.83 g
Bring up to 1 L with double distilled water. Store refrigerated in the dark.

NLN medium (10x concentrated)

KNO_3	1.25 g
$MgSO_4 \cdot 7H_2O$	1.25 g
KH_2PO_4	1.25 g
$Ca(NO_3)_2 \cdot 4H_2O$	5 g
Sequestrene 330 Fe	0.4 g
Glutathione	0.3 g
L-serine	1 g
L-glutamine	8 g
Add 10 ml of vitamin stock	
Add 10 ml of micro nutrient stock	
Add 10 ml of KI stock	

Dissolve each compound before adding the next and bring up to 1 L with double distilled water. Freeze in 100 ml aliquots.
Note: 100 ml of 10x concentrate makes 1 L of medium (full strength).

Efficiency and Applications

B. napus is the most responsive of the *Brassica* species, however, the other *Brassica* species do respond with varying embryogenic frequencies (Table 2.30-1). Just as there are differences in embryogenic response among species, there are also genotypic differences in microspore culture response. Although not all *B. napus* will respond in culture, usually 1-5% of the microspores will undergo embryogenesis. Embryo yields of up to 10% have been reported with certain genotypes and careful selection of the donor buds (Kott *et al.*, 1988; Huang *et al.*, 1990). Frequencies of 50% embryogenesis have been predicted based on estimated embryo yields from single buds of a highly embryogenic line (Pechan and Keller, 1988). Spring types of *B. napus* tend to be more embryogenic than winter types. The genotype, Topas DH4079, is a highly embryogenic line, which has been used in many microspore embryogenesis experiments (protocol 2.28). Genotypic screening studies have also been conducted in *B. rapa* (Ferrie *et al.*, 1995), *B. oleracea* (Kuginuki *et al.*, 1999), *B. carinata* (Barro and Marti, 1999) and *B. juncea* (Lionneton *et al.*, 2001). In *B. rapa*, there are differences between plants within a genotype (Ferrie, unpublished). A highly embryogenic line has been identified in *B. rapa* (Ferrie *et al.*, 1995). This line, CV-2, can produce up to 400 embryos/bud (Table 2.30-1).

Table 2.30-1. Microspore culture response of several *Brassica* species

Species	Embryos/bud	Reference
B. carinata	325.6	Ferrie, unpublished
B. juncea	90.8	Ferrie, unpublished
B. napus ssp.		
oleifera	50% of microspores	(Pechan and Keller, 1988)
Rapid-cycling	49.0	(Aslam *et al.*, 1990)
B. nigra	26.5	Ferrie, unpublished
B. oleracea ssp.		
italica	887.0	(Takahata *et al.*, 1993)
gemmifera	1.2	(Duijs *et al.*, 1992)
sabauda	12.1	(Duijs *et al.*, 1992)
botrytis	0.2	(Duijs *et al.*, 1992)
fimbriata	4.2	(Duijs *et al.*, 1992)
capitata	180.9	(Kuginuki *et al.*, 1999)
B. rapa ssp.		
oleifera (CV-2)	407.6	Ferrie, unpublished
chinensis	57.0	(Cao *et al.*, 1994)
pekinensis	112.5	(Sato *et al.*, 1989)

In addition to genotype, donor plant conditions, developmental stage of the pollen grain, pretreatments, medium constituents and culture conditions can influence microspore embryogenesis of *Brassica*. The identification and optimization of these factors will result in the establishment of a reliable protocol for production of sufficient numbers of embryos required for a breeding programme.

The location and conditions under which the donor plants are grown can influence embryo yield. Effects of temperature and light regimes can also vary depending on the genotype, however, most *Brassica* species require a cool temperature (usually 10/5°C). Donor plants are usually grown in growth cabinets, but can also be grown in the field or glasshouse. Under glasshouse or field conditions, there can be a decrease in embryogenesis.

Embryogenic microspores of *B. napus* are usually at the mid to late uninucleate stage of development. It has been suggested that the presence of younger or older cells can actually be toxic to embryogenic cells. A correlation can be made between the developmental stage of the pollen grain and the size of the bud, but needs to be gauged for each genotype and growing conditions.

Pretreating the donor plants, floral organs, or anthers with growth regulators or environmental stress can influence embryogenesis. Pretreatments have not been used very much in *Brassica* species; however, treatments have included: cold, gamma irradiation and ethanol stress (Pechan and Keller, 1989).

The composition of the culture medium is an important factor influencing embryogenesis. The major constituents are carbohydrate source and concentration, nitrogen source and concentration, pH, and growth regulators. For the *Brassica* species, the basal medium used for washing is B5 (Gamborg *et al.*, 1968) and for culturing, NLN (Lichter, 1982). For microspore culture of *Brassica* species, sucrose concentrations of the culture medium are generally higher than for other species. This is usually 13% sucrose or 17% sucrose with a media change to 10% or 13% sucrose.

The *in vitro* culture conditions can also influence embryogenesis. The main factor is temperature although light and CO_2 levels also influence embryogenesis. A heat shock treatment of 32°C for 24–72 h is required for many *Brassica* species. However, there are other treatments (e.g. colchicine) that allow the heat shock to be eliminated, but these are not used routinely. The procedure from sowing donor plant seed to harvesting seed from double haploid plants can take up to nine months.

Patents

Currently, there are a few patents on the induction of embryogenesis from *Brassica* microspore culture. These primarily deal with the use of colchicine, anti-microtubule agents, anti-cytoskeleton agents or protein synthesis inhibitors for the induction of embryogenesis. The inclusion of these in the media would remove the need for a heat shock treatment.

Simmonds, D.H. and Newcombe, W. Induction of embryogenesis using cytoskeleton inhibitors or protein synthesis inhibitors. US5900375.

Simmonds, D.H., Newcombe, W., Zhao, J., and Gervais, C. Induction of embryogenesis from plant microspores. C. US6200808.

Simmonds, D.H. and Newcombe, W., Induction of embryogenesis and generation of doubled haploid plants using microtubule inhibitors. CA2145833.

References

Aslam, F.N., M.V. MacDonald, P. Loudon and D.S. Ingram, 1990. Rapid-cycling *Brassica* species: inbreeding and selection of *B. campestris* for anther culture ability. Ann.Bot. **65**: 557-566.

Barro, F. and A. Marti, 1999. Response of different genotypes of *Brassica carinata* to microspore culture. Plant Breed. **118**: 79-81.

Cao, M.Q., Y. Li, F. Liu and C. Dore, 1994. Embryogenesis and plant regeneration of pakchoi (*Brassica rapa* L. ssp. *chinensis*) via *in vitro* isolated microspore culture. Plant Cell Rep. **13**: 447-450.

Duijs, J.G., R.E. Voorrips, D.L. Visser and J.B.M. Custers, 1992. Microspore culture is successful in most crop types of *Brassica oleracea* L. Euphytica **60**: 45-55.

Ferrie, A.M.R., D.J. Epp and W.A. Keller, 1995. Evaluation of *Brassica rapa* L. genotypes for microspore culture response and identification of a highly embryogenic line. Plant Cell Rep. **14**: 580-584.

Gamborg, O.L., R.A. Miller and K. Ojima, 1968. Nutrient requirements of suspension cultures of soybean root cells. Exp.Cell Res. **50**: 151-158.

Huang, B., S. Bird, R.J. Kemble, D. Simmonds, W. Keller and B. Miki, 1990. Effects of culture density, conditioned medium and feeder cultures on microspore embryogenesis in *Brassica napus* L. cv. Topas. Plant Cell Rep. **8**: 594-597.

Kott, L.S., L. Polsoni and W.D. Beversdorf, 1988. Cytological aspects of isolated microspore culture of *Brassica napus*. Can.J.Bot. **66**: 1658-1664.

Kuginuki, Y., T. Miyajima, H. Masuda, K. Hida and M. Hirai, 1999. Highly regenerative cultivars in microspore culture in *Brassica oleracea* L.var. *capitata*. Breed.Sci. **49**: 251-256.
Lichter, R., 1982. Induction of haploid plants from isolated pollen of *Brassica napus*. Z.Pflanzenphysiol. **105**: 427-434.
Lionneton, E., W. Beuret, C. Delaitre, S. Ochatt and M. Rancillac, 2001. Improved microspore culture and doubled-haploid plant regeneration in the brown condiment mustard (*Brassica juncea*). Plant Cell Rep. **20**: 126-130.
Murashige, T. and F. Skoog, 1962. A revised medium for rapid growth and bioassays with tobacco tissue cultures. Physiol.Plant. **15**: 473-497.
Pechan, P.M. and W.A. Keller, 1988. Identification of potentially embryogenic microspores in *Brassica napus*. Physiol.Plant. **74**: 377-384.
Pechan, P.M. and W.A. Keller, 1989. Induction of microspore embryogenesis in *Brassica napus* L. by gamma irradiation and ethanol stress. *In Vitro* Cell.Dev.Biol. **25**: 1073-1074.
Sato, T., T. Nishio, and M. Hirai, 1989. Plant regeneration from isolated microspore cultures of Chinese cabbage (*Brassica campestris* spp. *pekinensis*). Plant Cell Rep. **8**: 486-488.
Takahata, Y., Y. Takani and N. Kaizuma, 1993. Determination of microspore population to obtain high frequency embryogenesis in broccoli (*Brassica oleracea* L.). Plant Tissue Culture Lett. **10**: 49-53.

2.31
Protocol for microspore culture in Brassica

M. Hansen
Agricultural University of Norway, Department of Horticulture and Crop Sciences, P.O Box 5022, N-1432 Ås. Norway

Introduction

We have been using microspore culture of Brassica at the Institute of Horticulture and Crop Sciences, Agricultural University of Norway, since 1989, our focus has been on *Brassica oleracea* (cabbage), *Brassica napus* (swede) and *Brassica rapa* (chinese cabbage and summer turnip rapeseed). The protocol presented here is based on over 12 years of experience. In the very beginning we visited Centre for Plant Breeding and Research, Wageningen, The Netherlands, the Horticultural Research International, Wellesbourne, UK and University of Guelph, Canada to discuss an optimal protocol for microspore culture in *Brassica* and our protocol includes components from all three places.

Protocol

Essential equipment and tools for microspore culture in Brassica
- growth room for donor plants
- laminar flow cabinet
- centrifuge
- incubators
- light microscope
- inverted microscope
- pH meter
- equipment for filter sterilization of media
- tea baskets for disinfecting flower buds
- jars with autoclaved water (the kind used for jam, honey etc.)
- hand-held pipetter
- sterile 10 ml pipettes
- sterile pistons from 20 ml disposable syringes
- autoclaved 100 ml Erlenmeyer flasks supplied with a 45 µm nylon filter in a funnel and wrapped in aluminium foil
- sterile 100 ml beakers
- Petri dishes, 45x20 mm
- Petri dishes, 90x20 mm
- autoclaved filter paper.

Media
The microspore culture medium NLN is used for washing microspores as well as for culturing. During the last 6 years, we have used the ready-made powder purchased from Duchefa, the Netherlands. This medium consists of one bottle of micro and macro elements and one bottle of a vitamin mixture for preparation of one litre NLN medium. The two bottles are mixed with one litre of deionized water. In addition, 500 mg $Ca(NO_3)_2 \times 4H_2O$ and 130 g (13%) sucrose need to be added, and the final solution is adjusted to pH 6.0. The NLN medium is filter sterilized through a 0.22 µm filter unit. NLN medium with 13% sucrose is usually referred to as NLN-13 medium.

We use solid Gamborg B5 medium to germinate microspore derived embryos. Again, this medium is prepared from mixes purchased from Duchefa, the Netherlands. The powder for one litre medium is supplemented with 20 g sucrose (2%) and 8 g agar. After adjusting the final medium to one litre with deionized water, it is adjusted to pH 6.0 and autoclaved.

Donor plants and growth conditions
Plants of cabbage, swede and chinese cabbage are grown in plastic pots (20 cm diameter) under glasshouse conditions to the 10 leaf stage. We grow one plant in each pot. At this stage, plants are vernalized at 5°C for 90 days under a 16 h photoperiod with approximately 60 µmol m^{-2} s^{-1} photosynthetically active radiation. The vernalized plants are then grown to flowering in a growth room under a 16 h photoperiod with approximately 200 µmol m^{-2} s^{-1} photosynthetically active radiation at 15°C in the light and 10°C in the dark. Seeds of summer turnip rapeseed are sown directly in 20 cm plastic pots (five plants in each pot) and grown to flowering in the growth room under a 16 hr photoperiod with approximately 200 µmol m^{-2} s^{-1} photosynthetically active radiation at 15°C in the light and 10°C in the dark.

We use a commercial fertilized soil mix (macro and micro elements) that consists of 84% sphagnum peat moss, 10% sand and 6% clay. Plants are watered once a week with a liquid fertilizer (N:P:K – 20:10:20). We avoid using pesticides or fungicides because these treatments affect the responsiveness to microspore culture. To prevent mildew in the growth room it is important to keep relative humidity as low as possible. To keep the donor plants healthy, it is important to remove old yellowing leaves from plants.

Preparation of microspore culture suspension from 50 buds
Cellular activity in microspores
It is important that the metabolic activity of microspores is as low as possible from the time inflorescences are harvested from the donor plants until the microspore suspension is placed in the incubator for heat shock. The temperature should therefore be held as low as possible during all steps in the isolation of microspores. This is made possible if the buds are kept on ice during the measuring of buds and if the sterilizing solution, rinsing water and NLN-medium are kept at 4°C (fridge temperature). In addition, all the steps during the procedure should be done quickly and with no break.

Harvesting and measuring of buds
One of the most important factors for a successful microspore culture is the choice of buds. The inflorescences are harvested from the donor plants in a beaker placed in a polystyrene box with ice and brought to the lab. The operation from harvesting to the extraction of buds should be done as quickly as possible. Fifty buds of desirable bud length are selected and placed in a beaker on ice.

The buds for microspore culture should contain microspores at the mid uninucleate and late uninucleate stage. In the mid uninucleate state the microspores are round and have developed a thick exine with the nucleus in the centre. During development to the late uninucleate stage, the nucleus migrates to the cell wall and the exine becomes more yellow. To decide the correct size of microspores, anthers can be squeezed in a drop of water and examined under a light microscope. The distinction between early uninucleate and mid uninucleate can be seen by the thickness of the exine while the distinction between mid uninucleate and late uninucleate can be seen by the thick yellow exine. It is also possible to see the nucleus and locate its position.

Using the growing conditions described above, anthers of cabbage ('Hawke') contain microspores at the late-nuclear stage when the buds are 6.5–7.0 mm in length, swede ('Gry') when the buds are 4.5-5.5 mm, chinese cabbage ('Kasumi') when the buds are 2.5-3.0 mm and summer turnip seed rape (CR25) when the buds are about 2.0-2.5 mm.

Sterilization and rinsing
Buds are transferred to a tea basket and placed in the sterilizing solution (5 g Dichloro-isocyanuric acid dissolved in 300 ml water and supplied with 4 drops of Tween 20) for 6 minutes. The sterilizing solution is placed in the laminar flow cabinet and is at about 4°C. The tea basket should be stirred gently two or three times in the solution to avoid air bubbles. The tea baskets are transferred from the sterilizing solution to three jars in turn containing sterile deionized water at 4°C, for rinsing.

Macerating
The sterile buds are transferred from the tea basket to a 100 ml sterile beaker and 5 ml NLN medium is added. The microspores are released onto the medium by squeezing the buds gently with a piston taken from a 20 ml disposable syringe. If this operation is done too hard, microspores will be damaged and embryo yield will be lowered.

Filtering
Another 15 ml of NLN medium are added to the beaker and the contents of the beaker are poured through a funnel covered with two layers of a nylon mesh placed in a sterile 150 ml Erlenmeyer flask. The funnel is then washed by adding additional 20 ml of NLN medium.

This operation adds up to 40 ml filtrate that is divided among four 10 ml volume centrifuge tubes.

Centrifugation
Tubes are placed in a centrifuge and centrifuged at 190 g for 3 minutes. The grey green supernatant after the first spin is decanted away and 10 ml fresh NLN medium is added to the pellet. After a second centrifugation at 190 g for 3 minutes, the cloudy supernatant is decanted and 10 ml fresh NLN medium is added to the pellet. After the third and last spin at 190 g for 3 minutes the supernatant is clear.

Adjust concentration
After the last centrifugation the supernatant is decanted away and the pellet is resuspended in NLN medium. The desirable plating density is about 40,000–50,000 microspores per ml NLN medium. The density of microspores is determined using a haemocytometer. We have the experience from cabbage and swede that if we adjust the concentration to one bud to one ml NLN medium we will get a final microspore suspension of approximately 40,000 – 50,000 per ml medium.

Plating
The microspore suspension is dispensed 5 ml/plate (45x20 mm Petri dishes) using the handheld pipetter. The Petri dishes are wrapped with a single layer of Parafilm.

Incubation
Heat shock
The Petri dishes are put in an incubator for heat shock in dark. The optimal heat shock for embryo production varies between species and genotypes. Experiments have shown that optimal heat shock for the cabbage 'Hawke' is 30°C for 48 h while the cabbage 'Apex' produces most embryos if the heat shock is 32°C for 48 h. For most genotypes of turnip rape, 32.5°C for 48 h seems to be the best heat shock treatment.

Incubation at 24°C
After heat shock, plates are moved to an incubator or a tissue culture room at 24°C in darkness. Ten days after microspore isolation the plates are placed on a gyratory shaker (70 rpm, gentle shaking) at 24°C in the dark. The first embryos are visible by eye 10 days after isolation and they grow very fast when the shaking starts. After 21 days, the embryos are counted and the embryo quality is evaluated.

Procedure for germination of embryos
Medium
After 21 days, the old NLN-13 medium is removed from Petri dishes using a hand-held pipetter and fresh NLN-13 medium with 5 mg abscisic acid (ABA) per litre is added. Petri dishes are then placed on the gyratory shaker (70 rpm) at 24°C in dark for 24 h.

Desiccation on filter paper
After 24 h on the shaker, the NLN medium is removed and the embryos are transferred to new Petri dishes (90 mm x 20 mm) supplied with one layer of an autoclaved filter paper (Whatman no. 40). Each Petri dish can accommodate 200-300 embryos. Dishes are then sealed with one layer of Parafilm and placed in the tissue culture room (or an incubator) at 24°C in dark for 4 weeks. A very slow desiccation takes place. Fresh embryos taken from the NLN medium have water content of about 90%. After two weeks of desiccation on filter paper, the water content is about 50% and after four weeks is it about 10–15%. During the

Germination on Gamborg B5 medium
After 3-4 weeks of desiccation treatment, we begin the germination procedure by transferring embryos to 90x20 mm Petri dishes (15 embryos per dish) containing solidified Gamborg's B5 medium. The dishes are placed in tissue culture room at 22°C with a 16 h light period and a light intensity of 150 µmol $m^{-2} s^{-1}$. After the transfer to B5 medium, embryos regain the size they were before desiccation. In some cases the cotyledons turned greenish, but in the most cases the cotyledons retained the white/yellowish in colour. The embryos first form roots, but later (within 2-3 weeks) a normal shoot develops from the growing point in the centre of cotyledon (Figure 2.31-1).

Figure 2.31-2. Germinating desiccated embryos from turnip rapeseed on B5 medium It is tree weeks since the desiccated embryos were put on the medium and the plantlets are ready to be put into soil (Photo: Astrid Sivertsen).

Embryos by mail
Experiments have shown that desiccated embryos remain viable for several weeks after the start of desiccation. This procedure for desiccating of embryos has given us the opportunity of sending desiccated embryos from our laboratory to breeders at other locations. In research collaboration between our laboratory and scientists in Slovenia and the Netherlands, it has been possible to send desiccated embryos very simple through normal mail. We post embryos in sealed Petri dishes that are wrapped and packed into a sturdy parcel tolerant of rough handling.

Transfer to soil
When plantlets have developed a well-established root system, they are removed from the petri dishes and planted in Jiffy 7 pots. Pots are then placed in a tray covered with a transparent lid to maintain a high humidity and kept in an illuminated growth room at 20°C. Over the course of a week, the lids are lifted to allow more and more airflow. After two

weeks in the Jiffy7 pots, plantlets are transferred to 9 cm pots filled with the same soil used for donor plants, and pots are placed in the glasshouse.

Efficiency and Applications

There is a large variation in responses to microspore culture in the different *Brassica* species; there is also within species variation. In white cabbage (*B. oleracea* ssp. *capitata*), the most responsive variety we have tested is Hawke which in some experiments can produce an embryo yield approaching 10,000 embryos per 100 buds. Most varieties of swede (*B. napus* ssp. *rapifera*) seem to be responsive, and the Norwegian variety 'Gry' yields up to 10,000 embryos per 100 isolated buds. Responsiveness in *Brassica rapa*, generally, is lower than for cabbage and swede. In Chinese cabbage (*B. rapa* ssp *pekinensis*), the most responsive variety we have tested is 'Kasumi' (1,500 embryos per 100 buds). Summer turnip rape (*B. rapa* ssp. *oleifera*) is relatively recalcitrant to microspore culture. In a screen of 25 varieties of turnip rapeseed, the most responsive was the Canadian variety 'CR25' (1,400 embryos per 100 buds) and the Bangladeshi variety 'PT303' (2,000 embryos per 100 buds), but 18 varieties gave poor results and yielded less than 100 embryos per 100 isolated buds. In a selection of microspore derived lines from CR25, line 'HL14' yielded more than 14,000 embryos per 100 buds.

The desiccation of embryos has been a significant step forward in the production and deployment of microspore embryogenesis in Brassica species. Since we developed this protocol for germination the practical work with embryos has become much easier.

Selected references for microspore culture in Brassica

Gland, A., R. Lichter and H.G. Schweiger, 1988. Genetic and exogenous factors affecting embryogenesis in isolated microspore cultures of *Brassica napus* L. J.Plant Physiol. **132**: 613-617.
Guo, Y.-D. and S. Pulli, 1996. High-frequency embryogenesis in *Brassica campestris* microspore culture. Plant Cell Tiss.Org.Cult. **46**: 219-225.
Hansen, M., 2000. ABA treatment and desiccation of microspore-derived embryos of cabbage (*Brassica oleracea* ssp. *capitata* L.) improves plant development. J.Plant Physiol. **156**: 164-167.
Hansen, M. and K. Svinnset, 1993. Microspore culture of swede (*Brassica napus* ssp. *rapifera*) and the effects of fresh and conditioned media. Plant Cell Rep. **12**: 496-500.
Lichter, R., 1982. Induction of haploid plants from isolated pollen of *Brassica napus*. Z.Pflanzenphysiol. **105**: 427-434.
Rudolf, K., B. Bohanec and M. Hansen, 1999. Microspore culture of white cabbage, *Brassica oleracea* var. *capitata* L.: genetic improvement of non-responsive cultivars and effect of genome doubling agents. Plant Breed. **118**: 237-241.

2.32
Anther and microspore culture in tobacco

A. Touraev and E. Heberle-Bors
Vienna Biocenter, Institute of Microbiology & Genetics, Vienna University, Dr. Bohrgasse 9, A-1030, Vienna, Austria

Introduction

Serving as the oldest model species, several variants of tobacco anther and microspore culture have been developed over the years, depending on the timing of the inductive stress treatment and microspore isolation (Heberle-Bors *et al.*, 1996). A convenient, one-step procedure of anther culture in *Nicotiana tabacum* consists of a cold-treatment of excised flower buds, followed by anther excision and anther culture on a charcoal-containing medium. For microspore culture, sucrose and nitrogen starvation of isolated microspores in combination with a mild heat shock induces the formation of embryogenic pollen which, after transfer to a simple, sucrose and nitrogen-containing liquid medium, divide repeatedly and produce large numbers of embryos (sporophytic pathway, Touraev *et al.*, 1996). Under non-stress conditions in a rich medium isolated microspores develop into mature, fertile pollen (gametophytic pathway, Benito Moreno *et al.*, 1988; Touraev and Heberle-Bors, 1999). The usually haploid microspore embryos are transferred to a low-sucrose agar medium for germination and treated with colchicine for diploidization.

Protocol

Essential equipment and supplies
Equipment
- laminar air flow cabinet
- autoclave
- two incubators or growth rooms (25° and 33°C)
- growth room, 16 h light, 25°C
- microwave oven
- clinical centrifuges and centrifuge tubes
- growth chambers for donor plants (optional)
- light microscope with fluorescence lamp
- inverted microscope
- domestic fridge and freezer
- balance
- pH meter
- millipore (or Corning) filtration units (0.2 μm membrane)
- magnetic stirrer.

Glassware, culture vessels and miscellaneous items
- glass beakers (for 100-1000 ml), flasks (100-500 ml), funnels (10-1000 ml)
- 6-, 12- and 24-well plastic Petri dishes (tissue culture tested from Falcon or Corning)
- glass vials with cap (17 ml, 26 mm diameter, Merck), and 12 ml tipped and sterile plastic centrifugation tubes (Kabe Labortechnik, Nürnberg, Germany)
- magnetic rod (18 mm in length) that can move freely on bottom of the glass vial
- metal sieve (60 μm pore size, 30 mm diameter, Sigma) that fits the top of a 100 ml Erlenmayer flask with a wide neck
- Parafilm (American National Can.)
- Pasteur pipette (plastic, disposable from local producer)
- Gilson pipetman (for 20, 200 and 1000 μl)
- plastic Petri dishes (tissue culture treated, Falcon or Corning) 100x15; 100x20; 60x15 and 35x10 mm.

Preparations
Glassware and distilled water are sterilized by autoclaving at 120°C for 20 min. Concentrated media stocks containing salts and organic supplements can be prepared in advance and stored in a freezer for several months. Media for washing and culture of microspores must be filter sterilized.

Donor plants and growth conditions
Many tobacco genotypes and species, including *N. tabacum* and *Nicotiana rustica* L., have been shown to be good donor plants for microspore cultures. Emphasis should be on conditions for profuse flowering and high male fertility (good quality pollen). Here we give protocols established in our laboratory for *N. tabacum* variety 'Petit Havana SR1'. Donor plants are grown in a growth chamber (25°C, 16 h light) with regular supply of fertilizers and routine watering. Continuous flowering can be achieved by regular harvest of open flowers. Thus one generation of tobacco plants can be used for 4-6 months without significant decrease in microspore culture response.

Flower bud sterilization
1. Flower buds with an approximate length of 10-11 mm representing uninucleate microspores (which can be confirmed by DAPI staining) are excised. Diamidino-2-phenylindole (DAPI) is a specific fluorescent stain to visualize DNA. It can be purchased

as a ready to use solution from Partec (Germany) or as a powder from Sigma. The powder is dissolved in buffer (Partec solution, Germany) or in 50% ethanol. The nuclei of tobacco microspore, stained with DAPI, can be observed with a range of UV filter sets under a fluorescent microscope. Alternatively, staging can be performed by aceto carmine or aceto orcein staining, but the sensitivity of these stains is weaker than DAPI.

2. The buds are sterilized by dipping them into 70% ethanol for a maximum of 30 seconds. In general, surface sterilization of buds with 70% ethanol is sufficient, but sometimes sterilization in 5% sodium hypochlorite solution (Sigma) for 5- 10 min can be used after the ethanol treatment.

Flower bud pre-treatment for anther cultures
Flower buds are cold pre-treated for approximately 12 days at 7-8°C prior to culture by placing them in a refrigerator at a location where the indicated temperature can be kept constant. Buds are placed inside the 10 cm Petri dishes, sealed with Parafilm to maintain the humidity.

Anther isolation and culture
1. Tobacco anthers are gently squeezed out of the flower buds (through a cut made with a scalpel at the bud tips) by pressing from the base of the flower towards the ring of anthers close to the tip of the flower. The anthers can be squeezed directly onto the surface of an agar solidified (0.6 to 0.8% w/v) culture medium. They should be evenly distributed on the surface of the agar with sterilized forceps and placed with the filament side pointing to the agar. Manipulations should be kept to a minimum to avoid damage to the anther. Such damage results in early browning and often prevents plantlet formation. The anthers can also be squeezed out of the bud directly onto the surface of a liquid medium where they should float and not allowed to sink to the bottom of the dish. The four (if one anther has been used for staging) or five anthers of a single bud are cultured as a unit to avoid spread of contamination. The medium recommended is MS medium (Murashige and Skoog, 1962) supplemented with 2% (w/v) sucrose and 1% (w/v) activated charcoal.

2. The anther-containing Petri dishes are sealed with Parafilm and incubated for 1.5-2 months at 25°C in dim light.

3. Plantlets with cotyledons and primary roots are transferred onto an agar solidified MS medium (Murashige and Skoog, 1962) with 1% activated charcoal and 1% sucrose in a Petri dish or, as single plantlets, in a test tube and incubated in the light for further growth. Light is essential in this step for the production of chlorophyll and for normal growth. Continuous illumination from cool-white fluorescent lamps (300 lux) gives satisfactory results.

Anther cultures can be performed with any flowering tobacco plant available and without cold treatment of the excised flower buds. Yields will be lower though.

Microspore isolation
1. Under sterile conditions, anthers are squeezed out of the flower buds into a glass vial (17 ml), containing approximately 3 ml of medium B (Table 2.32-1). Medium B is simple, cheap, osmotically well balanced and used in our laboratory for all tobacco microspore isolation procedures. Depending on skill and isolation procedure used, the viability of isolated microspores should be 90-95% with this medium. Anthers from 10 buds are squeezed into a vial.

2. The medium, which might have the trace amounts of the sterilization agents, is removed and replaced by 6 ml fresh medium B.

3. A magnetic bar is placed into the vial and stirred for 2-3 min at maximum speed until the mixture of microspores/anther debris and medium becomes milky. The duration and time of stirring depends on the vial used and on the type of stirrer and can be easily optimized empirically. The magnet bar should move freely inside the glass vial. In case of small scale isolations (less than 5 buds) anther maceration with a glass rod can be used successfully and in case of large scale isolations (more than 50 buds) a Waring blender is the better choice. The isolation procedure using a Waring blender is essentially similar to wheat, barley or rapeseed isolation procedures described in this manual.

4. A suspension of released microspores and anther debris are collected with a Pasteur pipette and filtered through a 60 µm metal sieve. Uninucleate and binucleate tobacco microspores have average sizes of 20-23 µm, and the pore size of the sieve used for filtration can vary from 40 to 60 µm.

5. The filtrate is centrifuged for 2-3 min at 250 g. Microspores can resist centrifugation speed even higher than 600 g but high speed may affect viability of the microspores. The supernatant is removed together with the top green layer of the two-layered pellet, which contains anther wall debris, by using a 200 µl or 1000 µl Gilson pipette. The microspores form pellet much faster than the small anther debris, and therefore one can see two clear pellet layers (one greenish, one whitish) after centrifugation. The greenish layer can be removed easily after some practice.

6. The whitish pellet consisting of microspores is resuspended in 2-10 ml medium B and centrifuged again. This procedure should be repeated at least 2-3 times until there is no greenish pellet layer above the whitish pellet.

Table 2.32-1. B medium (Kyo and Harada, 1986)

Components	Concentration mg/L	Molarity (mM)
KCl	1,49	20
$CaCl_2$ x $2H_2O$	147	1
$MgSO_4$ x $7H_2O$	250	1
Mannitol	54,700	300
K_2HPO_4/KH_2PO_4		1
pH	7.0	

The medium is filter sterilized

Induction of microspore embryogenesis and regeneration of haploid tobacco plants
1. The final whitish pellet is resuspended in medium B (microspores from one flower bud in 2 ml medium B gives the optimal density for a good response) and poured into small Petri dishes (35x10 mm); 1.5 ml suspension per dish. Microspore density is an important parameter for optimal induction of embryogenesis. The calculation of optimal density for tobacco microspores is given by Garrido et al. (1991) and found to be: microspores from one bud to 2-4 ml medium or approximately 50,000 cells per ml.

2. Petri dishes are sealed with Parafilm and incubated at 33°C for 5-6 days in the dark. This is the key step to block the gametophytic pathway of development and divert

microspores to the sporophytic pathway. Embryogenic microspores at the end of the starvation treatment should have increased size and show a "star-like" phenotype under the inverted microscope with the nucleus being in a central position as compared to the starting stage. Cytoplasmic strands should radiate out from the nucleus through the vacuoles.

3. The microspore suspension is collected from Petri dishes and placed into 12 ml test tubes and pelleted by centrifugation at 250 g for 5 min. The formed embryogenic microspores are very sensitive to damage, and low speed must be used to pellet them gently.

4. The supernatant is discarded, the pellet is resuspended in AT3 medium (Table 2.32-2) and plated back to the original Petri dish in 1.0 ml per Petri dish. The Petri dishes are sealed with Parafilm and incubated for 1.5-2 months at 25°C in the dark.

5. The developing plantlets are transferred onto agar solidified MS medium with activated charcoal and 1% sucrose, and incubated in the light.

Table 2.32-2. AT3 medium (Touraev and Heberle-Bors, 1999)

Components	Concentration mg/L	Molarity (mM)
KNO_3	1,950	19
$(NH_4)_2SO_4$	277	2
KH_2PO_4	400	2.9
$CaCl_2 \times 2H_2O$	166	1.1
$MgSO_4 \times 7H_2O$	185	0.7
Fe-EDTA	10 ml from a 3.67 g/L stock solution	
B-5 vitamins	1 ml from 1000x stock solution	
MS microsalts	1 ml from 1000x stock solution	
MES*	1,950	10
Glutamine	1,256	8.5
Maltose	90,000	250
pH	6.2	

* 2-[N-morpholino] ethansulphonic acid. The medium is filter sterilized

Diploidization of tobacco haploids
Approximately 10-12% of plants regenerated *via* microspore embryogenesis are spontaneous diploids. In order to diploidize the remaining haploid plants, young plantlets are soaked in an aqueous solution of colchicine (Sigma, USA) (0.5% w/v) for 24-48 h. Following the colchicine treatment, plantlets are rinsed with distilled water and grown further until flowering. Ploidy can be determined by using flow cytometry measurements of the DNA content of the regenerated plants or by counting the chromosome numbers of mitotic cells stained by acetocarmine or DAPI.

References

Benito Moreno, R. M., F. Macke, M. T. Hauser, A. Alwen and E. Heberle-Bors, 1988. Sporophytes and male gametophytes from *in vitro* cultured, immature tobacco pollen. In: Sexual Reproduction in Higher Plants. Cresti, M., P. Gori and E. Pacini (Eds.) Springer, Heidelberg, New York, pp.137-142.
Garrido, D., B. Charvat, R.M. Benito Moreno, A. Alwen, O. Vicente and E. Heberle-Bors, 1991. Pollen culture for haploid plant production in tobacco. In: A Laboratory Guide for Cellular and

Molecular Plant Biology. Negrutiu,I. and G.Gharti-Chhetri (Eds.), Birkhäuser, Basel, pp. 59-69.

Heberle-Bors, E., E. Stöger, A. Touraev, V. Zarsky and O. Vicente, 1996. *In vitro* pollen cultures: progress and perspectives. In: Pollen Biotechnology. Gene Expression and Allergene Characterization. Mohapatra, S.S. and R.B. Knox (Eds.), Chapman and Hall, New York, pp. 85-109.

Kyo, M. and H. Harada, 1986. Control of the developmental pathway of tobacco pollen *in vitro*. Planta **168**: 427-432.

Murashige, T. and F. Skoog, 1962. A revised medium for rapid growth and bioassays with tobacco tissue cultures. Physiol.Plant. **15**: 473-497.

Touraev, A. and E. Heberle-Bors, 1999. Microspore embryogenesis and *in vitro* pollen maturation in tobacco. In: Methods in Molecular Biology. Vol. 111. Hall, R.D. (Ed.) Humana Press Inc., Totowa, NJ. pp. 281-291.

Touraev, A., A. Indrianto, I. Wratschko, O. Vicente and E. Heberle-Bors, 1996. Efficient microspore embryogenesis in wheat (*Triticum aestivum* L.) induced by starvation at high temperature. Sex.Plant Reprod. **9**: 209-215.

2.33
Haploid production of potatoes by anther culture

G.C.C. Tai and X.Y. Xiong
Potato Research Centre, Fredericton, New Brunswick, Canada

Introduction

The major cultivated species of the potato, *Solanum tuberosum* L. ssp. *tuberosum* is an autotetraploid ($2n=4x=48$). Other cultivated species often used in genetics and breeding are: the tetraploid species andigena (*S. tuberosum* L. spp. *andigena* Hawkes), the diploid species ($2n=2x=24$) stenotomum (*S. stemotomum* Juz. Et Buk.) and phureja (*S. phureja* Juz. Et Buk.). Chacoense (*S. chacoense* Bitt.), a wild diploid species, is often used in genetic studies and anther culture. There are a large number of wild species with ploidy levels ranging from diploid to hexaploid ($2n=6x=72$). Of the 176 species surveyed by chromosome counts, 73% are diploid, 15% are tetraploid, and 6% are hexaploid. The rest are hybrid species that are triploid or pentaploid. Several major gene banks maintain the world collections of potato species and varieteis. The German-Netherlands Potato Genebank in Braunschweig, Germany, and the USA Inter-regional Potato Introduction Project have published up-to-date inventories. Systematic evaluation of germplasm collections are carried out by the International Potato Center, the German-Netherlands Potato Genebank, and the USA Inter-regional Potato Introduction Project. The tetraploid varieties can be used to produce dihaploid progenies ($2n=2x=24$). Dihaploids provide many advantages for genetic analysis and breeding work. They are used to mate with cultivated or wild diploid species. The diploid hybrid progenies are screened for superior performance of agronomic traits and, more importantly, for their ability to produce unreduced $2n$ gametes. The $2n$ gametes are formed due to the failure of either the first or the second division during meiosis. The first division restitution (FDR) and second division restitution (SDR) gametes have fundamentally different consequences on genetic segregation. The FDR gametes are capable of maintaining about 80% of the heterozygosity of a parent and thus are beneficial in maintaining heterosis in the progenies when it is used to cross with other tetraploid parents. Hybrid progenies from $4x \times 2x$ matings are now produced in many breeding programs. Monohaploids ($2n=x=12$) of potato can also be obtained from the tetraploid varieteies. Two successive cycles of chromosome number reductions are involved. A tetraploid is reduced to a dihaploid. The dihaploid is further reduced to a monohaploid. Monohaploids are also obtained from diploid species and interspecific hybrids. Monohaploids are useful tools for critical genetic analysis of the diploid *S. tuberosum* species

(Jacobsen and Ramanna, 1994). Monoploids provide the cytological evidence of the basic chromosome number ($x=12$) of the potato genome. Monohaploid progenies are used to screen out undesirable and lethal genes and identify beneficial mutants. Doubling chromosomes of monohaploids lead to the production homozygous diploids and tetraploids. The hemizygous condition of monohaploids also facilitate mapping of molecular markers due to the simple (1:1) segregation ratio. Both di- and monohaploids can be produced by two methods: parthenogenesis and androgenesis. Production of dihaploids by parthenogenesis is a well established procedure and used in potato breeding to generate parents for $4x$ x $2x$ crosses. Anther or microspare culture appears to be the preferred procedure to obtain a large number of monohaploids because the number of microspores far exceeds the number of ovules in an ovary (Jacobsen and Ramanna, 1994).

Protocol

Donor plants and growth conditions
A great deal of research work on androgenesis of potatoes in the literature is on the diploid materials. The protocol described here is mainly based on the experimental reports using the diploids as anther donors. Anther donor plants are maintained as clones. They are multiplied from seed tubers or by stem cuttings in a greenhouse in the winter or in the field during the summer months. For greenhouse stock plants, tubers are planted in pots of appropriate size, e.g. 2 L peat pots containing composted soil. Plants are watered daily. Water soluble fertilizer with approximately 20:19:18, $N:P_2O_5:K_2O$ ratio is used once every two weeks. Greenhouse temperature is maintained between 17-20°C (night) and 23-26°C (day). In winter months, the natural day length was supplemented with a 1:3 combination of incandescent and cool white fluorescent lamps for a total light intensity of 1,000 $\mu E\ m^{-2}\ s^{-1}$ in a 16-18 h photoperiod. High humidity (70%) should be maintained in the greenhouse by wetting the greenhouse floor to reduce the flower abscission. Grafting the plants onto tomato root-stocks can prolong the flowering period.

Flower bud selection and treatment
The flower buds are harvested from selected plants when the microspores are at the uninucleate or early binucleate stages. This normally occurs when the buds are 4-6 mm long. Aceto carmine (4% carmine in 45% acetic acid) can be used to stain the nuclei in microspores to check on the developmental stages. The unopened flower buds are placed in plastic vials (5 ml) that have 1-2 mm holes punched in the sides. They are wrapped in moist paper towels and treated at 4-6°C in a refrigerator for 72 h. The buds are surface sterilized by immersion in 70% ethanol for 30 seconds followed by 15 minutes in 10% Javex-12 (commercial bleach containing 10.85% (w/w) sodium hypochlorite) and rinsed twice in sterilized distilled water. The buds are then aseptically dissected under a microscope to excise the anthers.

Anther culture and induction of embryos and calli
Excised anthers are placed in induction medium for the induction of embryos and calli. Three liquid and one semi-solid media are given in the Table 2.33-1 (Aziz et al., 1999). Fig. 2.33-1 illustrates the anther culture procedure as well as anther derived embryo and plantlets emerging from callus.

Procedure

Uhr85 (Uhrig, 1985) and MSU93 (Meyer *et al.*, 1993) liquid media
Fifty anthers per 25 ml of medium are placed in 50 ml Erlemeyer flasks, which are capped with non-absorbent cotton plugs and wrapped with aluminum foil. The cultures are shaken at 26°C with 120 rpm in an incubator. A lowlight intensity of approximately 1 $\mu E\ m^{-2}\ s^{-1}$ with 16 h day length is used. Embryos are harvested after 4 to 8 weeks.

S&V95 (Shen and Veilleux, 1995) liquid medium
Thirty anthers are distributed in 125 ml Delong culture flasks (Bellco Glass Co., Vineland, NJ) with 15 ml liquid medium. The flasks are covered with a Magenta 2-way cap (Magenta Plastics, Chicago, Ill.) and sealed with Parafilm. They are subjected to pre-incubation temperature at 35°C for 12 h and then incubated at 30°C for 16 h and 20°C for 8 h/day over 6 weeks. Cultures are maintained in the dark on a shaker with 120 rpm. Embryos are harvested after 4-6 weeks.

RVP95 semi-solid medium (Rokka *et al.*, 1995)
20 ml medium is distributed in 90 mm (diameter) sterile plastic Petri dishes. Thirty anthers are placed on the medium and the Petri dishes are sealed with Parafilm. The cultures are maintained at 28°C for 4 weeks and then decreased at 24°C, with light intensity of $50\mu E m^{-1} s^{-1}$ and 16 h daylength. Embryos are formed after 8 weeks.

pH of all media are adjusted to pH5.8 and are sterilized by autoclaving. Out of four induction media listed in the Table 2.33-1 we would recommend S&V95 induction medium, which gave the best results in our work.

Table 2.33-1. Composition of media for anther culture of potatoes

Constituents	Induction media				Morphogenic medium (mg/L)	Plantlet medium (mg/L)
	Uhr85 (mg/L)	MSU93 (mg/L)	S&V95 (mg/L)	RVP95 (mg/L)		
MS basal salts[1]	2,160	2,160	2,160	4,310	2,160	2,160
Sucrose	60,000	60,000	60,000	60,000	20,000	20,000
Charcoal	500	500	2,500	5,000	-	-
Wheat starch	-	-	-	30,000	-	-
Agar	-	-	-	-	8,000	8,000
myo-Inositol	50	50	100	-	-	50
Thiamine HCL	0.2	0.2	0.4	-	0.2	0.2
Ascorbic acid	-	-	-0	200	-	-0
Glutamine	-	7.3	-	-	-	-
Asparagine	-	6.6	-	-	-	-
Lysine HCL	-	-	-	10	-	-
IAA	0.1	0.1	0.1	-	-	-
BA	2.5	1	2.5	3	-	-
GA$_3$	-	-	-	-	0.1	-

[1] MS basal salts (Murashige and Skoog, 1962) - Sigma M5524

Induction of morphogenesis
The harvested embryos and/or calli from liquid medium are plated on a morphogenic medium in 60 mm (diameter) single well Petri dishes. The composition is listed in Table 2.33-1. The

medium is solidified with 8 g/L agar. The cultures are maintained in a 16 h day in a culture room and kept at temperature 24°C.

Culture of plantlets
Plantlets developed from the morphogenic medium are excised and maintained on a planlet medium (Table 2.33-1) in Magenta GA-7 vessel. The plantlet medium promotes root and shoots formation.

Checking ploidy level of plantlets
Ploidy level of plantlets can be checked by three methods: chromosome counting, chloroplast counting in the guard cells, and flow cytometry. The most accurate method is to count the chromosomes. Root tips of the plantlets are immersed in 0.002 M aqueous solution of 8-hydroxyquinoline or ice-water for 18-24 h. The root tips are then maintained at 4°C in pre-chilled fixative (1 part 85% lactic acid: 99 parts 30% ethanol) for at least 4 h. The tips are then hydrolysed for 6-7 min in 1N HCl at 60°C. After hydrolysis the material is blotted and flooded with Schiff's reagent for 30 minutes. On a glass slide, the root tips are stained with lacto-proprionic-orcein (0.9% orcein in 45% proprionic-lactic acid in equal parts). The tips are macerated, put under a cover slip and an even pressure applied. Chromosomes are counted under a microscope. For chloroplast counting a piece of the lower epidermis is removed and examined under light microscope. Plantlets with reduced ploidy have fewer numbers of chroloplasts in the guard cells than that of the donors.

Chromosome doubling
Young plants (25-30 cm tall) are used for colchicine treatment. Twenty-four hours before the colchicine is applied, the axillary buds inside the leaf axils, along the stem, are carefully removed by a razor blade or ground-down forceps. Leaf axils without axillary buds or pseudostiples are also removed. On average, six leaf axils per plant are kept. Some of the terminal and outmost lateral leaflets can be cut so that the stem could be enclosed in a polyethylene bag. Oblong pellet of absorbent cotton is firmly placed in a leaf axil. A strip of adhesive tape 3-4 mm wide is centered over the pellet and wrapped around the stem. The concentration of colchicine solution is between 0.25-0.5%. It is applied to the top of the cotton pellet with an eye dropper until the cotton is saturated. The plant is immediately enclosed in a polythylene bag and placed in a shady area in the greenhouse for a predetermined length of treatment. The bag and the pellets are then removed at the end of treatment. The plant is labelled and rinsed with water spray and grown on a greenhouse bench. The shoot grown from a treated sub-axillary meristem is cut when it reached 6-8 cm. It is treated with a rooting hormone (indole-3 butyric acid 0.8%, Stim-Root No.1, Plant Products Co. Ltd., Bramalea, Ontario, Canada), grown in moist vermiculite to root, and then transplanted into a small pot (7-10 cm) of soil to tuberize. The chromosomes are counted using root tips to determine if doubling had occurred.

Efficiency and Applications

Diploid genotypes originated from complex crosses of *S. tuberosum* dihaploids with diploid wild species were reported to generate on Uhr85 or MSU93 media, both monohaploid and diploid plants with an average rate of about 27 plants/100 anthers. Over 80% were diploid.

Diploid genotypes derived from crosses between dihaploid *S. andigena* and F_1 hybrids from diploid species (*S. phureja, S. berthaultii, S. chacoense, S. microdontum*) were used for anther culture on S&V95 liquid medium. The regeneration frequency ranged from 1-18% from both first and second harvests of the cultures. Anther culture of somatic hybrid between *S. tuberosum* dihaploid and the wild diploid species *S. brevident,* with the use of RVP95 semi-solid medium has generated about 1 shoot/100anthers.

Regenerants from anther culture vary in ploidy level. The diploid donors often yield between 20-60% monohaploids, most of the rest are diploids and some are tetraploids. This may be caused by chromosome doubling of reduced microspores during culture, embryogenesis of a somatic cell or androgenesis of unreduced microspores. Anther culturing of potatoes is genotype dependent. Another key factor determining the degree of success is the composition of the culture medium. Only a few of the successful media for anther culture of potatoes in the literature are listed here in the protocol. Other key factors are: the proper physiological condition of the anther donor plants, the proper developmental stage of microspores inside the anthers, the low (4-5°C for 72 h) or high (35°C for 12 h) temperature pretreatments, or the elevated incubation temperature (30/20°C) shock applied to the flower buds prior to anther culture, the incubation temperature and proper light (cool-white F20T12/cw, fluorescent lamps) during anther culture. There are other media for plant regeneration from anthers of diploid (Cappadocia *et al.*, 1984) and tetraploid potatoes (Tiainen, 1993). 2,4-D, zeatin, GA_3 and abscisic acid (ABA) are not effective in promoting embryogenesis. Auxin-lacking media or media with a low concentration (0.1 mg/L) of 3-indoleacetic acid (IAA) favour the induction of embryogenesis. Presence of 2,4-D in the normal sucrose-containing medium is required for callus initiation.

Benefits:
- Rapid production of homozygous lines may be equal to several generations of inbreeding.
- Reduction of ploidy level leads to the elimination of undesired and lethal genes from the progenies in a genetic population. Generation of monohaploids provide the most effective tool to achieve this.
- Anther-derived plants provide materials with reduced ploidy levels for mapping of molecular markers. Mapping of monohaploids, for example, avoids the problem of dominance in the segregation of gene markers.
- Dihaploids are used to cross with diploid species. Diploid hybrid species are then used in *4x* x *2x* mating in the potato breeding program after screening the diploid hybrids ($2n=2x=24$) for ability to produce *2n* gametes. These plants are used to cross with the tetraploid *S. tuberosum* cultivars ($2n=4x=48$) to produce *4x* progenies.

References

Aziz, A.N., J.E.A. Seabrook, G.C.C. Tai and H. De Jong, 1999. Screening diploid *Solanum* genotypes responsive to different anther culture conditions and ploidy assessment of anther-derived roots and plantlets. Amer.J.Potato Research **76**: 9-16.

Cappadocia, M., D.S.K. Cheng and R. Ludlum-Simonette, 1984. Plant regeneration from *in vitro* culture of anthers of *Solanum chacoense* Bitt. and interspecific diploid hybrids *S. tuberosum* L. x *S. chacoense* Bitt. Theor.Appl.Genet. **69**: 139-143.

Jacobsen, E. and M.S. Ramanna, 1994. Production of monohaploids of *Solanum tuberosum* L. and their use in genetics, molecular biology and breeding. In: Potato Genetics. Bradshaw, J.E. and G.R. Mackay (Eds.), CAB International, Cambridge, pp. 155-170.

Meyer, R., F. Salamini and H. Uhrig, 1993. Isolation and characterization of potato diploid clones generating a high frequency of monohaploid or homozygous diploid androgenic plants. Theor.Appl.Genet. **85**: 905-912.

Murashige, T. and F. Skoog, 1962. A revised medium for rapid growth and bioassays with tobacco tissue cultures. Physiol.Plant. **15**: 473-497.

Rokka, V.M., J.P.T. Valkonen and E. Pehu, 1995. Production and characterion of haploids derived from somatic hybrids between *Solanum brevidens* and *S. tuberosum* through anther culture. Plant Sci. **112**: 85-95.

Shen, L.Y. and R.E. Veilleux, 1995. Effect of temperature shock and elevated incubation temperature on androgenic embryos yield of diploid potato. Plant Cell Tiss.Org.Cult. **43**: 29-35.
Tiainen, T., 1993. The influence of hormones on anther culture response of tetraploid potato (*Solanum tuberosum* L.). Plant Sci. **88**: 83-93.
Uhrig, H., 1985. Genetic selection and liquid medium conditions improve the yield of androgenetic plants from diploid potatoes. Theor.Appl.Genet. **71**: 455-460.

2.34
Anther culture through direct embryogenesis in a genetically diverse range of potato (*Solanum*) species and their interspecific and intergeneric hybrids

V.-M. Rokka
MTT Agrifood Research Finland, Plant Production Research, Crops and Biotechnology, FIN-31600 Jokioinen, Finland

Introduction

The present anther culture procedure has been utilized successfully for a genetically diverse range of potato material including commercial varieties of cultivated potato (*Solanum tuberosum* ssp. *tuberosum*), its related wild species and various hybrids having wild *Solanum* spp. germplasm. Previously, cultivated potato has been described as a recalcitrant crop plant with regard to androgenesis, and progress in haploid production through direct embryogenesis using anther culture has been considerably limited. Therefore parthenogenesis through pollination with selected clones of *S. phureja* is still widely utilized for haploid production in potato. However, wild characteristics derived from the *S. phureja* pollinator and aneusomatic genomic constitutions in the resultant haploid lines are common when parthenogenesis (gynogenic pathway) is carried out for practical haploid production in potato. Isolated microspore culture in potato has yet to provide results for plant production. Therefore, *in vitro* androgenesis through direct embryogenesis in anther culture, the method described in the current article, is suggested for haploid production in potato and its relatives. The present anther culture protocol has been conducted on a genetically diverse range of *Solanum* spp. material. The method was originally introduced for dihaploid ($2n=2x=24$) production from tetraploid ($2n=4x=48$) varieties of cultivated potato (Rokka *et al.*, 1996), but since then it has also been utilized for: tetraploid breeding material with wild potato germplasm; dihaploid lines of cultivated potato; *S. acaule* ssp. *acaule* wild potato species (Rokka *et al.*, 1998); interspecific hybrids between A genome potato species (*S. tuberosum* and *S. acaule*); tetraploid and hexaploid interspecific hybrids between A and E genome species [*S. tuberosum* x *S. brevidens* hybrids, Rokka *et al.* (1995); *S. tuberosum* x *S. etuberosum* hybrids, V-M. Rokka, R. Thieme and T. Gavrilenko, unpublished results]; and intergeneric tomato (*Lycopersicon esculentum* + *Solanum etuberosum*) somatic hybrids (Gavrilenko *et al.*, 2001).

Protocol

Donor plants and growth conditions

Potato material from tubers
The anther donor plants are grown from tubers or from *in vitro* plantlets transferred to the greenhouse. Tubers sprouted in daylight at RT (room temperature) are planted in 21 cm diameter pots, and allowed to emerge in dry compost. Following emergence, plantlets are watered, and at each watering treated with 1% Superex 5 (Kekkilä) nutrient solution. Certificated tuber material is preferable as donor material for anther culture of potato to avoid contamination by endogenous bacteria and fungi during tissue culture. Field grown potato material can also be used for anther culture, but contamination during the tissue culture regime may be high.

Potato material from in vitro
In vitro propagated plant material is preferred as anther-donor plants when contamination of anther cultures is high or when the tubers are very small, weak, old or unavailable (some wild species and interspecific hybrids are non-tuberous). However, when *in vitro* material is used, the plants normally require about one extra month to produce flower buds. The number of buds developed also can be considerably fewer in such donor plants when compared to those grown from tuber derived donor plants. *In vitro* grown plantlets are multiplied using nodal sections on the following medium:

MS-20 medium
    Murashige and Skoog (MS) basal medium (Sigma, M 5519)
    Sucrose   20 g/L
    Casein hydrolysate (Merck)   100 mg/L
    NAA (BDH)   0.05 mg/L
    Agar (Biokar Diagnostics)   8 g/L
    pH   5.6

After propagation and rooting, the plantlets are transferred from test tubes to compost after carefully washing the agar from the roots. They are planted in wet compost in 8 cm diameter pots, sprayed with water and covered with plastic to maintain high humidity. Once the plantlets are well established, the plastic cover is removed. When the plants have grown to a height of about 5-10 cm in the pots, they are transplanted to 15 cm diameter pots and fed with 1% Superex nutrient solution to aid in the production of flower buds.

All the anther donor plants (from tubers and *in vitro* plantlets) are grown in 65-70% relative humidity under a photoperiod of 16 h. In summer the photoperiod does not need to be controlled, because certain genotypes may actually produce more flower buds under longer day length. Natural light can be supplemented with high pressure sodium lamps (Philips SON-T-AGRO 400) to reach light intensity of 240–360 $\mu E\ m^{-2}\ s^{-1}$. The temperature in the greenhouse is set at 12°C night and 17°C day. However, daytime temperature may vary, but should be maintained below 25°C. The plants should be maintained insect-free during development. However, insecticides must not be used, because they reduce flower bud formation and response to embryogenesis.

Analysis of the developmental stage of the microspores
When flower buds are harvested for anther culture, the microspores in the anthers must be at the uninucleate stage to achieve optimum embryo yields. It is also preferable that flower buds are collected in the morning for each experiment to encourage uniformity of the developmental stage.

The length of anthers can be measured with a ruler or millimetre paper. The buds should be harvested from *S. tuberosum* genotypes when they are of 4-6 mm in length, and anthers are 3-4 mm long. At this stage, petals are not or only slightly visible in most genotypes. In certain genotypes, such as *S. tuberosum* variety 'Torridon', the buds should be allowed to grow larger, as the developing rate of anthers is slower in this particular genotype. In *S. acaule*, smaller buds (3-4 mm in length) should be used for anther culture. Flower buds can be stored at 4°C for 1–3 days, but it is preferable to use freshly harvested buds for anther culture.

Staining of microspores
DAPI stock solution
    1 mg DAPI (4'6-diamidino-2-phenylindole) (Sigma)
    diluted in 1 ml dH_2O
    kept in the dark.

Staining buffer
    0.605 g Bistris (Sigma)
    5 g sucrose
    0.5 ml Triton-100
    adjusted to 100 ml with dH_2O.

Final staining solution
    1 µl DAPI stock solution diluted in 1 ml staining buffer.

Anthers are squashed using a small glass rod in a few drops of the staining solution and maintained for 10-15 min in the dark. The developmental stage of the microspores can then be analysed using a microscope with UV light and a 20x or 40x objective lens.

Anther culture
Flower buds are surface sterilized in 20% (2–2.5% active chlorine) sodium hypochlorite (Oy FF Chemicals AB) for 20 min, thereafter treated with 70% ethanol for 2 min and rinsed three times in sterile dH_2O.
    Anthers are isolated using a sharp scalpel (BB 511, No. 11, Aesculap) and tweezers (BD 329, Aesculap). Twenty anthers are transferred per plate onto anther culture medium.

Anther culture medium
    Murashige and Skoog basal medium (MS) (Sigma, M 5519)

Sucrose	60 g/L
L-cysteine (BDH)	50 mg/L
BA (Sigma)	3 mg/L (from a stock solution 1mg/ml)
Wheat starch[1] (Sigma)	60 g/L
Activated charcoal[2] (Merck)	5 g/L
Ascorbic acid[3] (Sigma)	200 mg/L
pH	5.6

[1] Wheat starch is gently boiled separately in 600 ml of H_2O in a water bath while continuously stirring for 1 h.

[2] MS basal medium, sucrose, L-cysteine and BA are diluted to 350 ml and the pH is adjusted with 1 N and 0.1 N KOH, and subsequently activated charcoal is added. The wheat starch solution is then added to the MS solution, and mixed with a glass rod. The volume is then adjusted to 1,000 ml. The medium is sterilized by autoclaving for 20 min.

[3] Ascorbic acid is added filter sterilized after autoclaving. About 20 ml of medium is poured into 60 mm Petri dishes.

After transferring the anthers onto the medium, the cultures are incubated under illumination of 40–85 µE m^{-2} s^{-1}, photoperiod of 16 h. The cultures are incubated at 28°C for four weeks. The number of embryos developed can then be assessed. After the four weeks period, the cultures are incubated at 24°C for a further four weeks.

After a total of eight weeks culture, the numbers of embryos and shoots developed can be assessed. The embryos, which did not develop shoots on the anther culture medium should be transferred to the regeneration medium.

Regeneration medium (Wenzel and Uhrig, 1981)
    Murashige and Skoog basal medium (MS) (Sigma, M 5519)
Sucrose	30 g/L
Coconut water (Sigma)	100 ml/L
Agar (Biokar Diagnostics)	8 g/L
Zeatin[1] (Sigma)	0.3 mg/L (from a stock solution 1 mg/ml)
pH	5.6

[1] Zeatin is filter sterilized and added after autoclaving.

A volume of about 10 ml of medium is poured into 30 mm Petri dishes. The cultures are maintained at 24°C, and the final shoot number is recorded 4–8 weeks after transferring to the regeneration medium. When regenerated, shoots are transferred onto MS-20 medium (see above) in test tubes for shoot elongation and root development at 20°C.

Ploidy level analyses

Ploidy level (reduced or unreduced) can be analysed with similar results using either chromosome counts or flow cytometric determinations, preferably prepared from *in vitro* material. Chromosome counts are analysed from root tips and nuclear DNA content analysed from leaf mesophyll tissue (Rokka *et al.*, 1995, 1998).

Chromosome counts
Chromosome counts are prepared from mitotic cells of root tips grown on MS-20 medium (see above) using the following protocol for root pre-treatment, fixation, hydrolysis and chromosome staining.

Root pretreatment
    0.29 g/L 8-hydroxyquinoline (Merck)
    dissolved in dH$_2$O at 60 °C for 2 h
    can be stored at RT for two weeks.
Roots, grown on MS-20 medium for 5-14 days until 1 cm in length, are pre-treated in the root pre-treatment solution at RT for 3-4 h. The length of pre-treatment may vary according to the genotype analysed. For instance *S. acaule* root tips are pre-treated exactly 4 h to obtain metaphase stage chromosomes visible by microscopy. After pre-treatment roots are transferred to fixation solution.

Fixation
    3.75 ml absolute ethanol + 1.25 ml glacial acetic acid (3:1) is prepared at the time of use.
Roots can be preserved in the fixation solution at 4°C for some weeks.

Hydrolysis
    1 N HCl is preheated to 60°C. The roots are hydrolysed in HCl at 60°C for 7 min.
After hydrolysis, roots are rinsed twice with dH$_2$O and transferred to Feulgen staining solution.

Feulgen staining solution (250 ml)
  0.9 g Fuchsin (Merck)
  4.8 g Na$_2$S$_2$O$_5$
  3.2 ml HCl
Adjust to 250 ml with dH$_2$O. Mix thoroughly. Wrap the bottle with aluminium foil and shake for 24 h. Add 2–2.5 g active charcoal and shake for 2 min. Then pass through filter paper until the solution is clear. Store in the dark at 4°C.

After staining in Feulgen solution, the roots are gently transferred onto a microscope slide and the root tip (length of 1 mm) is cut off using a sharp knife. The Feulgen solution surrounding the root tip is blotted off with a small piece of filter paper, and a drop of acetocarmine (1% carmine in 45% acetic acid) is added. The root tip is squashed in acetocarmine solution with a glass rod (do not let the tip dry up), covered with a cover slip and thereafter pressed by a thumb. The chromosome numbers are counted under a light microscope.

Flow cytometry
Ploidy level (nuclear DNA content analyses) can be determined from *in vitro* grown leaf mesophyll cells using the flow cytometry procedure of Arumuganathan and Earle (1991). About 50 mg of leaf tissue is required and an internal DNA content standard of chicken red blood cells can be used. The samples are stained with propidium iodide and measured with an argon-ion laser at a wavelength of 488 nm. About 1,000-5,000 isolated nuclei are sufficient for DNA content analyses.

Requirements of growing haploid potato plants
After ploidy level analyses, agar is washed from the roots and the plantlets are transferred to wet compost. They are planted in 8 cm diameter pots, sprayed with water and covered with plastic to maintain humidity. When plants have grown to a height of about 5-10 cm, they are transplanted to 15 cm diameter pots. Haploid potatoes are grown in the greenhouse until they have produced tubers. Some haploid genotypes may have poor tuber formation and require a growing period of several months. After harvesting, the tubers are allowed to dry for a couple days and then stored at 4°C until they are planted in the field. The dormancy requirements vary between genotypes. Some genotypes may have a prolonged period of dormancy (six months or more). When haploid lines contain wild species genetic material, the tubers cannot be stored for a long period at 4°C, because their dormancy is very short.

Efficiency and Applications

A relatively high androgenic response has been obtained using the present protocol compared to the previously published articles relating to potato anther culture. However, there are still a high number of potato cultivars that have not shown androgenic capacity. Of 48 tetraploid *S. tuberosum* ssp. *tuberosum* genotypes tested by Rokka *et al.* (1996), 33 produced embryos and 23 of these regenerated shoots. Subsequently, of 19 tetraploid *S. tuberosum* genotypes containing wild potato species germplasm, 11 produced shoots using anther culture. The highest estimated androgenic efficiency in tested *S. tuberosum* genotypes was obtained in potato variety 'Calgary' (24% of anthers cultured led to green shoot formation). A higher than 2% rate of shoot production was also shown in potato varieties 'Concorde', 'Helios', 'Petra', 'Pito', 'Rustica', Torridon, 'Van Gogh' and 'White Lady', but the efficiency often varies from one experiment to another (Rokka *et al.*, 1996).

Many wild potato species and their hybrids show a greater response to shoot formation than does *S. tuberosum*. In *S. acaule* accession 'PI 472655', 9% of anthers isolated formed green shoots. Less than 2% androgenic response was obtained in *S. acaule* acc. 'PI 208856' while *S. acaule* acc. 'PI 473515' did not show androgenic competence at all. Hexaploid ($2n=6x$) interspecific somatic hybrids of potato showed higher androgenic response than the tetraploid ($2n=4x$) somatic hybrids, but the trait was strongly genotype dependent (Rokka *et al.*, 1995). The presented anther culture method was also applied to amphihaploid production from amphidiploid hybrids containing the tomato genome resulting in the first successful direct embryogenesis in anther culture of tomatoes. A total of six shoots were obtained from the described *L. esculentum* + *S. etuberosum* somatic hybrids (Gavrilenko *et al.*, 2001).

The frequency of haploids obtained from anther culture of potatoes is also highly dependent on the genotype utilized. In many cases, haploids are difficult to recognize from anther derived plants having unreduced chromosome composition. Therefore, ploidy level determination using cytological or nuclear DNA content analyses are suggested. Specific tetraploid *S. tuberosum* cultivars, such as varieties Calgary, 'Cicero', 'Satu' and White Lady produce almost entirely regenerants with a reduced ploidy (dihaploids), but variety. Pito and Torridon mainly form tetraploid androgenic regenerants (Rokka *et al.*, 1996). In those specific cultivars that commonly regenerate into plants with unreduced ploidy, the first regenerated plants are mainly tetraploids. Thus, dihaploids seem to have a slower regeneration rate compared to the tetraploids.

Monohaploids ($2n=x=12$) are extremely difficult to obtain in potato. Testing secondary anther culture capacity in *S. tuberosum* did not result in monohaploids, but instead, in diploid, triploid or tetraploid plants. However, androgenic plants regenerated from interspecific hybrids have generally had a reduced ploidy level (Rokka *et al.*, 1995; Gavrilenko *et al.*, 2001).

Dihaploids of *S. tuberosum* obtained with anther culture can be utilized in practical potato breeding to produce intraspecific somatic hybrids. Haploid lines derived from interspecific and intergeneric somatic hybrids are also relevant research materials for inheritance of resistance traits expressed in the somatic hybrids (Rokka *et al.*, 1995) and for cytological determination of the transmission of parental species chromosomes into microspores and haploid plants (Gavrilenko *et al.*, 2001).

Acknowledgements

The author would like to acknowledge Ms. Jill Middlefell-Williams for the critical review of this manuscript.

References

Arumuganathan, K. and E.D. Earle, 1991. Estimation of nuclear DNA content of plants by flow cytometry. Plant Mol.Biol.Rep. **9**: 229-231.

Gavrilenko, T., R. Thieme and V.M. Rokka, 2001. Cytogenetic analysis of *Lycopersicon esculentum* (+) *Solanum etuberosum* somatic hybrids and their androgenetic regenerants. Theor.Appl.Genet. **103**: 231-239.

Rokka, V.M., C.A. Ishimaru, N.L.V. Lapitan and E. Pehu, 1998. Production of androgenic dihaploid lines of the disomic tetraploid potato species *Solanum acaule* ssp. *acaule*. Plant Cell Rep. **18**(1-2): 89-93.

Rokka, V.M., L. Pietilä and E. Pehu, 1996. Enhanced production of dihaploid lines via anther culture of tetraploid potato (*Solanum tuberosum* ssp. *tuberosum*) clones. Am.Potato J. **73**: 1-12.

Rokka, V.M., J.P.T. Valkonen and E. Pehu, 1995. Production and characterization of haploids derived from somatic hybrids between *Solanum brevidens* and *S. tuberosum* through anther culture. Plant Sci. **112**: 85-95.

Wenzel, G. and H. Uhrig, 1981. Breeding for nematode and virus resistance in potato via anther culture. Theor.Appl.Genet. **59**: 333-340.

2.35
Potato haploid technologies

M.J. De, Maine
Scottish Crop Research Institute, Invergowrie, Dundee, DD2 5DA, UK

Introduction

The term 'dihaploid' is used to describe a plant which contains the haploid number of somatic chromosomes but which contains two genomes per somatic cell. The common potato *Solanum tuberosum* L. has 48 chromosomes per somatic cell ($2n=4x=48$), so that the number of chromosomes in the egg cell and in haploid somatic cells is 24. Haploids of *S. tuberosum* have been the subject of widespread study because they promised a means of investigating the genetic architecture of the world's most commercially important broad-leaved crop. As the common potato is tetraploid, genetic investigations using it involve much more complex numbers than those arising from the use of diploid plants. Dihaploids offered a means of breeding the common potato at the diploid level. It is important to note, however, that the tetraploid state is generally regarded as the physiologically most efficient ploidy level for potato. Tetraploid *S. tuberosum* potatoes have commercial yields that far outweigh those of their dihaploid counterparts. It is therefore necessary to have technologies that can both generate haploid plants and regain the tetraploid chromosome level.

Protocol

Induction of haploids from tetraploid potatoes by means of inter-species crosses

Dihaploids can be produced by inter-ploidy crosses where the $4x$ female is *S. tuberosum* and the $2x$ pollinator is *S. phureja*. The $4x$ female produces an embryo sac containing an egg cell and two endosperm nuclei, all with the genetic constitution $1n=2x=24$. The $2x$ pollinator produces two sperms ($1n=1x=12$). During normal fertilization in potato one sperm fuses with an egg cell to produce a zygote and the remaining sperm fuses with the secondary endosperm nucleus, itself formed by fusion of the two haploid (primary) endosperm nuclei just prior to fertilization. If this was a $4x$ x $4x$ cross the result would be a $4x$ zygote and a $6x$ endosperm.

In a $4x$ x $2x$ cross, this would be expected to result in a $3x$ zygote and a $5x$ endosperm nucleus. This can occur but a $6x$ endosperm, formed by both x sperms fusing with the secondary endosperm nucleus ($4x$) can result and the resulting endosperm tissue is able to support development of an embryo from an unfertilized egg cell ($2x$). *S. tuberosum* dihaploids from $4x$ x $2x$ crosses therefore are matrilineal. The pollinator does not contribute to the embryo except for a very small amount of DNA, which has been detected in some dihaploids.

A further consequence of $4x$ x $2x$ crosses is $4x$ hybrids with $6x$ endosperm. These are caused by meiotic chromosome doubling in the diploid parent, to form $2x$ gametes that then carry out normal fertilization of the $2x$ egg and $4x$ endosperm nuclei in the tetraploid parent.

The production of dihaploid seed is low and steps must be taken to maximise the yield from the potato flowers. The female has a significant effect on dihaploid induction but the pollinator has an even greater effect. The choice of pollinator is critical to success and, using an efficient pollinator, most *S. tuberosum* plants that flower will yield some dihaploids. The most efficient pollinators for dihaploid induction are the IVP series of *S. phureja*, bred at the Agricultural University, Wageningen, Netherlands Those which have given the best results are IVP35, IVP48 and IVP101. They can be obtained directly from Wageningen or the Scottish Crop Research Institute. These have also been bred to be homozygous dominant for purple embryo spot marker. The spot will show up on all seeds whose embryos possess a genome from the pollinator. As such spotted seeds cannot be dihaploid, they can be discarded without further testing.

Flower production

Dihaploid induction is a rare event and seeds are produced at a low frequency. This can average 20% of all seeds with only 7% of total seeds giving rise to viable dihaploid plants. These figures are given as a guide, as dihaploid frequency can vary greatly between $4x$ genotypes. Photosynthate source-sink relationships can be manipulated to encourage flowering. The over-riding demand for photosynthate after tuber initiation is tuber filling at the expense of all other potential sinks. More flowers can be stimulated to form by removing tubers. There are two main methods of reducing the tuber sink.

Grafting potato scions onto tomato rootstocks

Any variety of tomato will do but F_1 hybrid varieties are more vigorous. First, plant potato tubers of the genotype required for crossing. A month later sow the tomato seeds singly in 100 mm pots filled with good compost. Tomato seed leaves and the first pair of true leaves will open out, about two weeks after seed germination. By this time, the potato should have produced plenty of shoots, which can be used as scions. Carry out a top graft by cutting off the top of the tomato shoot just below the first pair of true leaves, and above the seed leaves. Then slit the stem at right angles to the axis of the seed leaves.

The next stage, which is preparation of the scion, should be carried out as quickly as possible to minimise moisture loss from tomato and the scion. Cut a young stem, about 100 mm long, from the potato plant. Trim off the lower leaves and pare the cut end to a wedge approximately the same length as the slit you have made in the tomato stem. Insert the scion wedge into the slit and bind in place with water-proof tape such as grafting tape or strips of laboratory film (such as Nesco film). Place the graft in a cool place. If necessary, spray with water from time to time. The scion will almost certainly wilt but should recover the following day. After two weeks, put the grafts on the laboratory bench and water and fertilize as any other plant. High potash liquid fertilizer should help flower production once flowers have initiated. The presence of the tomato rootstock means that potato tubers cannot be formed and flowers become the main sink. Twist off tomato shoots as they appear, allowing only potato stems to grow. Keep the graft well-staked as the graft union is a major weakness.

Picking off potato tubers as they form
The simplest way to remove the tuber sink is to pull the tubers off the plant as they form. To do this, the plant is raised up as it grows, to expose the stolons that form at the base of the stem, but allow the roots to penetrate deep into the culture medium. A good way to do this is to grow potatoes on bricks. Place a house-brick on a bed of compost in a glasshouse and place a long plant support next to it, making sure it is well secured at the top as this stake will have to support what can be a huge top-growth. For the same reason use a soil-containing compost (such as John Innes Number 2) to improve root anchorage. Also, put a label with the potato's number or name on the stake. Place a large potato tuber or several small ones on the brick and cover with soil. Water as usual over the following weeks. After about one month shoots will emerge from the tuber. As they grow, train them up the stake. From time to time check the base of the plant for tuber formation. Pull off the tubers and discard them or store if necessary for further use. As the roots penetrate the compost below the brick, pull some of the compost off the tuber to expose the stolons a little and make it easier to check for tubers. Each week pull off a little more compost. The plants may well wilt at first but will recover. When the roots are well into the lower compost and there is a sizeable top-growth, the remaining compost can be washed off the tuber and the brick with a garden hose. Continue to harvest tubers as long as they are formed. In compensation for the loss of the major sink (the tubers) top-growth will accelerate and flowers will be produced in increased numbers compared to conventionally grown plants

Pollination

Emasculation
The female (*S. tuberosum*) flower must be emasculated before its anthers dehisce as many tetraploid *S. tuberosum* plants are self-fertile. To avoid self-pollination it is best to emasculate before the bud opens but when the petals have their final colour. Green buds are too young to emasculate and the flower will usually drop before setting seed. Use a mounted needle or fine forceps to bend each anther back on its filament and break it off, or grasp the filament at the base and pluck it out of the flower. As dihaploid induction is a rare event, target several whole trusses of flowers for emasculation and pollination with the inducer.

Pollination
One or two days after emasculation, the petals should have begun to open up and the flower is ready for pollination. Extract pollen from the inducer (eg. *S. phureja* IVP35, 48 or 101) by running a flat Borrodaile needle down the inside of each of the four longitudinal pollen sacs on an anther. Wipe the pollen on to the stigma of the emasculated flower. Alternatively shake anthers in a Petri dish to extract the pollen and dip into this with the needle. It is possible to store pollen in a freezer along with silica gel as a desiccant. It will retain some efficacy for several months, but always use pollen that is as fresh as possible since its germinability reduces with storage time. An efficient method of extraction that can be used to provide dry pollen for storage or for rapid pollination is as follows:

Pull the anthers from harvested flowers and place them in a suitable-sized Petri dish (ensure you pick out any styles which may fall off into the dish as their stigmata produce a surprising amount of sugary liquid overnight to which the pollen will stick). Place the Petri dish, minus its lid in a larger container with a layer of silica gel at the bottom. Seal the larger container and leave it on a laboratory bench overnight. The following day remove the Petri dish containing the dried anthers, put the lid on and shake for a minute. Flick the dried anthers out of the Petri dish with forceps and collect the extracted dry pollen with a small spatula or clean, dry finger and store in a labelled Eppendorf tube for storage or use. Sterilize the spatula or finger in absolute alcohol between genotypes.

The use of cut flowers
If glasshouse space is limited but field space is available, potatoes for crossing can be grown in field plots. These will give flowers for pollen and flowers on the stem for use as female parents in the glasshouse. Cut stems of flowers from the plants and trim off the bottom leaves, allowing at least one upper leaf to remain to lift water up the stem by transpiration. Place two or three flowering stems in each container of water (holding about 500 ml). Do not be tempted to put lots of stems into large containers as rotting is a common problem with this method and the more bunches can be kept separate to reduce its spread, the better. A little potassium permanganate solution can be added to keep bacteria at bay. Some experimentation on the actual amount to use is required to get a balance between killing bacteria and not killing the potato flowers. Regional differences in the levels of chlorination of available water can have an effect. Cutting flower stems appears to help stop flower drop. Cutting or even constricting the stem is thought to stop photosynthate draining out of the flower so that it is available for development of the berry (McLean and Stevensen, 1952).

Cut flower stems can be fed by adding a proprietary liquid feed to the water each week or 0.5 ml doses of Shive's three salt solution (see formula below) can be added to each 500 ml container. Again, experimentation is required to get a balance between plant feed being used to support the flowers and the control of bacterial rot.

Shive's three salt solution
KH_2PO_4 24.5 g/L
$Ca(NO_3)_2$ 8.5 g/L
$MgSO_4$ 1.8 g/L
$CuSO_4$, $MnSO_4$, $FeSO_4$ trace amounts
Add 0.5 ml to 500 ml water.

Identification of dihaploids
All the techniques for identifying dihaploids described below are useful and should be used together for best results. Chromosome counting is considered to be the best single test but it is easy to make mistakes in counting the potato's small chromosomes. The best possible test of a dihaploid, if it flowers, is whether it sets plenty of seed (i.e., hundreds of seeds per berry) in crosses with other known, highly fertile diploids (usually *S. phureja* used as a pollen parent).

As stated above, if one of the IVP series of dihaploid inducers is the pollinator, the purple embryo spot can be used to identify hybrid seed that can be discarded. Dihaploid seed will be spotless. This is the first stage of checking dihaploids. Observe plants as they grow because it is possible to get imperfectly expressed embryo spots in seeds. In the seedling the genes governing embryo spot also determine a purple nodal band on the stem. You can use this to discard hybrids at the seedling stage if you miss seeing the spot in the seed.

Incomplete or late emasculation can result in self-pollination of the *S. tuberosum* parent. The seed resulting from self-pollination will also be spotless. Dihaploids are expected to be slow-growing and tetraploids vigorous but, in practice, it can be difficult to distinguish between tetraploid selfs and dihaploids. Mutants may be common in some $4x$ selfed progenies and some dihaploids are surprisingly strong-growing. Take precautions to avoid selfs but also keep the seeds extracted from different berries separate in case you produce a mixture of selfed and non-selfed berries. The presence of spotted seeds in a berry is reassuring, indicating that you do not have selfs. Dihaploids, in common with naturally-occurring diploid species such as *S. phureja*, tend to have leaflets which are narrower, thinner and lighter green than those of tetraploid *S. tuberosum*.

Cytological tests for identifying dihaploids
Chloroplast counts
The chloroplast number of epidermal stomatal guard cells is an indicator of ploidy in potatoes (Frandsen, 1967). Strips of epidermis are peeled from the abaxial surface of a leaflet, placed in a drop of water and covered (but not squashed) with a cover-slip. The total number of chloroplasts in the pair of guard cells is counted. When carrying out chloroplast counts ensure that epidermis is available from a leaflet of the putative dihaploid parent for comparison. The ratio of chloroplast numbers per pair of stomatal guard cells of a dihaploid plant to one with double the somatic chromosome number is 0.67:1. A parent with chloroplast numbers averaging 21 will yield dihaploids averaging 14 chloroplasts per guard cell pair. Guard cell chloroplast numbers vary from stomata to stomata. There is an overlap in the range between ploidy levels. Count at least ten guard cell pairs, if possible twenty. There will be occasional stomata containing large numbers of chloroplasts. These outliers should be avoided.

Chromosome counts
Counting chloroplast numbers is easy and fast but it still may not be possible to be sure whether you have correctly identified a dihaploid, because of the overlap between the ranges seen in $2x$ and $4x$ plants. Also it should be noted that triploids can occur and may be very common in the offspring of some genotypes. It is difficult to distinguish a triploid from a diploid or a tetraploid. Counting chromosome numbers in somatic cells should give the required answer but potato has small chromosomes and the preparation of cytological specimens is critical to obtaining good counts.

Root tips are used as a source of somatic cells for chromosome counting. To obtain actively dividing root tips avoid over-watering and use young vigorously growing plants. Also avoid a compost containing sand as sand grains can adhere to the roots resulting in slide or cover slip breakages when the preparation is squashed. Harvest time is also important. It should preferably be at the early mid-growing season (May to June in the UK), sunny but not hot and late morning to early afternoon is best. Choose white, hairless roots and collect 3 to 6 roots per plant and put them in chilled water in Eppendorf tubes kept in ice. Store in a refrigerator overnight, then fix by transferring the root tips to a mixture of 3:1, ethyl alcohol: glacial acetic acid for 24 h. Macerate the root tips by transferring them to watch glasses of 5 M HCl at room temperature for 45 min. Rinse the roots three times in distilled water, then transfer them to watch glasses containing Schiff's Reagent. Put the watch-glasses in a box and it is important to seal the box with a lid after which they can be left until the roots develop a pink colour (up to 2 hours). Place a root on a microscope slide and cut off the tip about 1 mm from the end. Using a dissecting microscope, pull off and discard as much of the root-tip sheath as possible with a mounted needle. Add a little 45% aqueous acetic acid if necessary then squash the remaining ball of dividing cells under a cover slip and tap out gently to disperse the cells into a layer one cell thick. Squash the cells as much as possible. At this stage you are trying to spread the chromosomes in the cells. Examine the slide under a microscope for mitotic stages where the chromosomes are well separated and can be counted. Count chromosomes in three to six cells per preparation.

Chromosome doubling
There are two main methods of somatic cell chromosome doubling. One involves the use of colchicine and the other relies on spontaneous mitotic irregularities that occur during tissue culture.

Use of colchicine
Colchicine treatment can be carried out *in vivo* and *in vitro*. Colchicine is a toxic substance and care should be taken when working with it, such as dispensing the powder in a fume cupboard and wearing gloves when handling the solutions.

In vivo method
Dionne's method (Ross *et al.*, 1967) involves the application of colchicine solution to meristematic tissue in potato scions. For this to work you must first get rid of all actively growing meristematic stem tissue. This means you must remove or prevent the development of the stolons and all terminal and axillary buds. Use a dihaploid potato scion grafted onto a tomato root-stock to avoid the stolon meristems developing. Then all terminal and axillary buds must be removed and the tertiary meristems at the base of the pseudostipules at each node treated with colchicine. Shoots will grow out from the bases of the pseudostipules. These are rooted and ploidy-checked (De,Maine and Fantes, 1983).

In vitro method with colchicine treatment
If the appropriate facilities are available, a better colchicine method is to treat tertiary shoot meristems of aseptically-grown micropropagated plants *in vitro* (DeMaine and Simpson, 1999). Under sterile conditions cut the stem of a plant into pieces, each containing a node. Plate out on Basal Medium (i.e., Murashige and Skoog nutrient mixture, with micro and macro elements and vitamins, 3% sucrose, 0.8% agar, pH 5.6) in 9 cm diameter Petri dishes and allow the bud to produce a shoot 10-15 mm long, after about eight days. Excise the bud using a mounted needle and discard it. Next immerse the de-budded stem segments in a solution of 0.5% colchicine and place in a shaker for 24 or 48 h. After this time rinse off the solution with at least 6 changes of sterile, distilled water and transfer the nodes to Petri dishes of Basal Medium. Harvest the shoots that grow from the sub-axillary meristems when at least 25 mm in length and root them in fresh medium. Use root-tip chromosome counts to identify plants with doubled ploidy.

We found that this method was better than the following. By using a sub-lethal colchicine dose you obtain few products but what you obtain are most likely to be doubled. In the next method described you may obtain many plants but can have difficulty identifying dihaploid plants from the mass of material.

In vitro methods
Callus culture
Ploidy doubling occurs spontaneously during potato tissue culture. To induce doubling, callus must be produced and then shoots initiated from it. The shoots are excised from the callus, rooted and allowed to develop into plants *in vitro* before being transferred to compost in a glasshouse. There are several different culture methods (De,Maine and Simpson, 1999) and genotype has a significant effect on success. The medium for inducing callus from tissue explants is usually different from that which is used to stimulate shoot growth from callus. The tissue explants may be pieces of internodal stem, leaf discs or tuber discs.

Protoplast fusion
Another method of regaining the tetraploid level for *S. tuberosum* dihaploids is to fuse somatic cell protoplasts of one selected $2x$ genotype with those of another. The protocol for this, described by Cooper-Bland *et al.* (1994) has been used successfully at the Scottish Crop

Research Institute. Following protoplast fusion the resulting cells are placed on callus-forming medium. At this stage there could be chromosome doubling of unfused protoplasts so the tetraploid products which finally emerge will contain a mixture of hybrids and non-hybrids. The hybrids can only be identified if the putative parents carry different markers that can be seen in the end-products.

Sexual polyploidization
Diploid potatoes may form unreduced male and female gametes (Mendiburu *et al.*, 1974). The frequency varies with genotype and is usually less than 10%. However you can exploit natural selection for unreduced gametes by carrying out interploidy ($4x \times 2x$ and $2x \times 4x$) crosses. As there is a triploid block, most of the seed produced from interploidy crosses is tetraploid. Teteraploid hybrids can arise from carrying out the cross in either direction. There is usually a lack of male fertility in *S. tuberosum* dihaploids, so $2x \times 4x$ crosses are usually carried out. In $4x \times 2x$ crosses, as crossing is taking place within *S tuberosum*, there are usually no obvious markers to indicate that tetraploid offspring are hybrids or selfs, so meticulous care has to be taken with emasculation prior to pollination. Tetraploid offspring from a $2x \times 4x$ cross, on the other hand, will be true hybrids. It is possible to obtain tetraploid offspring from $2x \times 2x$ crosses but, because there is no selection pressure for unreduced gametes, these are very rare.

References

Cooper-Bland, S., M.J. De,Maine, M.L.M.H. Fleming, M.S. Phillips, W. Powell and A. Kumar, 1994. Synthesis of intraspecific somatic hybrids of *Solanum tuberosum*: assessments of morphological, biochemical and nematode (*Globodera pallida*) resistance characteristics. J.Exp.Bot. **45**: 1319-1325.

De,Maine, M.J. and J.A. Fantes, 1983. The results of colchicine treatment of dihaploids and their implications regarding efficiency of chromosome doubling and potato histogeny. Potato Research **26**: 289-294.

De,Maine, M.J. and G. Simpson, 1999. Somatic chromosome number doubling of selected potato genotypes using callus culture of the colchicine treatment of shoot nodes *in vitro*. Ann.Appl.Biol. **134**: 125-130.

Frandsen, N.O., 1967. Ploidy chimeras from haploid potato clones. Z.Pflanzenzüchtg. **57**: 123-145.

McLean, J.G. and F.J. Stevensen, 1952. Methods of obtaining seed on Russet Burbank and similar flowering varieties of potatoes. Am. Potato J. **29**: 206-211.

Mendiburu, A.O., S.J. Peloquin and D.W.S. Mok, 1974. Potato breeding with haploids and *2n* gametes. In: Haploids in Higher Plants. Kasha,K.J. (Ed.) University of Guelph, Guelph, pp. 249-258.

Ross, R.W., L.A. Dionne and R.W. Hougas, 1967. Doubling the chromosome number of selected *Solanum* genotypes. European Potato J. **10**: 37-52.

2.36
Anther culture of linseed (*Linum usitatissimum* L.)

K. Nichterlein
Plant Breeding and Genetics Section, Joint FAO/IAEA Division, International Atomic Energy Agency, Vienna, Austria

Introduction

Linum usitatissimum (L.) is grown as linseed (for oil) or flax (for fibre) and is one of the oldest cultivated plants in temperate regions. The oilseed types (linseed or oil flax) compared to the fibre types (flax) have shorter and thicker stems with more branches, whereas flax produces fewer capsules and smaller seeds. It is an autogamous species with about 5-10% cross-fertilization. Therefore, breeders mainly use pedigree selection, bulk breeding or progeny methods to develop new breeding lines and varieties. Traditional breeding methods are time consuming taking 10-15 years to develop improved varieties. Doubled haploid techniques offer an opportunity to accelerate breeding. In linseed, haploidy can be achieved through polyembryonic seedlings or through *in vitro* methods of anther or isolated microspore culture. Polyembryony was reported to be the first source of haploid plants in linseed. The diploid part of the twin is derived from the fertilized egg cell and the haploid probably from one of the synergids of the same embryo sac. Polyembryony is genetically determined and high-twinning genotypes with up to 8% twin seedlings were identified. Hybridization and subsequent selection resulted in an increased frequency of twinning seedlings up to 30%. In order to increase the efficiency of haploid plants independent from the hybridization with twinning lines, anther and microspore culture systems were developed. The original protocol for linseed anther culture and root regeneration (Nichterlein *et al.*, 1991; Nichterlein and Friedt, 1993) was further optimized through modifications of the induction medium, and anther pre-treatments (Chen *et al.*, 1998).

Protocol for anther culture

Donor plans and growth conditions

The protocol below has been developed for European oilseed flax varieties and successfully applied for Canadian oilseed flax varieties. Anther response is high and more reproducible when donor plants are grown in a controlled, stress-free environment, such as in a growth chamber. Before sowing, the growth chamber should be incubated at high humidity (95%) at 25°C for 24 h to stimulate germination of fungal spores present. Thereafter, the chamber is disinfected with a disinfecting detergent (Mucocit-F). Seed are treated with a fungicide (such as Arbosan®) before sowing. The plants are grown in pots in a soil mixture of steamed loess, steamed compost and Perlite (in ratio 5:3:1) with a water capacity of 55%. Each pot is provided with a collecting pan to avoid unequal loss of nutrients. Plants are fertilized three times weekly by using a liquid fertilizer 'Wuxal Super 8:8:6', containing N 8%, P_2O_5 8%, K_2O_5 6% and the following nutrients in mg/L: Fe 190, B 101, Co 4, Cu 82, Zn 61, Mn 161, Mo 10. Plants are grown at 16 h photoperiod with 28,000–30,000 lux (at plant height) provided by high-pressure mercury lamps (Philips HPI-T and Philips SON-T) under 14/8°C day/night regime, 70% humidity and circulating air (Nichterlein et al., 1991). Plants should be kept disease free. In the case of mildew infection, the plants are sprayed with wetted sulphur. The use of systemic fungicides should be avoided; they can negatively affect the *in vitro* response of anthers/microspores.

If a controlled environment is not available, plants of well responding genotypes can be cultivated in pots outside, with weekly application of liquid fertilizer and maintained as stress-free, as possible. However, the overall anther response is expected to be much lower than from plants grown under controlled conditions.

Collection of buds and pre-selection of suitable anthers

Flower buds are cut before flowering, when the microspores are in the late uninucleate stage, preferably from plants which have not yet started flowering (Nichterlein et al., 1991). Depending on the genotype under the proposed culture conditions, bud formation begins 8-10 weeks from sowing. The buds of European oilseed flax varieties under the growth conditions described above, with a diameter of 1.3-2.3 mm, have most of their microspores in the uninucleate stage. However, when using other varieties and/or growth conditions of donor plants, the suitable bud size of each genotype should be established by a cytological examination at the microspore stage. A continuous supply of anthers can be organized by sowing new donor plants every two weeks.

Cytological examination of microspores

Anthers are isolated from buds of different diameters and fixed in the solution of ethanol and propionic acid (3:1). They are carefully squeezed with a lancet on a microscope slide and the fixation solution is allowed to dry. One drop of the iron-alum-hematoxylin solution is added and after one minute a cover slip is applied to the preparation. Heat slightly to intensify the stain, until the stain turns brown and examine under light microscopy. The microspore stage of different bud sizes is assessed. Buds 1.3-2.3 mm in diameter, with the majority of microspores at the late uninucleate stage (the nucleus is located close to the microspore wall), are selected for anther culture.

Hematoxylin staining procedure after Kindiger and Beckett, (1985)
Solution I: Dissolve 2 g hematoxylin in 100ml of 50% propionic acid. The solution can be kept for 1 week in a dark bottle in the refrigerator.
Solution II: Dissolve 0.5 g ferric ammonium sulfate ($FeNH_4(SO_4)_2$) in 100 ml of 50% propionic acid. Mix both solutions in a ratio of 1:1 for cytological examination of microspores.

In vitro protocol
Various steps of the protocol such as sterilization of buds, inoculation of anthers, transfer of cultures to fresh and new culture media have to be carried out in a laminar flow bench to avoid contamination of cultures.

Preparation of culture media
The induction medium is a modified MS medium with reduced ammonium nitrate content 165 g/L, thiamine HCl 10 mg/L, glutamine 750 mg/L, 2,4 D 2 mg/L and NAA 1 mg/L, 6% sucrose and 0.4% agarose (Table 2.36-1). The agarose together with macro and micro nutrients are autoclaved at 121°C for 15 min, other components are filter sterilized (0.2 μm cellulose acetate filter). Ten ml of medium are poured into each sterile plastic Petri dish 60x15 mm (Greiner Labortechnik, Germany or Falcon 3002, Bectin Dickinson, Oxnard, CA, or equivalent).

The regeneration medium is a modified N6 medium with glutamine 375 mg/L, asparagine 250 mg/L, serine 125 mg/L and zeatin 1 mg/L (Table 2.36-1). The macro and micro salts and Gelrite is autoclaved at 121°C for 15 min whereas the other components are filter sterilized. Either sterilized glass jars (250 ml, Weck Glaswerk, No. 100, Bonn, Germany) each containing 50 ml of medium or sterile Magenta vessels (76x76x102 mm) containing 40 ml of medium can be used.

The shoot growth medium is prepared in the same way as the regeneration medium except it is free of growth regulators. The root induction medium is a modified MS medium, with reduced sucrose content, IAA 0.1 mg/l, high Gelrite concentration, and activated charcoal (Table 2.36-1). It is autoclaved and poured into sterilized test tubes (25x150 mm, 10 ml medium/tube).

Sterilization of flower buds
Use pre-sterilized glass Petri dishes and forceps for this procedure. Surface sterilize flower buds with microspores at the late uninucleate stage by immersion in 96% ethanol for 30 seconds and then in 2% sodium hypochlorite solution for 5 minutes, followed by rinsing three times in sterilized deionized water.

Anther culture
Isolate the anthers using a binocular microscope using sterilized forceps and scalpels and a pre-sterilized Petri dish: open the buds, remove filaments and plate the anthers flat on the induction medium with a density of 10 anthers per Petri dish (60x15 mm).

Induction of callus and organogenesis
1. Incubate anthers in the dark for one day at 35°C (Chen *et al.*, 1998). This pre-treatment increases regeneration efficiency in some genotypes, others do not respond to this treatment. Incubate anthers for the remaining period at 25°C. Transfer calli with a diameter of 1-3 mm onto the regeneration medium and cultivate at 27/24°C day/night under 16 h photoperiod in a lit incubator or a culture room with fluorescent light: mixture 1:1 of Philips TL-D 58 W/36 and Osram L 58W/21 or Sylvania cool white 40 W with an irradiance of 4,000 lux. Maintain these culture conditions during the entire *in vitro* period.

2. Transfer calli after induction of shoots onto the shoot growth medium – the same basal medium but without growth regulators. This medium will stimulate shoot elongation and shoots of 3-5 cm length can be cut for rooting.

Table 2.36-1. Composition of linseed anther culture media (Nichterlein et al., 1991; Nichterlein and Friedt, 1993)

Media components	Induction medium (mg/L)	Regeneration medium (mg/L)	Shoot growth medium (mg/L)	Root induction medium (mg/L)
Macro salts				
NH_4NO_3	165	-	-	1,650
KNO_3	1,900	2,830	2,830	1,900
$CaCl_2 \times 2H_2O$	440	166	166	440
$MgSO_4 \times 7H_2O$	370	185	185	370
KH_2PO_4	170	400	400	170
$(NH_4)_2SO_4$	-	463	463	-
Micro salts				
H_3BO_3	6.2	1.6	1.6	6.2
$MnSO_4 \times H_2O$	16.9	3.3	3.3	16.9
$ZnSO_4 \times 7H_2O$	8.6	1.5	1.5	8.6
KJ	0.83	0.8	0.8	0.83
$Na_2MoO_4 \times 2H_2O$	0.25	-	-	0.25
$CuSO_4 \times 5H_2O$	0.025	-	-	0.025
Iron source				
$FeSO_4 \times 7H_2O$	27.8	27.8	27.8	27.8
Na_2EDTA	37.3	37.3	37.3	37.3
Other components				
myo-Inositol	100	-	-	100
Thiamine HCl	10 [a]	1.0	1.0	0.4
Pyridoxine HCl	-	0.5	0.5	-
Nicotinic acid	-	0.5	0.5	-
Ca-Pantothenate	-	0.5	0.5	-
Glycine	-	2.0	2.0	-
Glutamine	750	375	375	-
Asparagine	-	250	250	-
Serine	-	125	125	-
2,4 D	2 [a]	-	-	-
BAP	1 [a]	-	-	-
Zeatin	-	1	-	-
IAA	-	-	-	0.1
Sucrose	60,000	-	-	10,000
Maltose	-	30,000	30,000	-
Silver nitrate	-	-	-	1
Seaplaque agarose	4,000	-	-	-
Gelrite	-	4,000	4,000	6,000
Activated charcoal	-	-	-	1,000*
pH	5.5	5.8	5.8	5.5

*modified by Chen et al., 1998.

3. Transfer calli without shoots every four weeks onto fresh regeneration medium until shoots appear, then follow step 2. Calli not responding with shoot regeneration can be discarded after the third subculture.

Rooting and acclimatization of plants
1. Cut shoots when they reach 3-5 cm and transfer them onto the rooting medium - 1 shoot per tube, 15x150 mm.

2. Transfer rooted shoots into vermiculite for two weeks under an inverted glass jar, then transfer to soil and culture for two weeks under plastic bags in a glasshouse. Alternatively, transfer directly to soil in a mist chamber for one week and then into a growth chamber or glasshouse.

Checking of ploidy level of anther culture derived plantlets
Haploid plantlets often can be recognized by their thinner stems, smaller leaves and at later stages by sterility. For young plants chromosome counting is the most reliable method of assessing the ploidy. In order to stimulate cell divisions, a fertilizer is applied to plants 4-6 days before sampling. Root tips of *in vitro* derived plants are arrested for 3-4 h in 2 mM aqueous solution of 8-hydroxyquinoline. Thereafter they are fixed in ethanol-propionic acid 3:1 (v/v) for 3 h and then stained using 2% orceine propionic acid.

Chromosome doubling
Spontaneous doubling occurs in anther culture at a rate of about 40%. Colchicine treatment may not therefore be necessay. However, for genotypes with low response and lack of sufficient spontaneous doubled haploids, the following procedure can be applied using the uptake of colchicine through roots. Cuttings can be taken to clonally propogate a number of plants for colchicines treatment. Plantlets of 10 cm in size are taken out of soil. The roots are carefully washed and cut back to a length of 4 cm. The roots of plantlets are submerged in a colchicine solution (0.05% colchicine plus 1.5% DMSO) for 8 h, after this they are washed in hand warm tap water for 2.5 h. The roots are cut to a size of 1 cm, the plantlets transferred into fresh soil.

Differentiation of spontaneous doubled haploids from somatic regenerants
Somatic regenerants from hybrids will segregate in the following generation and if hybrid parents have distinctive differences, the progeny of the regenerated plant shows segregation whereas the progeny of the spontaneous doubled haploid does not segregate. Segregating progenies can be discarded. Alternatively, the microspore-derived plants can be identified after successful transfer to soil using molecular markers (Chen *et al.*, 1999)

Efficiency and Applications

The androgenesis in linseed anther and microspore culture predominantly involves the organogenesis pathway in which plant regeneration is achieved through an intermediate callus phase. The controlled growth conditions of donor plants are important for good response in anther culture (Nichterlein *et al.*, 1991). Anther culture response is also genotype dependent affecting callus formation and shoot regeneration. Callus formation and shoot regeneration is relatively high in anther and microspore cultures of the European cultivar 'Atalante' and its hybrids. For Canadian genotypes, various hybrids of the cultivar 'AC McDuff' and some breeding lines also respond well to anther culture. Anther culture efficiency is defined here as the number of anthers producing shoots/100 cultured anthers, the overall response is the number of anthers producing anther derived lines. A culture efficiency and overall response of three F_1 hybrids of Atalante or an Atalante derived doubled haploid line (DH 57-1), was 3.3 to 6.8% and 1.8-4%, respectively (Nichterlein *et al.*, 1991; Friedt *et al.*, 1995). For 44 Canadian genotypes including varieties and advanced breeding lines, the anther efficiency fluctuated from 0.1% to 10% (Chen *et al.*, 1999). Using the original protocol (Nichterlein *et al.*, 1991) and the European hybrids (one parent being the variety

Atalante) as donor plants, the frequency of spontaneous doubled-haploids was 36%. About 50% of the anther culture derived lines were of somatic origin as shown by segregation of their progenies (Friedt et al., 1995). With the modified induction protocol and Canadian hybrids (one parent being the variety AC McDuff) the frequency of microspore derived plants was about 55% from which 38% showed spontaneous chromosome doubling (Chen et al., 1999). Increasing the sucrose content of the induction medium from 6 to 15%, increased the percentage of microspore derived plants from 67 to 94%, but not the overall response due to the lower number of calli forming shoots on the high sucrose medium. With regard to the overall response, there is also a beneficial effect of pre-culturing anthers on a high osmotic medium (15% sucrose) for a period of time, before transferring them on a lower sucrose concentration (Chen et al. 1998).

Microspore culture allows the isolation and culture of haploid cells exclusively, i.e. without the possible contamination by diploid, non-gametic tissue. A protocol is available for microspore culture of linseed and doubled haploid plants can be produced (Nichterlein and Friedt, 1993). In a study comparing anther culture and microspore culture of linseed, it was concluded that the response to anther culture is higher than microspore culture with respect to callus formation. However, the calli from microspore cultures generally show better shoot differentiation (Steiss et al., 1998).

Anther culture has been used for the development of breeding lines. A breeding programme was started to combine high linolenic and high oil content with valuable agronomic traits and various breeding populations were produced. The comparison of small populations of doubled-haploid lines (43) with F_5 pedigree lines (26) of three cross combinations showed a superior performance of some pedigree lines concerning seed yield and disease resistance, whereas in the doubled haploid population the lines with increased oil and linolenic acid content were found (Steiss et al., 1998).

References

Chen, Y., E. Kenaschuk and P. Dribnenki, 1998. High frequency of plant regeneration from anther culture in flax, Linum usitatissimum L. Plant Breed. **117**: 463-467.

Chen, Y., E. Kenaschuk and P. Dribnenki, 1999. Response of flax genotypes to doubled haploid production. Plant Cell Tiss.Org.Cult. **57**: 195-198.

Friedt, W., C Bickert and H. Schaub, 1995. In vitro breeding of high-linolenic, doubled-haploid lines of linseed (Linum usitatissimum L.) via androgenesis. Plant Breed. **114**: 322-326

Kindiger, B. and J.B. Beckett, 1985. A hematoxylin staining procedure for maize pollen grain chromosomes. Stain Technology **60**: 265-269.

Nichterlein, K. and W. Friedt, 1993. Plant regeneration from isolated microspores of linseed (Linum usitatissimum L.). Plant Cell Rep. **12**: 426-430.

Nichterlein, K., H. Umbach and W. Friedt, 1991. Genotypic and exogenous factors affecting shoot regeneration from anther callus of linseed (Linum usitatissimum L.). Euphytica **58**: 157-164.

Steiss, R., A. Schuster and W. Friedt, 1998. Development of linseed for industrial purpose via pedigree-selection and haploid technique. Industrial Crops and Products **7**: 303-309.

2.37
Doubled haploid production of sugar beet
(*Beta vulgaris* L.)

E. Wremerth Weich and M.W. Levall
Syngenta Seeds AB, P.O. Box 302, S-261 23 Landskrona, Sweden

Introduction

The attempts to produce haploid sugar beet plants have embraced miscellaneous techniques, such as natural polyembryony, induction with irradiated pollen, crosses with polyploid plants or wild species, and anther and microspore culture. Most resulted in limited success, but by using a gynogenetic approach Hosemans and Bossoutrot (1983) were able to report successful doubled haploid production. Since then sugar beet breeding companies, as well as public institutes, have been involved in optimizing the ovule culture method. Several publications have presented studies supporting modifications with beneficial effects, reviewed by Pedersen and Keimer (1996). In this paper a robust protocol well suited for high-throughput DH-production of sugar beet is presented.

Protocol

Definitions
Clone: in the protocol, one embryo gives rise to one clone, and the clone consists of about 5 ramets (plants).
Embryo frequency: total number of embryos/total number of ovules.
Shoot frequency: total numbers of shoots/total number of ovules.
Rooting frequency: total number of rooted plants/total number of plants on rooting medium.
Doubling efficiency: total number of DH plants/total number of colchicine-treated plants.

Basic essential equipment
In vitro
- dissecting-microscope (x20), with extensive depth of field
- growth cabinet
- growth chamber (for tissue culture)

Analysis
- light microscope (10x100)

Other
- glasshouse, for donor material and DH-plants
- facilities for storage of disposables, chemicals, waste

Optional
- 1 medium storage cabinet/room (12°C)
- walk-in climate chambers for whole plants.

Donor plants and growth conditions
Differences in embryo response between the lines generally range from 1-15%. A variation of similar magnitude is also evident between different plants within lines (Fig. 2.37-1a,b). Grow the donor plants under controlled conditions in a growth room or glasshouse.

Recommendations
- temperature: 17±3 °C
- lighting: a metal halide light-source e.g. GE Kolorarc 400 W (Hungary) and a photon flux of more than 200 μmol m^{-2} s^{-1}, is beneficial
- light period: a 16/8 h day/night is advisable
- soil: use a commercial soil with a pH close to 7, e.g. based on peat + 10% clay granules, 10% sand, 5% Leca-granules + 1.7 kg/m^3 NPK (11:5:18) + 5 kg/m^3 ground lime and 1 kg/m^3 ground dolomite-lime
- grow plants in ca. 2 dm^3 pots (diameter 16,5 cm top) or similar.
- fertilization: frequent fertilization, e.g. twice a day with solution Bv0 (Table 2.37-1), upon irrigation improves the development and quality of the explant-material.

Note: A pathogen free environment is essential, as e.g. sucking insects increase the risk of endogenous contaminants.

Table 2.37-1. Bv0 solution for fertilization of donor plants and other glasshouse material

Elements	mg/L
N	224
P	43
K	298
Mg	17
Ca	124
S	35
Na	24
Fe	5.8
Mn	1.3
Cu	0.09
Zn	0.38
B	0.38
Mo	0.042
pH	6.1±0.1
Conductivity	2.0 mS

Harvest of inflorescences
Harvest takes place before anthesis in early flowering stage. Due to the within-line variation, it is important to harvest material from several plants, as the reliability of the calculated mean increases with increased number of plants, samples from ≥ 4 donor plants are recommended (Fig. 2.37-1b).
1. Start by harvesting the branches with inflorescences from the donor plant. Remove the flower buds that already have opened. Keep careful track of the seed numbers and individual plant identification.

2. Place the inflorescence branches inside conventional plastic bags. Moisten the inflorescences with a minimal amount of water (5-10 ml), and keep them refrigerated at 8±2°C. The inflorescence containing plastic bags may be stored in the refrigerator for about 1 week.

Isolation of flower buds
1. Separate the flower clusters from the inflorescence by a sharp cut. Every cluster may contain 1-5 flower buds.

2. Remove and discard open flowers and the stem parts between buds.

3. Transfer the buds to a moist filter paper, to avoid desiccation.

Surface sterilization of buds
Use commercial bleach, e.g. 70% Klorin, or 3% hypochlorite as the sterilizing agent.
1. Add a surplus of the 70% bleach to the beaker and leave for 5 minutes. A final volume of 50 ml is recommended.
Note: From this point and onwards all work is carried out under sterile conditions in a laminar flow.

2. Pour off the commercial bleach and rinse with several volumes of sterile double distilled water (ddH$_2$O). Apply gentle shaking and pour off the water after every rinse.

3. After removal of the final rinse, store the buds under sterile conditions and prevent the buds from drying-out by leaving a minimal amount of residual water in the container. Storage at 8±2°C in refrigerator is possible for 1-2 days.

Isolation of ovules
1. Isolate about 200 ovules from each plant.

2. Excise the ovules under a microscope.

3. Transfer the ovules to 5 cm Petri dishes with medium Bv1 (Table 2.37-2). If possible place about 20 ovules in every Petri dish and label with the identification code.

4. Stack the Petri dishes, with an empty dish on the top, and transfer to dry plastic bags (Fig. 2.37-2b). Pierce ca. 20 holes, using a forceps, in the bag unit to improve aeration, and mark the bags with the identification codes.

5. The material in the bag units is grown in darkness at 30±2°C.
Note: The Fig. 2.37-2a shows the approximate position of the opaque ovule. The bag units facilitate scale-up and reduce the risk of mix-ups.

Subculture of embryos
The embryos become visible after 30-70 days (Fig. 2.37-2c).
1. Transfer the embryos to tubes or containers with shoot induction medium Bv2 (Table 2.37-2).

2. Grow the embryos under 50±10 μmol m^{-2} s^{-1} fluorescent light e.g. Osram Biolux (Germany), and 16/8 h day/night at 24±2°C.

Subculture of shoots
Every 3-4 week, embryos and new shoots are transferred to fresh medium Bv2.
1. Remove old or dead tissue. Discard callus and dead embryos. Make a slightly wedge shaped cut at the shoot base to improve contact with medium. Separate the shoots and promote apical dominance. The shoots should be vigorous and not hyperhydric.

2. Transfer the shoots to Bv2.
Note: The shoots originating from the same embryo can be considered as shoots (ramets) from the same clone. Multiply every clone to about 5 ramets as soon as possible (Fig. 2.37-2d), to reduce the risk of loosing clones due to infections or poor rooting.

Root induction
1. Transfer the vigorous shoots to growth containers with pre-rooting medium Bv3 (Table 2.37-2), and subsequently to rooting medium Bv4 (Table 2.37-2).

2. Subculture less frequently, every 6-8 week, to avoid disturbing roots.
Note: It is essential that the shoots have a well developed apical dominance before initiating root formation.

Table 2.37-2. Bv1-Bv4 media used for subsequent steps in sugar beet ovule culture

	Media components	Bv1 embryo induction medium	Bv2 shoot induction and propagation medium	Bv3 pre-rooting medium	Bv4 rooting medium
A	MS powder (basal medium)	full strength	full strength	full strength	half strength 1/2
	Agarose	5.8 g/L	-	-	-
	Agar	-	9 g/L	9 g/L	9.5 g/l
B	Sucrose	80 g/L	20 g/L	-	-
	Sugar	-	-	30 g/L	30 g/L
C	BA	1.33 µM	-	-	-
	2,4-D	0.23 µM	-	-	-
	Kinetin	-	0.93 µM	2.32 µM	-
	NAA	-	0.54 µM	-	-
	IBA	-	-	2.46 µM	24.6 µM
pH		5.8	5.8	5.8	5.8

Suppliers: MS-powder: SIGMA M-5519 or equal; Agarose: Sigma A-6013 or equal; Agar: Sigma 4550 or equal; Sucrose: high grade, p.a.; Sugar: Conventional household grade, i.e. about 99.9% sucrose.

Preparing Bv1 medium (a final volume of 1 liter)
- Autoclave component A and B, separately, at 1.1 bar for 20 minutes
- Mix A and B after sterilization
- Add component C to AB by filter sterilization at 50±5°C
- Allow cooling to 35±5°C before dispensing in small Petri dishes
  - Component A: dissolve the MS powder in 800 ml ddH$_2$O
    adjust the pH to 5.8 with 0.5 M KOH
    add 5.8 g agarose
  - Component B: dissolve 80 g sucrose in 150 ml ddH$_2$O
  - Component C: 300 µl BA from stock solution (1mg/L in EtOH)
    50 µl 2,4-D from stock solution (1mg/L in EtOH)
    30 ml ddH$_2$O.

If preparation from stock solutions is preferred, component A may be prepared according to the medium concentrations designated in Murashige and Skoog (1962).

Preparing Bv2 medium (a final volume of 1 liter)
- Component A and B are autoclaved separately at 1.1 bar for 20 minutes
- Mix A and B after sterilization
- Add component C to AB by filter sterilization at 50±5°C
- Allow cooling to 35±5°C before dispensing in tubes or growth containers
  - Component A: dissolve MS-powder in 800 ml ddH$_2$O
    adjust the pH to 5.8 with 0.5 M KOH; add 9 g agar
  - Component B: dissolve 20 g sucrose in 180 ml ddH$_2$O
  - Component C: 20 µl kinetin from stock solution (1mg/L in EtOH)
    10 µl NAA from stock-solution (1mg/L in EtOH)
    20 ml ddH$_2$O

Preparing Bv3 medium (a final volume of 1 liter)
Component A and B are autoclaved separately at 1.1 bar for 20 minutes
- Mix A and B after sterilization
- Add component C to AB by filter sterilization at 50±5°C
- Allow cooling to 35±5°C before dispensing in tissue culture containers

    Component A: dissolve MS powder in 800 ml ddH_2O
                       adjust the pH to 5.8 with 0.5 M KOH
                       add 9.0 g agar
    Component B: dissolve 30 g sugar in 180 ml ddH_2O
    Component C: 50 µl kinetin from stock solution (1mg/L)
                       50 µl IBA from stock solution (1mg/L)
                       20 ml ddH_2O

Preparing Bv4 medium (a final volume of 1 liter)
- Component A and B are autoclaved separately at 1.1 bar for 20 minutes
- Mix A and B after sterilization
- Add component C to AB by filter-sterilization at 50±5°C
- Allow cooling to 35±5°C before dispensing in tissue culture containers

    Component A: dissolve ½ MS-powder in 800 ml ddH_2O
                       adjust the pH to 5.8 with 0.5 M KOH
                       add 9.0 g agar
    Component B: dissolve 30 g sugar in 180 ml ddH_2O
    Component C: 500 µl IBA from stock-solution (1 mg/L)
                       20 ml ddH_2O

Transfer to glasshouse and acclimation
Note: At this point the material returns to *in vivo* conditions.

Once the root has emerged, transfer the shoots to glasshouse conditions with a high relative humidity (about 90%). The plants may need special attention during the first weeks of acclimation. Metal halide lamps are a suitable light source, but avoid too high photon flux densities in the early stages. After acclimation secure good development by using glasshouse or growth chamber. Growth conditions such as 24±2°C, light >200 µmol m^{-2} s^{-1} and 16/8 h day/night are recommended. A similar fertilization as described for donor plants (Bv0) may be used.

Chromosome number determination
1. 1-3 weeks after transfer to glasshouse, select a leaf close to the shoot apical meristem.

2. Remove the leaf and transfer to 2 mM 8-hydroxyquinoline solution and leave for 2.5 h.

3. Fix the material in Carnoy's solution and keep in refrigerator at 8±2°C for at least 24 h.

4. Hydrolyze the material in 1M HCl at 60°C for 5 minutes, and rinse by soaking in water for 5 min.

5. Transfer the hydrolyzed tissue to a watch glass with lacto-propionic orcein stain for 15-20 min.

6. Use a dissecting-microscope, cut and transfer the lower part of the leaf to a slide. Add one drop of lacto-propionic orcein and mount with a cover slip.

7. Squash gently, but firmly, on the cover slip and if necessary, allow to stain for a few more min. Remove excess fluid with a filter paper.

8. Determine the chromosome number under a microscope (Bosemark and Bormotov, 1971).

Note: The plantlets should be vigorous, show apical dominance and have a well developed root system. The meristematic activity is higher just after dawn, and to improve the analysis it is advisable to harvest the leaves in the early morning. After fixation (step 3), the material may be stored in the freezer at -30±5°C for several years.

Preparation of Carnoy's solution
Mix ethanol and acetic acid [3:1]
    Acetic acid [100%]
    Ethanol [96%]

Preparing Lacto-propionic orcein (100 ml of the solution)
- Mix equal volumes of lactic acid and propionic acid
- Add 2 g orcein to 100 ml of the lacto-propionic mix
- Dilute to 45% with ddH$_2$O

Chromosome doubling

1. Rinse the roots under running water and remove the outer part (approx.1 mm) of the root tips as this will facilitate transport of colchicine to the meristematic region of the shoot apex.

2. Transfer the plant to a 0.2% solution of colchicine + 0.25% DMSO and leave for 5 h (these chemicals are harmful and care is needed in their handling). Make sure that the roots are fully submerged during the whole period.

3. Rinse the roots with water.

4. Re-pot in soil and leave the plants to adapt and recover for several weeks in a glasshouse or growth chamber under acclimation conditions.

Note: The plant stage is non critical as long as no taproot has developed. However, treatment directly after receiving the chromosome number counts, at the 6-leaf stage is recommended. Normally the efficiency well exceeds 90%, however when setting up the protocol the first time, a confirmation of the doubling by a second determination is recommended.

Re-growth, vernalization, bolting and flowering

1. After recovery, transfer the plants to vernalization under 5±1°C and 60±20 µmol m^{-2} s^{-1} metal halide light, e.g. GE Kolorarc 400 W (Hungary), and 16/8 h day/night, for 14 weeks.

2. Transfer the plants to acclimation under 12±1°C at 60±20 µmol m^{-2} s^{-1} (light source as above) and 16/8 h day/night for 2 weeks.

3. Finally, transfer the plants to a glasshouse or growth chamber under 20±4°C with light >200 µmol m^{-2} s^{-1} (light source as above) and 16/8 h day/night.

4. After bolting and inflorescence development, the plants are isolated by paper bags to secure self-pollination.

Fertility determination - optional
1. Harvest a minute amount of pollen and desiccate for 15-30 minutes.

2. Dust-out the pollen on the pollen germination medium and leave under humid conditions (RH of about 85%).

3. Evaluate the pollen germination after 15-45 minutes.

Preparation of pollen germination medium (modified after Glenk et al., 1969)
- Dissolve the agar by gentle heating or in water-bath, and add the sucrose and the boric acid.
- Spread a thin layer of medium on microscopy slides, and allow cooling off.
- Leave for 15 minutes in a high-humidity chamber, before use.
    Components: 30% sucrose in ddH_2O
                  0.05% H_3BO_3 (pH6.4)
                  3% agar (Agar Agar Kadoya)
- pH 5.4-5.7, adjusted with HCl

Note: It is very important to use a high quality agar or the germination will be inhibited. Glenk *et al.* (1969) used 5% gelatin instead of agar. Over-heating or autoclaving when dissolving the gelling agent will inhibit germination, due to the degradation of sucrose.

Seed maturation
Seed maturation takes place in the isolation bags, and mature seed is available 10-12 weeks after anthesis. Harvest the whole inflorescence, remove stalk parts, thresh the material and transfer the seed to storage, in darkness at 15°C, RH 50%, for later use.

Efficiency and Applications

The overall time required in setting-up a reproducible production line is 2 years, minimum. Approximately 7 months are required to produce good starting material, another 6 months for the *in vitro* period and additionally 9-12 months to produce doubled haploid seed. The number of ovules that can be excised during one working day differs, depending on the appearance of the donor plant, e.g. a monogerm inflorescence is easier to work with than a multigerm. In addition, the genotype and line variation affects the calculations of a general estimate of efficiency. However, a skilled technician may harvest and excise about 1,000 ovules per day. The embryo response varies between 1-15% and the conversion to shoot cultures is generally 40%. Considering a rooting efficiency of about 95% and a chromosome doubling success rate above 90% gives the following rough prediction. Assuming 2,000 ovules/week x 5% x 40% x 95% x 90% results in about 30 clones at the end of the period. Since the estimation includes the propagation of ramets and support functions, the annual output per full-time employee is more than 1,000 clones. This compares favourably with doubled haploid productions systems of many other crops.

Different groups use various basal media e.g. N6, PGo and MS. Results (unpublished) indicate that the different basal media are not significantly different in respect to the embryo response, however show a non-significant positive effect of MS medium, which also displayed less genotypic variation. Since MS-components are easily accessible, it appears to be a good choice when setting up a production laboratory.

In a publication Lux *et al.* (1990) reported improved embryo development by replacing sucrose with maltose. In our laboratory, Schenk *et al.* (1991) showed the degradation of sucrose, during heat sterilization together with medium components, to give rise to fructose species that are detrimental to tissue culture growth. Therefore it is recommended to follow a special protocol for medium preparation, designed to avoid the creation of these inhibiting compounds.

Pedersen and Keimer (1996) discussed position effects, such as the position of flower buds on the inflorescence or the positioning of ovules in relation to the medium. In addition, seasonal variations in the embryo response may be observed, particularly if the culture conditions of the donor plants are not optimized.

The spontaneous doubling is rather low in sugar beet ovule culture, at about 5%, and chromosome doubling needs to be induced artificially. The chromosome doubling protocol (described above) consistently results in close to 100% doubling. However, as the use of colchicine needs careful handling, in respect of personal safety, alternative compounds have been desired and evaluated by several groups. Some labs have investigated the effect of doubling during the *in vitro* step (Hansen *et al.*, 1994; Gürel *et al.*, 2000). Compounds such as colchicine, oryzalin, trifluralin or amiprophos-methyl (APM) have been evaluated. So far, the success rate for doubling *in vitro* has been lower compared to doubling *in vivo*, 30-60% and 90-100%, respectively. However, reported data suggest that APM might offer a safer alternative (Hansen, personal communication), but the approach appears to need optimization to be applicable for commercial use. Moreover, the use of APM under *in vivo* conditions needs thorough investigation as doubling *in vitro* may negatively affect the embryo response.

The DH technique is presently used for the production of homozygous parental lines, in the molecular marker-assisted breeding and production of elite inbreds. Future use may include material for F_1 hybrid production.

References

Bosemark, N. O. and V. E. Bormotov, 1971. Chromosome morphology in a homozygous line of sugar beet. Hereditas **69**: 205-211.

Glenk, H.O., G. Blaschke and K.L. Barocka, 1969. Untersuchung zur Variabilität des Pollenschlauchwachstums bei Pollen di- und tetraploider Zuckerrüben. I. Bedingungen zur Keimung von *Beta*-Pollen *in vitro*. Theor.Appl.Genet. **39**: 197-205.

Gürel, S., E. Gürel and Z. Kaya, 2000. Doubled haploid plant production from unpollinated ovules of sugar beet (*Beta vulgaris* L.). Plant Cell Rep. **19**: 151-159.

Hansen, A.L., C. Plever, H.C. Pedersen, B. Keimer and S.B. Andersen, 1994. Efficient *in vitro* chromosome doubling during *Beta vulgaris* ovule culture. Plant Breed. **112**: 89-95.

Hosemans, D. and D. Bossoutrot, 1983. Induction of haploid plants from *in vitro* culture of unpollinated beet ovules (*Beta vulgaris* L.). Z. Pflanzenzüchtg. **91**: 74-77.

Lux, H., L. Herrmann and C. Wetzel, 1990. Production of haploid sugar beet (*Beta vulgaris* L.) by culturing unpollinated ovules. Plant Breed. **104**: 177-183.

Murashige, T. and F. Skoog, 1962. A revised medium for rapid growth and bioassays with tobacco tissue cultures. Physiol.Plant. **15**: 473-497.

Pedersen, H.C. and B. Keimer, 1996. Haploidy in sugar beet (*Beta vulgaris* L.). In: *In Vitro* Haploid Production in Higher Plants. Vol. 3. Jain, S.M., S.K. Sopory and R.E. Veilleux (Eds.) Kluwer Academic Publishers, Dordrecht, pp. 17-36.

Schenk, N., K.C. Hsiao, and C.H. Bornman, 1991. Avoidance of precipitation and carbohydrate breakdown in autoclaved plant tissue culture media. Plant Cell Rep. **10**: 115-119.

2.38
Asparagus microspore and anther culture

D.J. Wolyn and B. Nichols
Department of Plant Agriculture, University of Guelph, Guelph, Ontario N1G 2W1, Canada

Introduction

Asparagus is a dioecious, perennial crop where male plants generally yield more than female plants. Consequently, an all-male hybrid can have improved productivity compared to one that is dioecious. Genotypes of males and females are *Mm* and *mm*, respectively, and all-male hybrids result from crossing females and supermales (*MM*). Supermales can be obtained from hermaphroditic flowers produced on male plants, however, the occurrence of perfect flowers is often infrequent and affected by genotype and environment. Anther or microspore culture, generating doubled haploid supermales from important genetic stocks, can enhance the breeding of all-male hybrid cultivars. Asparagus hybrids are produced by crossing clonally propagated parental lines Development of homozygous lines through traditional breeding is difficult and time consuming due to the dioecious nature of the crop. Many successful hybrids have resulted from intermating heterozygous females with supermales derived from a single generation of selfing; these cultivars can be somewhat heterogeneous. Through *in vitro* culture, homozygous, doubled-haploid supermales and females have been produced, fostering hybridization between heterozygous females and homozygous supermales, or both homozygous females and supermales. The latter type of cross has resulted in the timely production of homogeneous F_1 hybrids with superior performance (Falavigna *et al.*, 1999). Critical steps in asparagus microspore culture are: the production of healthy, dormant, one-year-old crowns, growth of plants from these crowns in a controlled environment, cold pre-treatment of flower buds containing a high frequency of microspores at the late uninucleate stage of development, media components, incubation conditions, and culture vessel. Micropsore-derived calli are produced in culture, from which shoots are usually regenerated. A low frequency of embryos can also be recovered from the calli. Microspores shed in culture are used exclusively because multiple attempts to culture mechanically isolated microspores have been unsuccessful. Cold pre-treatment of flowers with microspores at the late uninucleate stage of development for 7 days induced symmetrical cell divisions, the frequencies of which were correlated with callus production (Peng *et al.*, 1997). Empirical analyses have also been published, identifying optimal media and conditions (Peng *et al.*, 1997; Peng and Wolyn, 1999). An inherent problem in microspore culture is variability of response; individual experiments may produce no calli. Preliminary research suggests the use of Transwell plates decreases the frequency of non-responsive experiments when compared to tissue culture dishes.

Protocol - Asparagus microspore culture

Donor plants and growth conditions
The need to grow plants in a controlled environment for efficient microspore culture is complicated by the perennial nature of the crop. Success has been achieved with the production of healthy one-year-old crowns in the field, the crowns are then transplanted and placed in a growth chamber to yield flowers. Optimal results occur only when completely dormant crowns are dug from the field.

Production of dormant one-year-old crowns
1. Plants are produced from seeds or *in vitro* micropropagated clones in the glasshouse for 10 to11 weeks in the spring.

2. Seedlings are space planted in the field, 50 cm apart within rows, 1.5 m between rows.

3. The crowns are dug in the autumn when are fully dormant. Plants are assumed to be dormant when the fern has senesced and buds no longer elongate. Poor culture response is observed if crowns are dug too early.

4. Crowns are stored at 4°C until use. They are placed in paper bags with wood shavings and the paper bags placed in plastic bags with two ventilation holes. Microspore culture has been successful with healthy crowns stored for up to six months.

5. When required, the crowns are planted in 15 L pots containing Promix BX soil mix and established in a growth chamber at 25/19°C (day/night) with a 16 h photoperiod provided by a combination of cool white and Gro Lite WS fluorescent bulbs (5:1), 200 µmol m^{-2} s^{-1} at soil level. Plants are fertilized weekly with liquid fertilizer (20N:8P:20K). Flower buds are produced after 2 to 3 weeks of growth.

Shed microspore culture
1. Flower buds are collected when microspores are at the late-uninucleate stage (e.g. 2.1 to 2.4 mm in length for variety 'Guelph Millennium') and placed immediately on ice. The relationship between microspore stage and bud size may depend on genotype and plant size, and should be verified with acetocarmine or DAPI staining. The buds can be collected three to four times over a period of 7 to 10 days for each planting and are pooled from 4-10 plants to obtain sufficient numbers of microspores to conduct experiments.

2. Store buds at 4°C for 1 week in 100x15 mm Petri dishes sealed with Parafilm.

3. Surface sterilize the buds in 1.6% sodium hypochlorite for 7 min and then rinse three times for 5 min in sterile deionized water.

4. Store sterilized buds on ice in a 100x15 mm Petri dish until dissection. Remove only the number of buds required to provide sufficient anthers to fill one culture well.

5. Dissect anthers from flowers, making sure no filaments remain, as they callus quickly in culture.

Culture of microspores
Microspores can be cultured in 24 mm diameter Transwell - Clear plates (Corning 3450) or in 35x10 mm tissue culture dishes (Corning 2500). We have found that use of Transwell - Clear

plates reduced the frequency of non-responsive experiments, however, tissue culture dishes are less expensive.

Transwell plates
1. Place 75 anthers per well in 24 mm diameter Transwell - Clear plates filled according to the manufacturer's recommendations (1.5 ml of culture medium in the upper chamber and 2.6 ml in the lower chamber) with FHA microspore culture medium (Table 2.38-1) (Peng et al., 1997; Peng and Wolyn, 1999).

2. Incubate at 34°C in darkness for 10 days.

3. Remove the anthers and replace medium in the lower compartment. Cell division can be determined at this time.

4. Incubate the cultures in darkness with gentle shaking. Replace medium in lower compartment at weekly intervals.

Tissue culture dishes
1. Place 125 anthers in a 35x10 mm tissue culture dish containing 1.5 ml of FHA medium.

2. Incubate at 34°C in darkness for 10 days.

3. Remove the anthers and replenish the medium by adding 0.5 ml of fresh FHA medium. Cell division frequency can be determined at this time.

4. Incubate the cultures in darkness with gentle shaking. Maintain medium at approximately 1.5 ml with weekly replenishments.

Shoot regeneration
1. After 4 weeks of culture on the shaker, remove microcalli that are larger than 0.5 mm in diameter and transfer to CG medium (Table 2.38-1) (Peng and Wolyn, 1999) for further growth. The microcalli can be collected over 4 to 6 weeks.

2. Incubate the cultures in darkness at 30°C for 8 weeks.

3. Transfer calli 3-5mm in size to WHZ medium (Table 2.38-1) and culture at 25°C under continuous cool white fluorescent light (50 μmol m^{-2} sec^{-1}) for shoot regeneration.

Propagation of plants
1. Cut the shoots into short segments containing one or two axillary buds and place horizontally, in magenta boxes, on shoot proliferation medium (Table 2.38-1). Axillary shoots which fail to root after 8 weeks should be cut into short segments and subcultured on shoot proliferation medium. This procedure is repeated until roots are produced. Shoots and rooted plants are cultured at 25°C, with a 16 h photoperiod, under cool white fluorescent light (50 μmol m^{-2} sec^{-1}).

2. The plantlets are transferred to root growth medium (Table 2.38-1). After one or two subcultures on root growth medium at 6 week intervals, the plantlets should be strong enough to transplant into soil.

3. Rooted plants can be micropropagated (cloned) by splitting the crown and culturing on root growth medium.

Table 2.38-1. Composition of media used for asparagus microspore culture

Media components	FHA microspore culture medium (mg/L)	CG calli proliferation medium (mg/L)	WHZ shoot regeneration medium (mg/L)	Shoot proliferation medium (mg/L)	Root growth medium (mg/L)
Macro salts					
KNO_3	1,900	1,900	1,900	1,900	1,900
NH_4NO_3	1,650	1,650	1,650	1,650	1,650
KH_2PO_4	170	170	170	170	170
$MgSO_4 \times 7H_2O$	370	370	370	370	370
$CaCl_2 \times 2H_2O$	440	440	440	440	440
$NaH_2PO_4.2H_2O$	-	-	-	155	-
Iron source					
$FeSO_4 \times 7H_2O$	27.8	27.8	27.8	27.8	27.8
Na_2EDTA	37.3	37.3	37.3	37.3	37.3
Micro salts					
$MnSO_4 \times 4H_2O$	22.3	22.3	22.3	22.3	22.3
H_3BO_3	6.2	6.2	6.2	6.2	6.2
$ZnSO_4 \times 4H_2O$	8.6	8.6	8.6	8.6	8.6
$CoCl_2 \times 6H_2O$	0.025	0.025	0.025	0.025	0.025
$CuSO_4 \times 5H_2O$	0.025	0.025	0.025	0.025	0.025
$Na_2MoO_4 \times 2H_2O$	0.25	0.25	0.25	0.25	0.25
KI	0.83	0.83	0.83	0.83	0.83
Other components					
Thiamine HCl	0.1	0.1	0.1	1	0.1
Nicotinic acid	0.5	0.5	0.5	5	0.5
Pyroxidine HCl	0.5	0.5	0.5	5	0.5
myo-Inositol	100	100	100	90	90
Glycine	2	2	2	-	2
Casein hydrolysate	500	500	-	-	-
Glutamine	800	800	-	-	-
Yeast extract	2,000	-	-	-	-
Sucrose	60,000	30,000	30,000	60,000	60,000
NAA	2	2	-	0.1	0.1
BA	1	1	-	-	-
Zeatin	-	-	2	-	-
Kinetin	-	-	-	0.1	0.1
Adenine hemisulfate	-	-	-	18.4	-
Ancymidol	-	-	-	2.5	4
Agar	-	7,500	7,500	7,000	7,000
pH	5.9	5.9	5.9	5.9	5.9

Protocol - Asparagus anther culture

Important steps for asparagus anther culture are: collection of flowers with microspores at the uninucleate stage of development from healthy field-grown plants; subculturing calli or embryoids derived from gametophytic cells while avoiding sporophytic contamination; media components; and culture conditions. Calli are commonly produced, from which shoots are regenerated, however, embryos may also be regenerated from calli or directly from microspores (Tsay, 1996; Falavigna et al., 1999).

Anther culture I, Tsay (1996)
1. Collect flower buds with microspores at the uninucleate stage from field grown plants.

2. Sterilize buds in 0.5% sodium hypochlorite for 10 min and rinse several times with sterile water.

3. Remove anthers aseptically from flowers and culture on induction medium (Table 2.38-2).

Table 2.38-2. Composition of media used for asparagus anther culture I, Tsay, (1996)

Media components	Induction medium (mg/L)	Germination medium (mg/L)	Differentiation medium (mg/L)
Macro salts			
KNO_3	950	1,900	1,900
NH_4NO_3	825	1,650	1,650
KH_2PO_4	85	170	170
$NaH_2PO_4 \times 2H_2O$	-	-	170
$MgSO_4 \times 7H_2O$	185	370	370
$CaCl_2 \times 2H_2O$	220	440	440
Iron source			
$FeSO_4 \times 7H_2O$	27.8	27.8	27.8
Na_2EDTA	37.3	37.3	37.3
Micro salts			
$MnSO_4 \times 4H_2O$	11.15	22.3	22.3
H_3BO_3	3.1	6.2	6.2
$ZnSO_4 \times 4H_2O$	4.3	8.6	8.6
$CoCl_2 \times 6H_2O$	0.0125	0.025	0.025
$CuSO_4 \times 5H_2O$	0.0125	0.025	0.025
$Na_2MoO_4 \times 2H_2O$	0.125	0.25	0.25
KI	0.415	0.83	0.83
Other components			
Thiamine HCl	0.4	0.1	1
Nicotinic acid	0.5	0.5	5
Pyroxidine HCl	0.5	0.5	5
myo-Inositol	100	100	100
Glycine	1	2	-
Sucrose	60,000	30,000	25,000
Adenine sulphate dihydrate	-	-	40
Difco-Bacto malt extract	-	-	500
NAA	2	-	1
BA	1	-	0.5
Difco-agar	8,000	6,000	6,000
pH	5.7	5.7	5.7

4. Incubate at 26°C in darkness for callus and embryo induction.

5. Transfer embryos to MS basal medium with no growth regulators for germination (Table 2.38-2).

6. Select callus from split anthers, avoid callus from filament and anther wall.

7. Transfer calli to differentiation medium (Table 2.38-2) (Murashige et al., 1972) and incubate at 26°C in fluorescent light with a 16 h photoperiod (20 µmol m^{-2} sec^{-1}; equivalent to 1500 lux).

8. Transfer non-regenerative calli after two months to MS medium without hormones for organ differentiation.

Anther culture II, Falavigna et al. (1999)
1. Collect flower buds with microspores at the uninucleate stage from field grown plants.

2. Sterilize buds in 10% calcium hypochlorite for 10 min and rinse four times with sterile water.

3. Dissect anthers aseptically and plate on A2 medium (Table 2.38-3).

4. Incubate at 25°C with a 16 h photoperiod under grolux lamps (50 µmol m^{-2} sec^{-1}; equivalent to 1,500 lux).

5. After 25 to 40 days transfer embryos, 0.5 to 1mm in diameter, to T1 medium (Table 2.38-3) for adventitious shoot development.

6. Micropropagate shoots on 85 medium (Table 2.38-3).

7. Transfer regenerating shoots to root on 102 medium (Table 2.38-3).

Chromosome doubling
Several methods have been reviewed (Tsay, 1996). Two are outlined below.

Method A, Dore (1976)
1. Culture meristems on propagation medium containing 0.5 to 0.1 g/L colchicine for 2 to 6 days. Propagation medium can be that described above for microspore culture (Propagation of Plants - Table 2.38-1).

2. Transfer to medium without colchicine for 2 to 6 days. Continue subculture for plantlet development.
Genotype response is variable, some lines are non-responsive and efficiency may not be greater than 10%.

Method B, Tsay (1996)
1. Apply 1.2% colchicine in lanolin to plantlet shoot tips *in vivo*. Chromosome doubling produces noticeable stem thickening.

2. Culture the doubled-haploid stems in rooting medium [MS medium (½ MS salts) plus 50 g/L sucrose, 6 g/L agar, 0.1 mg/L NAA, 0.1 mg/L kinetin, 0.13 mg/L ancymidol, pH 5.7] for one month.

3. Culture plants for an additional month in the above medium (liquid with filter paper support) for fast growth.
Efficiencies of 21-97% have been reported.

Table 2.38-3. Composition of media used for asparagus anther culture II, Falavigna et al. (1999)

Media components	A2 medium for embryo/callus induction (mg/L)	T1 medium for adventitious shoot development (mg/L)	85 medium for shoot micropropagation (mg/L)	102 medium for root growth (mg/L)
Macro salts				
KNO_3	1,900	1,900	1,900	1,900
NH_4NO_3	1,650	1,650	1,650	1,650
KH_2PO_4	170	170	170	170
$MgSO_4 \times 7H_2O$	370	370	370	370
$CaCl_2 \times 2H_2O$	440	440	440	440
Iron source				
$FeSO_4 \times 7H_2O$	27.8	27.8	27.8	27.8
Na_2EDTA	37.3	37.3	37.3	37.3
Micro salts				
$MnSO_4 \times 4H_2O$	22.3	22.3	22.3	22.3
H_3BO_3	6.2	6.2	6.2	6.2
$ZnSO_4 \times 4H_2O$	8.6	8.6	8.6	8.6
$CoCl_2 \times 6H_2O$	0.025	0.025	0.025	0.025
$CuSO_4 \times 5H_2O$	0.025	0.025	0.025	0.025
$Na_2MoO_4 \times 2H_2O$	0.25	0.25	0.25	0.25
KI	0.83	0.83	0.83	0.83
Other components				
Thiamine HCl	0.1	0.1	0.1	0.1
Nicotinic acid	0.1	0.1	0.1	0.1
Pyroxidine HCl	0.1	0.1	0.1	0.1
Calcium panthotenate	0.1	0.1	0.1	0.1
myo-Inositol	100	100	100	100
Biotin	0.01	-	-	-
Sucrose	20,000	20,000	30,000	60,000
Glucose	20,000	-	-	-
2,4-D	0.5	-	-	-
NAA	0.1	-	0.2	0.2
BA	0.5	0.3	0.2	0.1
Kinetin	-	-	0.1	-
Ancymidol	-	-	-	0.5
Microagar Duchefa	6,000	6,000	6,000	6,000
pH	5.7	5.7	5.7	5.7

Chromosome Counts

1. Collect root tips and place immediately in water on ice and store overnight at 4°C.

2. Fix in Carnoy's solution (3:1 ethanol /acetic acid) for 3 to 24 h.

3. Remove fixative and rinse with 70% ethanol. Store in 70% ethanol until required for staining.

4. Soak root tips in 1 N HCl at 60°C for 10 min.

5. Rinse with distilled water.

6. Soak in 45% glacial acetic acid at 60°C for 10 min.

7. Rinse with distilled water. Roots may be held in distilled water temporarily.

8. Stain with modified carbol fuchsin (0.3 g basic fuchsin, 10 ml 70% ethanol, 90 ml 5% phenol, 11.5 ml glacial acetic acid, 11.5 ml 37% formaldehyde) for 30 to 60 min.

9. Place root tip on a slide in a drop of glycerol.

10. Cover with cover slip and squash.

11. Observe with a light microscope at 1,000x magnification.

Evaluation of doubled haploids

Important traits for a successful asparagus hybrid include long-term yield, quality and disease resistance; male plants used for doubled haploid production should be superior for these characteristics. Doubled haploid populations will segregate for important traits, allowing the best genotypes to be selected for hybrid development. Evaluation varies based on individual breeding objectives and priorities. A scheme adapted from Falavigna *et al.* (1999) is presented below as one example.

1. Plant doubled haploid clones, 5 to 20 per diploid genotype, in soil naturally infested with *Fusarium* spp.

2. Testcross vigorous male clones to confirm gametophytic origin (to check against any derived from contaminating sporophytic cells). Males (*Mm*) will segregate 1 male: 1 female when crossed to a female clone while a supermale (*MM*) will produce only male progeny.

3. Observe for three years and make selections based on percentage of living plants, vigour, stalk diameter, fertility and any other trait important for the local breeding programme.

Efficiency and Applications

Among 12 genotypes tested in microspore culture, nine were responsive and produced 1.5 to 150 calli per100 anthers (mean = 40). Five to 28% of calli regenerated shoots. Experiments were sometimes inconsistent; the mean response for one genotype ranged from 24 to 375 calli per 100 anthers cultured, and regeneration rates have varied three-fold. Callus production and regeneration were not always correlated; some genotypes produced limited callus that regenerated well. Plants recovered were 49% haploid, 34% diploid, 4% triploid and 11% tetraploid (Peng and Wolyn, 1999).

In anther culture method, among the 85 cultivars tested by two research groups, conducting anther culture, genotypic variation ranged considerably, from near zero response to more than 50% of anthers producing calli. Tsay (1996) observed a higher response than Falavigna *et al.* (1999) but also sampled a narrower range of germplasm. Although regeneration rates differed among the researchers, the overall efficiencies of the methods, based on mean cultivar data, were similar; approximately one plant was produced for each 100 anthers cultured. The ploidy distribution of recovered plants was similar for the two groups: 3-8% haploid, 59-61% diploid, 5-11% triploid and 25-27% tetraploid.

Acknowledgments

The authors wish to thank Drs. A. Falavigna and H.S. Tsay for reviewing sections of the manuscript.

References

Dore, C., 1976. Doublement du stock chromosomique d'haploides d'Asperge per culture *in vitro* des meristemes en presence de colchicines. Ann.Amelior.Plantes **26**: 647-653.

Falavigna, A., P.E. Casali and A. Battaglia, 1999. Achievement of asparagus breeding in Italy. Acta Hort. **479**: 67-74.

Murashige, T., M.N. Shabed, P.M. Hasegawa, F.H. Takatori and J.B. Jones, 1972. Propagation of asparagus through shoot apex culture. I. Nutrient medium for formation of plantlets. J.Amer.Soc.Hort.Sci. **97**: 158-161.

Peng, M. and D.J. Wolyn, 1999. Improved callus formation and plant regeneration for shed microspore culture in asparagus (*Asparagus officinalis* L.). Plant Cell Rep. **18**: 954-958.

Peng, M., A. Ziauddin and D.J. Wolyn, 1997. Development of asparagus microspores *in vivo* and *in vitro* is influenced by gametogenic stage and cold treatment. *In Vitro* Cell Dev.Biol.-Plant **33**: 263-268.

Tsay, H.S., 1996. Haploidy in asparagus anther culture. In: *In Vitro* Haploid Production in Higher Plants. Jain, S.M., S.K. Sopory and R.E. Veilleux (Eds.) Kluwer Academic Publishers, Dordrecht, pp. 109-134.

2.39
In vitro gynogenesis induction and doubled haploid production in onion *(Allium cepa* L.*)*

L. Martínez
Plant Physiology Laboratory, Department of Biology, Agricultural College, National University of Cuyo, 5505, CC Nº7, Chacras de Coria, Mendoza, Argentina

Introduction

One of the goals in an onion breeding program is the production of F_1 hybrids. The main limitation of this scope is the long period required to produce inbred lines. The most time-consuming and labor-intensive aspect of developing these hybrids is the traditional inbreeding process that requires manual self-pollination necessary to generate homozygous parental lines. This process requires six or more generations of inbreeding to adequately establish stable lines that can be used in hybrid combinations. In many species *in vitro* methods have speed up the production of homozygous lines, as an alternative to the slower inbreeding process. Haploid plants can be obtained by anther, non-fertilized ovule, ovary or whole flower bud culture. The haploid chromosome number means that the effects of meiotic recombinations and recessive genes are expressed at the haploid plant level. Then, spontaneous or induced chromosome doubling permits the regeneration of doubled haploid (DH) homozygous material, with restored fertility that can be used in different breeding strategies. Because all the alleles of DH lines are fixed, selection for quantitative characters is often more reliable than in conventional populations.

Protocol

Donor plants and growth conditions
Research on gynogenesis in onion has been carried out in our laboratory since 1993 and has dealt with the long-day length Argentinean onion (*Allium cepa* L.) cultivar 'Valcatorce INTA', and one intermediate-day length population, 'Torrentina' (Martinez *et al.*, 2000). This work represents the first evaluation of these genotypes for unpollinated flower culture response.

Fifty bulbs, from the donor plants of each genotype, are planted at regular intervals from August to September in open field condition to obtain umbels continuously over a two month period.

Sterilization
Young flowers are removed from the umbels 3-5 days before anthesis.

Fisrt procedure
Flowers buds are surface-sterilized with 10% sodium hypochlorite solution containing a few drops of Tween 20 for 15 min, and then rinsed 3 times with sterile distilled water.

Second procedure
This previous step can be replaced by using 96% alcohol for 2 min followed by 16.6 g/L dichloroisocyanuric acid (sodium salt) with the addition of a few drops of Tween 20 for 8 min and then rinsed three times in sterile water.

After being washed, 30 flowers are plated on 90 mm Petri dishes and sealed with Parafilm (Fig. 2.39-1).

Flowers culture procedure and media
1. Flowers are cultured using 2 steps culture procedure. In the first step, flowers are inoculated on induction medium (Medium A-1) consisting of a mixture of Dunstan and Short macroelements (Dunstan and Short, 1977), Gamborg microelements (Gamborg *et al.*, 1968) and Murashige and Skoog (MS) vitamines (Murashige and Skoog, 1962), supplemented with 100 g/L sucrose (Table 2.39-1). Putrescine was added to the Medium A-1 at 2 mM concentration.

2. After 15 days, flowers are transferred to a fresh medium: Medium A-2 containing the same ingredients as Medium A-1 but supplemented with 0.1 mM spermidine instead of putrescine (Table 2.39-1) (second step).

3. Gynogenic embryos breaking through the ovary wall (Fig. 2.39-2) are transferred to embryo culture medium composed of macro, microelements and vitamins according to MS, without growth regulators and supplemented with 40 g/L sucrose (Medium B; Table 2.39-1) until complete plant development is reached.

4. After 40 days, the regenerated plantlets are removed and micropropagated by cutting the bulb into halves on Medium B supplemented with 2 mg/l NAA and 2 mg/l BA (Table 2.39-1).

All components of the media are adjusted to pH 5.8, solidified with 0.75% agar (MAG S.A, Argentinean brand) and sterilized by autoclaving at 121°C for 20 min. Putrescine and spermidine are dissolved in water and are added filter-sterilized to the autoclaved media, using a bacterial filter (pore size 0.22 µm).

Table 2.39-1. Inorganic and organic constituents of various culture media used in *in vitro* onion gynogenesis

Components	Media A		Media B	
	A-1 (mg/L)	A-2 (mg/L)	Embryo culture medium (mg/L)	Micropropagation medium (mg/L)
Macro elements	Dunstan and Short		Murashige-Skoog	
KNO_3	2,530	2,530	1,900	1,900
NH_4NO_3	320	320	1,650	1,650
$CaCl_2 \times 2H_2O$	150	150	440	440
$MgSO_4 \times 7H_2O$	247	247	370	370
$NH_4H_2PO_4$	230	230	-	-
$NaH_2PO_4 \times 2H_2O$	152	152	-	-
KH_2PO_4	-	-	170	170
$(NH_4)_2SO_4$	134	134	-	-
Micro elements	Gamborg		Murashige-Skoog	
KI	0.75	0.75	0.83	0.83
H_3BO_3	3	3	6.2	6.2
$MnSO_4 \times H_2O$	10	10	-	-
$MnSO_4 \times 4H_2O$	-	-	22.3	22.3
$ZnSO_4 \times 7H_2O$	2	2	8.6	8.6
$Na_2MoO_4 \times 2H_2O$	0.25	0.25	0.25	0.25
$CuSO_4 \times 5H_2O$	0.025	0.025	0.025	0.025
$CoCl_2 \times 6H_2O$	0.025	0.025	0.025	0.025
$FeSO_4 \times 7H_2O$	27.8	27.8	27.8	27.8
Na_2-EDTA	37.3	37.3	37.3	37.3
Vitamins	Musashige-Skoog		Murashige-Skoog	
Inositol	100	100	100	100
Nicotinic acid	0.5	0.5	0.5	0.5
Pyroxidine x HCl	0.5	0.5	0.5	0.5
Thiamine x HCl	0.1	0.1	0.1	0.1
Glycine	2.0	2.0	2.0	2.0
Growth regulators				
Putrescine (mM)[1]	2.0 mM	-	-	-
Spermidine (mM)[1]	-	0.1 mM	-	-
NAA	-	-	-	2.0
BA	-	-	-	2.0
Others				
Sucrose	100,000	100,000	40,000	40,000
Agar	7,500	7,500	7,500	7,500
pH	5.8	5.8	5.8	5.8

[1] Added by sterile filtration

Culture condition
- Petri dishes containing 30 ml of medium each are placed under a 16 h photoperiod (30 μmoles m^{-2} s^{-1}, supplied by Phillips cool-white fluorescent bulbs), at 25/20°C±2°C.

Ploidy level analysis
First Procedure: Chromosome stain
1. Chromosome numbers are determined in root-tip cells obtained from plantlets after treatment in 0.1% colchicine for 3 h.

2. Then, fix in 3:1 ethanol 96% glacial acetic acid.

3. Digest in 1 N HCl at 60°C for 8 min and squash in 45% acetic acid.

4. For microscopic inspection of the karyotype, the root tips are stained with 1% haematoxilin.

The chromosome counting in root tips does not necessarily reflect the situation in shoot tip.

Second Procedure: Flow cytometry
Nuclei are isolated from micropropagated plantlets using the following procedure:
1. 120 mg of healthy young leaves from micropropagated plantlets are chopped with a razor blade in 1 ml of ice-cold lysis and staining buffer (100 ml of this buffer contains: 181.7 mg Tris; 74.45 mg Na_2 EDTA; 596,5 g KCl; 116.9 mg NaCl; 100 µl Triton X-100; 5 mg RNase A and 100 mg propidium iodide).

2. Chopped leaves are filtered through a 50 µm mesh nylon filter.

3. Measurements are performed on flow cytometer Becton Dickinson equipped with on 488 nm argon laser lamp.

4. 2,000-5,000 nuclei are measured per sample.

5. Relative amount of DNA is scored by comparison of the G_1 and G_2 peaks of haploid and diploid samples using a diploid onion as external standard.

Chromosome doubling procedure
In vitro diploidization is carried out using colchicine as a genome doubling agent.
1. Whole micropropagated haploid plantlets are cut longitudinally into halves. Then, the sections are placed in a flask sealed with Parafilm and incubated with 0.25 g/l colchicine, filter sterilized, for 48 h. Following colchicine treatment, plantlets are placed on fresh Medium B plus 2 mg/l NAA and 2 mg/l BA, and cultured under a 16 h photoperiod (30 µmoles $m^{-2} s^{-1}$, supplied by Phillips cool-white fluorescent bulbs), at 25/20°C±2°C.

2. Ploidy is measured by flow cytometry on new young leaves.

3. Doubled haploid plantlets are transferred to greenhouse conditions.

4. Acclimatized plantlets are transferred to open field condition.

Efficiency and Applications

In onion, the use of the female gametophyte (gynogenesis) is an alternative way for production of haploids. Currently, it is the method of choice and has been reported to be successful by several authors (Bohanec *et al.*, 1995; Geoffriau *et al.*, 1997; Martinez *et al.*, 2000; Michalik *et al.*, 2000). Nonfertilized flower culture has proven to be the least laborious and most practical method for a large number of onion accessions (Geoffriau *et al.*, 1997). The results obtained so far, indicate that the effectiveness of gynogenesis depends on the flower donor genotype, developmental stage of the female gametophyte, the nutrient medium composition and culture conditions. Therefore, it is necessary to develop efficient protocols for regeneration and multiplication of gynogenic plantlets, as well as for the doubling of their genomes to obtain DH lines, especially when the genotypes used are considered to be average or low responsive materials. This could allow for the production of a sufficient number of genetically stable DHs plants from a wide range of genotypes.

Polyamines are normal plant growth regulators involved in all growth or developmental process in plants. The use of spermidine followed by putrescine treatment could permit an improvement in the production of embryos and haploid onion plants through *in vitro* gynogenesis (Martínez *et al.*, 2000). A protocol, using these growth regulators, is presented in this Manual.

Chromosome doubling during the *in vitro* stage is one of the main obstacles to achieve DH lines. Several doubling agents have been reported so far. The use of colchicine, as an antimitotic agent, is proposed in this protocol. Feulgen stain and flow cytometry are commonly used procedures for determining the DNA content of plant nuclei. The latter one, is a more expensive technology, but provides extremely rapid measurements in a reliable and accurate way.

Following the presented protocol, the Argentinean genotypes, Valcatorce INTA, and Torrentina can be considered as average responsive materials relative to the numbers of the gynogenic embryos and haploid plants produced. The achieved gynogenic embryo and haploid plants rates were 2.0 and 9.5, and 1.0 and 1.9 per 100 cultured flowers, for Valcatorce INTA and Torrentina, respectively. Chromosome doubling efficiency, using colchicine, was 66.6%.

References

Bohanec, B., M. Jakse, A. Ihan and B. Javornik, 1995. Studies of gynogenesis in onion (*Allium cepa* L.): induction procedures and genetic analysis of regenerants. Plant Sci. **104**: 215-224.

Dunstan, D.I. and K.C. Short, 1977. Improved growth of tissue cultures of the onion, *Allium cepa*. Physiol.Plant. **41**: 70-72.

Gamborg, O.L., R.A. Miller and K. Ojima, 1968. Nutrient requirements of suspension cultures of soybean root cells. Exp.Cell Res. **50**: 151-158.

Geoffriau, E., R. Kahane and M. Rancillac, 1997. Variation of gynogenesis ability in onion (*Allium cepa* L.). Euphytica, **94**: 37-44.

Martinez, L., C. Aguero, M. Lopez and C. Galmarini, 2000. Improvement of *in vitro* gynogenesis induction in onion (*Allium cepa* L.) using polyamines. Plant Sci. **156**: 221-226.

Michalik, B., A. Adamus and E. Nowak, 2000. Gynogenesis in Polish onion cultivars. J.Plant Physiol. **156**: 211-216.

Murashige, T. and F. Skoog, 1962. A revised medium for rapid growth and bioassays with tobacco tissue cultures. Physiol.Plant. **15**: 473-497.

2.40
Haploid induction in onion *via* gynogenesis

M. Jakše and B. Bohanec
University of Ljubljana, Biotechnical Faculty, Centre for Plant Biotechnology and Breeding, Jamnikarjeva 101, 1111 Ljubljana, Slovenia

Introduction

Onion (*Allium cepa* L.) is a biennial monocotyledoneus plant belonging to the Alliaceae family. Within the *Allium* species onion is the most economically important with a total world acreage of 2,7 million ha dry onions and production of over 46 million ton in 2001 (FAO, 2001).The demand for specific varieties adapted to local agroclimatic conditions is very high since onion is a photoperiodically sensitive plant and forms bulbs only after specific environmental conditions, which vary among genotypes. According to the breeding method, two types of onion varieties are cultivated: open pollinated and hybrid varieties. Hybrid varieties have been produced for over 50 years from elite inbred lines. Expected benefits from hybrid cultivars are: higher yield (expressed heterosis), improved uniformity and for the seed producers - protection of plant material. The characteristic for onion inbred lines is their relatively high heterozygosity resulting from limited (two to three) cycles of self-pollination. Doubled haploids provide an alternative strategy that offers, for the first time in onion breeding, complete homozygosity and phenotypic uniformity. Another advantage is a substantial reduction in the time required to produce inbred lines, considering that onion is a biennial plant that requires up to 10 years to obtain nearly homozygous inbred lines by conventional breeding. Haploid plants can be obtained from male or female gametic cells. However, as reviewed by Keller and Korzun (1996), large anther culture experiments in onion have failed. Flowering is induced by environmental factors causing the apex to cease production of leaf primordia and initiate an inflorescence. A single onion inflorescence (umbel) might consist of up to 2,000 or more flowers and some genotypes form more than one (even more than a dozen) flower stalks. Within the umbel, flowers open in successive order so that the blooming time of each umbel is over 10 days. Each flower consists of a single superior pistil with three locules having two ovules in each carpel. The nectaries are between the carpels of the ovary and the three inner stamens. According to Klein and Korzonek (1999), flower development can be divided into seven stages related to the umbel size and developmental stages of anthers and ovules. In variety 'Kutnowska' during first 3 stages, the umbel (smaller than 2 cm) is enveloped by a spathe. The beginning of meiosis in anthers starts in stage IV, with the subsequent meiotic divisions in stage V. Ovule development starts later than meiosis in the anthers, when flower buds are 3.5-4.0 mm long (predominantly in stage VI). Flowers in stage VI are actually used in majority of gynogenic haploid production procedures.

Protocol

Donor plants and growth condition

Flowers in stage VI (as defined by Klein and Korzonek, 1999) are generally used for inoculation. Choose genetically variable breeding material to obtain information on gynogenic ability of particular lines. Expect high variability of gynogenic ability between lines and between individual donor plants. Hybrids and A-lines possessing CMS-sterile cytoplasm should be avoided. We would recommend the publicly available inbred line B2923B with individual bulb responsiveness over 80%, which can be requested from M.J. Havey, USDA-ARS, 1575 Linden Drive, Madison, WI 53706, USA, as suitable standard with high gynogenic response. Mark and keep records of each donor plant within the line or population. Keep bulbs and try to induce flowering of the most responsive ones in the second growth year.

Conditions for donor plant growth, in particular temperature, are often of highest importance for success of haploid induction procedures. Place donor plants at bolting time in low temperature growth chambers. Optimal temperature regime should be about 14°C continuously, light intensity 50 µmol m^{-2} s^{-1}. Alternatively, keep donor plants in a cooled greenhouse and try to induce flowering as soon as possible to avoid high temperatures. Watering with systemic insecticide Confidor® (imidacloprid) at regular intervals is advisable since thrips are the major cause of contamination.

Flower bud collection

1. Two possible methods are used for flower bud collection: either the whole umbel is excised at the stage at which about 30% of the flower buds have reached the appropriate stage, or the buds are sheared off by scissors when they reach the appropriate stage, usually at two day intervals. The flowers are cut with minimal pedicle.

2. Sterilize flowers from each donor plant separately by a 10 min treatment in dichloroisocyanuric acid disodium salt (16.6 g/L) with a few drops of Tween 20. Agitate during treatment and wash three times with sterile water. Sterilized flowers might remain in water until placement to induction media. This disinfectant is superior to the solid organic chlorine and the unstable sodium hypochlorite, which may cause damage to the delicate tissue.

In vitro culture

1. Place 30 flowers in each 10 cm Petri dish containing induction medium consisting of BDS (Dunstan and Short, 1977) medium supplemented with 500 mg/L myo-Inositol, 200 mg/L proline, 2 mg/L BAP, 2 mg/L 2,4-D, 100 g/L sucrose and 7 g/L agar (Table 2.40-1). Seal the dishes with Parafilm. Place Petri dishes in a culture room with moderate intensity fluorescent white light illumination (16/8 h) at 21-25°C. Check for contamination during the first 2 weeks and repeat inoculation if necessary.

2. Embryo formation is expected between 60 to 180 days in culture (mostly around day 100). Note that at time of embryo emergence the majority of ovaries will turn from green to pale yellow in color. Some genotypes tend to form callus on flowers bases (region of nectaria) on induction medium. On BDS induction medium shoot formation from callus or other flower parts almost never happens, however such shoots need to be removed and not mixed with haploid regenerates.

3. Embryos are formed gradually and therefore only flowers forming an embryo (Fig. 2.40-1a) are removed from culture and embryos extracted. Check carefully other locules of removed flowers for smaller embryos that have not yet penetrated the ovary wall. Discard

abnormal embryos. Those are clearly distinguishable from somatic regenerants that on standard medium, on very rare occasions, proliferate at the flower base. Haploid embryos are complete bipolar structures with both shoot and root (Fig. 2.40-1b) and are visible when extracted from ovaries. It is important to detect them and transfer them to elongation medium or alternatively to genome doubling treatments soon after they emerge from ovaries. Petri dishes need to be scored at least once a week for embryo appearance.

Table 2.40-1. Composition of BDS induction medium

Medium components	(mg/L)
Macro nutrients	
KNO_3	2,530
$(NH_4)_2SO_4$	134
$MgSO_4 \times 7H_2O$	247
$NH_4H_2PO_4$	230
NH_4NO_3	320
$NaH_2PO_4 \times 2H_2O$	172
$CaCl_2 \times 2H_2O$	150
Micro nutrients	
$MnSO_4 \times 4H_2O$	13.2
$ZnSO_4 \times 7H_2O$	2
H_3BO_3	3
KJ	0.75
$CuSO_4 \times 5H_2O$	0.039
$Na_2MoO_4 \times 2H_2O$	0.25
$CoCl_2 \times 6H_2O$	0.025
Iron source	
$Na_2EDTA \times 2H_2O$	37.25
$FeSO_4 \times 7H_2O$	27.85
Vitamins	
Nicotinic acid	1
Pyridoxine HCl	1
Thiamine HCl	10
myo-Inositol	500
Other components	
Proline	200
BAP	2
2,4-D	2
Sucrose	100,000
Agar	7,000
pH	6.0

Genome doubling procedures

Reports on spontaneously doubled haploids during gynogenesis vary from 1% to 30%. In our studies, at least 90% of the regenerates are haploid. Our unpublished findings indicate that

2,4-D may be partially involved in the rate of "spontaneous" doubling, based on the observation that less than 3% were doubled when it was not in the induction media. One of the problems in genome doubling in onion is the inaccessibility of the apical meristem that is hidden in the bulb (bottom part of the plant). The majority of published approaches for genome doubling have exposed longitudinally sliced bottom parts of *in vitro* grown plantlets. Such plantlets need first to be grown from embryos to appropriate stage so that the bottom parts are large enough to be sliced and the apices exposed to antimitotic agents added in media. However, we usually use the embryo stage for doubling treatments.

Figure 2.40-1. a) Gynogenic embryos arising from the top of the ovary as loop structures; b) Haploid embryos at chromosome doubling treatment.

Protocol for chromosome doubling at embryo stage
1. Embryos of the same genotype are placed in Petri dishes (multiwell plates, etc. Fig. 2.40-1b) for genome doubling treatment. Colchicine and oryzalin are more often used as antimitotic agents in a wide range of concentrations, since an optimal concentration is difficult to define. On one hand low concentrations or short treatment duration are not efficient enough and on the other hand high concentrations or long lasting treatments (3 days or more) have a tendency to cause severe damage to plantlets or partly induce tetraploidization. We propose 1-2 day treatment of embryos on liquid medium (½ strength BDS + 30 g/L glucose) supplemented with 50 μM APM (amiprofos-methyl).

2. After treatment, place embryos in individual test tubes (150x25 mm) containing elongation medium consisting of ½ strength BDS medium, 30 g/L glucose and 7 g/L agar. Plantlets develop from embryos in approximately 6 weeks. At this stage a portion of the embryos will fail to develop. Transplant only rooted plantlets after removing agar in water and acclimatize them in garden flower potting mix and keep them under high relative humidity.

3. Part of the third or fourth leaf is used for ploidy analysis by flow cytometry. Note, if flow

cytometer is not available at the station, leaves should be wrapped in moistened paper tissues and mailed to laboratory with flow cytometric facility that can be reached within 3-4 days.

4. Doubled haploid and mixaploid genotypes (30-50% expected) are further grown to maturity. As soon as the bulb neck becomes soft and foliage is starting to collapse, bulbs should be pulled from soil and dried. Vernalized bulbs are induced to flower in the next season and, typically, selfed or alternatively intercrossed. At flowering, male fertility might be checked using ordinary acetocarmine staining of pollen grains. It can be expected that depending on the genotype only a proportion of doubled haploid plants will produce seeds.

Efficiency and Applications

For plant breeding purposes the induction of doubled haploid lines needs to be efficient, relatively easy, and as much as possible genotype insensitive. To obtain such high demands, the regenerants should be formed in high frequencies, should be easy to double from haploid to the doubled haploid level, and the plantlets produced should be easily acclimatized. The double haploids generated should maintain their genetic integrity and produce fertile seed. To estimate doubled haploid frequencies, Bohanec and Jakse (1999) analyzed 39 onion accessions from Europe, North America and Japan. Two European and three Japanese accessions produced no embryos. The highest gynogenic yield was obtained from North American cultivars and inbred lines. Two inbred lines and one F_1 hybrid produced up to 22.6 embryos per 100 cultured flowers

The outlined procedure for haploid induction and the constitution of media have been used in our laboratory for five years with about 3-8 thousand embryos induced each year. The procedure was optimized to require minimal input with optimal output. There are several alternatives to the proposed haploid induction scheme. In particular, chromosome doubling can be postponed until plantlets are formed (about 2 months). Such *in vitro* grown plantlets might be either micropropagated (basically according to Kahane *et al.*, 1992) or sliced basal parts exposed to genome doubling treatments. This alternative approach has both positive and negative aspects. Using micropropagation, valuable lines are multiplied, however micropropagation requires addition of cytokinins, which might cause hyperhydricity and loss of regenerants. Prolonged *in vitro* treatments increase risk of contamination and the genome doubling procedure of sliced shoot bases is not necessarily more efficient than treatments of embryos. Another procedure might be to check for ploidy level while still in *in vitro* culture. In this case, only the shoot apices of haploid plants will be treated with antimitotic drugs, as described above.

References

Bohanec, B. and M. Jakse, 1999. Variation in gynogenic response among long-day onion (*Allium cepa* L.) accessions. Plant Cell Rep. **18**: 737-742.
Dunstan, D.I. and K.C. Short, 1977. Improved growth of tissue cultures of the onion, *Allium cepa*. Physiol.Plant. **41**: 70-72.
FAO, 2001. FAO Statistical Database,*http://apps.fao.org*.
Keller, E.R. and L. Korzun, 1996. Haploidy in onion (*Allium cepa* L.) and other *Allium* species. In: *In Vitro* Haploid Production in Higher Plants. Vol. 3. Jain, M.S., S.K. Sopory and R.E. Veilleux (Eds.) Kluwer Academic Publishers, Dordrecht. pp. 51-75.
Klein, M. and D. Korzonek, 1999. Flower size and developmental stage of *Allium cepa* L. umbels. Acta Biol. Cracoviensia Series Botanica, **41**: 185-192.
Kahane, R., M. Rancillac and B.T. Delaserve, 1992. Long-term multiplication of onion (*Allium cepa*. L.) by cyclic shoot regeneration *in vitro*. Plant Cell Tiss.Org.Cult. **28**: 281-288.

2.41
In vitro androgenesis in apple

M. Höfer
Federal Centre for Breeding Research on Cultivated Plants, Institute for Fruit Breeding, Pillnitzer Platz 3a, D-01326 Dresden, Germany

Introduction

Most temperate fruit trees are characterized by a long reproductive cycle and juvenile phase (both of several years), a tendency to allogamy and a large tree size. Fruit trees are generally highly heterozygous, outbreeding species, which are propagated asexually. For these reasons, their genetic improvement by conventional methods is time-consuming and limited by space required for field experiments. The production of haploids offers new possibilities for genetic studies and for an increased efficiency in selection. Successful *in vitro* approaches to induce haploids in apple (*Malus domestica* Borkh.; $2n=2x=34$) have been rather limited in comparison with other plant species, until recently (Höfer and Lespinasse, 1996). The methods generally used for haploid induction are: *in vitro* androgenesis by anther and microspore culture; *in vitro* gynogenesis by unfertilized ovule culture, and *in situ* parthenogensis by irradiated pollen followed by *in vitro* culture of immature embryos or cotyledons. Due to higher efficiency, a description of methods in this manual will be limited to the protocols of anther and microspore culture. Anther culture in apple was pioneered by Japanese scientists at the beginning of the 1970s. They induced calli capable of root formation. Subsequently, several working groups initiated haploid induction in apple by anther culture and the induction of embryogenesis. However, limited plant formation has been reported. For further details of protocols and results see the review by Höfer and Lespinasse (1996).

Protocol for anther culture

Donor plants and growth conditions
For anther culture of apple, flower buds are collected when micropsores are at the mid uninucleate stage of pollen development, before vacuole formation. Morphologically this stage is characterized by the beginning of petal emergence on the king flower, which is the most advanced in the development on an inflorescence (Fig. 2.41-1a).

The determination of the optimal stage of pollen development is an essential step to increase the efficiency of anther culture. However, routine identification of the stages of pollen development is difficult because of the gradients within inflorescences. Thus, an indirect method associated with a morphological indicator was developed. On the basis of correlations between the stages of pollen development and easily determinable macroscopic characters of the flower bud, it was possible to define anther length, independent on genotype, as a marker of the optimal stage of pollen development in apple. The defined length is dependent on the type of the donor material (see below) and should be tested for your own genotypes and donor material by staining with aceto carmine or other stain.

To extend the anther culture period throughout the year, flower buds are taken from cut bud wood forced to flower under different temperatures (room temperature; 12 h 16°C in light and 12 h 12°C in the dark) and directly from trees growing in the orchard. Using this method, the flowers are available at almost any time from January to April, just until three weeks before blossom in Central Europe.

Pretreatment
The requirement of a cold pretreatment should be tested for each genotype. Single buds are stored at 4°C for 1 to 2 weeks in the refrigerator, the buds are held so that only the cut under part of the buds are in contact with water (this is done to prevent phenols accumulating).

In vitro procedure
1. Surface sterilize buds by soaking in 0.1% mercuric hypochloride for one minute (take care, mercuric chloride is toxic), separate the single flowers and isolate the anthers. Use only the ten bigger anthers of the outside anther circle of the flower (Fig. 2.41-1b).

2. The defined length for anthers of buds from forced bud wood should be 1.0 to 1.2 mm and 1.4 mm for anthers of buds taken from field-grown trees.

3. Incubate the 10 anthers per flower in a Petri dish (60 mm in diameter) with solid induction medium and place the anther cultures in a controlled environment chamber.

4. Cultures should be checked every 8 weeks for a period of 9 months.

5. Transfer embryos only to the optimized regeneration medium in flasks (Fig. 2.41-1c) and replenish the medium during the regeneration phase every 4 weeks (Fig. 2.41-1d). Note that the regeneration process can take 12 months or longer.

6. New regenerated shoots should be transferred to the proliferation medium (Fig. 2.41-1e).

7. The first evaluation of the regenerated lines is done during the proliferation phase and includes determination of the ploidy and investigation of the homozygosity by isoenzyme and microsatellite marker analysis.

8. The transfer of shoots to *in situ* conditions can be carried out by *in vitro* rooting and subsequent acclimatization (Fig. 2.41-1f) or by grafting *in vitro* shoots directly onto trees in the orchard.

Culture media
- The culture medium for the induction process is a solid MS basic medium with 50 g/L sucrose. According to the optimization experiments two phytohormone combinations can be used for all experiments: IBA 0.2 mg/L and kinetin 0.2 or 0.5 mg/L supplemented with 4 mg/L IAA, 1 mg/L GA_3 and 20 mg/L adenine sulphate, pH 5.7.
- The regeneration medium is a solid MS basic medium with 0.1 mg/L IBA, 0.1 mg/L thidiazuron, 1 mg/L GA_3 and 30 g/L sucrose, pH 5.7.
- The proliferation medium is also a solid MS basic medium with 0.2 mg/L IBA, 0.5 mg/L BA, 1 mg/L GA_3 and 30 g/L sucrose, pH 5.8.
- For *in vitro* rooting, a MS basic medium with 0.3 mg/L IBA and 20 g/L sucrose is used pH 5.7.

Conditions for in vitro culture
For embryo induction, the Petri dishes are incubated at 27°C in the dark, whereas the regeneration process and the proliferation phase take place in a light/dark regime of 16 h under 3000 lux/ 8 h dark, at 23°C. For rooting, we use the same culture regime as described above, but with the first four days in darkness.

Protocol for microspore culture

This was developed from the anther culture protocol.

Donor material
For microspore culture, the same donor material and methods of extending the incubation period are used as described for anther culture. The difference compared to the anther culture is that the flower buds are collected when micropsores are at the late uninucleate stage of development, after vacuole formation.

Pretreatment
Buds are given the following cold pre-treatment: 4°C for 1 or 2 weeks depending on the known responses of genotype and donor material in anther culture.

In vitro procedure
1. Surface sterilize buds by soaking in 0.1% mercuric hypochloride for one minute (take care, this is toxic), separate the single flowers and isolate the anthers.

2. Only the ten bigger anthers of the outside anther circle of the flower are used (Fig. 2.41-1b).

3. The anther lengths are: 1.4–1.6 mm for buds from forced bud wood, and 1.8 mm for anthers of buds taken from field-grown trees.

4. Collect and stir the anther in medium B (see below) with a magnetic stirrer for 4 min at 250 rpm.

5. Filter the crude microspore population through a 30 µm filter.

6. Wash the filtrate four times with the same medium by centrifugation at 220 g for 5 min.

7. The microspore pellet obtained after the final centrifugation step is resuspended in medium B and cultured for 2 or 3 days at 4°C, this is a starvation and cold pre-treatment (Fig. 2.41-2a).

8. After pre-treatment the microspores are collected, centrifuged and resuspended in induction medium in a 4-well plate (6x6 cm; Fig. 2.41-2b).
   Note: Microspore viability should be determined by staining with fluorescein diacetate after isolation and starvation. Responsive cultures have a viability of 60–80%.

9. The density of microspores should be adjusted to about $10\text{-}15 \times 10^5$ microspores/ml.

10. Cultures should be checked every 8 weeks for 9 months (Fig. 2.41-2 c,d).

11. Embryos and embryo like structures are transferred to the regeneration medium as they develop (Fig. 2.41-2e).

12. All following steps, the regeneration (Fig. 2.41-2f), the proliferation and the transfer to *in situ* conditions are according to the protocol for anther culture.

Culture media
- The isolation and starvation medium B contains 1.49 g/L KCl, 0.12 g/L $MgSO_4$, 0.11 g/L $CaCl_2$, 0.14 g/L KH_2PO_4 and 54.7 g/L mannitol, at pH 7.0
- One basal medium with 2 different sugar concentrations is used for the induction of apple microspore embryogenesis for all tested genotypes: modified N6 macro minerals [1950 mg/L KNO_3, 277 mg/L $(NH_4)_2SO_4$, 400 mg/L KH_2PO_4, 166 mg/L $CaCl_2$ x $2H_2O$, 185 mg/L $MgSO_4$ x $7H_2O$) plus MS micro minerals, B5 vitamins, 1,256 mg/L glutamine and 1,950 mg/L MES [2-(N-morpholino) ethanesulfonic acid] (Touraev *et al.*, 1996) with 90 g/L or 120 g/L maltose, pH 6.2.
- The regeneration, proliferation and rooting media are those given for anther culture.

Conditions for in vitro culture
For embryo induction, the 4-well plates are incubated at 27°C in the dark, whereas the regeneration process and the proliferation phase take place in a light/dark regime of 16 h under 3000 lux/8 h dark, at 23°C. For rooting, we use the same conditions, but the first four days are in darkness.

Evaluation of the lines induced by *in vitro* androgenesis and transfer to *in situ* conditions

The first step of an evaluation should be the determination of the ploidy level with the use of flow cytometer. Previous results have shown a distribution of the ploidy level in regenerating tissues from haploid to tetraploid, an investigation of the homozygosity is therefore necessary.

Investigations of the homozygosity state can be done by isozyme analysis (Höfer and Grafe, 2000) and by analysis of simple sequence repeat markers (Höfer *et al.*, 2002). According to these analyses all lines induced by *in vitro* androgenesis in our laboratory are homozygous.

Since not all lines respond to one method, both *in vitro* rooting and grafting of *in vitro* shoots are used to transfer lines to *in situ* conditions. After one year, the lines can be grafted on rootstocks in the nursery and can be available for a comprehensive evaluation compared to the donor cultivar.

Efficiency and Applications

In our laboratory, the refinement of *in vitro* androgenesis in apple was initiated using anther culture. Experiments were carried out with different genotypes from the apple breeding programme at Dresden-Pillnitz, Germany including cultivars and breeding clones. The first step of assessment, the efficiency of embryo induction, was strongly dependent on genotype and reached maximum rates of 15%. Compared to other plant species, the first embryos appear relatively late, after 3 months. Unfortunately, in apple, no source of material from plants cultivated under defined conditions exists. Differentiation of flower buds occurs in the year prior to their maturation and depends strongly on environmental conditions. Therefore, anthers probably have different endogenous hormone levels from one season to another. After transferring embryos to an optimized regeneration medium, many embryos increased in size, changed colour and developed the first cotyledons. Plant regeneration could only be observed after adventitious shoot formation, where adventitious buds develop directly from the primary embryos or after a phase of secondary embryogenesis taking 6-12 months. The conversion frequency depends strongly on the quality of the primary embryos, and on the donor genotype, but could reach a maximum of 52%.

Although anther culture is much simpler in terms of handling procedures, microspore culture has several important advantages. Firstly, the formation of calli and embryos that often form from somatic tissues of the anther is avoided. Secondly, there is direct access to the microspores, which speeds up the optimization of culture conditions. A protocol for wheat (Touraev *et al.*, 1996) was used as a basis for microspore embryogenesis, which was carried through to plant formation in apple (Höfer *et al.*, 1999). By varying the parameters of stress pretreatment and culture conditions, embryo induction rates of up to 35% can be achieved. After starvation of microspores in combination with a cold treatment, the first sporophytic mitoses were observed in carbohydrate-containing media after 3-5 days of culture. A small percentage of microspores continued repeated cell divisions and formed multicelluar structures. After 8-12 weeks, the first embryos could be observed. The regeneration process is similar to that of androgenic embryos *via* apple anther culture.

Androgenesis has been successfully induced *in vitro* and a reproducible protocol for plant regeneration from embryos in apple has been developed. At present, the first 24 androgenic lines induced *via* anther or microspore culture exist as grafts on different rootstocks in the orchard and are available for evaluation of morphology, resistance traits and fruit quality.

Acknowledgments
I am grateful to Prof. E. Heberle-Bors and to Dr. A. Touraev for teaching me about wheat and tobacco microspore culture methods during a short-term scientific mission of the COST-Action 824 (Microspore embryogenesis).

References

Höfer, M., A. Gomez, E. Aguiriano, J.A. Manzanera, and M.A. Bueno, 2002. Analysis of SSR markers in homozygous lines of apple. Plant Breed. **121**: 159-162.
Höfer, M. and Ch. Grafe, 2000. Preliminary evaluation of doubled haploid-material in apple. Acta Hort. **538**: 587-592.
Höfer, M. and Y. Lespinasse, 1996. Haploidy in apple. In: *In Vitro* Haploid Production in Higher Plants. Jain, M.S., S.K. Sopory and R.E. Veilleux (Eds.) Kluwer Academic Publishers, Dordrecht. pp. 259-274.
Höfer, M., A. Touraev, and E. Heberle-Bors, 1999. Induction of embryogenesis from isolated apple microspores. Plant Cell Rep. **18**: 1012-1017.

Touraev, A., A. Indrianto, I. Wratschko, O. Vicente, and E. Heberle-Bors, 1996. Efficient microspore embryogenesis in wheat (*Triticum aestivum* L.) induced by starvation at high temperature. Sex.Plant Reprod. **9**: 209-215.

2.42
Doubled haploid production in poplar

S.B. Andersen
The Royal Veterinary and Agricultural University, Department of Agricultural Sciences, Section Plant Breeding and Crop Science, Thorvaldsensvej 40, DK-1871 Frederiksberg C, Denmark

Introduction

Various species of poplar (*Populus*) are cultivated for their high production of wood for paper pulp, energy and for park trees. Like other tree species they have a long generation time and show strong depression of vigour if forced to inbreed. In addition, poplar species are dioecious with separate male and female plants, which further complicates traditional inbreeding for increased homozygosity. Chromosome doubled haploids in these species have future potential both as a means to improve traditional breeding approaches and as an important tool for genetic studies. Successful anther culture was first reported in the genus *Populus* by Wang *et al.* (1975) who described the formation of haploid plants *via* organogenesis from pollen calli. The methods using relatively high concentrations of growth hormones in the anther induction medium have been reported successful in inducing microspore derived callus and subsequent plant regeneration in a number of *Poplar* species (Uddin *et al.*, 1988; Wu and Nagarajan, 1990; Stoehr and Zsuffa, 1990). These methods, however, also induce the formation of embryogenic callus derived from somatic anther culture. The present protocol (Baldursson *et al.*, 1993) induces microspore embryogenesis directly in three different *Populus* species (*P. maximowiczii, P. balsamifera* and *P. trichocarpa*), with frequencies of 1-10 embryos per 1,000 cultured anthers. On average, 30-75% of these embryos can be grown successfully into plants and established in soil. For anther culture in most other species, the genotype of the donor plant has a strong effect on results with known protocols. Genotype differences in anther culture response have also been demonstrated in poplars (Uddin *et al.*, 1988). It is therefore important to test the response of a number of trees for each species to identify particularly responsive genotypes.

Protocol

Donor plants and growing conditions
Because poplars are generally large trees by the time they flower, it is normally not possible to grow the donor plants for anther culture, except in their natural environment, outside in fields, forests and parks. Therefore, twigs with flower buds from male trees must be collected during the winter period while dormant. The twigs can be stored with their cut ends in water for up to three weeks in darkness at 4°C temperature before being forced to flower at 21-24°C with 20-40 W/m^2 white fluorescent light.

Anthers isolation
Pollen development within catkins is synchronised and entire catkins with almost all microspores at the mid to late unicellular stage of development can be isolated shortly after they have started to emerge from the bud scales. The developmental stage of microspores within the catkins can be checked cytologically by squash preparations of a few anthers stained with aceto carmine acid.

Entire catkins can be surface sterilized by brief dipping in 70% ethanol with an added drop of Tween 80 per litre, followed by 10 min in 7% calcium hypochlorite and three rinses with sterile water. Anthers are isolated under aseptic conditions, e.g. in a laminar flow bench, from sterilized catkins using a pair of sterile forceps.

Anther plating and culture
For embryo induction, isolated anthers are placed on MS-1 medium (Table 2.42-1), at a density of approximately 100 anthers per 9 cm Petri dish with 25-30 ml medium. Cultures are incubated at 27°C in darkness for 6-8 weeks for embryo formation. Embryos emerging from cultured anthers are transferred to fresh MS-1 medium and incubated in light (15 W/m^2) for 10 days at 27°C for maturation, before plant regeneration.

Regeneration
For shoot regeneration, mature embryos are sub-cultured on WPM medium with 1.5 µM BA (Table 2.42-1) and transferred to fresh medium every two weeks. The cultures are maintained at 27°C with continuous white fluorescent light (15 W/m^2). Adventitious shoots are formed after 3-5 subcultures. Upon subculture on the same WPM medium, regenerated shoots will form axillary shoots, which can be divided into new clumps of shoots for clonal multiplication. Alternatively, shoots may form roots on WPM medium with 0.2 µM of IBA. Rooted shoots are established in a 1:1:1 mixture of peat, vermiculite and perlite substrate in a mist chamber (96% relative humidity) for 2-3 weeks. Plants can be subsequently grown in a glasshouse and subsequently planted into the field.

Chromosome doubling
Plants regenerated from anther cultures of *Populus* species generally have a high frequency of spontaneous chromosome doubling. In most cases, 60-70% of plants obtained have a diploid chromosome number, which means that artificial chromosome doubling of the plants is generally not necessary.

Media
MS-1 medium for induction of embryos from cultivated anthers is modified Murashige and Skoog's medium with concentration of ammonium nitrate reduced 10 times, with 6% maltose instead of sucrose, supplemented with 5.1 mM L-glutamine, 5 µM 6-benzylaminopurine (BA) and solidified with 0.3% gelrite (Kelco). The WPM (woody plant medium) is the medium described by Lloyd and McCown (1981) containing 2% sucrose, 0.1% gelrite (Kelco) and 0.3% Difco Bacto agar, but without growth regulators.

Table 2.42-1. Composition of media used for poplar anther culture

Media components	MS-1 induction medium (mg/L)	WPM shoot regeneration medium (mg/L)	WPM root formation medium (mg/L)
Macro elements			
KNO_3	1,900	-	-
NH_4NO_3	165	400	400
K_2SO_4	-	990	990
$CaCl_2 \times 2H_2O$	440	96	96
KH_2PO_4	170	170	170
$MgSO_4 \times 7H_2O$	370	370	370
$Ca(NO_3)_2 \times 4H_2O$	-	556	556
Micro elements			
$ZnSO_4 \times 7H_2O$	8.6	8.6	8.6
KI	0.83	-	-
$MnSO_4 \times H_2O$	22.3	22.3	22.3
H_3BO_3	6.2	6.2	6.2
$Na_2MoO_4 \times 2H_2O$	0.25	0.25	0.25
$CoCl_2 \times 6H_2O$	0.025	-	-
$CuSO_4 \times 5H_2O$	0.025	0.025	0.025
Iron source			
Na_2EDTA	37.3	37.3	37.3
$FeSO_4 \times 7H_2O$	27.8	27.8	27.8
Vitamins			
Thiamine HCl	0.4	1	1
Nicotinic acid	0.5	0.5	0.5
Pyrodixine HCl	0.5	0.5	0.5
myo-Inositol	100	100	100
Other componenets			
Glycine	-	2	2
L-glutamine	5.1 mM	-	-
Maltose	60,000	-	-
Sucrose	-	20,000	20,000
IBA	-	-	0.2 µM
BA	5 µM	1.5 µM	-
Gelrite	3,000	1,000	1,000
Difco Bacto agar	-	3,000	3,000
pH	5.6	5.6	5.6

References

Baldursson, S., P. Krogstrup, J.V. Norgaard and S.B. Andersen, 1993. Microspore embryogenesis in anther culture of three species of *Populus* and regeneration of dihaploid plants of *Populus trichocarpa*. Can.J.For.Res. **23**(9): 1821-1825.

Lloyd, A.D. and B. McCown, 1981. Commercially-feasible micropropagation of mountain laurel, *Kalmia latifolia*, by use of shoot-tip culture. Comb.Proc.Int.Plant Propag.Soc. **30**: 421-427.

Stoehr, M.U. and L. Zsuffa, 1990. Induction of haploids in *Populus maximowiczii* via embryogenic callus. Plant Cell Tiss.Org.Cult. **23**(1): 49-58.

Uddin, M.R., M.M.J. Meyer and J.J. Jokela, 1988. Plantlet production from anthers of eastern cottonwood (*Populus deltoides*). Can.J.For.Res. **18**: 937-941.

Wang, C., Z. Chu, and C. Sun, 1975. Induction of pollen plants of *Populus*. Acta Bot.Sinica, **17**: 56-62.

Wu, K. and P. Nagarajan, 1990. Poplars (*Populus* spp.): *in vitro* production of haploids. In: Biotechnology in Agriculture and Forestry. Vol. 12. Haploids in Crop Improvement. Bajaj, Y.P.S. (Ed.) Springer-Verlag, Berlin, pp. 215-236.

2.43
Oak anther culture

M. A. Bueno and J. A. Manzanera[1]
INIA. Ctra de la Coruña Km 7,5, 28040 Madrid, Spain
[1] IMIA. Ctra N-II, Km 38, 28800 Alcalá de Henares, Madrid, Spain

Introduction

Oaks (*Quercus* spp.) are widely distributed trees of temperate regions, and are of major importance in forestry. Most species are used for wood production, although there are a few interesting peculiarities, such as cork production from the cork oak. The cork oak industry is of economic and social relevance in rural areas with scarce alternative resources. Oaks are medium- to large-sized trees, reaching sexual maturity at 10-20 years old, depending on the species. The chromosome number is 24 and the genome size is 1.9 pg/nucleus. Oaks present some difficulties for traditional breeding, e.g., allogamy, a high degree of heterozygosity and long life cycles. Furthermore, they are mainly seed propagated, and vegetative propagation is in general very difficult. Propagation of cork oak adult trees by cuttings is not possible, and other methods, such as grafting or layering, are only used on a small scale. Some of these problems may be overcome by production of doubled haploids. Doubled haploids are mainly produced by androgenesis in either anther or microspore cultures. Woody species in general, and forest trees in particular, have been shown to be extremely recalcitrant in anther culture, with few exceptions. Jörgensen (1988) reported preliminary work in oak and showed induced embryogenesis in anther culture by using plant growth regulators in the medium. Recently, the induction of embryogenesis in anther culture has been approached by the application of specific stress conditions, such as heat shock, starvation or a combination of both treatments. These stress treatments are the main stimuli for gametic embryogenesis and the induction of sporophytic development.

Protocol

Plant material
Branches bearing catkins are collected from trees of different origins every week during May, which is flowering period of cork oak in Spain. A sample of anthers from the catkins is squashed in aceto carmine [4% (w/v) carmine in 45% (v/v) acetic acid] to determine the developmental stage of the microspores or pollen grains. Microspore development is highly asynchronous, as anthers containing all developmental stages, from tetrads to late bicellular pollen grains, can be observed within the same catkin. This is also observed in other oak species.

Pretreatment
The cut branches are given pieces of moist cotton wool, wrapped in aluminium foil and kept in the dark at 4°C for one week.

In vitro procedure
1. Catkins between 0.5 and 1 cm in length are collected and sterilized by immersion in 96% ethanol for 30 s, and then placed in 2% sodium hypochlorite for 20 min.

2. Anthers are isolated under aseptic conditions.

3. Anthers are then plated into Petri dishes (12 cm diameter, ca 100 anthers per plate), on a solid induction medium (Table 2.43-1).

4. Anthers are cultured in the dark for 5 days at 33°C and then transferred to 25°C.

5. After 20 days, embryo formation is observed in early responsive anthers. In all cases, embryos grew from the interior of the anthers, breaking through the degenerating anther walls (Fig. 2.43-1a). More slowly developing embryo cultures are also produced, these can take up to 10 months to produce embryos.

6. One month later, all the embryos obtained are isolated and transferred to plates containing proliferation medium (Table 2.43-1), where the embryos are clonally propagated (Fig. 2.43-1b).

7. The clones can be maintained for 6 months under these conditions, by monthly sub-culturing into fresh medium.

8. Individual cotyledonary embryos are vernalized for 8 weeks at 4°C, in the darkness, and then cultured at 25°C.

9. The embryos are transferred to test tubes with sterile vermiculite and liquid regeneration medium (Table 2.43-1), where shoot development and plantlet regeneration are achieved, this takes about one month.

10. *In vitro* plants are transferred to pots containing a mixture of soil, perlite and vermiculite (1:1:1) and placed in a mist tunnel in a glasshouse for *ex vitro* acclimation.

Culture medium
The induction medium contains macro nutrients of Sommer *et al.* (1975), MS micro nutrients and cofactors, 3% (w/v) sucrose and 1% (w/v) activated charcoal, at pH 5.6, and is solidified with 0.8% (w/v) agar. The proliferation medium is a modification of the induction medium

without activated charcoal and supplemented with 0.5 g/L glutamine. The regeneration medium is a modification of the induction medium without activated charcoal and agar and supplemented with 0.5 mg/L BA.

Table 2.43-1. Composition of media used in cork oak anther culture

Media ingredients	Induction medium (mg/L)	Proliferation medium (mg/L)	Regeneration medium (mg/L)
Macro nutrients (Sommer)			
KNO_3	1,000	1,000	1,000
$(NH_4)_2SO_4$	200	200	200
$MgSO_4 \times 7H_2O$	250	250	250
$NaH_2PO_4 \times 2H_2O$	129.5	129.5	129.5
KCl	300	300	300
$CaCl_2 \times 2H_2O$	150	150	150
Micro nutrients (MS)			
$MnSO_4 \times H_2O$	16.9	16.9	16.9
$ZnSO_4 \times 7H_2O$	8.6	8.6	8.6
H_3BO_3	6.2	6.2	6.2
KI	0.83	0.83	0.83
$CuSO_4 \times 5H_2O$	0.025	0.025	0.025
$Na_2MoO_4 \times 2H_2O$	0.25	0.25	0.25
$CoCl_2 \times 6H_2O$	0.025	0.025	0.025
Iron source (MS)			
$Na_2EDTA \times 2H_2O$	37.2	37.2	37.2
$FeSO_4 \times 7H_2O$	27.8	27.8	27.8
Vitamins (MS)			
Nicotinic acid	0.5	0.5	0.5
Pyridoxine HCl	0.5	0.5	0.5
Thiamine·HCl	0.1	0.1	0.1
myo-Inositol	100	100	100
Others			
Glycine	0.2	0.2	0.2
Glutamine [1]	-	500	-
Ascorbic acid	0.2	0.2	0.2
Sucrose	30,000	30,000	30,000
BA	-	-	0.5
Charcoal	10,000	-	-
Agar	8,000	8,000	-
pH (before autoclaving)	5.6	5.6	5.6

[1] Filter sterilize, add to medium at temperature below 40 °C.

In vitro conditions

For embryo induction, anthers are cultured in the dark for 5 days at 33°C and then transferred to 25°C, whereas the proliferation and regeneration phases take place with a photoperiod of 16 h light/8 h dark and a photon flux density of 100 µmol m^{-2} s^{-1} provided by Osram cool-white 18 W fluorescent lamps, until roots have formed.

Evaluation of anther derived embryos

The totipotency of plant cells opens the possibility of induction of embryogenesis from cells of the anther wall (somatic embryogenesis). These somatic embryos should be differentiated from androgenic embryos by chromosome counting. However, the spontaneous induction of doubled haploids is also possible. Therefore, the ploidy level alone is insufficient and must be analyzed in conjunction with genetic markers to confirm the origin of the embryos.

Chromosome counts
The diploid chromosome number of oaks is $2n=24$. The haploid embryos have $n=12$. A sample of embryos is fixed in glacial acetic acid : ethanol (1:3, v/v) after pre-treatment with 2 mM hydroxyquinoline for 2-4 h. The samples are then washed in distilled water, hydrolyzed in 5 M HCl for 30 min at room temperature (22°C), washed again and incubated for 1 to 2 h in Feulgen solution until staining of meristematic regions is visible. Finally, each sample is squashed on a microscope slide in a drop of aceto-carmine. Chromosome preparations are examined under a microscope with oil-immersion (100x).

Flow cytometry
Small pieces of material (about 50 mg) are chopped with a razor blade in the presence of 200 µl extraction buffer (Nuclei Extraction Buffer, Partec). Two minutes later, 1.5 ml staining buffer are added, containing 0.2 mM Tris-HCl (pH 7.5), 4 mM $MgCl_2$, 0.5% (v/v) Triton X-100 and 4 µg/ml 4,6-diamidino-2-phenylindole (DAPI). The sample is filtered through a 30 µm filter, recovered in a vial, and placed in a Partec flow cytometer for DNA quantification.

To avoid possible differences in the efficiency of DNA staining due to sample preparation, the available material most closely related to these embryogenic cultures, i.e. a somatic embryogenic callus culture originally induced from immature zygotic embryos of cork oak, is used as a standard (Bueno et al., 2000b). Nuclei isolated from the diploid standard are passed through the flow cytometer, and the G1 DNA peak (2C) is set at channel number 250 (Fig. 2.43-1c).

Nuclei prepared from one of the anther cultures showed a G1 peak with lower DNA content than the standard, appearing at about channel 125, and a very small G2 peak in channel 250, suggesting that the sample contained only a very small fraction of actively dividing cells (Fig. 2.43-1d). This shows that the embryos formed in the anther are indeed haploid, i.e. of microspore origin, which can be confirmed by chromosome counting. Ninety eight cultures out of 108 showed a G1 peak with half the DNA content of the standard and were therefore haploid (90.7%), while diploid DNA amounts were found for ten cultures (7.4%) and 1.9% were a mixture of haploid and diploid cells.

Molecular markers
Co-dominant molecular markers are more suitable for tree identification, genotypic characterisation, heterozygosity evaluation and determination of the ploidy level in anther culture induced embryos because they allow to distinguish between homozygous and heterozygous individuals (Bueno et al., 2000a). For this purpose, DNA markers such as simple sequence repeats (microsatellites) are ideal. Simple sequence repeats (SSR) localized in *Q. petraea* can be PCR-amplified using the same primers in other oaks (Gomez et al., 2001).

The DNA is obtained from leaves of the parent tree using GENECLEAN® kit (BIO 101), and used a control. Embryonic tissue for DNA extraction is sampled from haploid embryo cultures as described by Gomez et al. (2001). Three *Quercus petraea* microsatellite loci, $(GA)_n$ repeats, are PCR amplified with the primers SsrQpZAG15, SsrQpZAG46 and SsrQpZAG110. Each 25 µl amplification reaction contains: 20 ng of genomic DNA, 0.2 µM of fluorescently labelled forward primer and unlabelled reverse primer (Progenetic), 200 µM each dNTPs, 50 mM KCl, 10 mM Tris-HCl (pH 9), 2.5 mM $MgCl_2$ and 0.5 U of Taq-DNA polymerase (Ecogen). Fluorescent labelled PCR products are separated and analyzed in a semiautomated sequencer (ABI-Prism, Perkin-Elmer). Standards are used for length determination of alleles. Our results indicate that the embryos have multiple microspore origins. In the few cases where embryos were diploid, only one allele per locus was found, confirming the hypothesis of a spontaneous duplication of the haploid genome.

Efficiency

Our laboratory reported the induction of haploid embryos from anthers of cork oak subjected to heat shock treatment in a simple agar medium without growth regulators (Bueno *et al.*, 1997). We reported an efficient androgenic method, obtaining 14% embryogenic anthers with 90.7% of them producing haploid embryos, 7.4% were spontaneously doubled and 1.9% mixaploid (having haploid and diploid cells). The microspore origin of the anther-derived embryos was confirmed by chromosome counting, flow cytometry and genetic markers. An application to patent this method has been presented (P-200001953).

References

Bueno, M.A., M.D. Agundez, A. Gomez, M.J. Carrascosa and J.A. Manzanera, 2000a. Haploid origin of cork oak anther embryos detected by enzyme and RAPD gene markers. Int.J.Plant Sci. **161**(3): 363-367.

Bueno, M.A., A. Gomez, M. Boscaiu, J.A. Manzanera and O. Vicente, 1997. Stress-induced formation of haploid plants through anther culture in cork oak (*Quercus suber*). Physiol.Plant. **99**: 335-341.

Bueno, M.A., A. Gomez and J.A. Manzanera, 2000b. Somatic and gametic embryogenesis in *Quercus suber* L. In: Somatic Embryogenesis in Woody Plants. Jain, S.M., P.K. Gupta and R.J. Newton (Eds.) Kluwer Academic Publisher, Dordrecht, pp.479-508.

Gomez, A., B. Pintos, E. Aguiriano, J.A. Manzanera and M.A. Bueno, 2001. SSR markers for *Quercus suber* tree identification and embryo analysis. J.Hered. **92**(3): 290-292.

Jorgensen, J., 1988. Embryogenesis in *Quercus petrea* and *Fagus sylvatica*. J.Plant Physiol. **132**: 638-640.

Sommer, H.E., C.L. Brown, and P.P. Kormanik, 1975. Differentiation of plantlets in longleaf pine (*Pinus palustris* Mill.) tissue cultured *in vitro*. Bot.Gaz. **136**: 196-200.

2.44
Haploids and doubled haploids in *Citrus* ssp.

M.A. Germanà
Istituto di Ricerca per la Genetica degli Agrumi. C.N.R. c/o Facolta di Agraria. Viale delle Scienze, 11. 90128 Palermo, Italy

Introduction

Citrus breeding is based either on conventional (hybridization, selection, mutation) or biotechnological methods, the latter employing *in vitro* tissue culture, regeneration from protoplasts, somatic hybridization, *in vitro* mutant selection, genetic transformation and haploid/doubled haploid production. All cultivated forms of *Citrus* spp. and related genera (*Poncirus, Fortunella, etc.*) are diploid with a monoploid number of chromosomes ($n=x=9$). Triploid and tetraploid forms of *Citrus* spp. also exist. Haploid plants, with a gametophytic set of chromosomes in the sporophyte, have potential use in mutation research, selection, genetic analysis and genetic transformation. The possibility of obtaining triploid somatic hybrids (important for the seedlessness of their fruits) by fusion between haploid and diploid protoplasts is an important application of haploidy in *Citrus* spp. breeding. Doubled haploids are also important in genome mapping and in exploring gametoclonal variation. Haploids can be induced in woody plants mainly through two approaches: gynogenesis, where they arise from the female gamete (*in situ* parthenogenesis induced by irradiated pollen and followed by *in vitro* culture of embryos, *in situ* or *in vitro* parthenogenesis induced by triploid pollen followed by *in vitro* culture of embryos), and androgenesis where they are regenerated from the male gametes (anther culture, isolated microspore culture). Although in our laboratory haploid plantlet regeneration through gynogenesis in *C. clementina* Hort. ex Tan., variety 'Nules', has been induced by *in vitro* pollination with the pollen of 'Oroblanco', a triploid grapefruit-type (triploidy of pollen, like irradiation, does not hinder pollen germination, but prevents pollen fertilization and stimulates the development of haploid embryoids from ovules (Germana and Chiancone, 2001), the experimental procedure in this Manual will be limited to the protocol of anther culture because of its proven efficiency. In particular, the protocol given is for anther culture of several varieties (Nules, 'SRA 63' and 'Monreal') of *C. clementina* Hort. ex Tan, which is currely the only *Citrus* species that responds well to androgenesis. In addition to genotypic effects, physiological conditions of donor plants and seasonal dependency are important factors for anther culture in *Citrus* and other woody plants (Germana *et al.*, 2000a). Although a lot of research has been carried out on gametic embryogenesis in *Citrus* spp. and their relatives, not much of it has been successful. For further details regarding *Citrus* haploidy, see the review by Germanà (1997).

Protocol for anther culture

Growth conditions
Floral buds are collected from the donor plants growing in the field, at the end of February to the end of April in Italy, depending on the season.

Anther culture
The stage of pollen development is commonly determined by staining one or more anthers per bud with aceto carmine, Schiff's reagent, or DAPI staining. Usually, the stage of pollen development is tested in one anther per floral bud size by the acetic carmine method. Anthers are squashed in 1% aceto carmine in 45% acetic acid for observation under an optical microscope to determine the uninucleate stage of pollen development, which is the most responsive for clementine androgenesis. DAPI fluorescent staining has also been used. The flower bud size at the uninucleate microspore stage for *C. clementina* is about 3.5-4.0 mm in length.

1. After pretreatment (4°C for 8-15 days in the dark), the buds are surface sterilized by immersion for 3 min in 70% (v/v) ethanol, followed by immersion in sodium hypochlorite solution (about 1.5% active chlorine in water containing a few drops of Tween 20) for 15-20 min, and finally rinsed three times for 5 minutes each with sterile distilled water.

2. Petals are aseptically removed with forceps, and anthers are carefully dissected and placed onto the culture medium. Anthers from 3-4 flower buds (60-80 anthers) are placed in each Petri dish (60 mm of diameter).

3. Petri dishes are incubated at 27±1°C for fifteen days in the dark, and then placed under cool white fluorescent lamps (Philips TLM 30W/84) with a photosynthetic photon flux density of 35 μmol m^{-2} s^{-1} and a 16 h photoperiod. Observations of the cultures are performed for the following ten months, every two weeks.

4. Embryogenic calli are transferred to the multiplication medium. Embryoids are isolated and placed in the germination medium. Subculturing is done every month.

5. Characterization of regenerants.

6. *In vitro* grafting of small shoots (2-3 mm) onto etiolated 20-day old 'Troyer' citrange seedlings.

7. *In vivo* acclimation.

Culture media
The basal medium used in *Citrus clementina* anther culture is N6 medium (Chu, 1978), supplemented with Nitsch and Nitsch (1969) vitamins, 18 g/L lactose, 9 g/L galactose, 5% coconut water (Sigma), 500 mg/L casein, 200 mg/L L-glutamine, 0.5 mg/L biotin, 500 mg/l ascorbic acid (Table 2.44-1). The following growth regulators (mg/L): NAA 0.02 + 2,4-D 0.02 + kinetin 1.0 + BA 0.5 + zeatin 0.5 + thidiazuron 0.1 + GA$_3$ 0.5, are added to the culture medium before autoclaving. The pH is adjusted to 5.8 with 1 N KOH before autoclaving (20 min, 120°C). Agar (0.8% of washed agar from Sigma) is added as a gelling agent.

The highly embryogenic haploid callus is multiplied on MS medium supplemented with 5% sucrose, 0.02 mg/L NAA and 0.8 % agar. The cultures maintain embryogenic potential for several years. Embryoids are germinated in Petri dishes with MS medium containing 3% (w/v) sucrose, 1 mg/L GA$_3$, 0.01 mg/L NAA and 0.75% (w/v) agar (germination medium), and they are later transferred to Magenta boxes (Sigma V8505) or to test tubes.

Table 2.44-1. Culture media composition for *Citrus* androgenesis

Media components	Anther culture medium (mg/L)	Callus multiplication medium (mg/L)	Embryoid germination medium (mg/L)
Macro salts			
$(NH_4)_2SO_4$	463	-	-
KNO_3	2,830	1,900	1,900
$CaCl_2$	125.3	-	-
$CaCl_2 \times 2H_2O$	-	440	440
NH_4NO_3	-	1,650	1,650
KH_2PO_4	400	170	170
$MgSO_4$	90.3	-	-
$MgSO_4 \times 7H_2O$	-	370	370
Iron source			
FeNaEDTA	36.7	-	-
$Na_2EDTA \times 2H_2O$	-	37.3	37.3
$FeSO_4 \times 7H_2O$	-	27.8	27.8
Micro salts			
H_3BO_3	1.6	6.2	6.2
$CoCl_2 \times 6H_2O$	-	0.025	0.025
$CuSO_4 \times 5H_2O$	-	0.025	0.025
KI	0.8	0.83	0.83
$MnSO_4 \times H_2O$	3.33	-	-
$MnSO_4 \times 4H_2O$	-	22.3	22.3
$Na_2MoO_4 \times 2H_2O$	-	0.25	0.25
$ZnSO_4 \times 7H_2O$	1.5	8.6	8.6
Vitamins			
myo-Inositol	100	100	100
Nicotinic acid	5	0.5	0.5
Pyridoxine HCl	0.5	0.5	0.5
Thiamine HCl	0.5	0.1	0.1
Biotin	0.5	-	-
Folic acid	0.5	-	-
Other components			
Coconut water (Sigma)	50 ml	-	-
Casein hydrolysate	500	-	-
Glycine	2	2	2
Sucrose	-	50,000	30,000
Lactose	18,000	-	-
Galactose	9,000	-	-
L-Glutamine	200	-	-
Ascorbic acid	500	-	-
NAA	0.02	0.02	0.01
2,4-D	0.02	-	-
Kinetin	1	-	-
BA	0.5	-	-
Zeatin	0.5	-	-

Media components	Anther culture medium (mg/L)	Callus multiplication medium (mg/L)	Embryoid germination medium (mg/L)
Thidiazuron	0.1	-	-
GA_3	0.5	-	1
Agar	8,000	8,000	7,500
pH	5.8	5.8	5.8

Plant recovery and hardening
After 2-3 months of culture, anthers start to produce calli or embryoids. Most of the calli are not morphogenic, but many of them appear highly embryogenic and they maintain this potential for many years. In the best conditions, more than 4% of cultured anthers can produce embryoids or embryogenic calli. One embryogenic callus with a high regenerative potential can give rise up to fifty embryoids at the second subculture.

Plantlet formation from cultured anthers may occur either directly through embryogenesis of microspores or indirectly through organogenesis or embryogenesis of microspore derived callus. The embryoids develop normally through the globular, heart, torpedo and cotyledonary stages and often produce secondary embryoids. Very high conversion rate (85-95%) of the well-structured embryoids have been observed. Grafting *in vitro* of homozygous small shoots (2-3 mm) onto etiolated 20-day old 'Troyer' citrange seedlings is necessary to improve the survival of plantlets transferred to *in vivo* conditions. After 3-4 months, the grafted plantlets are washed with sterile water to remove medium from their roots and then transferred to sterilized pots containing peat moss, sand and soil in the ratio 1:1:1 for the acclimation phase. The new scions obtained are later grafted onto 2-year old sour orange or *C. macrophylla* seedlings. However, they show a more compact habit and a decreased vigour, with significantly smaller leaves, shorter internodes and more thorns when compared to the heterozygous parent of the same age of grafting (Germana *et al.*, 2000b).

Characterization of regenerants
Chromosome numbers are counted in root tip cells of regenerated embryos and plantlets, using the standard Feulgen technique. The explants are pretreated with 0.05% (w/v) aqueous solution of colchicine for 2 h at room temperature, fixed overnight in 3:1 (v/v) ethanol : glacial acetic acid, and stored in 70% ethanol until viewing.

Chromosome counts carried out on root apices of embryos and of plantlets obtained from *in vitro* androgenesis of clementine show the haploid set of chromosomes ($n=x=9$) (Germana *et al.*, 1994a). During culture, haploid calli spontaneously diploidize, producing doubled haploid (DH) embryoids and plantlets (Germana, 1997), and sometimes the presence of a triploid number of chromosomes in calli cells is also observed.

Because of the spontaneous diploidization of the haploid calli, cytological analysis cannot always identify androgenic plants and isozyme analyses have been employed to decide the gametic origin of calli and plantlets (Germana *et al.*, 1994b). Isozyme techniques provide a method to distinguish between androgenetic and somatic tissue when the enzyme is heterozygotic in the diploid condition of the donor plant and the regenerants show lack of one of the alleles.

To identify the origin of calli, embryoids and plantlets obtained, their crude extracts are analyzed using two enzyme systems: phosphoglucoisomerase (PGI) and phosphoglucomutase (PGM), as reported by Grosser *et al.* (1988). Numbering for isozymes (PGI-1) and lettering for different allozymes are the same as used by Torres *et al.* (1978).

Lines of *C. clementina* are heterozygous for PGI-1 and PGM. According to Torres *et al.*, (1978), the heterozygous clementine parent is FI (F = allele which specifies fast migration toward the anode enzyme; I = intermediate) in PGM, and WS (W = allele which specifies an enzyme migrating faster than F; S = allele which specifies a slowly migrating enzyme) in PGI. For analysis of calli and leaves obtained from anther culture, the presence of a single band is retained in the homozygous state. With one or two exceptions out of more than one hundred samples analyzed, both enzyme systems confirmed the androgenic nature of the regenerants.

References

Chu, C.C., 1978. The N6 medium and its applications to anther culture of cereal crops. In: Proc. Symp.Plant Tissue Cult. Science Press, Beijing, pp. 45-50.

Germana, M.A., 1997. Haploidy in Citrus. In: *In Vitro* Haploid Production in Heigher Plants. Vol. 5. Jain, S.M., S.K. Sopory and R.E. Veilleux (Eds.) Kluwer Academic Publishers, Dordrecht, pp. 195-217.

Germana, M.A. and B. Chiancone, 2001. Gynogenetic haploids of *Citrus* after *in vitro* pollination with triploid pollen grains. Plant Cell Tiss.Org.Cult. **66**: 59-66.

Germana, M.A., F.G. Crescimanno, F. de Pasquale, and W.Y. Ying, 1994a. Androgenesis in 5 cultivars of *Citrus limon* L. Burm. f. Acta Hort. **300**: 315-324.

Germana, M.A., F.G. Crescimanno, and A. Motisi, 2000a. Factors affecting androgenesis in *Citrus clementina* Hort. ex. Tan. Adv.Hort.Sci. **14**(2): 43-51.

Germana, M.A., F.G. Crescimanno, G.R. Recupero, and M.P. Russo, 2000b. Preliminary characterization of several doubled haploids of *Citrus clementina* cv. Nules. Acta Hort. **535**: 183-190.

Germana, M.A., Y.Y. Wang, M.G. Barbagallo, G. Iannolino, and F.G. Crescimanno, 1994b. Recovery of haploid and diploid plantlets from anther culture of *Citrus clementina* Hort. ex Tan. and *Citrus reticulata* Blanco. J.Hortic.Sci. **69**(3): 473-480.

Grosser, J.W., F.G. Jr. Gmitter, and J.L. Jr. Chandler, 1988. Intergeneric somatic hybrid plants from sexually incompatible woody species: *Citruis sinensis* and *Severina disticha*. Theor.Appl.Genet. **75**: 397-401.

Nitsch, J.P. and C. Nitsch, 1969. Haploid plants from pollen grains. Science **163**: 85-87.

Torres, A.M., R.K. Soost, and U. Diedenhofen, 1978. Leaf isozyme as genetic markers in *Citrus*. Am.J.Bot. **65**: 869-881.

3
Published doubled haploid protocols in plant species

M. Maluszynski, K.J. Kasha[1] and I. Szarejko[2]
Plant Breeding and Genetics Section, Joint FAO/IAEA Division, P.O. Box 100, Vienna, Austria;
[1]*Department of Plant Agriculture, Biotechnology Division, University of Guelph, Guelph, Ontario Canada N1G 2W1;* [2]*Department of Genetics, University of Silesia, Jagiellonska 28, 40-032 Katowice, Poland*

Forty-four protocols of doubled haploid production (DH) are presented in this Manual. They are related to at least 33 plant species, not counting protocols on production of DH from interspecific hybrids. The protocols provide procedures for production of DH in such major crops as wheat, maize, barley, rapeseed or potato and also in other crop species where this technology is more advanced, or in other words, where a relatively high frequency of doubled haploids was obtained. However, there have been several approaches to develop doubled haploids in many crops or plant species. We are presenting below the list of publications related to DH production in 226 additional plant species. This list does not include protocols of plant species published in Chapter 2 of this Manual. This means that efforts have been undertaken to develop doubled haploids in more than 250 plant species as the presented list was built on the basis of our own collection of related publications and as a result of searching in publicly available databases. It was impossible to cite all published information. We limited references to those which carried protocols or more recent information on DH production. In the cited papers, presented protocols did not always lead to the regeneration of haploid or doubled haploid plants. However, even negative results can be very helpful for someone who is initiating this work. We hope that publication of this list will help in further application of DH technology in crop improvement and basic research.

Table 3-1. List of plant species with available publications on production of doubled haploids
(a = anther culture, g = gynogenesis, i = interspecific or intergeneric crosses, irr = radiation, m = microspore culture, o = ovary or ovule culture, spont = spontaneous)

Species	Method	Reference
Aconitum carmichaeli	a	Hatano *et al.*, 1987
Actinidia spp.	a, g, irr	Fraser and Harvey, 1986; Pandey *et al.*, 1990
Aesculus carnea	a	Marinkovic and Radojevic, 1992
Aesculus hippocastanum	a	Jorgensen, 1991; Radojevic *et al.*, 2000
Agropyron spp.	a	Marburger and Wang, 1988; Chekurov and Razmakhnin, 1999
Albizzia lebbek	a	De and Rao, 1983; Gharyal *et al.*, 1983b
Allium giganteum	a	Inagaki *et al.*, 1994
Allium sativum	a	Suh and Park, 1986
Allium schoenoprasum	o	Kim *et al.*, 1998
Anemone canadensis	a	Johansson *et al.*, 1990
Anemone spp.	a	Johansson, 1986
Annona squamosa	a	Nair *et al.*, 1983
Antirrhinum majus	a	Sharma and Babber, 1990
Arabidopsis thaliana	a	Avetisov, 1976a, b; Amos and Scholl, 1978
Arachis hypogaea	a	Bajaj *et al.*, 1981; Banks, 1988; Willcox *et al.*, 1990; Bansal *et al.*, 1991
Arachis villosa	a	Bajaj *et al.*, 1981
Atriplex glauca	a	Kenny and Caligari, 1996
Atropa belladonna	a	Bajaj, 1978; Mazzolani *et al.*, 1979
Avena sterilis	a	Kiviharju *et al.*, 1997; Kiviharju and Pehu, 1998
Begonia x hiemalis	a	Khoder *et al.*, 1984
Boswellia serrata	a	Prakash *et al.*, 1999
Brassica carinata	a	Yadav *et al.*, 1988
Brassica rapa ssp. *chinensis*	m	Cao *et al.*, 1994
Brassicoraphanus	a	Lee and Yoon, 1987
Cajanus cajan	a, m	Bajaj *et al.*, 1980; Bajaj and Gosal, 1987; Fougat *et al.*, 1992; Kaur and Bhalla, 1998; Narasimham and Kishor, 1999; Vishukumar *et al.*, 2000
Camellia japonica	a, m	Pedroso and Pais, 1993, 1997
Camellia sinensis	a	Seran *et al.*, 1999
Capsicum annuum	a, m, spont.	Sibi *et al.*, 1979; Dumas-de-Vaulx *et al.*, 1981; Morrison *et al.*, 1986; Munyon *et al.*, 1989; Maheswary and Mak, 1993; Kristiansen and Andersen, 1993; Lefebvre *et al.*, 1994; Li and Ye, 1995; Mityko *et al.*, 1995; Gonzalez-Melendi *et al.*, 1995; Qin and Rotino, 1995a,b; Regner, 1996; Hwang *et al.*, 1998; Baracaccia *et al.*, 1999
Capsicum frutescens	a	Wu and Zhang, 1986
Carica papaya	a	Tsay and Su, 1985
Carthamus tinctorius	a	Prasad *et al.*, 1991
Cassia siamea	a	Gharyal *et al.*, 1983a
Catharanthus roseus	a	Abou-Mandour *et al.*, 1979; Kim *et al.*, 1994

Species	Method	Reference
Chrysanthemum spp.	a, g	Watanabe, 1977
Cicer arietinum	a	Khan and Ghosh, 1983; Bajaj and Gosal, 1987; Huda *et al.*, 2001
Cichorium spp.	a	Guedira *et al.*, 1989
Cichorium intybus	m	Theiler-Hedtrich and Hunter, 1995
Citrus aurantium	a	Hidaka *et al.*, 1982
Citrus celmentina	a	Germana *et al.*, 1994, 2000a,b
Citrus limon	a	Germana *et al.*, 1992
Citrus madurensis	a	Ling *et al.*, 1988
Citrus reticulata	a	Germana *et al.*, 1994
Citrus sinensis	a, g	Hidaka, 1984; Koltunow *et al.*, 1995
Citrus spp.	a, g, irr	Chen, 1985; Gmitter and Moore, 1986; Ling *et al.*, 1988; DeLange and Vincent, 1988; Geraci and Starrantino, 1990; Starrantino and Caponnetto, 1990; Ikeda *et al.*, 1993; Germana, 1997
Clausena excavata	a	Froelicher and Ollitrault, 2000
Cocos nucifera	a	Thanh-Tuyen and de Guzman, 1983a,b; Monfort, 1985
Coffea arabica	m	Neuenschwander and Baumann, 1995
Coffea canephora	spont, g	Couturon, 1982; Lashermes *et al.*, 1994a,b; Muniswamy and Sreenath, 2000
Coffea spp.	a, spont.	Raghuramulu and Prakash, 1996
Corchorus olitorius	m	Ali and Jones, 2000
Crepis capillaris	a	Sacristan, 1971; Slusarkiewicz-Jarzina and Zenkteler, 1979
Crepis tectorum	irr.pollen	Gerassimowa, 1936
Crotalaria pallida	a	Debata and Patnaik, 1983
Cucumis melo	a, g, i, irr	Dumas-de-Vaulx, 1979; Dryanovska and Ilieva, 1983; Sauton and Dumas-du-Vaulx, 1987; Savin *et al.*, 1988; Cuny *et al.*, 1992; Ficcadenti *et al.*, 1999
Cucumis sativus	a, g, irr	Lazarte and Sasser, 1982; Denissen and Den Nijs, 1987; Le Deunff and Sauton, 1994
Cucurbita moschata	g	Kwack and Fujieda, 1988
Cucurbita pepo	g	Dumas-de-Vaulx and Chambonnet, 1986; Metwally *et al.*, 1998a,b
Cyclamen persicum x *Cyclamen purpurascens*	a	Ishizaka, 1998
Cynara scolymus	a, g	Motzo and Deidda, 1993
Dactylis glomerata	a	Songstad and Conger, 1988; Christensen *et al.*, 1997; Caredda and Clement, 1999
Datura ferox	a	Padmanabhan *et al.*, 1977
Datura innoxia	a, m	Nitsch and Norreel, 1973a,b; Collins *et al.*, 1974; Dunwell and Sunderland, 1976; Schieder, 1976; Blaschke *et al.*, 1978; Iankulov *et al.*, 1979; Krumbiegel, 1979; Atanasov *et al.*, 1980; Forche *et al.*, 1981; Tyagi *et al.*, 1981; Badea and Raicu, 1982

Species	Method	Reference
Datura metel	a, m	Sangwan and Camefort, 1984; Babbar and Gupta, 1984; Babbar and Gupta, 1986a,c; Babbar and Gupta, 1990
Daucus carota	a, m	Matsubara *et al.*, 1995; Tyukavin *et al.*, 1999
Dianthus caryophyllus	a, o	Mosquera *et al.*, 1999; Sato *et al.*, 2000
Digitalis lanata	a	Badea *et al.*, 1985; Diettrich *et al.*, 2000
Digitalis obscura	a	Perez-Bermudez *et al.*, 1985
Digitalis purpurea	a	Corduan, 1976
Dolichos biflorus	a	Sinha and Das, 1986
Elaeis guineensis	a	Teixeira *et al.*, 1994
Ephedra foliata	g	Singh and Konar, 1979
Eragrostis tef	a	Tefera *et al.*, 1999
Eucalyptus spp.	a	Sommer and Wetzstein, 1984
Euphorbia pulcherima	a	Rudramuniyappa and Annigeri, 1985
Euphoria longan	a	Yang and Wei, 1984
Fagopyrum esculentum	a, g	Zheleznov, 1976; Bohanec, 1997
Fagus sylvatica	a	Jorgensen, 1991
Feijoa sellowiana	a	Canhoto and Cruz, 1993
Festuca arundinacea	a	Kasperbauer and Buckner, 1979; Zare *et al.*, 1999
Festuca arundinacea x Lolium multiflorum	a	Zwierzykowski *et al.*, 1998, 1999
Festuca pratensis	a	Rose *et al.*, 1987a,b
Festuca pratensis x Lolium multiflorum		Lesniewska *et al.*, 2001
Foeniculum vulgare	a, m	Matsubara *et al.*, 1995
Fragaria spp.	a, i, g	Rowe, 1974; Xue *et al.*, 1981; Jelenkovic *et al.*, 1984; Quarta *et al.*, 1992; Hennerty and Sayegh, 1996
Fragaria x ananassa	a, i	Li *et al.*, 1988; Rose *et al.*, 1993; Svensson and Johansson, 1995; Owen and Miller, 1996
Gerbera jamesonii	g	Preil *et al.*, 1977; Sitbon, 1981; Meynet and Sibi, 1984; Ahmin and Vieth, 1986; Cappadocia *et al.*, 1988; Honkanen *et al.*, 1992; Miyoshi and Asakura, 1996; Tosca *et al.*, 1999
Gingko biloba	m	Laurain *et al.*, 1993
Glycine max	a	Crane *et al.*, 1982; Jian *et al.*, 1986; Hildebrand *et al.*, 1986; Zhuang *et al.*, 1991; Kadlec *et al.*, 1991; Hu *et al.*, 1996; Kaltchuk-Santos *et al.*, 1997
Gossypium arboreum	a	Mehetre, 1984; Bajaj and Gill, 1989
Gossypium barbadense	spont.	Mehetre and Thombre, 1981; Mehetre, 1984; Bajaj and Gill, 1997
Gossypium hirsutum	a, o	Meredith *et al.*, 1970; Mehetre and Thombre, 1981, 1982; Pallares, 1984; Barrow, 1986; Stelly *et al.*, 1988; Zhou *et al.*, 1989; Bajaj and Gill, 1997

Species	Method	Reference
Gossypium spp.	a, o, m	Turaev and Shamina, 1986, 1993; Contolini and Menzel, 1987; Stelly et al., 1988; Bajaj and Gill, 1997; Khushwinder et al., 1998
Guizotia abyssinica	a	Makhmudov, 1978; Chaudhari, 1979; Mahill et al., 1984; Sarvesh et al., 1993; Adda et al., 1994; Kavi Kishor et al., 1997
Haemanthus katherinae	a	Zhou et al., 1986
Helianthus annuus	a, i, m, o, irr	Mezzarobba and Jonard, 1986; Hongyuan et al., 1986; Gelebart and San, 1987; Gurel et al., 1991; Thengane et al., 1994; Coumans and Zhong, 1995; Zhong et al., 1995; Nurhidayah et al., 1996; Badigannavar and Kuruvinashetti, 1996; Todorova et al., 1997; Friedt et al., 1997; Saji and Sujatha, 1998; Todorova and Ivanov, 1999
Helianthus annuus x H. smithii	a	Nenova et al., 1992
Helianthus annuus x H. eggerttii	a	Nenova et al., 1992
Helianthus mollis	a	Nenova et al., 2000
Helianthus salicifolius	a	Nenova et al., 2000
Helianthus smithii	a	Nenova et al., 2000
Hemerocallis fulva	a	Zhou, 1989
Hevea brasiliensis	a, o	Chen et al., 1988; Jayasree et al., 1999
Hieracium pilosella	a	Bicknell and Borst, 1996
Hordeum bulbosum	a	Gudu et al., 1993; Kihara et al., 1994
Hordeum marinum	i	Jorgensen and Bothmer, 1988; Kihara et al., 1994
Hordeum murinum	a, i	Gaj and Gaj, 1985; Wang et al., 1993; Kihara et al., 1994
Hordeum secalinum	i	Gaj and Gaj, 1985
Hordeum spontaneum	a, i	Simpson and Snape, 1980; Piccirilli and Arcioni, 1991; Kintzios and Fischbeck, 1994
Hyoscyamus muticus	a	Wernicke et al., 1979; Strauss et al., 1981; Fankhauser et al., 1984
Hyoscyamus niger	a	Corduan, 1975; Wernicke et al., 1979; Reynolds, 1985; Raghavan and Nagmani, 1989
Iochroma warscewiczii	a	Canhoto et al., 1990
Ipomoea batatas	a	Tsay et al., 1982; Mukherjee et al., 1991
Larix spp.	o	Nagmani and Bonga, 1985; Von-Aderkas and Bonga, 1988; Von-Aderkas et al., 1990
Lesquerella fendleri	a	Tomasi et al., 1999
Lilium davidii	o	Gu and Cheng, 1983
Lilium spp.	a, o	Vassileva-Dryanovska, 1966; Prakash and Giles, 1986; Miki-Hirosige et al., 1988; Qu et al., 1988; Han et al., 1997, 1999; van den Bulk and van Tuyl, 1997; Arzate-Fernandez et al., 1998;
Litchi chinensis	a	Fu and Tong, 1983

Species	Method	Reference
Lolium multiflorum x Festuca arundinacea	a	Rose *et al.*, 1987a; Pasakinskiene *et al.*, 1997; Humphreys *et al.*, 1998; Zare *et al.*, 1999
Lolium temulentum		Rose *et al.*, 1987a,b
Lupinus albus	a, m	Ormerod and Caligari, 1994
Lupinus polyphyllus	a	Sator *et al.*, 1982
Lycium spp.	a	Fan *et al.*, 1982
Lycopersicon esculentum	a, m	Gulshan *et al.*, 1981; Zagorska *et al.*, 1986; Sink and Reynolds, 1986; Varghese and Yadav, 1986; Evans and Morrison, 1989; Summers, 1997; Shtereva *et al.*, 1998
Lycopersicon esculentum + Solanum etuberosum	a	Gavrilenko *et al.*, 2001
Lycopersicon peruvianum	a	Ramulu, 1982
Lycopersicon pimpinellifolium	g, irr	Nishiyama and Uematsu, 1967
Malus prunifolia	a	Wu, 1981
Manihot esculenta	a	Liu and Chen, 1978; Abraham *et al.*, 1995
Medicago sativa	a, g	Bingham, 1969; Bingham and Gillies, 1971; Tanner *et al.*, 1988, 1990; Ray and Bingham, 1989; Zagorska and Dimitrov, 1995; Skinner and Liang, 1996; Zagorska *et al.*, 1997
Melandrium album	a, g	Veuskens *et al.*, 1992; Mol, 1992; Paulikova and Vagera, 1993
Mentha spp.	a	Van-Eck and Kitto, 1990
Morus alba	g	Thomas *et al.*, 1999
Morus indica	a	Jain *et al.*, 1996
Musa spp.	a	Dallos and Galan-Sauco, 1998; Assani *et al.*, 2003
Nephelium spp.	a	Imelda *et al.*, 1988
Nicotiana attenuata	a	Collins and Sunderland, 1974
Nicotiana knightiana	a	Collins and Sunderland, 1974
Nicotiana raimondii	a	Collins and Sunderland, 1974
Oenothera hookeri	a	Martinez and de Halac, 1995
Olea europaea	a	Perri *et al.*, 1994
Oryza perennis	a	Wakasa and Watanabe, 1979
Oryza spp. (wild species)	a	Tang *et al.*, 1998
Oryza sativa x O. glaberima	a	Woo *et al.*, 1983
Paeonia albiflora	a	Lee *et al.*, 1992
Panax quinquifolius	a	Du *et al.*, 1986
Papaver somniferum	a	Dieu and Dunwell, 1988
Parthenium spp.	a	Rudramuniyappa, 1985
Passiflora edulis	a	Tsay *et al.*, 1984
Pelargonium roseum	a	Kato *et al.*, 1980
Pelargonium zonale	a	Pol'kheim, 1972
Peltophorum pterocarpum	a	Rao and De, 1987
Pennisetum americanum	a	Nitsch *et al.*, 1982
Pennisetum glaucum	a, m	Bui-Dang-Ha and Pernes, 1982; Le Thi *et al.*, 1994; Choi *et al.*, 1997; Caredda and Clement, 1999
Pennisetum purpureum	a	Haydu and Vasil, 1981

Species	Method	Reference
Pennisetum spp.	a, i, g	Robert *et al.*, 1989; Busso *et al.*, 1995; Caredda and Clement, 1999
Petunia spp.	a, g, m	Singh and Cornu, 1976; Kusum and Maheshwari, 1977; Bajaj, 1978; Krumbiegel, 1979; Gupta, 1983; Hanson, 1984; Babbar and Gupta, 1984; DeVerna and Collins, 1984; Raquin, 1985; Raquin *et al.*, 1989; Jain and Bhalla-Sarin, 1997
Phaseolus vulgaris	a	Peters *et al.*, 1977; Munoz and Baudoin, 1994
Physalis ixocarpa	a	Bapat and Wenzel, 1982
Picea sitchensis	g	Baldursson *et al.*, 1993b
Pisum sativum	a	Gupta, 1975
Poncirus trifoliata	a	Hidaka *et al.*, 1979
Populus trichocarpa	a, g	Baldursson *et al.*, 1993a
Populus maximowiczii	a	Kim *et al.*, 1986; Stoehr and Zsuffa, 1990
Populus glandulosa	a	Kim *et al.*, 1983
Populus deltoides	a	Uddin *et al.*, 1988
Populus nigra	a	Wang *et al.*, 1975
Populus spp.	a	Wang *et al.*, 1975; Sommer and Wetzstein, 1984; Hyun *et al.*, 1986; Kiss *et al.*, 2001
Primula vulgaris	a	Piper *et al.*, 1986
Prunus avium	a	Höfer and Hanke, 1990; Long *et al.*, 1994
Prunus persica	a	Todorovic *et al.*, 1992; Hammerschlag, 1993
Pseudotsuga menziesii	g	Livingston, 1971
Psidium guajava	a	Babbar and Gupta, 1986b
Psophocarpus tetragonolobus	a, m	Pal, 1983; Rao *et al.*, 1986
Punica granatum	a	Moriguchi *et al.*, 1987
Pyrus communis	g, irr	Braniste *et al.*, 1984; Sniezko and Visser, 1987; Bouvier *et al.*, 1993
Quercus petraea	a	Jorgensen, 1991
Quercus suber	a	Bueno *et al.*, 1997
Raphanus sativus	m	Lichter, 1989; Takahata *et al.*, 1996
Ribes spp.	a	Sankina and Sankin, 1988
Ricinus communis	a	Jelenkovic *et al.*, 1980
Rosa spp.	g, m	Meynet *et al.*, 1994
Rosa elliptica	m	Wissemann *et al.*, 1998
Rosa micrantha	m	Wissemann *et al.*, 1998
Saccharum officinarum	a	Liu *et al.*, 1980; Fitch and Moore, 1996
Saccharum spontaneum	a	Fitch and Moore, 1984, 1996; Hinchee *et al.*, 1984
Salvia sclarea	a	Bugara *et al.*, 1986
Saintpaulia ionantha	a	Hughes *et al.*, 1975; Weatherhead *et al.*, 1982; Bhaskaran *et al.*, 1983; Radojevic *et al.*, 1985
Scilla indica	a	Chakravarty and Sen, 1989
Sesamum indicum	a	Ranaweera and Pathirana, 1992
Setaria italica	a	Ban *et al.*, 1971
Sinapis alba	a	Jain *et al.*, 1989
Sinocalamus latiflora	a	Tsay *et al.*, 1990; Caredda and Clement, 1999

Species	Method	Reference
Solanum acaule ssp. acaule	a	Rokka et al., 1998
Solanum bulbocastanum	a	Lysenko and Sidorov, 1985
Solanum carolinense	a	Reynolds, 1990
Solanum chacoense	a	Hermsen, 1969; Cappadocia and Ahmim, 1988; Birhman et al., 1994; Rivard et al., 1994
Solanum dulcamara	a	Binding and Mordhorst, 1984
Solanum melongena	a, m	Isouard et al., 1979; Misra et al., 1983; Hinata, 1986; Borgel and Arnaud, 1986; Tuberosa et al., 1987; Sanguineti et al., 1990; Quarta et al., 1992; Rotino, 1996; Miyoshi, 1996
Solanum phureja	a	Veilleux et al., 1985; Pehu et al., 1987; Owen et al., 1988; Teten Snider and Veilleux, 1994; Paz and Veilleux, 1999
Solanum surattense	a	Sinha et al., 1979
Solanum torvum	a	Jaiswal and Narayan, 1981
Solanum viarum	a	Debata and Patnaik, 1988
Sorbus domesticus	a	Arrillaga et al., 1995
Sorghum bicolor	a, m	Rose et al., 1986a,b; Elkonin et al., 1993; Kumaravadivel and Sree Rangasamy, 1994; Sairam and Seetharama, 1996; Liang et al., 1997; Can et al., 1998; Can and Yoshida, 1999
Sorghum vulgare	a	Schertz, 1963
Streptocarpus hybridus	a	Wolff et al., 1986
Taeniatherum caput-medusae	i	Frederiksen, 1989
Theobroma cacao	g, o, irr.poll	Dublin, 1978; Lanaud et al., 1988; Falque et al., 1992; Falque, 1994
Thuja gigantea	a	Pol'kheim, 1972
Trifolium spp.	a	Tomes and Peterson, 1981
Trifolium rubens	g	Ponitka and Slusarkiewicz-Jarzina, 1987
Triticum monococcum	a	Tan and Halloran, 1982
Triticum ventricosum	i	Fedak, 1983
Tritordeum	a	Barcelo et al., 1994
Tulip spp.	a, o, m	van den Bulk and van Tuyl, 1997; Custers et al., 1997
Vaccinium spp.	a	Smagula and Lyrene, 1984
Vicia faba	a	Hesemann, 1980
Vigna unguiculata	a	Arya and Chandra, 1989
Viola odorata	o	Wijowska et al., 1999
Vitis latifolia	a	Salunkhe et al., 1999
Vitis riparia	a	Mozsar and Sule, 1994
Vitis rupestris	a	Altamura et al., 1992
Vitis vinifera	a, o	Rajasekaran and Mullins, 1985; Mauro et al., 1986; Emershad et al., 1989; Cersosimo et al., 1990; Bensaad et al., 1996; Faure et al., 1996a,b; Torregrosa, 1998; Hollo and Misik, 2000
Zingiber officinale	a	Samsudeen et al., 2000

References

Abou-Mandour, A.A., S. Fischer and F.C. Czygan, 1979. Regeneration of intact plants from haploid and diploid callus cells of *Catharanthus roseus*. Z.Pflanzenphysiol. **91**: 83-88.
Abraham, A., P.N. Krishnan and S. Seeni, 1995. Induction of androgenesis, callus formation and root differentiation in anther culture of cassava (*Manihot esculenta* Crantz). Indian J.Exp.Biol. **33**(3): 186-189.
Adda, S., T.P. Reddy and P.B. Kavi Kishor, 1994. Androclonal variation in niger (*Guizotia abyssinica* Cass). Euphytica **79**: 59-64.
Ahmin, M. and J. Vieth, 1986. Regeneration of haploid plants of *Gerbera jamesonii* by ovules cultivated *in vitro*. Can.J.Bot. **64**: 2355-2357.
Ali, M.A. and J.K. Jones, 2000. Microspore culture in *Corchorus olitorius*: effect of growth regulators, teperature and sucrose on callus formation. Indian J.Exp.Biol. **38**(6): 593-597.
Altamura, M.M., A. Cersosimo, C. Majoli and M Crespan, 1992. Histological study of embryogenesis and organogenesis from anthers of *Vitis rupestris* du Lot cultured *in vitro*. Protoplasma **171**(3/4): 134-141.
Amos, J.A. and R.L. Scholl, 1978. Induction of haploid callus from anthers of four species of *Arabidopsis*. Z.Pflanzenphysiol. **90**: 33-43.
Arrillaga, I., V. Lerma, P. Perez-Bermudez and J. Segura, 1995. Callus and somatic embryogenesis from cultured anthers of service tree (*Sorbus domestica* L.). HortSci. **30**(5): 1078-1079.
Arya, I.D. and N. Chandra, 1989. Organogenesis in anther-derived callus culture of cowpea (*Vigna unguiculata* (L.) Walp). Current Science **58**(5): 257-259.
Arzate-Fernandez, A.-M., T. Nakazaki and T. Tanisaka, 1998. Production of diploid and tetraploid interspecific hybrids between *Lilium concolor* and *L. longiflorum* by *in vitro* ovary slice culture. Plant Breed. **117**: 479-484.
Assani, A., F. Bakry, F. Kerbellec, R. Haicour, G. Wenzel and B. Foroughi-Wehr, 2003. Production of haploids from anther culture of banana [*Musa balbisiana* (BB)]. Plant Cell Rep. **21**: 511-516
Atanasov, A.I., M. Abadzhieva and V. Becheva, 1980. Study on organogenesis ability in somatic tissue cultures of *Datura innoxia* Mill. with haploid diploid and tetraploid origin. Genet.Sel. **13**: 104-112.
Avetisov, V.A., 1976a. Production of haploids from anther cultures and isolated protoplast cultures of *Arabidopsis thaliana* (L.) Heynh. Sov.Genet. **12**: 408-413.
Avetisov, V.A., 1976b. Production of haploids under in vitro culture of *Arabidopsis thaliana* (L.) Heynh. anthers and isolated protoplasts. Genetika **12**: 17-25.
Babbar, S.B. and S.C. Gupta, 1984. Pathways in pollen sporophyte development in anther cultures of *Datura metel* and *Petunia hybrida*. Beitr.Biol.Pflanzen. **59**: 475-488.
Babbar, S.B. and S.C. Gupta, 1986a. Effect of carbon source on *Datura metel* microspore embryogenesis and the growth of callus raised from microspore-derived embryos. Biochem.Physiol.Pflanz.BPP. **181**: 331-338.
Babbar, S.B. and S.C. Gupta, 1986b. Induction of androgenesis and callus formation in *in vitro* cultured anthers of a myrtaceous fruit tree (*Psidium guajava* L.). Botanical Magazine **99**(1053): 75-83.
Babbar, S.B. and S.C. Gupta, 1986c. Putative role of ethylene in *Datura metel* microspore embryogenesis. Physiol.Plant. **68**: 141-144
Babbar, S.B. and S.C. Gupta, 1990. Phasic requirement of coconut milk for *Datura metel* microspore embryogenesis. Phytomorphol. **40**(1/2): 53-57.
Badea, E., M. Iordan and A. Mihalea, 1985. Induction of androgenesis in anther culture of *Digitalis lanata*. Revue Roumaine de Biologie, Biologie Vegetale **30**(1): 63-71.
Badea, E. and P. Raicu, 1982. In vitro anther and pollen culture in haploids (n=12) and triploids (3n=36) of *Datura innoxia* Mill. Rev.Roum.Biol.Ser.Biol.Veg. **27**: 163-165.
Badigannavar, A.M. and M.S. Kuruvinashetti, 1996. Callus induction and shoot bud formation from cultured anthers in sunflower (*Helianthus annuus* L.). Helia **19**(25): 39-46.

Bajaj, Y.P.S., 1978. Effect of super-low temperature on excised anthers and pollen-embryos of *Atropa, Nicotiana* and *Petunia*. Phytomorphol. **28**(2): 171-176.

Bajaj, Y.P.S. and M.S. Gill, 1989. Pollen-embryogenesis and chromosomal variation in anther culture of a diploid cotton (*Gossypium arboreum* L.). SABRAO J. **21(1)**: 57-63.

Bajaj, Y.P.S. and M.S. Gill, 1997. *In vitro* induction of haploidy in cotton. In: *In Vitro* Haploid Production in Higher Plants. Jain,M.S., S.K.Sopory and R.E.Veilleux (Eds.) Kluwer Academic Publishers, Dordrecht, pp.165-174.

Bajaj, Y.P.S. and S.S. Gosal, 1987. Pollen embryogenesis and chromosomal variation in cultured anthers of chickpea. Int.Chickpea Newsletter **17**: 12-13.

Bajaj, Y.P.S., A.K. Ram, K.S. Labana and H. Singh, 1981. Regeneration of genetically variable plants from the anther-derived callus of *Arachis hypogaea* and *Arachis villosa*. Plant Sci.Lett. **23**(1): 35-39.

Bajaj, Y.P.S., H. Singh and S.S. Gosal, 1980. Haploid embryogenesis in anther cultures of pigeon-pea (*Cajanus cajan*). Theor.Appl.Genet. **58**: 157-159.

Baldursson, S., P. Krogstrup, J.V. Norgaard and S.B. Andersen, 1993a. Microspore embryogenesis in anther culture of three species of *Populus* and regeneration of dihaploid plants of *Populus trichocarpa*. Can.J.For.Res. **23(9)**: 1821-1825.

Baldursson, S., J.V. Norgaard and P. Krogstrup, 1993b. Factors influencing haploid callus initiation and proliferation in megagametophyte cultures of Sitka spruce (*Picea sitchensis*). Silvae Genetica **42**(2-3): 79-86.

Ban, Y., T. Kokubu and Y. Miyaji, 1971. Production of haploid plant by anther-culture of *Setaria italica*. Kagoshima Univ.Fac.Agr.Bull. **21**: 77-81.

Banks, D.J., 1988. Characteristics of a rare, monosomic peanut (*Arachis hypogaea* L.- -Leguminosae), with implications for haploidy discovery. Am.J.Bot. 97-98.

Bansal, U.K., G. Bassi, S.S. Gosal and D.R. Satija, 1991. Induction of pollen embryogenesis and cytological variability in *Arachis hypogaea* L. through anther culture. Indian J.Genet.Plant Breed. **51** (1): 125-129.

Bapat, V.A. and G. Wenzel, 1982. *In vitro* haploid plantlet induction in *Physalis ixocarpa* Brot. through microspore embryogenesis. Plant Cell Rep. **1**: 154-156.

Baracaccia, G., C. Tomassini and M. Falcinelli, 1999. Further cytological evidence on the androgenesis pathway in pepper (*Capsicum annuum* L.). J.Genet.and Breed. **53**(3): 251-254.

Barcelo, P., A. Cabrera, C. Hagel and H. Lörz, 1994. Production of doubled-haploid plants from tritordeum anther culture. Theor.Appl.Genet. **87**(6): 741-745.

Barrow, J.R., 1986. The conditions required to isolate and maintain viable cotton (*Gossypium hirsutum* L.) microspores. Plant Cell Rep. **5**: 405-408.

Bensaad, Z.M., M.J. Hennerty and T.D Roche, 1996. Effects of cold pretreatment, carbohydrate source and gelling agents on somatic embryogenesis from anthers of *Vitis vinifera* L. cvs. 'Regina' and 'Reichensteiner'. Acta Hort. **440**: 504-509.

Bhaskaran, S., R.H. Smith and J.J. Finer, 1983. Ribulose bisphosphate carboxylase activity in anther-derived plants of *Saintpaulia ionantha* Wendl. Shag. Plant Physiol. **73**: 639-642.

Bicknell, R.A. and N.K. Borst, 1996. Isolation of reduced genotypes of *Hieracium pilosella* using anther culture. Plant Cell Tiss.Org.Cult. **45**(1): 37-41.

Binding, H. and G. Mordhorst, 1984. Haploid *Solanum dulcamara* L.: shoot culture and plant regeneration from isolated protoplasts. Plant Sci.Lett. **35**: 77-79.

Bingham, E.T., 1969. Haploids from cultivated alfalfa *Medicago sativa* L. Nature **221**: 865-866.

Bingham, E.T. and C.B. Gillies, 1971. Chromosome pairing, fertility, and crossing behavior of haploids of tetraploid alfalfa, *Medicago sativa* L. Can.J.Genet.Cytol. **13**: 195-202.

Birhman, R.K., S.R. Rivard and M. Cappadocia, 1994. Restriction Fragment Length Polymorphism analysis of anther-culture-derived *Solanum chacoense*. HortSci. **29(3)**: 206-208.

Blaschke, J.R., E. Forche and K.H. Neumann, 1978. Investigations on the cell cycle of haploid and diploid tissue cultures of *Datura innoxia* Mill. and its synchronization. Planta **144**: 7-12.

Bohanec, B., 1997. Haploid induction in buckwheat (*Fagopyrum esculentum* Moench). In: *In Vitro* Haploid Production in Higher Plants. Vol. 4. Jain,S.M., S.K.Sopory and R.E.Veilleux (Eds.) Kluwer Academic Publishers, Dordrecht, pp.163-170.

Borgel, A. and M. Arnaud, 1986. Progress in eggplant breeding, use of haplomethod. Capsicum Newsletter **5**: 65-66.
Bouvier, L., Y.X. Zhang and Y. Lespinasse, 1993. Two methods of haploidization in pear, *Pyrus communis* L.: greenhouse seedling selection and *in situ* parthenogenesis induced by irradiated pollen. Theor.Appl.Genet. **87**: 229-232.
Braniste, N., A. Popescu and T. Coman, 1984. Producing and multiplication of *Pyrus communis* haploid plants. Acta Hortic. **161**: 147-150.
Bueno, M.A., A. Gomez, M Boscaiu, J.A. Manzanera and O. Vicente, 1997. Stress-induced formation of haploid plants through anther culture in cork oak (*Quercus suber*). Physiol.Plant. **99**: 335-341.
Bugara, A.M., L.V. Rusina and S. A. Reznikova, 1986. Embryoidogenesis in anther culture of *Salvia sclarea*. Fiziol.Biokhim.Kult.Rast. **18**: 381-386.
Bui-Dang-Ha, D. and J. Pernes, 1982. Androgenesis in pearl millet. I. Analysis of plants obtained from microspore culture. Z.Pflanzenphysiol. **108**(4): 317-327.
Busso, C.S., C.J. Liu, C.T. Hash, J.R. Witcombe, K.M. Devos, J.M. J. de Wet and M.D. Gale, 1995. Analysis of recombination rate in female and male gametogenesis in pearl millet (*Pennisetum glaucum*) using RFLP markers. Theor.Appl.Genet. **90**: 242-246.
Can, N.D., S. Nakamura, T.A.D. Haryanto and T. Yoshida, 1998. Effects of physiological status of parent plants and culture medium composition on the anther culture of sorghum. Plant Prod.Sci.Kyoto **1**(3): 211-215.
Can, N.D. and T. Yoshida, 1999. Combining ability of callus induction and plant regeneration in sorghum anther culture. Plant Prod.Sci.Kyoto **2**(2): 125-128.
Canhoto, J. M. and G. S. Cruz, 1993. Induction of pollen callus in anther cultures of *Feijoa sellowiana* Berg. (Myrtaceae). Plant Cell Rep. **13**(1): 45-48.
Canhoto, J.M., M. Ludovina, S. Guimaraes and G.S. Cruz, 1990. *In vitro* induction of haploid, diploid and tetraploid plantlets by anther culture of *Iochroma warscewiczii* Regel. Plant Cell Tiss.Org.Cult. **21**(22): 171-177.
Cao, M.Q., Y. Li, F. Liu and C. Dore, 1994. Embryogenesis and plant regeneration of pakchoi (*Brassica rapa* L. ssp. *chinensis*) via *in vitro* isolated microspore culture. Plant Cell Rep. **13**: 447-450.
Cappadocia, M. and M. Ahmim, 1988. Comparison of two culture methods for the production of haploids by anther culture in *Solanum chacoense*. Can.J.Bot. **66**: 1003-1005.
Cappadocia, M., L. Chretien and G. Laublin, 1988. Production of haploids in *Gerbera jamesonii* via ovule culture: influence of fall versus spring sampling on callus formation and shoot regeneration. Can.J.Bot. **66**: 1107-1110.
Caredda, S. and C. Clement, 1999. Androgenesis and albinism in Poaceae: influence of genotype and carbohydrates. In: Anther and Pollen. From Biology to Biotechnology. Clement, C., E. Pacini and J.-C.Audran (Eds.) Springer-Verlag, Berlin, pp.211-228.
Cersosimo, A., M Crespan, G. Paludetti and M.M. Altamura, 1990. Embryogenesis, organogenesis and plant regeneartion from anther culture in *Vitis*. Acta Hort. **280**: 307-314.
Chakravarty, B. and S. Sen, 1989. Regeneration through somatic embryogenesis from anther explants of *Scilla indica* (Roxb.) Baker. Plant Cell Tiss.Org.Cult. **19**(1): 71-75.
Chaudhari, H.K., 1979. The production and performance of doubled haploids of cotton. Bull.Torrey Bot.Club. **106**: 123-130.
Chekurov, V.M. and E.P. Razmakhnin, 1999. Effect of inbreeding and growth regulators on the *in vitro* androgenesis of wheatgrass, *Agropyron glaucum*. Plant Breed. **118**(6): 571-573.
Chen, Z.G., 1985. A study on induction of plants from citrus pollen. Fruit.Var.J. **39**: 44-50.
Chen, Z.H., W.B. Li, L.H. Zhang, X. Xu and S.J. Zhang, 1988. Production of haploid plantlets in cultures of unpolinated ovules of *Hevea brasiliensis* Muell.-Arg. In: Somatic Cell Genetics of Woody Plants. Ahuja, M. R. (Ed.), Kluwer Academic Publishers, Dordrecht, pp.39-44.
Choi, B.H., R.Y Park and R.K. Park, 1997. Haploidy in pearl millet [*Pennisetum glaucum* (L.) R.Br.]. In: *In Vitro* Haploid Production in Higher Plants. Vol. 4. Jain, S. M., S. K. Sopory and R. E. Veilleux (Eds.) Kluwer Academic Publishers, Dordrecht, pp.171-179.

Christensen, J. R., E. Borrino, A. Olesen and S. B. Andersen, 1997. Diploid, tetraploid, and octoploid plants from anther culture of tetraploid orchard grass, D*actylis glomerata* L. Plant Breed. **116**: 267-270.

Collins, G.B., J.M. Dunwell and N. Sunderland, 1974. Irregular microspore formation in *Datura innoxia* and its relevance to anther culture. Protoplasma **82**: 365-378.

Collins, G.B. and N. Sunderland, 1974. Pollen-derived haploids of *Nicotiana knightiana*, *Nicotiana raimondii*, and *Nicotiana attenuata*. J.Exp.Bot. **25**: 1030-1039.

Contolini, C.S. and M.Y. Menzel, 1987. Early development of duplication-deficiency ovules in upland cotton. Crop Sci. **27(2)**: 345-348.

Corduan, G., 1975. Regeneration of anther-derived plants of *Hyoscyamus niger* L. Planta **127**: 27-36.

Corduan, G., 1976. Isozyme variation as an indicator for the generative or somatic origin of anther-derived plants of *Digitalis purpurea* L. Z.Pflanzenphysiol. **76(1)**: 47-55.

Coumans, M. and D. Zhong, 1995. Doubled haploid sunflower (*Helianthus annuus*) plant production by androgenesis: fact or artifact? 2. *In vitro* isolated microspore culture. Plant Cell Tiss.Org.Cult. **41(3)**: 203-209.

Couturon, E., 1982. Obtaining naturally occurring haploids of *Coffea canephora* Piere by grafting of embroys. Cafe,Cacao,Tee. **26**: 155-160.

Crane, C.F., W.D. Beversdorf and E.T. Bingham, 1982. Chromosome pairing and associations at meiosis in haploid soybean (*Glycine max*). Can.J.Genet.Cytol. **24**: 293-300.

Cuny, F., B. Dumas de Vaulx, B. Longhi and R. Siadous, 1992. Analyse des plantes de melon (*Cucumis melo* L) issues de croisements avec du pollen irradie a differentes doses. Agronomie **12**: 623-630.

Custers, J.B.M., E. Ennik, W. Eikelboom, J.J.M. Dons and M.M. Van Lookeren Campagne, 1997. Embrogenesis from isolated microspores of tulip; towards developing F_1 hybrid varieties. Acta Hort. **430**: 259-266.

Dallos, M.P. and V. Galan-Sauco, 1998. Pollen and anther culture in *Musa* spp. Acta Hort. **490**: 493-497.

De, D.N. and P.V.L Rao, 1983. Androgenetic haploid callus of tropical leguminous trees. In: Plant Cell Culture in Crop Improvement. Sen, S. K. and K. L. Giles (Eds.) Plenum Press, New York, pp.469-474.

Debata, B.K. and S.N. Patnaik, 1983. *In vitro* culture of anther of *Crotalaria pallida* Ait. for induction of haploid. Indian J.Exp.Biol. **21**: 44-46.

Debata, B.K. and S.N. Patnaik, 1988. Induction of androgenesis in anther cultures of *Solanum viarum* Dunal. J.Plant Physiol. **133(1)**: 124-125.

DeLange, J.H. and A.P. Vincent, 1988. Studies on *Citrus* pollination using gamma-irradiated pollen. S.Afr.J.Bot. **54(3)**: 257-264.

Denissen, C.J.M. and A.P.M. Den Nijs, 1987. Effects of gamma irradiation on *in vitro* pollen germination of different *Cucumis* species. Euphytica **36**: 651-658.

DeVerna, J.W. and G.B. Collins, 1984. Maternal haploids of *Petunia axillaris* (Lam.) B.S.P. via culture of placenta attached ovules. Theor.Appl.Genet. **69**: 187-192.

Diettrich, B., S. Ernst, and M. Luckner, 2000. Haploid plants regenerated from androgenic cell cultures of *Digitalis lanata*. Planta Med. **66(3)**: 237-240.

Dieu, P. and J.M. Dunwell, 1988. Anther culture with different genotypes of opium poppy (*Papaver somniferum* L.): effect of cold treatment. Plant Cell Tiss.Org.Cult. **12(3)**: 263-271.

Dryanovska, O.A. and I.N. Ilieva, 1983. *In vitro* anther and ovule cultures in muskmelon (*Cucumis melo* L.). Comptes Rendus de l'Academie Bulgare de Sciences **36(8)**: 1107-1110.

Du, L. G., Q.Q. Shao and A.S. Li, 1986. Somatic embryogenesis and plant regeneration from anther culture of *Panax quinquifolius* (ginseng). Newslet.Int.Plant Biotech.Network. **6**: 9.

Dublin, P., 1978. Diploidized haploids and production of fertile homozygous genotypes in cultivated cacao trees (*Theobroma cacao*). Cafe,Cacao,Tee **22**: 275-284.

Dumas-de-Vaulx, R., 1979. Obtaining haploid plants in the melon (*Cucumis melo* L.) after pollination by *Cucumis ficifolius* A. Rich. C.R.Hebd.Seances Acad.Sci.Ser.D.Sci.Nat. **289**: 875-878.

Dumas-de-Vaulx, R. and D. Chambonnet, 1986. Obtention of embryos and plants from *in vitro* culture of unfertilized ovules of *Cucurbita pepo*. In: Genetic Manipulation in Plant Breeding.

Horn, W., C.J. Jensen, W. Oldenbach and O. Schieder (Eds.) Walter de Gruyter and Co, Berlin, pp.295-297.

Dumas-de-Vaulx, R., D. Chambonnet and E. Pochard, 1981. *In vitro* culture of pepper (*Capsicum annuum* L.) anthers: high rate plant production from different genotypes by +35 degrees Celsius treatments haploid and diploid plants, cultivars. Agron.Sci.Prod.Veg.Environ. **1**: 859-864.

Dunwell, J.M. and N. Sunderland, 1976. Pollen ultrastructure in anther cultures of *Datura innoxia*. III. Incomplete microspore division. J.Cell Sci. **22**: 493-501.

Elkonin, L.A., T.N. Gudova, A.G. Ishin, and U.S. Tyrnov, 1993. Diploidization in haploid tissue cultures of sorghum. Plant Breed. **110**: 201-206.

Emershad, R.L., D.W. Ramming and M.D Serpe, 1989. *In ovulo* embryo development and plant formation from stenospermic genotypes of *Vitis vinifera*. Aust.J.Bot. **76**(3): 397-402.

Evans,D.A. and R.Morrison. (1989) Tomato anther culture. (US 4 835 339).

Falque, M., 1994. Pod and seed development and phenotype of M_1 plants after pollination and fertilization with irradiated pollen in cacao (*Theobroma cacao* L.). Euphytica **75**: 19-25.

Falque, M., A.A. Kodia, O. Sounigo, A.B. Eskes, and A. Charrier, 1992. Gamma-irradiation of cacao (*Theobroma cacao* L.) pollen: Effect of pollen viability, germination and mitosis on fruit set. Euphytica **64(3)**: 167-172.

Fan, Y.H., S.Y. Zang and J.F. Zhao, 1982. Induction of haploid plants in *Lycium chinense* Mill. and *Lycium barbarum* by anther culture. I. Chuan. Hereditas **4**: 25-26.

Fankhauser, H., F. Bucher and P.J. King, 1984. Isolation of biochemical mutants using haploid mesophyll protoplasts of *Hyoscyamus muticus*. IV. Biochemical characterisation of nitrate non-utilizing clones. Planta **160**: 415-421.

Faure, O., J. Aarrouf and A Nougarede, 1996a. Ontogenesis, differentiation and precocious germination in anther-derived somatic embryos of grapevine (*Vitis vinifera* L.): embryonic organogenesis. Ann.Bot. **78**(1): 29-37.

Faure, O., J. Aarrouf and A Nougarede, 1996b. Ontogenesis, differentiation and precocious germination in anther-derived somatic embryos of grapevine (*Vitis vinifera* L.): proembryogenesis. Ann.Bot. **78**(1): 23-28.

Fedak, G., 1983. Haploids in *Triticum ventricosum* via intergeneric hybridization with *Hordeum bulbosum*. Can.J.Genet.Cytol. **25**: 104-106.

Ficcadenti, N., S. Sestili, S. Annibali, M. di Marco and M. Schiavi, 1999. *In vitro* gynogenesis to induce haploid plants in melon *Cucumis melo* L. J.Genet.and Breed. **53**(3): 255-257.

Fitch, M.M. and P.H. Moore, 1984. Production of haploid *Saccharum spontaneum* L. Comparison of media for cold incubation of panicle branches and for float culture of anthers. J.Plant Physiol. **117**: 169-178.

Fitch, M.M.M. and P.H. Moore, 1996. Haploids of sugarcane. In: *In Vitro* Haploid Production in Higher Plants. Jain, S.M., S.K. Sopory and R.E. Veilleux (Eds.) Kluwer Academic Publishers, Dordrecht, pp.1-16.

Forche, E., R. Kibler and K.H. Neumann, 1981. The influence of developmental stages of haploid and diploid callus cultures of *Datura innoxia* on shoot initiation. Z.Pflanzenphysiol. **101**: 257-262.

Fougat, R.S., A.R. Pathak and P.S. Bharodia, 1992. Regeneration of haploid callus from anthers of pigeonpea. Gujarat Agr.Univ.Res.J. **17**(2): 151-152.

Fraser, L.G. and C.F. Harvey, 1986. Somatic embryogenesis from anther-derived callus in two *Actinidia* species. Sci.Hortic. **29**: 335-346.

Frederiksen, S., 1989. Chromosome elimination in a hybrid between *Taeniatherum caput-medusae* and *Hordeum bulbosum*. Hereditas **110**: 87-88.

Friedt, W., T. Nurhidayah, T. Rocher, H. Kohler, R Bergmann and R. Horn, 1997. Haploid production and application of molecular methods in sunflower (*Helianthus annuus* L.). In: *In Vitro* Haploid Production in Higher Plants. Vol. 5. Jain, S.M., S.K. Sopory and R.E. Veilleux (Eds.) Kluwer Academic Publishers, Dordrecht, pp.17-35.

Froelicher, Y. and P. Ollitrault, 2000. Effects of the hormonal balance on *Clausena excavata* androgenesis. Acta Hort. **535**: 139-146.

Fu, L.F. and D.Y. Tong, 1983. Induction pollen plants of litchi tree (*Litchi chinensis* Soon). Acta Genet.Sin. **10**(5): 369-374.
Gaj, M. and M.D. Gaj, 1985. Dihaploids of *Hordeum murinum* L. and *H. secalinum* Schreb. from interspecific crosses with *H. bulbosum* L. Barley Genet.Newsl. **15**: 33-34.
Gavrilenko, T., R. Thieme and V.-M. Rokka, 2001. Cytogenetic analysis of *Lycopersicon escullentum* (+) *Solanum etuberosum* somatic hybrids and their androgenetic regenerants. Theor.Appl.Genet. **103**: 231-239.
Gelebart, P. and L.H. San, 1987. Production of haploid plants of sunflower (*Helianthus annuus* L.) by *in vitro* culture of nonfertilized ovaries and ovules. Agron.Sci.Prod.Veg.Environ. **7**: 81-86.
Geraci, G. and A. Starrantino, 1990. Attempts to regenerate haploid plants from " *in vitro*" cultures of citrus anthers. Acta Hort. **280**: 315-320.
Gerassimowa, H., 1936. Experimantell erhaltene haploide Pflanze von *Crepis tectorum* L. Planta. **25**: 696-702.
Germana, M.A., 1997. Haploidy in Citrus. In: *In Vitro* Haploid Production in Heigher Plants. Vol. 5. Jain, S.M., S.K. Sopory and R.E. Veilleux (Eds.) Kluwer Academic Publishers, Dordrecht, pp.195-217.
Germana, M.A., F.G Crescimanno, F. de Pasquale and W.Y. Ying, 1992. Androgenesis in 5 cultivars of *Citrus limon* L. Burm. f. Acta Hort. **300**: 315-324.
Germana, M.A., F.G Crescimanno and A. Motisi, 2000a. Factors affecting androgenesis in *Citrus clementina* Hort. ex. Tan. Adv.Hort.Sci. **14**(2): 43-51.
Germana, M.A., F.G Crescimanno, G.R. Recupero and M.P. Russo, 2000b. Preliminary characterization of several doubled haploids of *Citrus clementina* cv. Nules. Acta Hort. **535**: 183-190.
Germana, M.A., Y.Y. Wang, M.G. Barbagallo, G. Iannolino and F.G. Crescimanno, 1994. Recovery of haploid and diploid plantlets from anther culture of *Citrus clementina* Hort. ex Tan. and *Citrus reticulata* Blanco. J.Hortic.Sci. **69**(3): 473-480.
Gharyal, P.K., A. Rashid and S.C. Maheshwari, 1983a. Androgenic response from cultured anthers of a leguminous tree, *Cassia siamea* Lam. Protoplasma **118**: 91-93.
Gharyal, P.K., A. Rashid and S.C. Maheshwari, 1983b. Production of haploid plantlets in anther cultures of *Albizzia lebbeck* L. Plant Cell Rep. **2**: 308-309.
Gmitter, F.G. Jr. and G.A. Moore, 1986. Plant regeneration from undeveloped ovules and embryogenic calli of Citrus: embryo production, germination, and plant survival. Plant Cell Tiss.Org.Cult. **6**(2): 139-147.
Gonzalez-Melendi, P., P.S. Testillano, P. Ahmadian, B. Fadon, O. Vicente and M.C. Risueno, 1995. *In situ* characterization of the late vacuolate microspore as a convenient stage to induce embryogenesis in *Capsicum*. Protoplasma **187**(1-4): 60-71.
Gu, Z.P. and K.C. Cheng, 1983. *In vitro* induction of haploid plantlets from unpollinated young ovaries of lily and its embryological observations (*Lilium davidii*). Acta Bot.Sinica **25**: 24-28.
Gudu, S., J.D. Procunier, A. Ziauddin and K. J. Kasha, 1993. Anther culture derived homozygous lines in *Hordeum bulbosum*. Plant Breed. **110**: 109-115.
Guedira, M., T. Dubois-Tylski, J. Vasseur and J. Dubois, 1989. Direct somatic embryogenesis obtained from anther culture of *Cichorium* (Asteraceae). Can.J.Bot. **67**: 970-976.
Gulshan, T.M. Varghese and D.R. Sharma, 1981. Studies on anther cultures of tomato - *Lycopersicon esculentum* Mill. Biol.Plant. **23**(6): 414-420.
Gupta, P.P., 1983. Microspore-derived haploid, diploid and triploid plants in *Petunia violacea* Lindl. Plant Cell Rep. **2**: 255-256.
Gupta, S., 1975. Morphogenetic response of haploid callus tissue of *Pisum sativum* (var. B22). Indian Agriculturist **19**(4): 11-21.
Gurel, A., K. Nichterlein and W. Friedt, 1991. Shoot regeneration from anther culture of sunflower (*Helianthus annuus*) and some interspecific hybrids as affected by genotype and culture procedure. Plant Breed. **106**(1): 68-76.
Hammerschlag, F.A., 1993. Factors affecting the frequency of callus formation among cultured peach anthers. HortSci. **18(2)**: 210-211.
Han, D.S., Y. Niimi and M Nakano, 1997. Regeneration of haploid plants from anther cultures of the Asiatic hybrid lily 'Conecticut King'. Plant Cell Tiss.Org.Cult. **47**: 153-158.

Han, D.S., Y. Niimi and M. Nakano, 1999. Production of doubled haploid plants through colchicine treatment of anther-derived haploid calli in Asiatic hybrid lily 'Connecticut King'. J.Japanese Soc.Hortic.Sci. **68**(5): 979-983.

Hanson, M.R., 1984. Anther and pollen culture. Monogr.Theor.Appl.Genet. **9**: 138-150.

Hatano, K., Y. Shoyama and I. Nishioka, 1987. Somatic embryogenesis and plant regeneration from the anther of *Aconitum carmichaeli* Debx. Plant Cell Rep. **6**(6): 446-448.

Haydu, Z. and I.K. Vasil, 1981. Somatic embryogenesis and plant regeneration from leaf tissues and anthers of *Penisetum purpureum* Schum. Theor.Appl.Genet. **59**(5): 269-273.

Hennerty, M.J. and A.J. Sayegh, 1996. Polyhaploidy in strawberry. In: *In Vitro* Haploid Production in Higher Plants. Jain, S.M., S.K. Sopory and R.E. Veilleux (Eds.) Kluwer Academic Publishers, Dordrecht, pp.231-260.

Hermsen, J.G.T., 1969. Induction of haploids and aneuhaploids in colchicine-induced tetraploid *Solanum chacoense* Bitt. Euphytica **18**: 183-189.

Hesemann, C.U., 1980. Haploid cells in calluses from anther culture of *Vicia faba*. Z.Pflanzenzüchtg. **84**: 18-27.

Hidaka, T., 1984. Induction of plantlets from anthers of 'Trovita' orange (*Citrus sinensis* Osbeck). J.Japanese Soc.Hortic.Sci. **53**(1): 1-5.

Hidaka, T., Y.Yamada and T. Shichijo, 1979. *In vitro* differentiation of haploid plants by anther culture in *Poncirus trifoliata* (L.) Raf. Japan.J.Breed. **29**: 248-254.

Hidaka, T., Y.Yamada and T. Shichijo, 1982. Plantlet formation by anther culture of *Citrus aurantium* L. Induction of haploid plants. Japan.J.Breed. **32**: 247-252.

Hildebrand, D.F., G.C. Phillips and G.B. Collins, 1986. Soybean [*Glycine max* (L.) Merr.]. In: Biotechnology in Agriculture and Forestry 2. Crops I. Bajaj, Y.P.S. (Ed.) Springer-Verlag, Berlin,pp.283-308.

Hinata, K., 1986. Egg Plant (*Solanum melongena* L.). In: Biotechnology in Agriculture and Forestry 2. Crop I. Bajaj, Y.P.S. (Ed.) Springer-Verlag, Berlin, pp.363-370.

Hinchee, M.A.W., A. dela Cruz and A. Maretzki, 1984. Development and biochemical characteristics of cold-treated anthers of *Saccharum spontaneum*. J.Plant Physiol. **115**(4): 271-284.

Hollo, R. and S. Misik, 2000. Investigations of grape anther culture aiming haploid plant production. Acta Hort. **528**: 347-350.

Hongyuan, Y., Z. Chang, C. Detian, Y. Hua, W. Yan and C. Xiaoming, 1986. *In vitro* culture of unfertilized ovules in *Helianthus annuus* L. In: Haploids of Higher Plants *In Vitro*. Hu, H. and H. Yang (Eds.) China Academic Publishers, Beijing, pp.182-191.

Honkanen, J., A. Aapola, P. Seppanen, T. Tormala, J.C de Wit, L.J.M. Stravers and H.F. Esendam, 1992. Production of doubled haploid gerbera clones. Acta Hort. **300**: 341-346.

Höfer, M. and V. Hanke, 1990. Induction of androgenesis *in vitro* in apple and sweet cherry. Acta Hort. **280**: 333-336.

Hu, C.Y., G.C. Yin, M. Helena and B Zanettini, 1996. Haploid of soybean. In: *In vitro* Haploid Production in Higher Plants. Vol. 3. Jain, S.M., S.K. Sopory and R.E. Veilleux (Eds.) Kluwer Academic Publishers, Dordrecht, pp.377-395.

Huda, S., R. Islam, M.A. Bari and M. Asaduzzaman, 2001. Anther culture of chickpea. Int.Chickpea and Pigeonpea Newslet. **8**: 24-26.

Hughes, K. W., S. L. Bell and J. D. Caponetti, 1975. Anther-derived haploids of the African violet. Can.J.Bot. **53**: 1442-1444.

Humphreys, M.W., A.G. Zare, I. Pasakinskiene, H. Thomas, W.J. Rogers and H.A. Collin, 1998. Interspecific genomic rearrangements in androgenic plants derived from a *Lolium multiflorum x Festuca arundinacea* (2n=5x=35) hybrid. Heredity, **80**: 78-82.

Hwang, J.K., K.Y. Paek and D.H. Cho, 1998. Breeding resistant pepper lines (*Capsicum annuum* L.) to bacterial spot (*Xanthomonas campestris* pv. *vesicatoria*) through anther culture. Acta Hort. **461**: 301-307.

Hyun, S.K., J.H. Kim, E.W. Noh and J.I. Park, 1986. Induction of haploid plants of *Populus* species. In: Plant Tissue Culture and Its Agricultural Applications. Withers, L.A. and P.G. Alderson (Eds.) Butterworths, London, pp.413-418.

Iankulov, I.K., M.D. Abadzhieva and A.I. Atanasov, 1979. Investigation of experimentally obtained haploid from *Datura innoxia* Mill. Dokl.Bulg.Akad.Nauk **32**: 213-216.

Ikeda, M., T. Masuda, T. Kotobuki, T. Yoshioka, T. Nakanishi and M. Yoshida, 1993. Artificial control of polyembryogenesis and production of hybrids by chronic gamma irradiation on *Citrus* ovule *in vitro*. Bull.Natl.Inst.Agrobiol.Resour. **8**: 25-46.

Imelda, M., S.H.A. Lubis and S. Sastrapradja, 1988. Anther culture of rambutan (*Nephelium* sp.). Annales Bogorienses, New Series **1**(1): 7-9.

Inagaki, N., H. Matsunaga, M. Kanechi and S. Maekawa, 1994. *In vitro* micropropagation of *Allium giganteum* R. 2. Embryoid and plantlet regeneration through the anther culture of *Allium giganteum* R. Sci.Rep.Fac.Agric.Kobe.Univ. **21**(1): 23-30.

Ishizaka, H., 1998. Production of microspore-derived plants by anther culture of an interspecific F_1 hybrid between *Cyclamen persicum* and *C. purpurascens*. Plant Cell Tiss.Org.Cult. **54**: 21-28.

Isouard, G., C. Raquin and Y. Demarly, 1979. Haploid and diploid plants obtained by culture *in vitro* from eggplant anthers (*Solaum melongena*). C.R.Hebd.Seances Acad.Sci.Ser.D.Sci.Nat. **288**: 987-989.

Jain, A.K., A. Sarkar and R. K Datta, 1996. Induction of haploid callus and embryogenesis in *in vitro* cultured anthers of mulberry (*Morus indica*). Plant Cell Tiss.Org.Cult. **44**(2): 143-147.

Jain, R.K., U. Brune and W. Friedt, 1989. Plant regeneration from *in vitro* cultures of cotyledon explants and anthers of *Sinapis alba* and its implications on breeding of crucifers. Euphytica **43**(1-2): 153-163.

Jain, S.M. and N. Bhalla-Sarin, 1997. Haploidy in Petunia. In: *In Vitro* Haploid Production Production in Higher Plants. Vol. 5. Jain, S.M., S.K. Sopory and R.E. Veilleux (Eds.) Kluwer Academic Publishers, Dordrecht, pp.53-71.

Jaiswal, V.S. and P. Narayan, 1981. Induction of pollen embryoids in *Solanum torvum* Swartz. Haploid induction. Curr.Sci. **50**: 998-999.

Jayasree, P.K., M.P. Asokan, S. Sobha, L. S. Ammal, K. Rekha, R. G. Kala and R. Jayasrree, 1999. Somatic embryogenesis and plant regeneration from immature anthers of *Hevea brasiliensis* (Muell.) Arg. Curr.Sci. **76**(9): 1242-1245.

Jelenkovic, G., O. Shifriss and E. Harrington, 1980. Association and distribution of meiotic chromosomes in a haploid of *Ricinus communis* L. Cytologia **45**: 571-577.

Jelenkovic, G., M.L. Wilson and P.J. Harding, 1984. An evaluation of intergeneric hybridization of *Fragaria* spp. x *Potentilla* spp. as a means of haploid production. Euphytica, **33**: 143-152.

Jian, Y.Y., D.P. Liu, X.M. Luo and G.L. Zhao, 1986. Studies on induction of pollen plants in *Glycine max* (L.) Merr. Chiang.su.Nung.Yeh.Hseuh.Pao.J.Agric.Sci. **2**: 26-30.

Johansson, L.B., 1986. Effects of activated charcoal, cold treatment and elevated CO_2-concentrations on embryogenesis in anther cultures. In: Genetic Manipulation in Plant Breeding. Horn, W., C.J. Jensen, W. Oldenbach and O. Schieder (Eds.), Walter de Gruyter and Co, Berlin, pp. 257-264.

Johansson, L.B., E. Calleberg, and A. Gedin, 1990. Correlation between activated charcoal, Fe-EDTA and other organic media ingredients in cultured anthers of *Anemone canadensis*. Physiol.Plant. **80**(2): 243-249.

Jorgensen, J., 1991. Androgenesis in *Quercus petraea*, *Fagus sylvatica*, and *Aesculus hippocastanum*. NATO.Adv.Sci.Inst.Ser.Ser.A.Life Sci. **210**: 323-354.

Jorgensen, R. B. and R. Bothmer, 1988. Haploids of *Hordeum vulgare* and *H. marinum* from crosses between the two species. Hereditas **108**: 207-212.

Kadlec, M., J. Suchomelova, V. A Smirnov, and S. L. Nikolajevna , 1991. Anther culture in soybean. Soybean Genet.Newsl. **18**: 121-124.

Kaltchuk-Santos, E., J.E. Mariath, E. Mundstock, C.-Y. Hu, and M.H. Bodanese-Zanettini, 1997. Cytological analysis of early microspore divisions and embryo formation in cultured soybean anthers. Plant Cell Tiss.Org.Cult. **49**: 107-115.

Kasperbauer, M.J. and R.C. Buckner, 1979. Haploid plants from anthers of *Festuca arundinacea* cultured with nurse tissue. Agron.Abstr. pp. 66.

Kato, M., T. Suga, and S. Tokumasu, 1980. Effect of 2,4-D and NAA on callus formation and haploid production in anther culture of *Pelargonium roseum*. Mem.Coll.Agric.Ehime.Univ. **24**: 199-207.

Kaur, P. and J.K. Bhalla, 1998. Regeneration of haploid plants from microspore culture of pigeon pea (*Cajanus cajan* L.). Indian J.Exp.Biol. **36**(7): 736-738.

Kavi Kishor, P.B., T.P. Reddy, A. Sarvesh, and G. Venkatesham, 1997. Haploidy in niger (*Guizotia abyssinica* Cass). In: *In Vitro* Haploid Production in Higher Plants. Jain, S.M., S.K. Sopory and R.E. Veilleux (Eds.) Kluwer Academic Publishers, Dordrecht, pp.37-51.

Kenny, L. and P.D.S. Caligari, 1996. Androgenesis of the salt tolerant shrub *Atriplex glauca*. Plant Cell Rep. **15**: 829-832.

Khan, S.K. and P.D. Ghosh, 1983. *In vitro* induction of androgenesis and organogenesis in *Cicer arietinum* L. Current Sci. **52**(18): 891-893.

Khoder, M., P. Villemur, and R. Jonard, 1984. Obtainment of monoploid and triploid plants by *in vitro* androgenesis in *Begonia x hiemalis* Fotsch cv. (A). Bull.Soc.Bot.Fr.Lett.Bot. **131**: 43-48.

Khushwinder, S., B.S. Sandhu, and S.S. Gosal, 1998. Anther culture response in cotton. Annals Biol.Ludhiana **14**(1): 11-13.

Kihara, M., K. Fukuda, H. Funatsuki, I. Kishinami, and Y. Aida, 1994. Plant regeneration through anther culture of tree wilde species of *Hordeum* (*H. murinum, H. marinum,* and *H. bulbosum*). Plant Breed. **112**: 244-247.

Kim, C.K., J.Y. Oh, and J.D. Chyng, 1998. Plant regeneration of Korean native Chinese chive by unpollinated ovule culture. J.Korean Soc.Hortic.Sci. **39**(6): 693-696.

Kim, J.H., H.K. Moon, and J.I. Park, 1986. Haploid plantlet induction through anther culture of *Populus maximowiczii*. Yon'gu.Pogo.Res.Rep.Inst.For.Genet. 116-121.

Kim, J.H., E.W. Noh, and J.I. Park, 1983. Haploid plantlets formation through anther culture of *Populus glandulosa*. Yon'gu.Pogo.Res.Rep.Inst.For.Genet. 93-98.

Kim, S.W., N.H. Song, K.H Jung, S.S. Kwak, and J.R. Liu, 1994. High frequency plant regeneration from anther-derived cell suspension cultures via somatic embryogenesis in *Catharanthus roseus*. Plant Cell Rep. **13**(6): 319-322.

Kintzios, S. and G. Fischbeck, 1994. Anther culture response of *Hordeum spontaneum* - derived winter barley lines. Plant Cell Tiss.Org.Cult. **37**: 165-170.

Kiss, J., M. Kondrak, O. Torjek, E. Kiss, G. Gyulai, K. Mazik Tokei, and L.E. Heszky, 2001. Morphological and RAPD analysis of poplar trees of anther culture origin. Euphytica **118**: 213-221.

Kiviharju, E. and E. Pehu, 1998. The effect of cold and heat pretreatments on anther culture response of *Avena sativa* and *A. sterilis*. Plant Cell Tiss.Org.Cult. **54**: 97-104.

Kiviharju, E., M. Puolimatka, and E. Pehu, 1997. Regeneration of anther-derived plants of *Avena sterilis*. Plant Cell Tiss.Org.Cult. **48**: 147-152.

Koltunow, A.M., K. Soltys, N. Nito, and S. McClure, 1995. Anther, ovule, seed, and nuclear embryo development in *Citrus sinensis* cv. Valencia. Can.J.Bot. **73**(10): 1567-1582.

Kristiansen, K. and S.B. Andersen, 1993. Effects of donor plant temperature, photoperiod, and age on anther culture response of *Capsicum annuum* L. Euphytica **67**(1/2): 105-109.

Krumbiegel, G., 1979. Response of haploid and diploid protoplasts from *Datura innoxia* Mill. and *Petunia hybrida* L. to treatment with X-rays and a chemical mutagen. Environ.Exp.Bot. **19**: 99-103.

Kumaravadivel, N. and S.R. Sree Rangasamy, 1994. Plant regeneration from sorghum anther cultures and field evaluation of progeny. Plant Cell Rep. **13**: 286-290.

Kusum, M. and S.C. Maheshwari, 1977. Enhancement by cold treatment of pollen embryoid development in *Petunia hybrida*. Z.Pflanzenphysiol. **85**(2): 177-180.

Kwack, S.N. and K. Fujieda, 1988. Somatic embryogenesis in cultured unfertilized ovules of *Cucurbita moschata*. J.Japanese Soc.Hortic.Sci. **57**(1): 34-42.

Lanaud, C., P. Lachenaud, and O. Sounigo, 1988. Behavior in the growth of doubled hapolids of cacao plants (*Theobroma cacao*). Can.J.Bot. **66**: 1986-1992.

Lashermes, P., E. Couturon, and A Charrier, 1994a. Combining ability of doubled haploids in *Coffea canephora* P. Plant Breed. **112**: 330-337.

Lashermes, P., E. Couturon, and A. Charrier, 1994b. Doubled haploids of *Coffea canephora*: development, fertility and agronomic characterictics. Euphytica **74**: 149-157.

Laurain, D., J. Tremouillaux-Guiller, and J.-C. Chenieux, 1993. Embryogenesis from microspores of *Ginkgo biloba* L., a medicinal woody species. Plant Cell Rep. **12**: 501-505.

Lazarte, J.E. and C.C. Sasser, 1982. Asexual embryogenesis and plant development in anther culture of *Cucumis sativus* L. HortSci. **17**(1): 88.

Le Deunff, E. and A. Sauton, 1994. Effect of parthenocarpy on ovule development in cucumber (*Cucumis sativus* L.) after pollination with normal and irradiated pollen. Sex.Plant Reprod. **7**: 221-228.

Le Thi, K., R. Lespinasse, S. Siljak-Yakovlev, T. Robert, N. Khalfallah, and A. Sarr, 1994. Karyotypic modifications in androgenetic plantlets of pearl millet, *Pennisetum glaucum* (L.) R. Brunken: occurrence of B chromosomes. Caryologia **47**: 1-10.

Lee, B.K., J.A. Ko, and Y.S. Kim, 1992. Studies on the thidiazuron of anther culture of *Paeonia albiflora*. J.Korean Soc.Hortic.Sci. **33**(5): 384-395.

Lee, S.S. and Y.J. Yoon, 1987. Anther culture of *X Brassicoraphanus*. Cruciferae Newsl. **12**: 68.

Lefebvre, V., A. Palloix, C Caranta, and E. Pochard, 1994. Construction of an interspecific integrated linkage map of pepper using molecular markers and doubled-haploid progenies. Genome **38**: 112-121.

Lesniewska, A., A. Ponitka, A. Slusarkiewicz-Jarzina, E. Zwierzykowska, Z. Zwierzykowski, A.R. James, H. Thomas, and M.W. Humphreys, 2001. Androgenesis from *Festuca pratensis x Lolium multiflorum* amphidiploid cultivars in order to select and stabilize rare gene combinations for grass breeding. Heredity **86**: 167-176.

Li, C.L. and B.J. Ye, 1995. Successful development of new sweet (hot) pepper cultivars by anther culture. Acta Hort. **402**: 442-444.

Li, S.Z., W.P. Wu, Z.H. Zhang, and D.Y. Wang, 1988. Study on anther culture of strawberry (*Fragaria ananassa*). Genet.Manipulation Crops.Newsl. **4**(1): 52-62.

Liang, G.H., X. Gu, G.L. Yue, Z.S. Shi, and K.D. Kofoid, 1997. Haploidy in sorghum. In: *In Vitro* Haploid Production in Higher Plants. Jain, S.M., S.K. Sopory and R.E. Veilleux (Eds.) Kluwer Academic Publishers, Dordrecht, pp.149-161.

Lichter, R., 1989. Efficient yield of embryoids by culture of isolated microspores of different *Brassicaceae* species. Plant Breed. **103**: 119-123.

Ling, J.T., M. Iwamasa, and N. Nito, 1988. Plantlet regeneration by anther culture of Calamondin (*Citrus madurensis* Lour.). In: Citriculture. Proc. Sixth Int. Citrus Congress, Middle East. Vol. 1. Balaban Publishers, Rehovot, pp.251-256.

Liu, M.C. and W.H. Chen, 1978. Organogenesis and chromosome number in callus derived from cassava anthers. Can.J.Bot. **56**(10): 1287-1290.

Liu, M.C., W.H. Chen, and L.S. Yang, 1980. Anther culture in sugarcane. I. Structure of anther and its pollen grain development stages. Taiwan Sugar. **27**(3): 86-91.

Livingston, G.K., 1971. Experimental studies on the induction of haploid parthenogenesis in Douglas-fir and the effects of radiation on the germination and growth of Douglas-fir pollen. Ph.D. Thesis, Univ.of Washington, Seattle, pp.189.

Long, C.M., C.A. Mullinix, and A.F. Iezzoni, 1994. Production of a microspore-derived callus population from sweet cherry. HortSci. **29**(11): 1346-1348.

Lysenko, E.G. and V.A. Sidorov, 1985. Obtainment of androgenous *Solanum bulbocastanum* haploids and mesophyll protoplast culture. Cytol.Genet. **19**: 35-37.

Maheswary, V. and C. Mak, 1993. The influence of genotypes and environments on induction of pollen plants for anther culture of *Capsicum annuum* L. AsPac.J.Mol.Biol.Biotechnol. **1**(1): 43-50.

Mahill, J.F., J.N. Jenkins, J.C.J. McCarty, and W.L. Parrott, 1984. Performance and stability of doubled haploid lines of upland cotton derived via semigany. Crop Sci. **24**: 271-277.

Makhmudov, T., 1978. Use of haploids in cotton breeding. Khlopkovodstvo pp. 31-32

Marburger, J.E. and R.R.C. Wang, 1988. Anther culture of some perennial triticeae. Plant Cell Rep. **7**: 313-317.

Marinkovic, N. and L. Radojevic, 1992. The influence of bud length, age of the tree and culture media on androgenesis induction in *Aesculus carnea* Hayne anther culture. Plant Cell Tiss.Org.Cult. **31**(1): 51-59.

Martinez, L.D. and N.I. de Halac, 1995. Organogenesis of anther-derived calluses in long-term cultures of *Oenotera hookeri* de Vries. Plant Cell Tiss.Org.Cult. **42**: 91-96.

Matsubara, S., N. Dohya, and K. Murakami, 1995. Callus formation and regeneration of adventitious embryos from carrot, fennel and mitsuba microspores by anther and isolated microspore cultures. Acta Hort. **392**: 129-137.

Mauro, M.C., C. Nef, and J. Fallot, 1986. Stimulation of somatic embryogenesis and plant regeneration from anther culture of *Vitis vinifera* cv. Cabernet-Sauvignon. Plant Cell Rep. **5**: 377-380.

Mazzolani, G., G. Pasqua, and B. Monacelli, 1979. Conditions for the development of haploid plants from *in vitro* cultures of pollen grains *Atropa belladonna*, *Nicotiana tabacum*, tobacco. Ann.Bot.Rome. **38**: 107-117.

Mehetre, S.S., 1984. Analysis of chromosome pairing in haploids of cotton (*Gossypium* spp.). Indian J.Agric.Res. **18**: 49-53.

Mehetre, S.S. and M.V. Thombre, 1981. Meiotic studies in the haploids (2n=2X=26) of the tetraploid cottons (2n=4X=52) *Gossypium hirsutum*, *Gossypium barbadense*. Proc.Indian Natl.Sci.Acad.Part.B.Biol.Sci. **47**: 516-518.

Mehetre, S.S. and M.V. Thombre, 1982. Cytomorphology of haploid Gossypium hirsutum X Gossypium anomalum Cotton. Indian J.Genet.Plant Breed. **42**: 144-149.

Meredith, W.R., R.R. Bridge, and J. F. Chism, 1970. Relative performance of F_1 and F_2 hybrids from doubled haploids and their parent varieties in upland cotton, *Gossypium hirsutum* L. Crop Sci. **10**: 295-298.

Metwally, E.I., S.A. Moustafa, B.I. El-Sawy, and T.A. Shalaby, 1998a. Haploid plantlets derived by anther culture of *Cucurbita pepo*. Plant Cell Tiss.Org.Cult. **52**: 171-176.

Metwally, E.I., S.A. Mustafa, B.I. El-Sawy, S.A. Haroun, and T.A. Shalaby, 1998b. Production of haploid plants from *in vitro* culture of inpollinated ovules of *Cucurbita pepo*. Plant Cell Tiss.Org.Cult. **52**(3): 117-121.

Meynet, J., R. Barrade, A. Duclos, and R. Siadous, 1994. Dihaploid plants of roses (*Rosa x hybrida*, cv. 'Sonia') obtained by parthenogenesis induced using irradiated pollen and *in vitro* culture of immature seeds. Agronomie **2**: 169-175.

Meynet, J. and M. Sibi, 1984. Haploid plants from *in vitro* culture of unfertilized ovules in *Gerbera jamesonii*. Z.Pflanzenzüchtg. **93**: 78-85.

Mezzarobba, A. and R. Jonard, 1986. Effects of the stage of isolation and pretreatments on *in vitro* development of cultivated sunflower anthers (*Helianthus annuus* L.). C.R.Acad.Sci.Ser.III.Sci.Vie. **303**(5): 181-186.

Miki-Hirosige, H., S. Nakamura, and I. Tanaka, 1988. Ultrastructural research on cell wall regeneration by cultured pollen protoplasts of *Lilium longiflorum*. Sex.Plant Reprod. **1**: 36-45.

Misra, N.R., T.M. Varghese, N. Maherchandani, and R.K. Jain, 1983. Studies on induction and differentiation of androgenic callus of *Solanum melongena* L. In: Plant Cell Culture in Crop Improvement. Sen, S.K. and K.L. Giles (Eds.) Plenum Press, New York pp.465-468.

Mityko, J., A. Andrasfalvy, G. Csillery, and M. Fari, 1995. Anther-culture response in different genotypes and F_1 hybrids of pepper (*Capsicum annuum* L.). Plant Breed. **114**(1): 78-80.

Miyoshi, K., 1996. Callus induction and plantlet formation through culture of isolated microspores of eggplant (*Solanum melongena* L.). Plant Cell Rep. **15**(6): 391-395.

Miyoshi, K. and N. Asakura, 1996. Callus induction, regeneration of haploid plants and chromosome doubling in ovule cultures of pot gerbera (*Gerbera jamesonii*). Plant Cell Rep. **16**: 1-5.

Mol, R., 1992. *In vitro* gynogenesis in *Melandrium album*: from parthenogenetic embryos to mixoploid plants. Plant Sci. **81**(2): 261-269.

Monfort, S., 1985. Androgenesis of coconut: embryos from anther culture. Z.Pflanzenzüchtg. **94**: 251-254.

Moriguchi, T., M. Omura, N. Matsuta, and I. Kozaki, 1987. *In vitro* adventitious shoot formation from anthers of pomegranate. HortSci. **22**: 947-948.

Morrison, R.A., R.E. Koning, and D.A. Evans, 1986. Anther culture of an interspecific hybrid of Capsicum. J.Plant Physiol. **126**: 1-9.

Mosquera, T., L.E. Rodriguez, A. Parra, and M. Rodriguez, 1999. *In vitro* adventive regeneration from carnation (*Dianthus caryophyllus*) anthers. Acta Hort. **482**: 305-308.

Motzo, R. and M. Deidda, 1993. Anther and ovule culture in globe artichoke. J.Genet.and Breed. **47**(3): 263-266.

Mozsar, J. and S. Sule, 1994. A rapid method for somatic embryogenesis and plant regeneration from cultured anthers of *Vitis riparia*. Vitis. **33**(4): 245-246.

Mukherjee, A., M. Unnikrishnan, and N.G. Nair, 1991. Callus induction, embryogenesis and regeneration from sweet potato anther. J.Root Crops. **17**: 302-304.

Muniswamy, B. and H.L. Sreenath, 2000. Plant regeneration through somatic embryogenesis from anther cultures of C x R cultivar of coffee. Journal of Plantation Crops **28**(1): 61-67.

Munoz, L.C. and J.P. Baudoin, 1994. Influence of the cold pretreatment and the carbon source on callus induction from anthers in Phaseolus. Annu.Rep.Bean.Improv.Coop. **37**: 129-130.

Munyon, I.P., J.F. Hubstenberger, and G.C. Phillips, 1989. Origin of plantlets and callus obtained from chile pepper anther cultures. In Vitro Cell Dev.Biol.J.Tissue Cult.Assoc. **25**: 293-296.

Nagmani, R. and J.M. Bonga, 1985. Embryogenesis in subcultured callus of *Larix decidua*. Can.J.For.Res. **15**: 1088-1091.

Nair, S., P.K. Gupta, and A.F. Mascarenhas, 1983. Haploid plants from *in vitro* anther culture of *Annona squamosa* Linn. Plant Cell Rep. **2**: 198-200.

Narasimham, M. and P.B.K. Kishor, 1999. Induction of androgenesis in pigeonpea - possible reason for low response of pollen to tissue culture conditions. In: Plant Tissue Culture and Biotechnology: Emerging Trends. Univerities Press Ltd., Hyderabad, India pp.232-239.

Nenova, N., M. Cristov, and P. Ivanov, 2000. Anther culture regeneration from some wild *Helianthus* species. Helia **23**(32): 65-72.

Nenova, N., P. Ivanov, and M. Christov, 1992. Anther culture regeneration of F_1 hybrids of *Helianthus annuus* x *H. smithii* and *H. annuus* x *H. eggerttii*. In: Proceedings of the 13th International Sunflower Conference. Vol. 2. Pisa. pp.1509-1514.

Neuenschwander, B. and T.W. Baumann, 1995. Increased frequency of dividing microspores and improved maintenance of multicellular microspores of *Coffea arabica* in medium with coconut milk. Plant Cell Tiss.Org.Cult. **40**(1): 49-54.

Nishiyama, I. and S. Uematsu, 1967. Radiobiological studies in plants - XII. Embryogenesis following X-irradiation of pollen in *Lycopersicon pimpinellifolium*. Radiation Bot. **7**: 481-489.

Nitsch, C., S. Andersen, M. Godard, M.G. Neuffer and W.F. Sheridan, 1982. Production of haploid plants of *Zea mays* and *Pennisetum* through androgenesis. In: Variability in Plants Regenerated from Tissue Culture. Earle, E.D. and Y. Demarly (Eds.), Praeger Publishers, New York, pp.69-91

Nitsch, C. and B. Norreel, 1973a. Effect of a thermic shock on the embryogenic power of the pollen of *Datura innoxia* cultivated in the anther or isolated from the anther. Acad.Sci Paris C.R.Ser.D. **276**: 303-306.

Nitsch, C. and B. Norreel, 1973b. Factors favoring the formation of androgenetic embryos in anther culture. In: Genes Enzymes and Populations. International Latin American Symposium, Cali. pp.129-144.

Nurhidayah, T., R. Horn, T. Röcher, and W. Friedt, 1996. High regeneration rates in anther culture of interspecific sunflower hybrids. Plant Cell Rep. **16**: 167-173.

Ormerod, A.J. and P.D.S. Caligari, 1994. Anther and microspore culture of *Lupinus albus* in liquid culture medium. Plant Cell Tiss.Org.Cult. **36**: 227-236.

Owen, H.R. and A.R. Miller, 1996. Haploid plant regeneration from anther cultures of three North American cultivars of strawberry (*Fragaria x ananassa* Duch.). Plant Cell Rep. **15**: 905-909.

Owen, H.R., R.E. Veilleux, F.L. Haynes, and K.G. Haynes, 1988. Photoperiod effects on 2n pollen production, response to anther culture, and net photosynthesis of a diplandrous clone of *Solanum phureja*. Am.Potato.J. **65**: 131-139.

Padmanabhan, C., M. Gurunathan, G. Pathmanabhan, and G. Oblisami, 1977. Induction of haploid plants from anther culture in *Datura ferox* L. Madras Agric.J. **64**: 542-543.

Pal, A., 1983. Isolated microspore culture of the winged bean, *Psophocarpus tetragonolobus* (L) DC - growth, development and chromosomal status. Indian J.Exp.Biol. **21**: 597-603.

Pallares, P., 1984. First results from '*in vitro*' culture of unfertilized cotton ovules (*Gossypium hirsutum* L.). Cotton et Fibres Tropicales **39**(4): 145-152.

Pandey, K.K., L. Przywara, and P.M. Sanders, 1990. Induced parthenogenesis in kiwifruit (*Actinidia deliciosa*) through the use of lethally irradiated pollen. Euphytica **51**: 1-9

Pasakinskiene, I., K. Anamthawat-Jonsson, M.W. Humphreys, and R.N. Jones, 1997. Novel diploids following chromosome elimination and somatic recombination in *Lolium multiflorum* x *Festuca arundinaceae* hybrids. Heredity **78**: 464-469.

Paulikova, D. and J. Vagera, 1993. *In vitro* induced androgenesis in *Melandrium album*. Biol.Plant. **35**(4): 645-647.
Paz, M.M. and R.E. Veilleux, 1999. Influence of culture medium and *in vitro* conditions on shoot regeneration in *Solanum phureja* monoploids and fertility of regenerated doubled monoploids. Plant Breed. **118**(1): 53-57.
Pedroso, M.C. and M.S. Pais, 1997. Anther and microspore culture in *Cammellia japonica*. In: *In Vitro* Haploid Production in Higher Plants. Vol. 5. Jain, S.M., S.K. Sopory and R.E. Veilleux (Eds.) Kluwer Academic Publishers, Dordrecht, pp.89-107
Pedroso, M.C. and S. Pais, 1993. Regeneration from anthers of adult *Camellia japonica* L. *In Vitro* Cell Dev.Biol.-Plant. **29P**(4): 155-159.
Pehu, E., R.E. Veilleux, and K.W. Hilu, 1987. Cluster analysis of anther-derived plants of *Solanum phureja* (Solanaceae) based on morphological characteristics. Am.J.Bot. **74**: 47-52.
Perez-Bermudez, P., M.J. Cornejo, and J. Segura, 1985. Pollen plant formation from anther cultures of *Digitalis obscura*. Plant Cell Tiss.Org.Cult. **5**: 63-68.
Perri, E., M.V. Parlati, R. Mule, and A.S. Fodale, 1994. Attempts to generate haploid plants from *in vitro* cultures of *Olea europaea* L. anthers. Acta Hort. **356**: 47-50.
Peters, J.E., O.J. Crocomo, W.R. Sharp, E.F. Paddock, I. Tegenkamp, and T. Tegenkamp, 1977. Haploid callus cells from anthers of *Phaseolus vulgaris*. Phytomorphol. **27**: 79-85.
Piccirilli, M. and S. Arcioni, 1991. Haploid plants regenerated via anther culture in wild barley (*Hordeum spontaneum* C. Kock). Plant Cell Rep. **10**(6/7): 237-276.
Piper, J.G., B. Charlesworth, and D. Charlesworth, 1986. Breeding system evolution in *Primula vulgaris* and the role of reproductive assurance. Heredity **56**: 207-217.
Pol'kheim, F., 1972. On the problem of selecting for breeding mutation chimeras and mutants in haploids of *Pelargonium zonale* Kleiner Liebling and *Thuja gigantea gracilis*. In: Eksperimental'nyi Mutagenez v Selektsii pp.199-221.
Ponitka, A. and A. Slusarkiewicz-Jarzina, 1987. Induction of gynogenesis in selected plant species from the family Papilionaceae. Genetica Polonica **28**(3): 239-242.
Prakash, D.V.S.S.R., S. Chand, and P.B.K. Kishor, 1999. *In vitro* response from cultured anthers of *Boswellia serrata* Roxb. In: Plant Tissue Culture and Biotechnology: Emerging Trends. Universities Press Ltd., Hyderabad, India, pp. 226-231.
Prakash, J. and K.L. Giles, 1986. Production of doubled haploids in oriental lilie. In: Genetic Manipulation in Plant Breeding. Horn, W., C.J. Jensen, W. Oldenbach and O. Schieder (Eds.) Walter de Gruyter & Co, Berlin, pp.335-337.
Prasad, B.R., M.A. Khadeer, P. Seeta, and S.Y. Anwar, 1991. *In vitro* induction of androgenic haploids in safflower (*Carthamus tinctorius* L.). Plant Cell Rep. **10**(1): 48-51.
Preil, W., W. Huhnke, M. Engelhardt, and M. Hoffmann, 1977. Haploids in *Gerbera jamesonii* from in vitro cultured capitulum explants. Z.Pflanzenzüchtg. **79**: 167-171.
Qin, X. and G.L. Rotino, 1995a. Anther culture of several sweet and hot pepper genotypes. Acta Hort. **402**: 313-316.
Qin, X. and G.L. Rotino, 1995b. Chloroplast number in guard cells as ploidy indicator of *in vitro*-grown androgenic pepper plantlets. Plant Cell Tiss.Org.Cult. **41**(2): 145-149.
Qu, Y., M.C. Mok, D.W.S. Mok, and J.R. Stang, 1988. Phenotypic and cytological variation among plants derived from anther cultures of *Lilium longiflorum*. *In Vitro* Cell Dev.Biol.J.Tissue Cult.Assoc. **24**: 471-476.
Quarta, R., D. Nati, and F.M. Paoloni, 1992. Strawberry anther culture. Acta Hort. **300**: 335-339.
Radijevic, L., N. Marinkovic, and S. Jevremovic, 2000. Influence of the sex of flowers on androgenesis in *Aesculus hippocastanum* L. anther culture. *In Vitro* Cell Dev.Biol.-Plant. **36**(6): 464-469.
Radojevic, L., L. Vapa, K. Borojevic, and J. Joksimovic, 1985. Plant regeneration in *Saintpaulia ionantha* Wendl. anther cultures. Savrem.Poljoprivreda **33**: 485-491.
Raghavan, V. and R. Nagmani, 1989. Cytokinin effects on pollen embryogenesis in cultured anthers of *Hyoscyamus niger*. Can.J.Bot. **67**: 247-257.
Raghuramulu, Y. and N.S. Prakash, 1996. Haploidy in coffee. In: *In Vitro* Haploid Production in Higher Plants. Vol. 3. Jain, S.M., S.K. Sopory and R.E. Veilleux (Eds.) Kluwer Academic Publishers, Dordrecht, pp.349-363.

Rajasekaran, K. and M.G. Mullins, 1985. Somatic embryo formation by cultured ovules of Cabernet Sauvignon grape: effects of fertilization and of the male gameticide toluidine blue. Vitis 24: 151-157.
Ramulu, K., 1982. Genetic instability at the S-locus of *Lycopersicon peruvianum* plants regenerated from *in vitro* culture of anthers: generation of new S-specificities and S-allele reversions. Heredity 49(3): 319-330.
Ranaweera, K.K.D.S. and R. Pathirana, 1992. Optimization of media and conditions for callus induction from anthers of sesame cultivar MI 3. J.Natn.Sci.Coun.Sri Lanka, **20(2)**: 309-316.
Rao, I.U., I.V. Ramanuja, and M. Narasimham, 1986. Induction of androgenesis in the *in vitro* grown anthers of winged bean (*Psophocarpus tetragonolobus*). Phytomorphol. **36**: 111-116.
Rao, P.V.L. and D.N. De, 1987. Haploid plants from *in vitro* anther culture of the leguminous tree, *Peltophorum pterocarpum* (DC) K. Hayne (Copper pod). Plant Cell Tiss.Org.Cult. **11**(3): 167-177.
Raquin, C., 1985. Induction of haploid plants by *in vitro* culture of *Petunia* ovaries pollinated with irradiated pollen. Z.Pflanzenzüchtg. **94**: 166-169.
Raquin, C., A. Cornu, E. Farcy, D. Maizonnier, G. Pelletier, and F. Vedel, 1989. Nucleus substitution between *Petunia* species using gamma ray-induced androgenesis. Theor.Appl.Genet. **78**: 337-341.
Ray, I.M. and E.T. Bingham, 1989. Breeding diploid alfalfa for regeneration from tissue culture. Crop Sci. **29**: 1545-1548.
Regner, F., 1996. Anther and microspore culture in Capsicum. In: *In Vitro* Haploid Production in Higher Plants. Vol. 3. Jain, S.M., S.K. Sopory and R.E. Veilleux (Eds.) Kluwer Academic Publishers, Dordrecht, pp.77-89.
Reynolds, T.L., 1985. Ultrastructure of anomalous pollen development in embryogenic anther cultures of *Hyoscyamus niger*. Am.J.Bot. **72**: 44-51.
Reynolds, T.L., 1990. Interactions between calcium and auxin during pollen androgenesis in anther cultures of *Solanum carolinense* L. Plant Sci. **72**(1): 109-114.
Rivard, S.R., M.K. Saba-El-Leil, B.S. Landry, and M. Cappadocia, 1994. RFLP analyses and segregation of molecular markers in plants produced by in vitro anther culture, selfing, and reciprocal crosses of two lines of self-incompatible *Solanum chacoense*. Genome **37**: 775-783.
Robert, T., A. Sarr, and J. Pernes, 1989. Haploid selections in pearl millet (*Pennisetum typhoides* (Burm.) Stapf et Hubb.): temperature effect. Genome **32**: 946-952.
Rokka, V.M., C.A. Ishimaru, N.L.V. Lapitan, and E. Pehu, 1998. Production of androgenic dihaploid lines of the disomic tetraploid potato species *Solanum acaule* spp. *acaule*. Plant Cell Rep. **18**(1-2): 89-93.
Rose, J.B., J.M. Dunwell, and N. Sunderland, 1986a. Anther culture of *Sorghum bicolor* (L.) Moench. I. Effect of panicle pretreatment, anther incubation temperature and 2,4-D concentration. Plant Cell Tiss.Org.Cult. **6**: 15-22.
Rose, J.B., J.M. Dunwell, and N. Sunderland, 1986b. Anther culture of *Sorghum bicolor* (L.) Moench. II. Pollen development *in vivo* and *in vitro*. Plant Cell Tiss.Org.Cult. **6**: 23-31.
Rose, J.B., J.M. Dunwell, and N. Sunderland, 1987a. Anther culture of *Lolium temulentum*, *Festuca pratensis* and *Lolium* x *Festuca hybrids*. I. Influence of pretreatment, culture medium and culture incubation conditions on callus production and differentiation. Ann.Bot. **60**: 191-201.
Rose, J.B., J.M. Dunwell, and N. Sunderland, 1987b. Anther culture of *Lolium temulentum*, *Festuca pratensis* and *Lolium* x *Festuca hybrids*. II. Anther and pollen development *in vivo* and *in vitro*. Ann.Bot. **60**: 203-214.
Rose, J.B., R.P. Jones, and D.W. Simpson, 1993. Anther culture and intergeneric hybridization of *Fragaria x ananassa*. Adv.Strawb.Res. **12**: 59-64.
Rotino, G.L., 1996. Haploidy in eggplant. In: *In Vitro* Haploid Production in Higher Plants. Vol. 3. Jain, S.M., S.K. Sopory and R.E. Veilleux (Eds.) Kluwer Academic Publishers, Dordrecht, pp.115-141.
Rowe, P.R., 1974. Methods of producing haploids: parthenogenesis following interspecific hybridization. In: Haploids in Higher Plants. Advances and Potential. Kasha, K.J. (Ed.) University of Guelph, Guelph. pp.43-52.

Rudramuniyappa, C.K., 1985. A histochemical study of developing sporogenous tissue and periplasmodial tapetum in the anther of *Parthenium* (Compositae). Cytologia **50**: 891-898.

Rudramuniyappa, C.K. and B.G. Annigeri, 1985. Histochemical observations on the sporogenous tissue and tapetum in the anther of Euphorbia. Cytologia **50**: 39-48.

Sacristan, M.D., 1971. Karyotypic changes in callus cultures from haploid and diploid plants of *Crepis capillaris* (L.) Waller. Chromosoma **33**: 273-293.

Sairam, R.V. and N. Seetharama, 1996. Androgenetic response of cultured anthers and microspores of sorghum. Int.Sorghum Millets Newsl. **37**: 69-71.

Saji, K.V. and M. Sujatha, 1998. Embryogenesis and plant regeneration in anther culture of sunflower (*Helianthus annuus* L.). Euphytica **103**: 1-7.

Salunkhe, C.K., P.S. Rao, and M. Mhatre, 1999. Plant regeneration via somatic embryogenesis in anther callus of *Vitis latifolia* L. Plant Cell Rep. **18**(7/8): 670-673.

Samsudeen, K., K.N. Babu, M. Divakaran, and P.N. Ravindran, 2000. Plant regeneration from anther derived callus cultures of ginger (*Zingiber officinale* Rosc.). J.Hortic.Sci.Biotech. **75**(4): 447-450.

Sanguineti, M.C., R. Tuberosa, and S. Conti, 1990. Field evaluation of androgenetic lines of eggplant. Acta Hortic. 177-181.

Sangwan, R.S. and H. Camefort, 1984. Cold-treatment related structural modifications in the embryogenic anthers of *Datura*. Cytologia **49**: 473-487.

Sankina, A.S. and L.S. Sankin, 1988. Characteristics of meiosis in the remote currant hybrid *Ribes nigrum* x *Ribes* "Holland Red" at the amphihaploid and the amphidiploid levels. Cytol.Genet. **22**: 12-16.

Sarvesh, A., T.P. Reddy, and P.B. Kavi Kishor, 1993. Embryogenesis and organogenesis in cultured anthers of an oil yielding crop niger (*Guizotia abyssinica* Cass). Plant Cell Tiss.Org.Cult. **35**: 75-80.

Sato, S., N. Katoh, H. Yoshida, S. Iwai, and M. Hagimori, 2000. Production of doubled haploid plants of carnation (*Dianthus caryophyllus* L.) by pseudofertilized ovule culture. Sci Hortic. **83**(3/4): 301-310.

Sator, C., G. Mix, and U. Menge, 1982. Investigations on anther culture of *Lupinus polyphyllus*. Landbauforsch.Volkenrode **32**: 37-42.

Sauton, A. and R. Dumas-du-Vaulx, 1987. Induction of gynogenetic haploid plants in muskmelon (*Cucumis melo* L.) by use of irradiated pollen. Agron.Sci.Prod.Veg.Environ. **7**: 141-147.

Savin, F., V. Decomble, M. Le-Couviour, and J. Hallard, 1988. The X-ray detection of haploid embryos arisen in muskmelon (*Cucumis melo* L.) seeds, and resulting from a parthenogenetic development induced by irradiated pollen. Rep.Cucurbit.Genet.Coop. pp. 39-42.

Schertz, K.F., 1963. Chromosomal, morphological, and fertility characteristics of haploids and their derivatives in *Sorghum vulgare* Pers. Crop Sci. **3**: 445-447.

Schieder, O., 1976. Isolation of mutants with altered pigments after irradiating haploid protoplasts from *Datura innoxia* Mill. with X-rays. Mol.Gen.Genet. **149**: 251-254.

Seran, T.H., K. Hirimburegama, W.K. Hirimburegama, and V. Shanmugarajah, 1999. Callus formation in anther culture of tea clones, *Camellia sinensis* (L.) O. Kuntze. J.Nat.Sci.Fund.Sri Lanka, **27**(3): 165-175.

Sharma, R. and S. Babber, 1990. *In vitro* studies of anther culture of *Antirrhinum majus*. Annals Biol.Ludhiana, **6**(2): 175-178.

Shtereva, L.A., N.A. Zagorska, B.D. Dimitrov, M.M. Kruleva, and H.K. Oanh, 1998. Induced androgenesis in tomato (*Lycopersicon esculentum* Mill). II. Factors affecting induction of androgenesis. Plant Cell Rep. **18**(312): 317.

Sibi, M., R. Dumas-de-Vaulx, and D. Chambonnet, 1979. Obtainment of haploid plants through *in vitro* androgenesis in sweetpepper (*Capsicum annuum* L.). Ann.Amelior.Plant. **29**: 583-606.

Simpson, E. and J.W. Snape, 1980. Haploid production in *Hordeum spontaneum* x *Hordeum bulbosum* crosses barley. Barley Genet.Newsl. **10**: 66-67.

Singh, I.S. and A. Cornu, 1976. Research on androgenetic Petunia haploids with gynogenetic cytoplasmic pollen sterility. Ann.Amelior.Plant. **26**: 565-568.

Singh, M.N. and R.N. Konar, 1979. *In vitro* induction of haploid roots and shoots from female gametophyte of *Ephedre foliata* Boiss. Beitr.Biol.Pflanzen. **55**: 169-177.

Sinha, R.R. and K. Das, 1986. Anther-derived callus of *Dolichos biflorus* L., its protoplast culture and their morphogenic potential. Curr.Sci. **55**(9): 447-452.
Sinha, S., R.P. Roy, and K.K. Jha, 1979. Callus formation and shoot bud differentiation in anther culture of *Solanum surattense*. Can.J.Bot. **57**(22): 2524-2527.
Sink, K.C. and J.F. Reynolds, 1986. Tomato (*Lycopesicon esculentum* L.). In: Biotechnology in Agriculture and Forestry 2. Crops I. Bajaj, Y.P.S. (Ed.) Springer-Verlag, Berlin, pp.319-344.
Sitbon, M., 1981. Production of haploid *Gerbera jamesonii* plants by *in vitro* culture of unfertilized ovules. Agron.Sci.Prod.Veg.Environ. **1**: 807-812.
Skinner, D.Z. and G.H. Liang, 1996. Haploidy in alfalfa. In: *In Vitro* Haploid Production in Higher Plants. Vol. 3. Jain, S.M., S.K. Sopory and R.E. Veilleux (Eds.) Kluwer Academic Publishers, Dordrecht, pp.365-375.
Slusarkiewicz-Jarzina, A. and M. Zenkteler, 1979. Cytological and embryological studies on haploids (n=3) of *Crepis capillaris* L. Bull.Soc.Amis.Sci.Lett.Poznan.Ser.D.Sci.Biol. 65-73.
Smagula, J.M. and P.M. Lyrene, 1984. Blueberry. Handb.Plant Cell Cult. **3**: 383-401.
Sniezko, R. and T. Visser, 1987. Embryo development and fruit-set in pear induced by untreated and irradiated pollen. Euphytica **36**: 287-294.
Sommer, H.E. and H.Y. Wetzstein, 1984. Hardwoods. Handb.Plant Cell Cult. **3**: 511-540.
Songstad, D.D. and B.V. Conger, 1988. Factors influencing somatic embryo induction from orchardgrass anther cultures. Crop Sci. **28**: 1006-1009.
Starrantino, A. and P. Caponnetto, 1990. Effect of cytokinins on embryogenic callus formation from undeveloped ovules of orange. Acta Hort. **280**: 191-194.
Stelly, D.M., J.A. Lee, and W.L. Rooney, 1988. Proposed schemes for mass-extraction of doubled haploids of cotton. Crop Sci. **28**: 885-890.
Stoehr, M. and L. Zsuffa, 1990. Genetic evaluation of haploid clonal lines of a single donor plant of *Populus maximowiczii*. Theor.Appl.Genet. **80**(4): 470-474.
Strauss, A., F. Bucher, and P. J. King, 1981. Isolation of biochemical mutants using haploid mesophyll protoplasts of *Hyoscyamus muticus*. Planta **153**: 75-80.
Suh, S.K. and H.G. Park, 1986. Studies on the anther culture of garlic (*Allium sativum* L.). I. Callus formation and plant regeneration. J.Korean Soc.Hortic.Sci. **27**: 89-95.
Summers, W.L., 1997. Haploid plantlet production in tomato. In: *In Vitro* Haploid Production in Higher Plants. Vol. 5. Jain, S.M., S.K. Sopory and R.E. Veilleux (Eds.) Kluwer Academic Publishers, Dordrecht, pp.219-231.
Svensson, M. and L.B. Johansson, 1995. Anther culture of *Fragaria x ananassa*: environmental factors and medium components affecting microspore divisions and callus production. J.Hortic.Sci. **69**(3): 473-480.
Takahata, Y., H. Komatsu, and N. Kaizyma, 1996. Microspore culture of radish (*Raphanus sativus* L.): influence of genotype and culture conditions on embryogenesis. Plant Cell Rep. **16**: 163-166.
Tan, B.H. and G.M. Halloran, 1982. Pollen dimorphism and the frequency of inductive anthers in anther culture of *Triticum monococcum*. Biochem.Physiol.Pflanz. **177**(2): 197-202.
Tang, K., X. Sun, Y. He, and Z. Zhang, 1998. Anther culture response of wild *Oryza species*. Plant Breed. **117**: 443-446.
Tanner, G.J., A.E. Moore, and P.J. Larkin, 1988. Reducing the ploidy of lucerne by anther culture or induced parthenogenesis. In: Ninth Australian Plant Breeding Conference, pp. 136.
Tanner, G.J., M. Piccirilli, A.E. Moore, P.J. Larkin, and S. Arcioni, 1990. Initiation of non-physiological division and manipulation of developmental pathway in cultured microspores of *Medicago* sp. Protoplasma. **158**(3): 165-175.
Tefera, H., F.J. Zapata-Arias, R. Afza, and A. Kodym, 1999. Response of tef genotypes to anther culture. AgriTopia. **14**(1): 8-9.
Teixeira, J.B., M.R. Sondahl, and E.G. Kirby, 1994. Somatic embryogenesis from immature inflorescences of oil palm. Plant Cell Rep. **13**: 247-250.
Teten Snider, K. and R.E. Veilleux, 1994. Factors affecting variability in anther culture and in conversion of androgenic embryos of *Solanum phureja*. Plant Cell Tiss.Org.Cult. **36**: 345-354.
Thanh-Tuyen, N.T. and E.V. de Guzman, 1983a. Formation of pollen embryos in cultured anthers of coconut (*Cocos nucifera* L.). Plant Sci.Lett. **29**(1): 81-88.

Thanh-Tuyen, N.T. and E.V. de Guzman, 1983b. Pollen development stages for coconut anther culture. Kalikasan, Phylippine J.Biol. **12**(1-2): 135-144.
Theiler-Hedtrich, R. and C.S. Hunter, 1995. Regeneration of dihaploid chicory (*Cichorium intybus* L. var.*foliosum* Hegi) via microspore culture. Plant Breed. **114**(1): 18-23.
Thengane, S.R., M.S. Joshi, S.S. Khuspe, and A.F. Mascarenhas , 1994. Anther culture in *Helianthus annuus* L., influence of genotype and culture conditions on embryo induction and plant regeneration. Plant Cell Rep. **13**: 222-226.
Thomas, T.D., A.K. Bhatanagar, M.K. Razdan, and S.S. Bhojwani, 1999. A reproducible protocol for the rediction of gynogenic haploids of mulberry, *Morus alba* L. Euphytica **110**(3): 169-173.
Todorova, M. and P. Ivanov, 1999. Induced parthenogenesis in sunflower: effect of pollen donor. Helia **22**(31): 49-56.
Todorova, M., P. Ivanov, P. Shindrova, M. Christov, and I. Ivanova, 1997. Doubled haploid production of sunflower (*Helianthus annuus* L.) through irradiated pollen-induced parthenogenesis. Euphytica. **97**: 249-254.
Todorovic, R.R., P.D. Misic, D.M Petrovic, and M.A. Mirkowic, 1992. Anther culture of peach cultivars 'Cresthaven' and 'Vesna'. Acta Hort. **300**: 331-333.
Tomasi, P., D.A. Dierig, R.A. Backhaus, and K.B. Pigg, 1999. Floral bud and mean petal length as morphological predictors of microspore cytological stage in lesquerella. HortSci. **34**(7): 1269-1270.
Tomes, D.T. and R.L. Peterson, 1981. Isolation of a dwarf plant responsive to exogenous GA3 from anther cultures of birdsfoot trefoil. Can.J.Bot. **59**(7): 1338-1342.
Torregrosa, L., 1998. A simple and efficient method to obtain stable embryogenic cultures from anthers of *Vitis vinifera*. Vitis. **37**(2): 91-92.
Tosca, A., L. Arcara, and P Frangi, 1999. Effect of genotype and season on gynogenesis efficiency in Gerbera. Plant Cell Tiss.Org.Cult. **59**(1): 77-80.
Tsay, H.S., J.Y. Hsu, T.P. Yang, and C.R. Yang, 1984. Anther culture of passion fruit (*Passiflora edulis*). J.Agric.Res.China. **33**: 126-131.
Tsay, H. S., P. C. Lai, L. J. Chen, X. S. Cai, B. Z. Lai, and L. Z. Chen, 1982. Organ regeneration from anther callus of sweet potato. J.Agr.Res.China. **31**(2): 123-126.
Tsay, H. S. and C. Y. Su, 1985. Anther culture of papaya (*Carica papaya* L.). Plant Cell Rep. **4**: 28-30.
Tsay, H.S., C.C. Yeh, and J.Y. Hsu, 1990. Embryogenesis and plant regeneration from anther culture of bamboo (*Sinocalamus latiflora* (Munro) McClure). Plant Cell Rep. **9**(7): 349-351.
Tuberosa, R., M.C. Sanguineti, and S. Conti, 1987. Anther culture of egg-plant (*Solanum melongena* L.) lines and hybrids. Genetica Agraria. **41**(3): 267-274.
Turaev, A.M. and Z.B. Shamina, 1986. Optimization of the medium for cotton anther culture. Sov.Plant Physiol. **33**: 439-444.
Turaev, A.M. and Z.B. Shamina, 1993. Cotton microspore culture in agarose. Sov.Plant Physiol. **40**(2): 276-279.
Tyagi, A.K., A. Rashid, and S.C. Maheshwari, 1981. Promotive effect of polyvinylpolypyrrolidone on pollen embryogenesis in *Datura innoxia*. Physiol.Plant. **53**(4): 495-406.
Tyukavin, G. B., N. A. Shmykova, and M. A. Monakhova, 1999. Cytological study of embryogenesis in cultured carrot anthers. J.Plant Physiol.(Russian). **46**(6): 767-773.
Uddin, M.R., M.M.J. Meyer, and J.J. Jokela, 1988. Plantlet production from anthers of eastern cottonwood (*Populus deltoides*). Can.J.For.Res. **18**: 937-941.
Van-Eck, J.M. and S.L. Kitto, 1990. Callus initiation and regeneration in *Mentha*. HortSci. **25**: 804-806.
van den Bulk, R.W. and J.M. van Tuyl, 1997. *In vitro* induction of haploid plants from the gametophytes of lily and tulip. In: *In Vitro* Haploid Production in Higher Plants. Vol.5. Jain, S.M., S.K. Sopory and R.E. Veilleux (Eds.) Kluwer Academic Publishers, Dordrecht, pp.73-88.
Varghese, T.M. and G. Yadav, 1986. Production of embryoids and calli from isolated microspores of tomato (*Lycopersicon esculentum* Mill.) in liquid media. Biol.Plant. **28**(2): 126-129.
Vassileva-Dryanovska, O.A., 1966. The induction of haploid embryos and tetraploid endosperm nuclei with irradiated pollen in *Lilium*. Hereditas **55**: 160-165.

Veilleux, R.E., J. Booze-Daniels, and E. Pehu, 1985. Anther culture of a 2n pollen producing clone of *Solanum phureja* Juz. & Buk. Can.J.Genet.Cytol. **27**: 559-564.
Veuskens, J., D. Ye, M. Oliveira, D.D. Ciupercescu, P. Installe, H.A. Verhoeven, and I. Negrutiu, 1992. Sex determination in the dioecious *Melandrium album* androgenic embryogenesis requires the presence of the X chromosome. Genome **35**: 8-16.
Vishukumar, U., M.S. Patil, and S.N. Nayak, 2000. Anther culture studies in pigeonpea. Karnataka J.Agri.Sci. **13**(1): 16-19.
Von-Aderkas, P. and J.M. Bonga, 1988. Formation of haploid embryoids of *Larix decidua*: early embryogenesis. Am.J.Bot. **75**: 690-700.
Von-Aderkas, P., K. Klimaszewska, and J.M. Bonga, 1990. Diploid and haploid embryogenesis in *Larix leptolepis*, *L. decidua*, and their reciprocal hybrids. Can.J.For.Res. **20**: 9-14.
Wakasa, K. and Y. Watanabe, 1979. Haploid plant of *Oryza perennis* (*spontanea* type) induced by anther culture. Japan.J.Breed. **29**(2): 146-150.
Wang, C.C., C.C. Chu, and C.S. Sun, 1975. The induction of *Populus nigra* pollen-plants. Acta Bot.Sinica **17**: 56-59.
Wang, X.H., P.A. Lazzeri, and H. Lörz, 1993. Regeneration of haploid, dihaploid and diploid plants from anther- and embryo-derived cell suspensions of wild barley (*Hordeum murinum*). J.Plant Physiol. **141**: 726-732.
Watanabe, K., 1977. Successful ovary culture and production of F_1 hybrids and androgenic haploids in Japanese *Chrysanthemum* species. J.Hered. **68**: 317-320.
Weatherhead, M.A., B.W.W. Grout, and K.C. Short, 1982. Increased haploid production in *Saintpaulia ionantha* by anther culture African violet. Sci.Hortic. **17**: 137-144.
Wernicke, W., H. Lörz, and E. Thomas, 1979. Plant regeneration from leaf protoplasts of haploid *Hyoscyamus muticus* L. produced via anther culture. Plant Sci.Lett. **15**: 239-249.
Wijowska, M., E. Kuta, and L. Przywara, 1999. In vitro culture of unfertilized ovules of *Viola odorata* L. Acta Biol.Cracoviensia **41**: 95-101.
Willcox, M.C., S.M. Reed, J.A. Burns, and J.C. Wynne, 1990. Microsporogenesis in peanut (*Arachis hypogaea*). Aust.J.Bot. **77**(10): 1257-1259.
Wissemann, V., C. Mollers, and F.H. Hellwig, 1998. Microspore culture in the genus Rosa, further investigations. Angew.Botanik **72**(1-2): 7-9.
Wolff, D.W., R.E. Veilleux, and C.J. Jensen, 1986. Evaluation of anther-derived *Streptocarpus* x *hybridus* and their progeny. Plant Cell Tiss.Org.Cult. **6**: 167-172.
Woo, S.-C., S.-W. Ko, C.-K. Wong, and X.-C. Wu, 1983. Anther culture of pollen plants derived from cross *Oryza sativa* L. x *O. glaberrima* Steud. Botanical Bull.Acad.Sinica **24**(1): 53-58.
Wu, H.N. and S.Z. Zhang, 1986. Effect of acridine yellow on development of anthers of *Capsicum frutescens* var. *longum* cultured *in vitro*. Chiang.su.Nung.Yeh.Hseuh.Pao.J.Agric.Sci. **2**: 34-39.
Wu, J.Y., 1981. Obtaining haploid plantlets of crabapple from anther culture *in vitro*. Yuan.I.Hsueh.Pao.Acta Hortic.Sin. **8**: 36.
Xue, G.R., K.W. Fei, and J. Hu, 1981. Induction of haploid plantlets of strawberry (*Fragaria orientalis*) by anther culture *in vitro*. Yuan.I.Hsueh.Pao.Acta Hortic.Sin. **8**: 9-14.
Yadav, R.C., P.K. Sareen, and J.B. Chowdhury, 1988. High frequency induction of androgenesis in Ethiopian mustard (*Brassica carinata* A. Br.). Cruciferae Newsl. **13**: 77.
Yang, Y.Q. and W.X. Wei, 1984. Induction of longan haploid plantlets from pollen cultured in certain proper media. Acta Genet.Sin. **11**(4): 288-293.
Zagorska, N. and B. Dimitrov, 1995. Induced androgenesis in alfalfa (*Medicago sativa* L.). Plant Cell Rep. **14**: 249-252.
Zagorska, N., B. Dimitrov, P. Gadeva, and P. Robeva, 1997. Regeneration and characterization of plants obtained from anther cultures in *Medicago sativa* L. *In Vitro* Cell Dev.Biol.-Plant. **33**(2): 107-110.
Zagorska, N.A., M.D. Abadjieva, and H.K. Oanh, 1986. Factors affectng callus and plant production in anther cultures of tomato. In: Genetic Manipulation in Plant Breeding. Horn, W., C.J. Jensen, W. Oldenbach and O. Schieder (Eds.) Walter de Gruyter & Co., Berlin. pp.361-363.
Zare, A.G., M. W. Humphreys, W.J. Rogers, and H.A. Collin, 1999. Androgenesis from a *Lolium multiflorum* x *Festuca arundinacea* hybrid to generate extreme variation for freezing-tolerance. Plant Breed. **118**: 497-501.

Zheleznov, A.V., 1976. Methods of obtaining parthenogenetic haploids in buckwheat. In: Apomiksis i Ego Ispol'zovanie v Selektsii pp.65-68.

Zhong, D., N. Michaux-Ferriére, and M. Coumans, 1995. Assay for doubled haploid (*Helianthus annuus*) plant production by androgenesis: fact or artifact? 1. *In vitro* anther culture. Plant Cell Tiss.Org.Cult. **41**(2): 91-97.

Zhou, C., 1989. Cell divisions in pollen protoplast culture of *Hemerocallis fulva* L. Plant Sci. **62**(2): 229-235.

Zhou, C., K. Orndorff, R.D. Allen, and A.E. DeMaggio, 1986. Direct observations on generative cells isolated from pollen grains of *Haemanthus katherinae* Baker. Plant Cell Rep. **5**: 306-309.

Zhou, S.Q., D.Q. Qian, and X.Y. Cao, 1989. Haploid breeding and its cytogenetics in cotton (*Gossypium hirsutum*). In: Review of Advances in Plant Biotechnology, 1985-88. Mujeeb-Kazi, A. and L.A. Sitch (Eds.) CIMMYT, IRRI, Mexico, Manila, pp.323-324.

Zhuang, X.J., C.Y. Hu, Y. Chen, and G.C. Yin, 1991. Embryoids from soybean anther culture. Soybean Genet.Newsl. **18**: 265.

Zwierzykowski, Z., A.J. Lukaszewski, A. Lesniewska, and B. Naganowska, 1998. Genomic structure of androgenic progeny of pentaploid hybrids, *Festuca arundinacea x Lolium multiflorum*. Plant Breed. **117**: 457-462.

Zwierzykowski, Z., E. Zwierzykowska, A. Slusarkiewicz-Jarzina, and A. Ponitka, 1999. Regeneration of anther-derived plants from pentaploid hybrids of *Festuca arundinacea x Lolium multiflorum*. Euphytica **105**(3): 191-195.

4.1
Doubled haploids in breeding

W.T.B. Thomas, B.P. Forster and B. Gertsson[1]
Scottish Crop Research Institute, Invergowrie, Dundee, DD2 5DA, UK; [1]Svalöf Weibull AB, Svalöv, SE-268 81, Sweden

Introduction

There are many factors influencing the deployment of doubled haploidy in breeding. Strategies for exploiting doubled haploids (DHs) vary within and among species and are dependent upon available technologies. Theoretical benefits and working protocols are not however sufficient to justify DH deployment in breeding. Breeders must also consider cost efficiency, the fixation of rare and useful alleles, the preservation of genetic variation in their lines, and distinctiveness, uniformity and stability (DUS) of their end products, new varieties. Doubled haploidy has both positive and negative effects in respect to these considerations. Rapeseed for instance, which is generally considered to be 2/3 inbreeding and 1/3 outcrossing, can be bred to produce homozygous or heterozygous varieties, the latter does not require DH production and is cheaper. However, strict DUS policies of some countries push production towards homozygous varieties and here DH production has advantages. Yield stability on the other hand favours more heterogeneous and heterozygous varieties in rapeseed. Many important traits in rapeseed breeding are controlled by rare alleles, which can be fixed using doubled haploidy. Breeders must maintain a reasonable level of diversity in their stocks in order to preserve rare alleles, which may become important in future breeding efforts.

Elite Crossing

Traditional breeding such as the pedigree programme for inbreeders takes 10-15 years to produce a variety. The time delay is costly and prevents breeders from responding rapidly to end users needs. In addition pedigree inbreeding (PI) has other drawbacks: in the early stages all individuals are unique and there is a biasing effect of dominance on the phenotype; non-competitive individuals are rejected and early selection is based on individuals grown in non-crop conditions without replication. Intensive selection cannot be made until lines approach homozygosity and sufficient seed is available for field trials. Single seed descent (SSD) has been developed to help speed up the development of homozygous lines, but also suffers from time delays and competitive interactions among plants. Doubled haploidy overcomes many of these problems. It is the fastest route to homozygosity, it is achieved in a single generation and can be performed at any generation in a breeding programme. DH production saves time, for instance it is possible to cross parental lines and conduct field trials of derived DH progeny within 2 years (Fig. 4.1-1). It is also possible to conduct some reliable selection for disease resistance in the glasshouse and therefore concentrate resource on the more promising lines. Since the DH lines are homozygous there are no dominance related effects and selection is therefore more reliable.

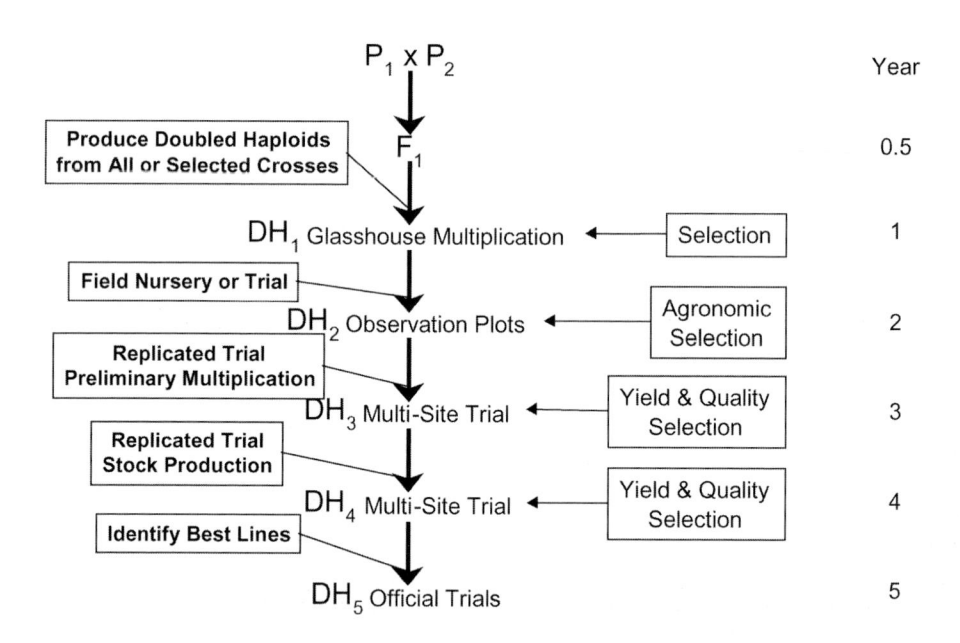

Stage	Process	Selection	Year
	$P_1 \times P_2$		
Produce Doubled Haploids from All or Selected Crosses	F_1		0.5
	DH_1 Glasshouse Multiplication	Selection	1
Field Nursery or Trial			
Replicated Trial Preliminary Multiplication	DH_2 Observation Plots	Agronomic Selection	2
Replicated Trial Stock Production	DH_3 Multi-Site Trial	Yield & Quality Selection	3
Identify Best Lines	DH_4 Multi-Site Trial	Yield & Quality Selection	4
	DH_5 Official Trials		5

Figure 4.1-1. Outline of development of doubled haploidy in a practical plant breeding programme.

Comparisons with other methods

The theoretical properties of progenies produced by DH and SSD were compared by Snape (1976), who concluded that there was no difference between the two methods in the absence of linkage. When linkage was present, SSD has the greater opportunity for recombination and therefore the frequency of recombinants is expected to be higher. One's choice of method would then depend upon whether one wished to preserve linkage blocks (F_1DH) or break them up (SSD or PI). A number of comparisons of populations produced by doubled haploidy, PI and/or SSD from a range of species, principally cereals, have been made (Table 4.1-1). All conclude that the adoption of doubled haploidy does not lead to any bias in populations and random DHs were even found to be comparable to selected lines produced by PI (Friedt *et al.*, 1986; Winzeler *et al.*, 1987). There is therefore no genetical reason not to adopt DH instead of pedigree inbreds in a breeding programme, provided that reliable protocols are available.

The number of lines involved in breeding programmes is another important consideration. For example, the potential number of homozygous lines in a cross is 2^n, in a population segregating for 100 genes, the possible number of homozygous lines is 1267 x 10^{27}. The number of homozygous genotypes is thus far too large to be handled by any breeding programme. In cases where segregating characters have high heritability PI can be efficient in cutting down the number of lines in early generations. In cross pollinated species gene frequencies can be manipulated by allowing out crossing and applying recurrent selection; this scheme prevents random fixation of lines and buffers negative effects which may result from excessive DH deployment.

The majority of breeders who could utilise DH continue, however, to rely upon variations of PI, possibly incorporating SSD. This trend is most marked in the commercial plant breeding sector, although there are examples of breeders who have converted programmes to doubled haploidy. The principal reasons for this reluctance is that DH production remains resource intensive. One has to accommodate within a team a person whose sole function is to produce doubled haploids and accept that the numbers produced will not be the same as, for instance, an F_2 population in a PI programme. There are, however, plenty of reports demonstrating that F_2 single plant selection is, at best, random and it is perhaps more realistic to compare population sizes at the F_3 generation, when doubled haploidy is competitive. Doubled haploid protocols are not as transferable as one might expect from some published studies and a number of breeders have tried to adopt the technique and failed to generate sufficient numbers of DHs to sustain a competitive programme. The human element is a key feature to success as is the ability to grow top quality donor plants. These factors should be borne in mind when debating the merits of utilising doubled haploids in breeding. The time advantage conveyed by doubled haploids is not that great for a spring crop if one adopts a shuttle breeding strategy, and especially if complete homozygosity is not needed or even not desired. It is possible to complete a crossing and selection cycle and enter a line into Official Trials in the same five-year time scale as that for doubled haploid shown in Fig. 4.1-1. Stock production in a breeding programme utilising DHs is potentially less resource-demanding but in practice requires the same amount of resources as a pedigree programme. This is because out-crossing can occur even in inbreeding species, requiring the same careful examination of plant progenies as a pedigree scheme. With these provisos, the ability to work with fixed inbred lines is a major advantage and it is probably worthwhile converting to doubled haploid if one has not had any success with a PI programme.

Table 4.1-1. Some comparisons of lines produced by doubled haploidy with other breeding methods

Crop	DH method	Comparison	Reference
Barley	anther culture	PI	Friedt et al., 1986
		SSD	Powell et al., 1992
		SSD	Bjornstad et al., 1993
	Hordeum bulbosum	SSD and PI	Park et al., 1976
		PI	Turcotte et al., 1980
		SSD	Choo et al., 1982
		SSD and PI	Powell et al., 1986
		SSD	Caligari et al., 1987
		SSD	Bjornstad et al., 1993
Brussels sprouts	anther culture	SSD	Kubba et al., 1989
Linseed	microspore culture	PI	Steiss et al., 1998
Maize	anther culture	SSD	Murigneux et al., 1993
Oilseed rape	microspore culture	SSD	Chen and Beversdorf, 1990
Rice	anther culture	SSD	Courtois, 1993
		PI	Martinez et al., 1996
Tobacco	anther culture	SSD	Schnell et al., 1980
		SSD	Jinks et al., 1985
		SSD and PI	Chung et al., 1992
	maternal DH	SSD and PI	Chung et al., 1992
Triticale	anther culture	SSD	Charmet and Branlard, 1985
Wheat	anther culture	PI	Winzeler et al., 1987
		SSD	Snape et al., 1992
		SSD	Mitchell et al., 1992
		SSD	Skinnes and Bjornstad, 1995
		SSD	Ma et al., 1999
	maize pollination	SSD and PI	Inagaki et al., 1998
		SSD	Ma et al., 1999

Varietys produced by doubled haploidy

The first variety produced *via* doubled haploidy was the barley "Mingo" in the late 1970's in Canada. Doubled haploidy now features in variety production of many crops (see other sections of this manual). We have found over 200 varieties that have been produced by deploying various DH methods (Table 4.1-2). This is an under-estimate of global DH varieties as our information from some parts of the world is scant, in addition breeders are not obliged to reveal methods and in some cases the information is sensitive. The vast majority of varieties derived from doubled haploidy are in barley (96), followed by rapeseed (47), wheat (20) and the rest. The most popular method of DH production varies from crop to crop. If we consider the top three species from Table 4.1-2, the most popular DH methods are wide crossing for barley, microspore culture for rapeseed and anther culture for wheat (although here wide crossing is also common). There is a trend from wide crossing to anther culture to microspore culture, which is promoted as more genotype independent tissue culture protocols are developed. The number of varieties produced by doubled haploidy remains relatively small compared to those produced by other means and it is arguable that even fewer have been major varieties in a country. For example, the winter wheat 'Savannah' is the fourth most popular variety in the UK but only has a 6% market share, although 'Xi19' at 2% in its first year of recommendation has the potential to be a major wheat variety, certainly in the UK. The blackleg resistant rapeseed variety 'Quantum' however proved highly successful in western Canada capturing 30% of the acreage in 1995 (Stringhem et al., 1995).

Table 4.1-2. List of varieties produced by doubled haploidy

Crop	Variety	Method	Country	Authority Reference/pers. com.
Asparagus	Andréas	DH female x DH supermale	France	Corriols et al., 1990
	Ringo		Italy	Falavigna et al., 1999
	Golia		Italy	Falavigna et al., 1999
	Guelph Millenium	female x DH supermale	Canada	D. Wolyn
	Argo		Italy	Falavigna et al., 1999
	Eros		Italy	Falavigna et al., 1999
	Gladio		Italy	Falavigna et al., 1999
Barley	Binalong	anther culture	Australia (NSW and Qld)	B. Read
	Tantangara		Australia (NSW)	B. Read
	Picard		Australia (SA and Vict.)	D. Moody
	Diplom		Czech Rep.	J. Weyen
	Merlot		Czech Rep., Germany	E. Laubach
	Tender		Denmark	Devaux et al., 1996
	Flag		France	E. Laubach
	Hamida		France	E. Laubach
	Ladoga		France	E. Laubach
	Lyric		France	Devaux et al., 1996
	Naomie		France	E. Laubach
	Annabel		Germany, Denmark, Hungary, Poland, Russia, Sweden	J. Weyen
	Anthere		Germany, France	E. Laubach
	Auriga		Germany,	J. Weyen
	Bayava	anther culture	Germany	E. Laubach
	Caprima		Germany	E. Laubach
	Carola		Germany, Belgium, France, Austria, Hungary, Czech Rep. Poland, UK	E. Laubach
	Clara		Germany, France, UK, Denmark	E. Laubach
	Danuta		Germany, Austria, Hungary	J. Weyen
	Henni		Germany Denmark, Spain, Netherlands, UK, Eire	E. Laubach
	Nelly		Germany, Czech Rep., Hungary	E. Laubach
	Nicola		Germany, France, UK, Denmark	E. Laubach
	Sarah		Germany, Belgium France	E. Laubach
	Traminer		Germany, Austria	E. Laubach

Crop	Variety	Method	Country	Authority Reference/pers. com.
	Uschi		Germany, France	E. Laubach
	Ursa		Germany	J. Weyen
	Verena		Germany	E. Laubach
	Viskosa		Germany, France	J. Weyen
	Zenobia		Germany	J. Weyen
	Justina		Norway, Poland	J. Weyen
	Bélin		Spain	L. Cistue
	Erika		Spain	J. Weyen
	Lola		Spain	L. Cistue
	Leonie		UK, Germany	E. Laubach
	Milena		UK, Eire	E. Laubach
	Beluga	*Hordeum bulbosum*	Canada	Devaux *et al.*, 1996
	Bronco		Canada	Devaux *et al.*, 1996
	Craig		Canada	Devaux *et al.*, 1996
	DB202		Canada	Devaux *et al.*, 1996
	H30-11		Canada	Devaux *et al.*, 1996
	HD87-12.1		Canada	Devaux *et al.*, 1996
	HD87-18.14		Canada	Devaux *et al.*, 1996
	Lester		Canada	Devaux *et al.*, 1996
	McGregor		Canada	Devaux *et al.*, 1996
	Mingo		Canada	Devaux *et al.*, 1996
	Ontario		Canada	Devaux *et al.*, 1996
	Prospect		Canada	Devaux *et al.*, 1996
	Rodeo		Canada	Devaux *et al.*, 1996
	Sandrina		Canada	Devaux *et al.*, 1996
	T081-009	*Hordeum bulbosum*	Canada	Devaux *et al.*, 1996
	T086-156		Canada	Devaux *et al.*, 1996
	T090-017		Canada	Devaux *et al.*, 1996
	T103-003		Canada	Devaux *et al.*, 1996
	TB891-6		Canada	Devaux *et al.*, 1996
	Winthrop		Canada	Devaux *et al.*, 1996
	Aberdeen		Denmark	Devaux *et al.*, 1996
	Bereta		Denmark	Devaux *et al.*, 1996
	Etna		Denmark	Devaux *et al.*, 1996
	Give		Denmark	Devaux *et al.*, 1996
	Loke		Denmark	Devaux *et a.l*, 1996
	Loma		Denmark	Devaux *et al.*, 1996
	Paloma		Denmark	Devaux *et al.*, 1996
	Perma		Denmark	Devaux *et al.*, 1996
	Pondus		Denmark	Devaux *et al.*, 1996
	Riga		Denmark	Devaux *et al.*, 1996
	Rima		Denmark	Devaux *et al.*, 1996
	Verona		Denmark	Devaux *et al.*, 1996
	Anka		France	Devaux *et al.*, 1996
	Douchka		France	Devaux *et al.*, 1996
	Gaelic		France	Devaux *et al.*, 1996
	Gotic		France	Devaux *et al.*, 1996
	Jerka		France	Devaux *et al.*, 1996
	Jing Zhuo		France	Devaux *et al.*, 1996
	Logic		France	Devaux *et al.*, 1996
	Lombard		France	Devaux *et al.*, 1996
	Michka		France	Devaux *et al.*, 1996

Crop	Variety	Method	Country	Authority Reference/pers. com.
	Moka		France	Devaux et al., 1996
	Tattoo		France	Devaux et al., 1996
	Vodka		France	Devaux et al., 1996
	ZF3642		France	Devaux et al., 1996
	Houshun		Japan	Furusho et al., 1999
	Gwylan		New Zealand	Devaux et al., 1996
	Valetta		New Zealand	Devaux et al., 1996
	KA7/3		Poland	Devaux et al., 1996
	Istok		Russia	Devaux et al., 1996
	Odesskill15		Russia	Devaux et al., 1996
	Preria		Russia	Devaux et al., 1996
	Doublet		UK	Devaux et al., 1996
	Pipkin		UK	Devaux et al., 1996
	Waveney		UK	Devaux et al., 1996
	Orca		USA (Oregon)	P. Hayes
	Strider		USA (Oregon)	P. Hayes
	Tango		USA (Oregon)	P. Hayes
	Sloop SA	microspore culture	Australia (SA)	D. Moody
	Jacinta		Denmark	A. Jensen
	Helium			A. Jensen
Brassicas (Brassica juncea)	Amulet	micropsore culture	Canada (Reg. Pending)	D. Males
	Arrid		Canada (Reg. Pending)	D. Males
Eggplant	Petra	F_1 from DH parent(s)	Spain	R. Dolcet
	Seven		Spain	R. Dolcet
	Cristal		Spain	R. Dolcet
	Milar		Spain	R. Dolcet
	Senegal		Spain	R. Dolcet
Melon	Cantasapo	F_1 from DH parent(s)	Spain	R. Dolcet
	Cantarino		Spain	R. Dolcet
	M1009		Spain	R. Dolcet
Pepper	Carisma	F_1 from DH parent(s)	Spain	R. Dolcet
Pepper	Tajo		Spain	R. Dolcet
Pepper	Pekin		Spain	R. Dolcet
Pepper	Olmo		Spain	R. Dolcet
Pepper	Alcor		Spain	R. Dolcet
Pepper	Aneto		Spain	R. Dolcet
Pepper	Beret		Spain	R. Dolcet
Pepper	Collado		Spain	R. Dolcet
Rapeseed	Hansen	unknown	Denmark	L. Sernyk
	Acropolis	unknown	Germany	G. Viden
	Licandy	unknown	Germany	G. Viden
	Mendel	DH parental line	Germany	M. Frauen
	Titan		Germany	M. Frauen
	44A53	microspore culture	Canada	J. Patel
	45A55		Canada	J. Patel
	45A71		Canada	J. Patel
	46A52		Canada	J. Patel
	46A54		Canada	J. Patel
	46A72		Canada	J. Patel
	46A73		Canada	J. Patel

Crop	Variety	Method	Country	Authority Reference/pers. com.
	46A74		Canada	J. Patel
	46A76		Canada	J. Patel
	46A77		Canada	J. Patel
	Acropolis		Canada	L. Kott
	Avalanche		Canada	L. Kott
	Dorothy		Canada	L. Kott
	Dynamite		Canada	L. Kott
	Explorer		Canada	L. Kott
	Griffin		Canada	L. Kott
	Hurricane		Canada	L. Kott
	Kastan		Canada	L. Kott
	Kingfisher		Canada	L. Kott
	LoLinda		Canada	L. Kott
	Phoenix		Canada	L. Kott
	Plumbshot		Canada	L. Kott
	Summit		Canada	L. Kott
	Tornado		Canada	L. Kott
	Hybrid male parent of SW-P Admire		Canada	G. Stringhem
	Conquest	microspore culture	Canada	G. Stringhem
	Dakota		Canada	G. Stringhem
	Hi-Q		Canada	G. Stringhem
	Kelsey		Canada	G. Stringhem
	Peace		Canada	G. Stringhem
	Q2		Canada	G. Stringhem
	Quantum		Canada	G. Stringhem
	Roper		Canada	G. Stringhem
	Bumper		Canada	D. Males
	SP Armada		Canada	D. Males
	Kosto		France	D. Louchart
	Pollen		France	D. Louchart
	Korall	spontaneous	Sweden	B. Gertsson
	Paroll		Sweden	B. Gertsson
	Haplona		UK	L. Sernyk
	Navajo		UK	L. Sernyk
	Saxon		UK	G. Viden
Rice	Hua Yu 1	unknown	China	-
	Hua Yu 2	unknown	China	-
	Hwaseong-byeosee	unknown	China	Moon et al., 1986
	Tanfeng 1	unknown	China	-
	Xin Xiou	unknown	China	-
	Zonghua 2	unknown	China	-
	Dorella	unknown	Italy	E. Lupotto
Tobacco	F211	unknown	China	-
	NC 744	unknown	China	-
	Tan Yuh 1	unknown	China	-
	Tan Yuh 2	unknown	China	-
	Tan Yuh 3	unknown	China	-
	LMAFC34	unknown	China	-
Triticale	Eleanor	unknown	Australia (NSW)	K. Cooper

Crop	Variety	Method	Country	Authority Reference/pers. com.
	Triathlon	anther culture	France	P. Devaux
	Tricolor		France	P. Devaux
Swede	Vigod	anther culture	Norway	M. Hansen
Wheat	Huapei 1	unknown	China	
	Lunghua 1	unknown	China	
	Korund	unknown	Germany	
	BR-43	anther culture	Brazil	A. Rosa
	OAC Montrose		Canada	T. Hunt
	Jinghua 1		China	W. Zhou
	Florin		France	De Buyser et al., 1987
	Rubens		France	P. Devaux
	GK Delibab		Hungary	J. Pauk
	GK Tunder		Hungary	J. Pauk
	MV Madrigal		Hungary, Ireland & UK	Z. Bed
	MV Szigma		Hungary	Z. Bed
	GK Szindbad		Hungary	J. Pauk
	SW Agaton		Sweden	S. Tuvesson
	Raspail	maize pollination	France	P. Devaux
	Napier		UK	T. Rhodes
	Solstice		UK	T. Rhodes
	Xi19		UK	T. Rhodes
	Marshal		UK & Ireland	T. Rhodes
	Savannah		UK, Ireland, Germany & France	T. Rhodes

Footnote: the authors would appreciate any additions, updates or corrections to this table.

DH and backcross conversion

As more knowledge is accumulated about the genetic control of economically important traits, it is becoming more feasible to target specific regions of the genome for introgression of novel characters. Previously, plant breeders utilized germplasm collections for easily recognized characters such as major disease resistance genes and then used either a backcrossing or a crossing and selection programme to introduce them with minimal residual donor genome. This process was lengthy and inefficient as deleterious effects of residual donor genome could persist for several cycles of crossing and selection. For example, the barley *Reg-La* mildew resistance from *Hordeum laevigatum* was first deployed in the varieties 'Vada' and 'Minerva' in the 1960's. Varieties carrying the gene had poor malting quality and it was not until the release of Doublet in the 1980's that the association was broken. Initially, it was thought that this was an association with the mildew resistance locus, later located on the long arm of chromosome 2H, but it was more likely to be due to partial resistance to leaf rust on chromosome 7H (Swanston, 1987).

The development of molecular markers means that such problems can be minimized or avoided in the future (Fig. 4.1-2). For example Chen et al., (1994), located barley stripe rust resistance loci on chromosomes 4H and 5H with RFLPs. These markers were then used to select BC_1 plants carrying the loci, which were then entered into doubled haploid production. DH plants homozygous for the resistance loci were identified by another round of selection with the markers. A final round of screening with random AFLP markers then identified those lines with minimum residual donor genome (Toojinda et al., 1998). The AFLP survey revealed that many of the BC_1DHs produced in this scheme had less than the theoretical average 25% donor genome and some approached the theoretical BC_3 donor genome content. Thus, with careful deployment of positive and negative selection for donor markers, one can rapidly derive BC_3 equivalent inbred lines of value (Fig. 4.1-3). By conducting a further round of backcrossing, either to the same or an alternative recipient parent, before producing DHs from selected BC_2 plants, one can reduce the donor genome contribution much more (Tanksley et al., 1989). One can rapidly introduce new genes into a variety using this combination of backcrossing, marker-assisted selection and doubled haploidy. It is easily possible to enter lines of an inbreeding species into official trials 5 years after the initial cross, irrespective of the crop habit and the dominance relationships of the trait (Fig. 4.1-3).

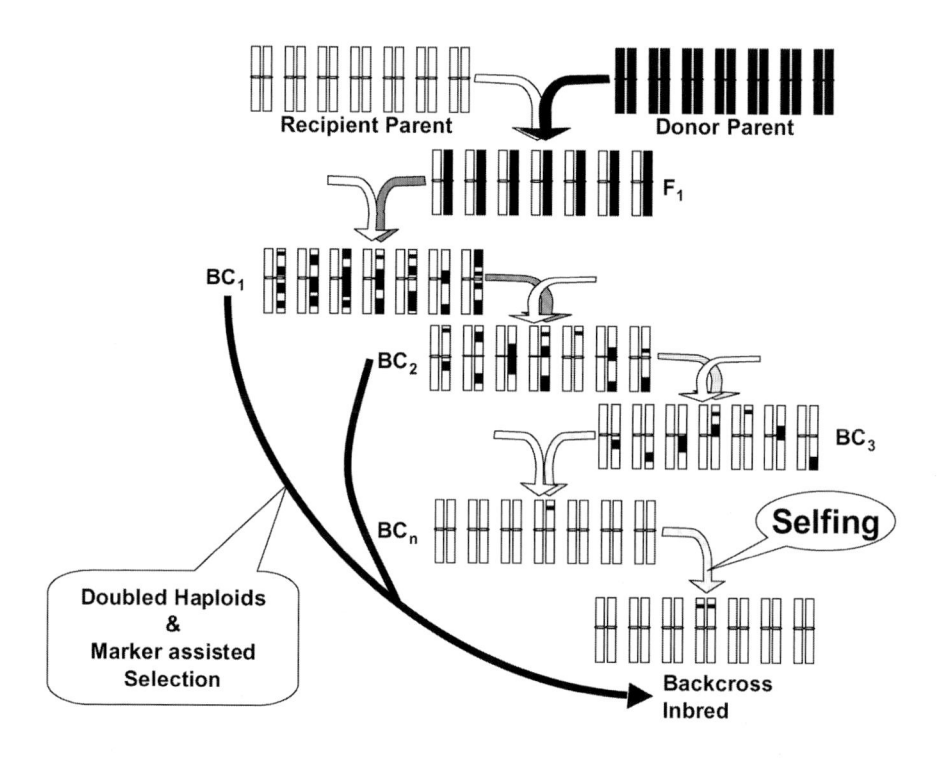

Figure 4.1-2. Development of doubled haploidy and Marker Assisted Selection in a rapid backcross conversion scheme.

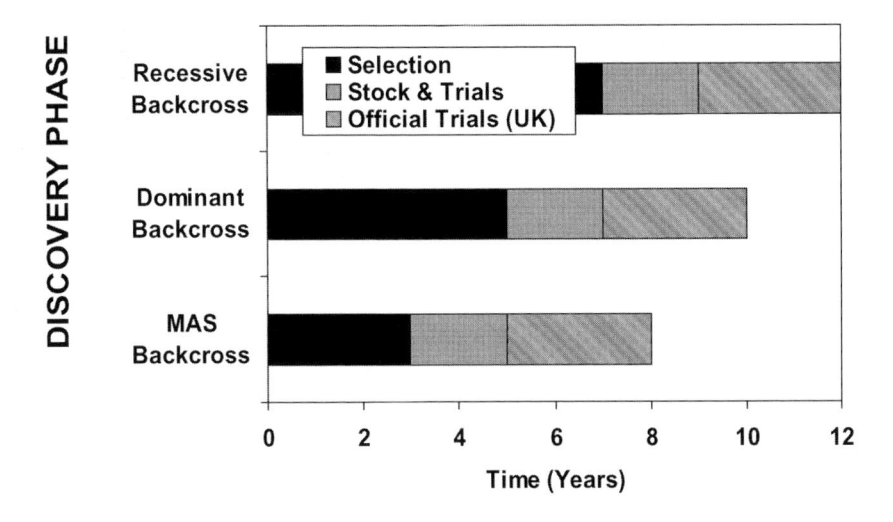

Figure 4.1-3. Comparison of time taken in DH and MAS backcross conversion scheme versus conventional backcrossing.

Acknowledgements
SCRI recieves grant-in-aid from the Scottish Executive Environment and Rural Affiars Department.

References

Bjornstad, A., H. Skinnes, and K. Thoresen, 1993. Comparisons between doubled haploid lines produced by anther culture, the *Horduem bulbosum* method and lines produced by single seed descent in barley. Euphytica. **66**: 135-144.
Caligari, P., D.S., W. Powell, J.L. Jinks, 1987. A comparison of inbred lines derived by doubled haploidy and single seed descent in spring barley (*Hordeum vulgare*). Annals Appl. Biol. **111**: 667-675.
Charmet, G., and G. Branlard, 1985. A comparison of anrogenetic doubled haploid and single seed descent lines in. Theor. Appl. Genet. **71**: 193-200.
Chen, F.Q., D. Prehn, P.M. Hayes, D. Mulrooney, A. Corey, and H. Vivar, 1994. Mapping genes for resistance to barley stripe rust (*Puccinia striiformis* f. sp. *hordei*). Theor. Appl. Genet. **88**: 215-219.
Chen, J.L. and W.D. Beversdorf, 1990: A comparison of traditional and haploid derived breeding populations of oilseed rape (*Brassica napus* L) for fatty acid composition of the seed oil. Euphytica. **51**: 59-65.
Choo, T. M., E. Reinbergs, and S. J. Park, 1982. Comparison of frequency distributions of doubled haploid and single seed descent lines in barley. Theor. Appl. Genet. **61**: 215-218.
Chung, Y.H., S.C. Lee, and D.U. Kim, 1992. Comparison of lines from anther and maternally-derived dihaploids, single-seed descent and bulk breeding method in flue-cured tobacco (*Nicotiana tabacum* L.). J. Korean Soc. Tob. Sci. **14**: 104-115.
Corriols, L., C. Doré, and C. Rameau, 1990. Commercial release in France of Andréas, the first asparagus all-male F1 hybrid. In: A. Falavigna and M. Schiavi (Eds.). Proc. VII International Asparagus Symposium. Acta Hort. **271**: 249-252.
Courtois, B., 1993: Comparison of single seed descent and anther culture derived lines of 3 single crosses of rice. Theor. Appl. Genet. **85**: 625-631.
De Buyser, J., Y. Henri, P. Lonnet, R. Hertzog, and A. Hespel, 1987. Florin: A doubled haploid wheat variety developed by the anther culture method. Plant Breed. **98**: 53-56.

Devaux, P., M. Zivy, A. Kilian, and A. Kleinhofs, 1996. Doubled haploids in barley. In: V International Oat Conference & VII International Barley Genetics Symposium. Eds: Scoles, G. and Rossnagel, B. 1, 213-222. 1996. University of Saskatchewan, Canada, University Extension Press.

Falavigna, A., P.E. Casali, and A. Battaglia, 1999. Achievement of asparagus breeding in Italy. In: B. Benson (Ed.). Proc. IX International Asparagus Breeding Symposium. Acta Hort. **479**: 67-74.

Friedt, W., J. Breun, S. Zuchner, and B. Foroughi-Wehr, 1986. Comparative value of androgenetic doubled haploid and conventionally selected spring barley lines. Plant Breed. **97**: 56-63.

Furusho, M., T. Baba, O. Yamaguchi, T. Yoshida, Y. Hamachi, R. Yoshikawa, K. Mizuta, and M. Yoshino, 1999. Breeding a new malting barley variety Houshun by the bulbosum method. Breed. Sci. **49**: 281-281.

Inagaki, M.N., G. Varughese, S. Rajaram, S., M. van Ginkel, and A. Mujeeb-Kazi, 1998. Comparison of bread wheat lines selected by doubled haploid, single-seed descent and pedigree selection methods. Theor. Appl. Genet. **97**: 550-556.

Jinks, J., L, M.K.U. Chowdhury, H.S. Pooni, 1985: Comparison of the inbred lines derived from a hybrid of tobacco (Burley x flue cured) by dihaploidy and single seed descent. Heredity. **55**: 127-133.

Kubba, J., B.M. Smith, D.J. Ockendon, A.P. Setter, C.P. Werner, M.J. Kearsey, 1989. A comparison of anther culture derived material with single seed descent lines in Brussels sprouts (*Brassica oleracea* var. *gemmifera*). Heredity. **63**: 89-95.

Ma, H., R.H. Busch, O. Riera-Lizarazu, H.W. Rines, and R. Dill-Macky, 1999. Agronomic performance of lines derived from anther culture, maize pollination and single-seed descent in a spring wheat cross. Theor. Appl. Genet. **99**: 432-436.

Martinez, C.P., F.C. Victoria, M.C. Amezquita, E. Tulande, G. Lema, and R.S. Zeigler, 1996. Comparison of rice lines derived through anther culture and the pedigree method in relation to blast (*Pyricularia grisea* Sacc) resistance. Theor. Appl. Gene. **92**: 583-590.

Mitchell, M.J., R.H. Busch, and H.W. Rines, 1992. Comparison of lines derived by anther culture and single seed descent in a spring wheat cross. Crop Sci. **32**: 1446-1451.

Moon, H.P., S.Y. Cho, Y.H. Son, B.T. Jun, M.S. Lim, H.C. Choi, N.K. Park, R.K. Park, and G.S. Chung, 1986. A new high-quality, high-yielding rice variety, Hwaseongbyeo, derived by anther culture. Res. Rep. Rural Devel. Admin. Crops, Korea Republic. **28**: 2, 27-33.

Murigneux, A., S. Baud, and M. Beckert, 1993. Molecular and morphological evaluation of doubled haploid lines in maize. 2. Comparison with single seed descent lines. Theor. Appl. Genet. **87**: 278-287.

Park, S.J., E.J. Walsh, E. Reinbergs, L.S.P. Song, K.J. Kasha, 1976. Field performance of doubled haploid barley lines in comparison with lines developed by the pedigree and single descent methods. Can. J. Plant Sci. **56**: 467-474.

Powell, W., P.D.S. Caligari, W.T.B. Thomas, 1986. Comparison of spring barley lines produced by single seed descent, pedigree inbreeding and doubled haploidy. Plant Breed. **97**: 138-146.

Powell, W., W.T.B. Thomas, D.M. Thompson, J.S. Swanston, and R. Waugh, 1992. Association between rDNA alleles and quantitative traits in doubled haploid populations of barley. Genetics. **130**: 187-194.

Schnell, R., J.E.A. Wernsman, L.G. Burk, 1980. Efficiency of single-seed-descent vs. anther-derived dihaploid breeding methods in tobacco. Crop Sci. **20**: 619-622.

Skinnes, H. and A. Bjornstad, 1995. Comparative trials with anther culture derived doubled haploids and single seed descent lines in wheat. Cereal Res.Commun. **23**: 267-273.

Snape, J.W., 1976. A theoretical comparison of diploidised haploid and single seed descent populations. Heredity. **36**: 275-277.

Snape, J.W., J.W. Ouyang, B.B. Parker, S. E. Jia, 1992. Evidence for genotypic selection in wheat during the development of recombinant inbred lines by anther culture and single seed descent. J. Genet. Breed. **46**: 167-172.

Steiss, R., A. Schuster, and W. Friedt, 1998. Development of linseed for industrial purposes via pedigree- selection and haploid-technique. Indust. Crops Prod. **7**: 303-309.

Stringhem, G.R., V.K. Bansel, M.R. Thiagarajah, D.F. Degenhardt, and J.P. Tewari, 1995. Development of an agronomically superior blackleg resistant canola variety *Brassica napus* L. using doubled haploidy. Can. J. Plant Sci. **75**: 437-439.

Swanston, J.S., 1987. The consequences for malting quality of *Hordeum laevigatum* as a source of mildew resistance in barley breeding. Annals Appl. Biol. **110**: 351-355.

Tanksley, S.D., N.D. Young, A.H. Paterson, and M.W. Bonierbale, 1989: RFLP mapping in plant breeding – new tools for an old science. Bio-Tech. **7**: 257-264.

Toojinda, T., E. Baird, A. Booth, L. Brocrs, P. Hayes, W. Powell, W. Thomas, H. Vivar, and G. Young, 1998. Introgression of quantitative trait loci (QTLs) determining stripe rust resistance in barley: an example of marker-assisted line development. Theor. Appl. Genet. **96**: 123-131.

Turcotte, P., C.A. St Pierre, K.M. Ho, 1980: Comparison between pedigree lines and doubled haploid lines of barley (*Hordeum vulgare* L.). [French]. Can. J. Plant Sci. **60**: 79-85.

Winzeler, H., J. Schmid, and P.M. Fried, 1987. Field performance of androgenetic doubled haploid spring wheat lines in comparison with lines selected by the pedigree system. Plant Breed. **99**: 41-48.

4.2
Doubled haploid mutant production

I. Szarejko
Department of Genetics, University of Silesia, Jagiellonska 28, 40-032 Katowice, Poland

Introduction

The use of haploid systems for mutant induction and selection has been listed among the most important applications of haploid technologies, since their development (Kasha, 1974). Haploid tissue can facilitate the generation of genetic variation and its identification. The haploid system provides several advantages for the application of mutation techniques in plant breeding and germplasm enhancement, however under the condition that an efficient methodology of doubled haploid production is available for a particular species. Many of these advantages arise from the fact that genotypes and genetic segregation ratios in DH populations are equivalent to those found in gametes. Some of the benefits of applying DH systems for induction and selection of mutants are:

- possibility to screen for both recessive and dominant mutants in the first generation after mutagenic treatment
- immediate fixation of mutated genotypes, which saves time in the production of pure mutant lines
- increased selection efficiency of desired mutants due to the gametic *versus* zygotic segregation ratios (1:1 *vs* 3:1, respectively) and the lack of chimerism
- possibility of applying *in vitro* selection methods at the haploid or doubled haploid level.

Conventional and DH mutagenesis

In conventional mutagenesis of seed propagated crops, selection of mutants is often initiated in the M_2 generation (Fig. 4.2-1). It should be noted, however, that due to the chimeric structure of M_1 plants, there is usually a deficit of recessive mutant phenotypes in the segregating progeny of M_1 plants. Additionally, mutagenic treatment induces more than one mutagenic event per initial cell that contributes to gamete formation, so beneficial mutations are often accompanied or masked by undesired ones. The selection in the M_2 generation is based on a single plant which makes it difficult to pick up mutants with quantitative changes. Taking these facts into consideration, Maluszynski et al., (2001) recommended postponing mutant selection to the M_3 generation, especially when screening for a quantitative trait, such as yield, disease and abiotic stress resistance or quality. In any case, the homozygosity test must be undertaken in the selfed progeny of a selected mutant. On average, the isolation of a homozygous mutant line in seed propagated crops takes 3-4 generations. Application of a DH system can shorten this process by half. All mutated traits are expressed in the first generation after mutagenic treatment, both before and after chromosome doubling. DH mutant lines selected on the basis of their phenotype are genetically fixed and will not segregate in the progeny. Due to gametic segregation and the lack of chimerism, the frequency of a desired mutant in a doubled haploid population will be much higher than in the M_2 generation. Higher selection response allows for the use of a smaller population size than in conventional mutagenesis. On the other hand, if mutant selection can be carried out *in vitro*, the populations of haploid cells or embryos can provide an extremely large mutagenized population, thus increasing the probability of identifying a rare mutation event.

Mutagenic treatment → Parent variety M_0 seeds (homozygous)

↓

Growth in good conditions, without selection → M_1 plants (chimeric, partially heterozygous)

↓

Selection of mutants on a single plant basis → M_2 generation (segregating for mutated individuals, deficit of mutants)

↓

Selection of mutant lines → M_3 generation (segregating for mutant lines) — Homozygosity test

↓

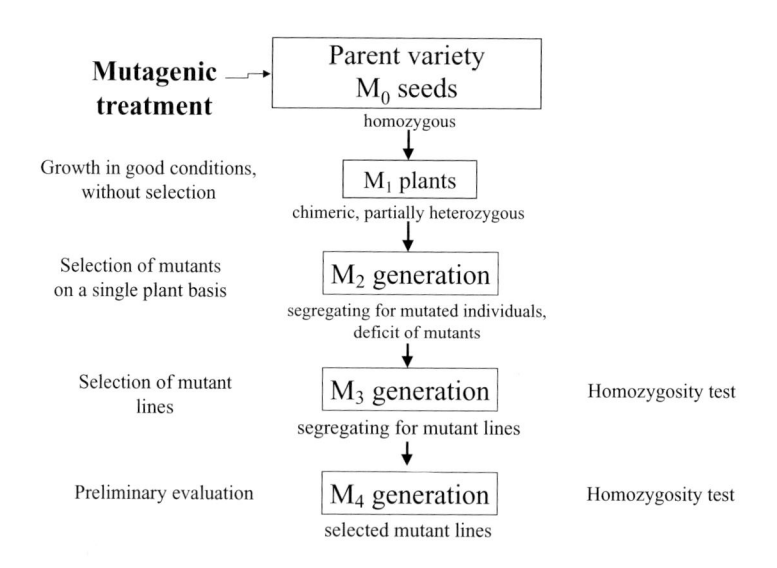

Preliminary evaluation → M_4 generation (selected mutant lines) — Homozygosity test

Figure 4.2-1. Development of mutant lines through conventional mutagenesis.

There are two different approaches to applying mutagenesis in combination with haploid *in vitro* culture (Maluszynski et al., 1996). One utilizes mutagenic treatment of haploid cells or tissues *in vitro*. In the alternative approach, gametes produced by M_1 plants originating from mutagenically treated seeds, are used as donor material for haploid culture. In both cases, mutants are obtained from haploid cells carrying a mutation in hemizygous stage. After chromosome doubling, a pure, genetically fixed mutant line is obtained. The main difference in both methodologies lies in the application of mutagenic treatment: to the haploid cell *in vitro*, or to the dormant seed *in vivo*. Examples of both approaches to produce DH mutant lines are presented below.

Inducing mutations in haploid cells *in vitro*
If available, isolated microspore culture is the best system for the application of *in vitro* mutagenesis and selection. The advantages of this approach were most fully demonstrated in rapeseed and other Brassicas, where several important mutants were isolated with the use of microspore technology (Kott *et al.*, 1996; Barro *et al.*, 2001). Mutagenic treatment is applied during the single cell stage, usually shortly after microspore isolation (Fig. 4.2-2). The detailed protocols of isolated microspore culture, including procedures for microspore extraction, *in vitro* microspore embryogenesis, embryo germination and chromosome doubling are presented in the other chapters of this Manual for many plant species. When employing these protocols for DH mutant production, one should consider several important factors that can affect the success of the whole procedure:

1. Mutagenic agent must be applied at the uninucleate stage of microspore development. It is usually done directly or shortly after microspore isolation. If microspores undergo the first nuclear division during pre-treatment or culture, before application of the mutagenic treatment, the regenerated haploid or doubled haploid plants can be chimeric and/or heterozygous. This is especially true for barley and wheat where most microspore derived regenerants are spontaneously doubled. As it was illustrated by Chen *et al.* (1984) and Kasha *et al.* (2001), fusion between two nuclei derived from the first mitotic division in microspores cultured *in vitro* can be the main mechanism responsible for diploidization.

2. The application of mutagenic treatment to microspores in culture can drastically decrease their survival, embryogenic ability and regeneration potential. Various mutagens can affect the microspore culture in a different ways. For example, in a study performed in rapeseed, UV light affected only embryo formation but not the regeneration potential of surviving embryos, whereas gamma radiation decreased both the frequency of embryos and the frequency of regenerated haploid plants (MacDonald *et al.*, 1991). Additionally, doses of a mutagen applied to cells *in vitro* must be several times lower than those used for dormant seed treatments. For these reasons it is always necessary to perform initial studies on microspore sensitivity to the mutagens that will be used for large scale treatments in culture. The results of sensitivity tests are often called 'killing curves' as they reflect the effect of each treatment in terms of embryo survival. The finally applied doses of mutagens should be adjusted according to the results of sensitivity tests and the desired frequency of induced mutations that are related to the objective of the mutation programme.

3. Both physical and chemical mutagens can be applied for microspore treatment *in vitro*. Chemical mutagenesis may be more difficult to handle, as it requires thorough washing of microspores after treatment to remove remaining residues of the mutagen. Nevertheless, chemical factors including alkylating agents (e.g. EMS, ENU, MNU) and sodium azide, were successfully utilized in microspore mutagenesis in Brassicas and barley (Table 4.2-1).

4. Isolated microspores are not the only haploid cells used for *in vitro* mutagenesis. Mutagenic treatments have also been applied to immature inflorescences, e.g. spikes or buds containing microspores in the stage proper for culture, for anthers prior to *in vitro*

incubation and for different explants originating from haploid cells, e.g., haploid calli, embryos or protoplasts. In all these cases the doses of mutagens need to be adjusted according to the sensitivity of material used for treatment (Table 4.2-2).

Figure 4.2-2. *In vitro* mutagenesis using isolated microspore culture.

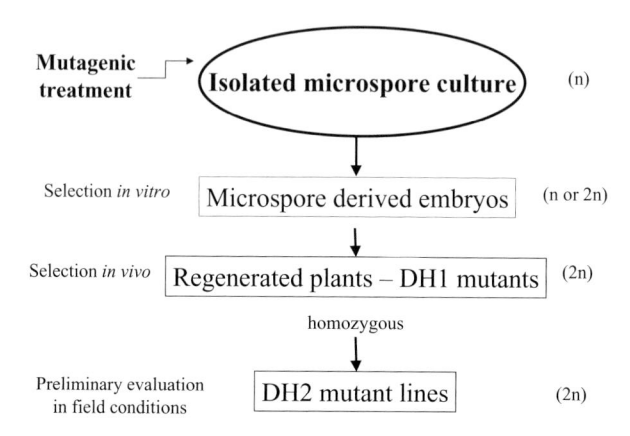

```
Mutagenic  ──►  ( Isolated microspore culture )   (n)
treatment

Selection in vitro   Microspore derived embryos   (n or 2n)
                              │
                              ▼
Selection in vivo    Regenerated plants – DH1 mutants   (2n)
                          homozygous
                              │
                              ▼
Preliminary evaluation    DH2 mutant lines   (2n)
in field conditions
```

Table 4.2-1. Mutagenic treatments applied to isolated microspore cultures

Species	Mutagen	Dose	Reference
Brassica napus	EMS		Beversdorf and Kott, 1987
	ENU	20 mM	Swanson *et al.*, 1988, 1989
	NaN$_3$		Polsoni *et al.*, 1988
	MNH (MNU)	0.2-1.5 mM, 3 h	Jedrzejaszek *et al.*, 1997
	gamma rays	5 Gy	Swanson *et al.*, 1988
		15 Gy	Beversdorf and Kott, 1987
		10-15 Gy	Beversdorf and Kott, 1987
		10-40 Gy	MacDonald *et al.*, 1991
	X-rays	10-40 Gy	MacDonald *et al.*, 1991
	UV	10-60 s*	MacDonald *et al.*, 1991
		15-30 s	Jedrzejaszek *et al.*, 1997
Brassica carinata	EMS	0.1-0.5%, 30 min	Barro *et al.*, 2001
Hordeum vulgare	NaN$_3$	10^{-4} M, 10^{-5} M, 1 h	Castillo *et al.*, 2001

EMS - ethyl methanesulfonate; ENU - N-ethyl-N-nitroso urea; MNH (MNU) - N-methyl-N-nitroso urea; NaN$_3$ - sodium azide; UV – ultra violet light; *dose rate: 33 erg mm^{-2} s^{-1}

Using *in vitro* mutagenesis at the haploid level, several important mutants have been isolated in rapeseed, barley and rice. Microspore mutagenesis has been most successful in rapeseed resulting in the identification of stable mutant lines. Among them there were DH mutants resistant to herbicides (Swanson *et al.*, 1988, 1989) and mutants having altered fatty acid composition or levels of glucosinolates (Kott *et al.*, 1996). Similarly, DH mutants with modified erucic acid content have been isolated from mutagenized microspore cultures of *Brassica carinata* (Barro *et al.*, 2001). In barley, treatments with sodium azide applied to freshly isolated microspores in culture resulted in induction of several morphological and developmental mutants, among them dwarf, *eceriferum*, *hexastichon*, and late or early heading (Castillo *et al.*, 2001). The frequency of DH lines with heritable changes reached 15.6%.

Besides isolated microspores, other haploid explants have been used successfully for DH mutant development (Table 4.2-2). After irradiation of microspore derived calli with gamma rays, several DH mutant lines were selected in rice, among them DH lines with earlier maturity, shorter stem and other important characters related to yield (Chen, 2001). Some of these DH mutant lines are under regional advanced rice breeding trials of Zhejiang Province. Recently, Lee and Lee (2002) reported on high frequency of rice mutants obtained from anther culture treated with EMS 10 and 20 days after anther inoculation. There were semi-dwarf forms, grain shape and glabrous mutants among DH lines obtained from these treatments.

Doubled haploid mutants from segregating gametes of M_1 plants

Another approach to mutant induction in DH systems omits the step of *in vitro* mutagenesis. In this approach M_1 plants obtained by mutagenic treatment of dormant seeds are used as donors of mutated gametes for haploid production (Fig. 4.2-3). The effectiveness of this system has been demonstrated in barley anther culture of M_1 donor plants (Szarejko *et al.*, 1995). However, when using M_1 plants for DH mutant induction, one should consider the several advantages and limitations of this approach:

1. Treatment of dormant seeds instead of microspores, anthers or inflorescences prior to culture allows the application of much higher doses of mutagen, which should result in the increased frequency of mutations. Additionally, the negative somatic effect of a mutagen that often drastically affects the efficiency of haploid induction, can be significantly reduced when mutagenic treatment is applied to seed embryos, many cell generations before meiosis of the M_1 donor plants. The doses of radiation or chemical mutagens used for seed treatments to obtain M_1 donor plants are usually within the range of doses applied to seeds using conventional mutagenesis. It is important, however, to examine the effect of mutagenic treatment on sterility of M_1 plants. Too high doses of mutagens, even when applied to seeds, can significantly decrease the viability of both male and female gametes, resulting in the partial sterility of M_1 plants.

2. The number of M_1 plants used as donors of gametes for DH mutant production should be sufficient enough to ensure the recovery of all desired mutations. Contrary to microspore mutagenesis *in vitro*, where only a few donor plants can supply hundreds of thousands of microspores, each of them being a potential target for mutagenic action, the number of mutations obtained in a segregating population of gametes produced by a single M_1 plant is more limited. In this case, the final number of DH mutants depends on the frequency of mutations per single nucleus and the number of 'genetically active' cells that are present in M_0 embryos and contribute, after mutagenic treatment, to the production of gametes. In cereals, samples of microspores taken from different sectors of about 50 M_1 plants can supply a mutagenized population large enough to assure the recovery of wide range of mutants.

3. Mutagenic treatment applied to dormant seeds will have relatively low somatic effect on the efficiency of haploid production from spikes or buds. For this reason, it is possible to utilise DH mutagenesis even in those species and cultivars that have not been reported as responsive in haploid cultures. M_1 plants have been used as donors of male gametophytes in anther culture of rice (Zapata and Aldemita, 1989) or female gametes in *Hordeum vulgare* x *H. bulbosum* crosses (Gaj and Maluszynski, 1989). Similarly, M_1 derived ovaries can be employed for gynogenic production of DH mutants in those species, where ovary or ovule culture is the most efficient way of haploid production.

Table 4.2-2. *In vitro* mutagenesis of haploid explants or plant organs that contain gametophytes

Species	Organ/explant treated	Mutagen	Dose	Reference
Brassica napus	flower buds before plating of anthers	gamma rays	10-50 Gy	MacDonald et al., 1986
	anthers	gamma rays	10-60 Gy	Jedrzejaszek et al., 1997
		N_f	10-16 Gy	Jedrzejaszek et al., 1997
	secondary embryoids	gamma rays	80-240 Gy	McDonald et al., 1988
	hypocotyl segments from microspore derived embryos	EMS	0.2, 0.25, 0.5%, 3-10 h	Shi et al., 1997
Datura innoxia	anthers	gamma rays	10 Gy	Sangwan and Sangwan, 1986
	haploid protoplasts	X-rays	2.5-15 Gy	Schieder, 1976
		MNNG	5-50 mg/L, 0.5 h	Krumbiegel, 1979
Hordeum vulgare	spikes before plating of anthers	gamma rays	1-10 Gy	Szarejko et al., 1995; Laib et al., 1996
	anthers	NaN_3	10^{-3}, 10^{-4} M, 6 h	Castillo et al., 2001
Malus x domestica	flower buds before plating of anthers	gamma rays	5-10 Gy	Zhang et al., 1992
Nicotiana plumbaginifolia	haploid protoplasts	gamma rays	13-23 J kg^{-1} 1-10 Gy	Sidorov et al., 1981 Nielsen et al., 1985
		UV	12.5 erg mm^{-2}s^{-1}, 40 s	Marion-Poll et al., 1988
			25 erg mm^{-2} s^{-1}	Sumaryati et al., 1992
Nicotiana sylvestris	haploid protoplasts	UV	32 erg mm^{-2} s^{-1}	Vunsh et al., 1982
Nicotiana tabacum	anthers	gamma rays	10 Gy	Sangwan and Sangwan, 1986
		EMS	0.005-0.01%, 1-2 h	Medrano et al., 1986
	haploid protoplasts	UV	32 erg mm^{-2} s^{-1}	Vunsh et al., 1982
Oryza sativa	panicles with microspores at the uninucleate stage	gamma rays	5-50 Gy 10-20 Gy	Hu, 1983 Chen, 1986
	anthers at early stage of culture	EI	0.5-1 ml/L, 20 h	Hu, 1983
		EMS	2-4 ml/L, 12 h	Hu, 1983
		MNNG	25-100 mg/L, 15 h	Hu, 1983
	anther culture at 0, 10 and 20 days after inoculation	EMS	0.5%, 6 h	Lee and Lee, 2002
	anthers after 35 days of culture and microspore-derived calli	gamma rays	10-50 Gy	Chen et al., 2001
Solanum tuberosum	inflorescences with PMC	gamma rays	2-30 Gy	Przewozny et al., 1980
		X-rays	10 Gy	Przewozny et al., 1980

Species	Organ/explant treated	Mutagen	Dose	Reference
		MMNG	0.05-2 mM, 3 h	Przewozny et al., 1980
		MNH	0.13-2 mM, 3 h	Przewozny et al., 1980
Triticum aestivum	spikes before plating of anthers	gamma rays	1-3 Gy	Ling et al., 1991

EI – ethylenimine; MNNG - N-methyl-N'-nitro-N-nitrosoguanidine; N_f - fast neutrons

The application of M_1 plants for DH mutant production has been successfully employed in barley and rice (Table 4.2-3). Using M_1 anther donor plants derived from seeds treated with chemical mutagens, Szarejko et al. (1995) obtained pure mutant lines of barley at the frequency as high as 25%. Such homozygous mutant lines have included various dwarf and semi-dwarf types, *eceriferum*, isozyme and chlorophyll mutants, and mutant lines with changes in quantitative traits related to yield. DH lines derived from M_1 generations of Peruvian barley varieties have been evaluated in the Altiplano region, at an elevation of 3,500-4,000 m above sea level, by researchers from the Cereal Programme, Agricultural University, LaMolina, Peru. Several promising barley lines have been identified and recommended for growth in these conditions (Gomez-Pando, 2002). Similarly, at the Central Agricultural Research Institute in Myanmar, several DH mutant lines with good yield parameters have been isolated from anther culture of rice using M_1 donor plants derived from seed irradiation. One of these lines, with yield parameters similar to the parent variety but maturing 19 days earlier, has been recommended for release as a variety after evaluation in large scale field experiments (Kyin San Myint, personal comm.). Khan et al. (2001) reported on promising DH mutant lines of spring wheat that were produced by anther culture from gamma irradiated F_1 plants. Some of these lines have been recommended for regional wheat trials in Punjab, Pakistan.

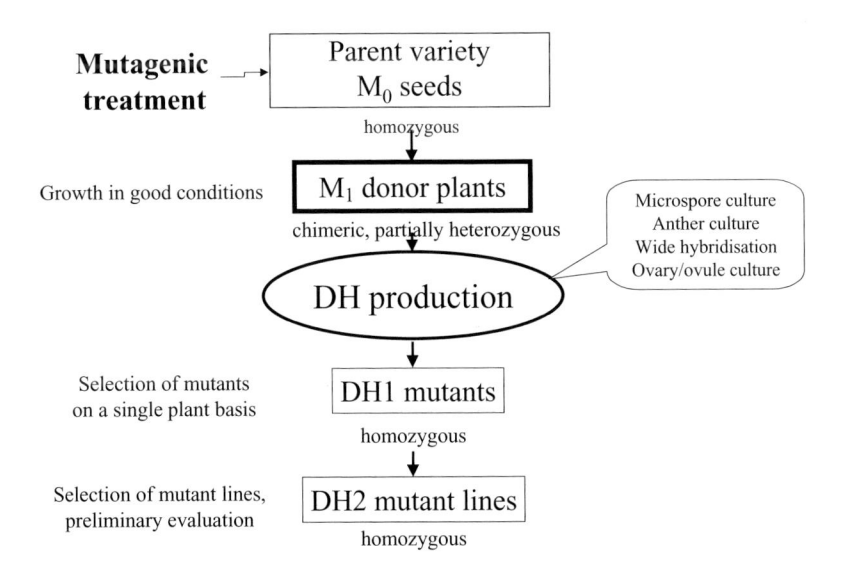

Figure 4.2-3. Development of mutant lines using seed mutagenesis and DH systems.

Table 4.2-3. Examples of mutagenic treatments applied to dormant seeds for production of M_1 plants used as donors for anther culture

Species	Mutagen	Dose	Reference
Hordeum vulgare	gamma rays	120 Gy	Szarejko et al., 1995
	MNH	2 x 0.5 mM, 3 h*	di-Umba et al., 1991; Szarejko et al., 1995
	NaN_3 and MNH	1.0 mM, 3 h and 0.5 mM, 3 h*	di-Umba et al., 1991
Oryza sativa	gamma rays	100-400 Gy	Aldemita and Zapata, 1991
		300-450 Gy	Kyin San Myint, personal comm.
	EMS	0.025 M EMS, 4 h	Wong et al., 1983
Triticum aestivum	gamma rays	150 Gy	Khan et al., 2001

*/ double treatment with 6 h inter-incubation germination period

In vitro selection

The efficient regeneration of plants from microspores provides an attractive system not only for mutagenesis but also for *in vitro* selection or identification of desired mutants. The selection can be carried out *in vitro* if the traits are expressed equally in the haploid cells or embryos and adult plants. The usefulness of *in vitro* selection has been best demonstrated in rapeseed, where several mutants resistant to diseases and herbicides have been recovered from mutagenized microspore populations (Kott, 1995, 1998; Palmer et al., 1996). The selective agent, e.g. herbicide is added to the culture medium at the concentration giving LD_{100} (Beversdorf and Kott, 1987). The herbicide used for selection can be incorporated into induction medium at two different times. First, immediately after the mutagenic treatment of microspores, so the selection is initiated at the microspore level (Swanson et al., 1988). Or secondly, it can be added into the medium when haploid embryos have fully developed, perhaps after 28 days (Beversdorf and Kott, 1987a; Polsoni et al., 1988). In both approaches, the surviving embryos are transferred onto regeneration medium, germinated and colchicine treated. The DH1 plants are harvested and the next generation (DH2) is checked for the level of herbicide tolerance in field conditions.

The main advantages of applying selection at the haploid cell/embryo level are: dealing with the extremely large populations of mutagenized individuals in a small space (in rapeseed several hundred or thousand embryos can be screened in a 10 cm diameter plate); immediate expression of recessive traits; homozygosity of selected mutants (after chromosome doubling DH lines are genetically fixed). Additionally, in rapeseed the phenomenon of secondary embryogenesis (Loh and Kim, 1992) can be used for cloning of mutated embryos surviving the selection pressure. The best examples of the successful application of *in vitro* mutagenesis/selection are rapeseed mutants resistant to widely used herbicides, sulfonoureas and imidazolines (Swanson et al., 1988, 1989). Single gene mutations of the enzyme acetohydrozyacid synthase (AHAS) being the primary action site of these herbicides, have been identified as a result of microspore mutagenesis. The mutated AHAS genes have been cloned and transferred into canola using transgenic approach, for further development of herbicide tolerant varieties (Huang, 1992).

Besides screening for disease or herbicide resistance, the microspore system in rapeseed offers another possibility of early selection. Since the processes of lipid biosynthesis in microspore-derived and zygotic embryos are similar, the cotyledons of microspore-derived embryos should contain fatty acids which are very similar in their composition to the ones found in the seeds (Wiberg et al., 1991; Chen and Beversdorf, 1991; Taylor et al., 1991). At the late cotyledonary stage, a single cotyledon of an embryo can be used for the determination of its fatty acid composition. After the analysis, the remaining part of the embryo can be either regenerated into a plant, or discarded. The advantage of non-destructive analysis makes microspore derived embryos an excellent target for early selection of rapeseed mutants with improved oil quality.

The successful application of this strategy has been demonstrated in rapeseed by isolation of DH lines carrying mutations responsible for the increased level of oleic acid from 60% to 80%, with the accompanying reduction of linolenic acid from 10% to 3-6% (Wong and Swanson, 1991). The high oleic acid canola has maintained the low levels of saturated palmatic and stearic acids (6%) compared with that of 16% in olive oil (Wong and Swanson, 1991), thus providing the additional value to consumers. Isolation of mutants with higher concentration of oleic acids and decreased levels of saturated fatty acids from mutagenized microspore culture, has been also reported by Turner and Facciotti (1990) and Huang (1992).

Microspore-derived embryos in rapeseed have proven to be amenable of early selection also for other compounds affecting seed quality. Cotyledons of microspore-derived embryos can be used for analysis of glucosinolate and synapine content (McClellan *et al.*, 1993; Kott *et al.*, 1996). Using the *in vitro* mutagenesis/selection system in microspore culture, several DH rapeseed mutants with low glucosinolate content have been identified (Kott, 1998).

Acknowledgments
The author gratefully acknowledges Prof. M. Maluszynski for his advice, valuable comments and the critical review of this manuscript.

References

Aldemita, R.R. and F.J. Zapata, 1991. Anther culture of rice: effects of radiation and media components on callus induction and plant regeneration. Cereal Res.Commun. **19(1-2)**: 9-32

Barro, F., J. Fernandez-Escobar, M. De La Vega and A. Martin, 2001. Doubled haploid lines of *Brassica carinata* with modified erucid acid content through mutagenesis by EMS treatment of isolated microspores. Plant Breed. **120**: 262-264.

Beversdorf, W.D. and L.S. Kott, 1987a. An *in vitro* mutagenesis/selection system for *Brassica napus*. Iowa State Journal of Research, **61(4)**: 435-443.

Beversdorf, W.D. and L.S. Kott, 1987b. An *in vitro* mutagenesis/selection system for *Brassica napus*. Iowa State J.Res. **61**: 435-443.

Castillo, A.M., L. Cistue, M.P. Valles, J.M. Sanz, I. Romagosa and J.L. Molina-Cano, 2001. Efficient production of androgenic doubled-haploid mutants in barley by the application of sodium azide to anther and microspore cultures. Plant Cell Rep. **20**: 105-111.

Chen, C.C., K.J. Kasha and A. Marsolais, 1984. Segmentation patterns and mechanisms of genome multiplication in cultured microspores of barley. Can.J.Genet.Cytol. **26**: 475-483.

Chen, J.L. and W.D. Beversdorf, 1991. Evaluation of microspore-derived embryos as models for studying lipid biosynthesis in seed of rapeseed (*Brassica napus* L.). Euphytica, **58**: 145-155.

Chen, Q.F., C.L. Wang, Y.M. Lu, M. Shen, R. Afza, M.V. Duren and H. Brunner, 2001. Anther culture in connection with induced mutations for rice improvement. Euphytica, **120**: 401-408

Chen, Y., 1986. The inheritance of rice pollen plant and its application in crop improvement. In: Haploids of Higher Plants *In Vitro*. Hu, H. and H. Yang (Eds.) China Acad.Publ. and Springer-Verlag, Beijing, pp.118-136

Gaj, M. and M. Maluszynski, 1989. Crossability of spring barley mutants with *Hordeum bulbosum*. In: Current Options for Cereal Improvement. Doubled Haploids, Mutants and Heterosis. Maluszynski, M. (Ed.) Kluwer Academic Publishers, Dordrecht, pp.203-210.

Gomez-Pando, L., 2002. Introduction of barley and other native mutant cultivars to Peruvian highlands. IAEA Technical Cooperation Report 2001. La Molina University, La Molina, Peru, pp.35

Hu, Z., 1983. Stimulating pollen haploid culture mutation in *Oryza sativa* subsp. Keng (*japonica*). In: Cell and Tissue Culture Techniques for Cereal Crop Improvement. Science Press, IRRI, Beijing, Manila, pp.291-301

Huang, B., 1992. Genetic manipulation of microspores and microspore-derived embryos. *In Vitro* Cell.Dev.Biol. **28**: 53-58

Jedrzejaszek, K., H. Kruczkowska, H. Pawlowska and B. Skucinska, 1997. Stimulating effect of mutagens on *in vitro* plant regeneration. MBNL. **43**: 10-11.

Khan, A.J., S. Hassan, M. Tariq and T. Khan, 2001. Haploidy breeding and mutagenesis for drought tolerance in wheat. Euphytica, **120**: 409-414

Kasha, K.J. (Ed.), 1974. Proc. 1st International Symposium on Haploids in Higher Plants: Advances and Potential. Univ. of Guelph, Guelph, pp.421

Kasha, K.J., T.C. Hu, R. Oro, E. Simion and Y. S. Shim, 2001. Nuclear fusion leads to chromosome doubling during mannitol pretreatment of barley (*Hordeum vulgare* L.) microspores. J.Exp.Bot. **52**(359): 1227-1238.

Kott, L., 1995. Production of mutants using the rapeseed doubled haploid system. In: Induced Mutations and Molecular Techniques for Crop Improvement. IAEA, Vienna, pp. 505-515.

Kott, L., R. Wong, E. Swanson and J. Chen, 1996. Mutation and selection for improved oil and meal quality in *Brassica napus* utilizing microspore culture. In: *In Vitro* Haploid Production in Higher Plants. Vol. 2. Jain, S.M., S.K. Sopory and R.E. Veilleux (Eds.) Kluwer Academic Publishers, Dordrecht, pp.151-167.

Kott, L.S., 1998. Application of doubled haploid technology in breeding of oilseed *Brassica napus*. AgBiotech News and Information, **10**(3): 69N-74N.

Krumbiegel, G., 1979. Response of haploid and diploid protoplasts from *Datura innoxia* Mill. and *Petunia hybrida* L. to treatment with X-rays and a chemical mutagen. Environ.Exp.Bot. **19**: 99-103.

Laib, Y., I. Szarejko, K. Polok and M. Maluszynski, 1996. Barley anther culture for doubled haploid mutant production. MBNL. **42**: 13-15.

Lee, J.H. and S.Y. Lee, 2002. Selection of stable mutants from cultured rice anthers treated with ethyl methane sulfonic acid. Plant Cell Tiss.Org.Cult. **71**: 165-171.

Ling, D.X., D.J. Luckett and N.L. Darvey, 1991. Low-dose gamma irradiation promotes wheat anther culture response. Aust.J.Bot. **39**: 467-474.

Loh, C.S. and G.K. Lim, 1992. The influence of medium components on secondary embryogenesis of winter oilseed rape, *Brassica napus* L. ssp. *oleifera* (Metzg.) Sinsk. New Phytol. **121**: 425-430

McClellan, D., L.S. Kott, W.D. Beversdorf and B.E. Ellis, 1993. Glucosinolate metabolism in zygotic and microspore-derived embryos of *Brassica napus* L. Plant Physiol. **141**: 153-159

MacDonald, M. V. and F. N. Aslam, 1986. The effect of gamma irradiation on buds of *Brassica napus* ssp. *oleifera* prior to anther culture. Cruciferae Newsl. **11**

MacDonald, M.V., I. Ahmad, J.O.M. Menten and D.S. Ingram, 1991. Haploid culture and *in vitro* mutagenesis (UV light, X-rays and gamma rays) of rapid cycling *Brassica napus* for improved resistance to disease. In: Plant Mutation Breeding For Crop Improvement. Vol.2. IAEA, Vienna, pp.129-138.

MacDonald, M.V., D.M. Newsholme and D.S. Ingram, 1988. The biological effects of gamma irradiation on secondary embryoids of *Brassica napus* ssp. *oleifera* (Metzg.) Sinsk., winter oilseed rape. New Phytol. **110**: 255-259.

Maluszynski, M., I. Szarejko and J. Maluszynska, 2001. Induced mutations in wheat. In: The World Wheat Book. Bonjean, A.P. and W.J. Angus (Eds.) Lavoisier Publishing, Londres, pp.939-977.

Maluszynski, M., I. Szarejko and B. Sigurbjörnsson, 1996. Haploidy and mutation techniques. In: *In Vitro* Haploid Production in Higher Plants. Vol. 1. Jain, S.M., S.K. Sopory and R.E. Veilleux (Eds.) Kluwer Academic Publishers, Dordrecht, pp. 67-93.

Marion-Poll, A., C. Missionier, J. Goujaud and M. Caboche, 1988. Isolation and characterization of valine-resistant mutants of *Nicotiana plumbaginifolia*. Theor.Appl.Genet. **75**: 272-277.

Medrano, H., E.P. Millo and J. Guerri, 1986. Ethyl-methane-sulphonate effects on anther cultures of *Nicotiana tabacum*. Euphytica, **35**: 161-168.

Nielsen, E., E. Selva, C. Sghirinzetti and M. Devreux, 1985. The mutagenic effect of gamma rays on leaf protoplasts of haploid and dihaploid *Nicotiana plumbaginifolia*, estimated by valine resistance mutation frequencies. Theor.Appl.Genet. **70**: 259-264

Palmer, C.E., W.A. Keller and P.G. Arnison, 1996. Utilization of *Brassica* haploids. In: *In Vitro* Haploid Production in Higher Plants. Jain, S.M., S.K. Sopory and R.E. Veilleux (Eds.) Kluwer Academic Publisher, Dordrecht, pp. 173-192.

Polsoni, L., L.S. Kott and W.D. Beversdorf, 1988. Large-scale microspore culture technique for mutation-selection studies in *Brassica napus*. Can.J.Bot. **66**: 1681-1685.
Przewozny, T., O. Schieder and G. Wenzel, 1980. Induced mutants from dihaploid potatoes after pollen mother cell treatment. Theor.Appl.Genet. **58**: 145-148.
Sangwan, R.S. and B.S. Sangwan, 1986. Effects des rayons gamma sur l'embryogenese somatique et l'androgenese chez divers tissus vegetaux cultives *in vitro*. In: Nuclear Techniques and *In Vitro* Culture for Plant Improvement. IAEA, Vienna, pp.181-185.
Schieder, O., 1976. Isolation of mutants with altered pigments after irradiating haploid protoplasts from *Datura innoxia* Mill. with X-rays. Mol.Gen.Genet. **149**: 251-254.
Shi, S.W., Y.M. Zhou, J.S Wu and H.L. Liu, 1997. EMS mutagenesis of microspore-derived embryogenic cultures of *Brassica napus*. MBNL. **43**: 8-9.
Sidorov, V., L. Menczel and P. Maliga, 1981. Isoleucine-requiring *Nicotiana* plant deficient in threonine deaminase. Nature, **294**: 87-88.
Sumaryati, S., I. Negrutiu and M. Jacobs, 1992. Characterization and regeneration of salt- and water-stress mutants from protoplast culture of *Nicotiana plumbaginifolia* (Viviani). Theor.Appl. Genet. **83**: 613-619.
Swanson, E.B., M.P. Coumans, G.L. Brown, J.D. Patel and W.D. Beversdorf, 1988. The characterization of herbicide tolerant plants in *Brassica napus* L. after *in vitro* selection of microspores and protoplasts. Plant Cell Rep. **7**: 83-87.
Swanson, E.B., M.J. Herrgesell, M. Arnoldo, D.W. Sippell and R.S.C. Wong, 1989. Microspore mutagenesis and selection: canola plants with field tolerance to the imidazolinones. Theor.Appl.Genet. **78**: 525-530.
Szarejko, I., J. Guzy, J. Jimenez Davalos, A. Roland Chavez and M. Maluszynski, 1995. Production of mutants using barley DH systems. In: Induced Mutations and Molecular Techniques for Crop Improvement. IAEA, Vienna, pp.517-530.
Taylor, D.C., N. Weber, D. Barton, E.W. Underhill, L.R. Hoffe, R.J. Weselake and M.K. Pomeroy, 1991. Triacylglycerol bioassembly in microspore-derived embryos embryos of *Brassica napus* cv. Reston. Plant Physiol. **97**: 65-79.
Turner, J. and D. Facciotti, 1990. High oleic acid *Brassica napus* from mutagenized microspores. In: Proc. 6th Crucifer Genetics Workshop. McFerson, J.R., S. Kresovich and S.G. Dwyer (Eds.) Geneva, NY. pp.24.
Umba, di-Umba., M. Maluszynski, I. Szarejko and J. Zbieszczyk, 1991. High frequency of barley DH-mutants from M_1 after mutagenic treatment with MNH and sodium azide. MBNL. **38**: 8-9
Vunsh, R., D. Aviv and E. Galun, 1982. Valine resistant plants derived from mutated haploid and diploid protoplasts of *Nicotiana sylvestris* and *Nicotiana tabacum*. Theor.Appl.Genet. **64**: 51-58.
Wiberg, E., L. Rahlen, M. Hellman, E. Tillberg, K. Glimelius and S. Stymne, 1991. The microspore-derived embryo of *Brassica napus* L. as a tool for studying embryo-specific lipid biogenesis and regulation of oil quality. Theor.Appl.Genet. **82**: 515-520.
Wong, R.S.C. and E. Swanson, 1991. Genetic modification of canola oil: high oleic acid canola. In: Fat and Cholesterol Reduced Food. Haberstroh, C. and C.E. Morris (Eds.) Gulf, Houston, Texas, pp.154-164.
Zapata, F.J. and R.R. Aldemita, 1989. Induction of salt tolerance in high-yielding rice varieties through mutagenesis and anther culture. In: Current Options for Cereal Improvemnt. Maluszynski,M. (Ed.) Kluwer Academic Publishers, Dordrecht, pp.193-202.
Zhang, Y.X., L. Bouvier and Y. Lespinasse, 1992. Microspore embryogenesis induced by low gamma dose irradiation in apple. Plant Breed. **108**: 173-176.

4.3
Barley microspore transformation protocol by biolistic gun

Y.S. Shim and K.J. Kasha
Dept. of Plant Agriculture, Biotech Division, University of Guelph, Guelph, ON, N1G 2W1, Canada

Introduction

Microspores are immature pollen and have the gametic number (*n*) of chromosomes. They can be induce to divide and form embryos or callus and, due to their large numbers, have been able to provide an efficient source of haploid or double haploid (DH) plants in some crops. Doubled haploids are genetically homozygous and produce pure breeding lines. Thus, if transgenes can be incorporated into the haploid microspore genome prior to DNA synthesis and chromosome doubling, the doubled haploids may also be homozygous for the transgenes. Thus, isolated microspores not only provide a good target for bombardment but also are readily amenable to transgene *in vitro* selection. Jähne *et al.* (1994) were the first to achieve plants homozygous for the transgenes using biolistic bombardment of barley microspores. Our lab has also studied this system and we have obtained transformed plants (Yao *et al*, 1997; Carlson *et al.*, 2001). However, our doubled haploids are heterozygous (hemizygous) for the transgenes. The difference in procedures appears to be the pretreatment used to induce embryogenesis and chromosome doubling. Jähne *et al.* (1994) used a 28 day cold pretreatment whereas we used a mannitol pretreatment either with cold (4°C) or room temperature for 4 days. We are presently cytologically investigating the possible reasons for these differences by following the development of the microspores during pretreatment In this paper we will describe the protocol for microspore bombardment and transformation currently used in our lab. For bombardment we used the Biolistic™ PDS-1000 devise from DuPont. The efficiency of transgenic selection by visual screening has been about 1 structure per 1 million isolated microspores which translates into only about 1 green plant recovered per 3 million isolated microspores that were bombarded (Carlson *et al.*, 2001). The essential detailed protocols for barley isolated microspore culture are found elsewhere in this manual (Kasha *et al.*) and will not be repeated here.

Spike collection and pretreatment

Spikes from donor barley plants are collected at mid- to late-uninucleate microspore stage (Kasha et al., 2001) when checked on a central floret near the middle of spike. Additional tillers are collected when they reach a similar stage of morphological development. This stage can vary from genotype to genotype. The tillers are collected in water and stored in a refrigerator at 4°C until (not longer than 4 days) sufficient spikes (5-10) are collected for blender isolation of microspores. Harvested tillers are defoliated before being wrapped in a damp cheesecloth and surface sterilized with 70% alcohol for 10 min. Spikes are then manually removed from the sheath and placed in large Petri dishes. We have used two pretreatments, namely:

1. Spikes in the Petri dishes are partially immersed in ice cold 0.3 M mannitol and kept in dark in a refrigerator for 4 days at 4°C.

2. Spikes are placed in the Petri dishes with 0.5 ml of sterile water at 4°C for 3 to 4 weeks.

Both treatments have been equally effective as pretreatments for induction of microspore embryogenesis.

Microspore osmotic treatment

Once the microspores have been isolated under cold conditions (see Kasha et al., Barley isolated microspore culture protocol), they are prepared for bombardment with a higher osmotic treatment. This can increase the receptiveness of target microspores to DNA coated particles and decrease susceptibility to cell damage from particle bombardment. Microspores on filter paper are placed on solidified FHG medium containing 0.5 M sorbitol plus mannitol. A drop of FHG liquid media (see chapter 6.1) containing 2mg/L PAA is added to the top. After 4 h at 25°C in the dark the microspores are bombarded and left there for another 16 h recovery period. The filter with the bombarded microspores in then transferred to solidified FHG induction medium for culture and placed in the dark at 28°C.

DNA preparation

The bombardment protocol requires purified plasmid DNA containing the transgenes at a high concentration (1µg/µl). A Wizard™ plus Minipreps (Promega) DNA isolation kit is used to produce tubes of concentrated clean transformation vector plasmid DNA recovered from DH5α E. coli cells (Gibco BRL). The details of this isolation are as follows:

1. Using sterile pipette tip, transfer a well-isolated colony into 3 ml LB media with antibiotic in culture tube. Cap loosely and incubate with shaking at 37°C overnight.

2. Transfer 1.5 ml of culture into a 1.5ml micro centrifuge tube and centrifuge at 12,000 g for 1 min.

3. Remove the medium by aspiration, leaving the pellet as dry as possible.

4. Resuspend the cells in 100 µl of ice-cold 50 mM glucose, 25 mM Tris-HCl pH 8.0, 10 mM EDTA. Pipette up and down to make sure that the pellet is completely suspended.

5. Add 200 µl of freshly prepared 0.2 N NaOH, 1% SDS. Mix by inversion and let sit on ice for 3 min at least.

6. Add 150 µl of ice-cold 3 M NaOAc (pH 5.2). Mix by inversion and leave on ice for 5 min at least.

7. Centrifuge at 12,000 g for 5 min. Transfer the clear supernatant to a fresh tube, avoiding the pellet, which tends to break up easily. Spin again if too much particulate matter remains in the supernatant.

8. Add 400 µl phenol : CIAA (1:1), vortex for 30 s, and centrifuge at 12,000 g for 1 min at room temperature. CIAA is a 24:1 mixture of chlorophorm : isoamyl alcohol.

9. Transfer the top aqueous phase to a fresh tube and add 800 µl ethanol (100%). Vortex and leave it at room temperature for 2 min, and keep at -80°C for 30 min, centrifuge 12,000 g for 5 min.

10. Decant the supernatant and add 400 µl ethanol to the pellet. Spin briefly, pour off the ethanol and allow the pellet to air dry in an inverted position for about 10 min.

11. Resuspend the pellet in 30 µl of TE buffer containing 20 µg/ml Rnase and leave overnight to dissolve the DNA at 4°C. Alternatively, this last step can be done by shaking or vortexing it for 1 h 30 min or 2 h.

Particle preparation
The protocol for gold particle preparation is modeled after Sanford *et al.* (1993).
1. Gold particles, 0.6 or 1.0 microns in diameter (Bio-Rad) are prepared in advance by weighing out 0.5 mg of gold per bombardment, vortexing it in a 1.5 ml microfuge tube containing 70% ethanol for 5 min and incubated for 15 min at room temperature.

2. After centrifugation, the ethanol is aspirated and the gold particles are cleaned with three subsequent dH_2O vortex/centrifuge washes. The gold particle pellets are then suspended in sterile 50% glycerol (50 µl for 3 mg gold) where they can remain at 4°C for up to 2 weeks.

3. On the day of bombardment, a 50 µl aliquot (sufficient for six bombardments) of gold in glycerol is transferred to a sterile microfuge tube.

4. While vortexing vigorously, 5 µl of 1 µg/µl vector DNA, 50 µl of 2.5 M $CaCl_2$ and 20 µl of 0.1 M spermidine are added to the gold. Continue to vortex this mixture for 3 min, then centrifuge to remove the liquid.

5. Two 70% ethanol washes and one 100% ethanol wash are performed without disturbing the pellet before the gold is suspended in 50 µl of 100% ethanol.

Macrocarrier preparation
In a sterile flow bench the gold labeled with the plasmid DNA is then added to the surface of Bio-Rad macrocarrier disks previously inserted into the macrocarrier holders. Continuously vortex the source tube during the removal of each 6 µl aliquot of gold as this is important in order to maximize uniform sampling. Once the macrocarriers are coated with the gold, they are placed into a sterile desiccation chamber, to evaporate the alcohol, and are held there until they are used.

Bombardment preparation
The bombardment parameters are similar to the ones used previously (Jähne *et al.*, 1994; Yao *et al.*, 1997). The parameters are: a 6 to 9 cm target distance, a 10 mm fly distance and a 6 mm gap distance. The vacuum in the chamber is lowered to 67.5 cm Hg and then the gun is fired using either 46 or 71 kg/cm^2 rupture disks.

Selection of transgenic plants

Transgenic embryos or calli can be selected from cultures by using resistance selection, either to herbicides or to an antibiotic resistance marker (Jähne et al., 1994; Yao et al., 1997) or with visual marker genes (Carlson et al., 2001). A visual marker such as GFP (green-fluorescent protein) can decrease the time spent to select transgenic plants. With the GFP marker, fluorescing microspores that undergo multistructure development can be monitored regularly with an epifluorescence microscope. However the fluorescing multicellular structures are often not visible during the first 10 days. Then, about 2-3 weeks after bombardment, the fluorescing structures or embryos can be isolated using a sterilized stereomicroscope with a blue fluorescent filter in a flow bench under dark conditions. The structures are scooped up and moved onto surface of the solidified FHG medium. This procedure optimizes the conditions for developing structures and it is easy to isolate these bigger fluorescing structures. After another 1-2 weeks the fluorescing structures are moved onto the solidified differentiation medium. At this time the structure may be isolated from any non-transgenic ones that came with the original transfer. Once shoots are formed the embryos are moved to regeneration media.

Acknowledgement

This research has been funded by NSERC while OMAFRA has provided both facilities and funding. The protocols are based mainly on the earlier references from our lab that are cited and the work of those authors is greatly appreciated.

References

Carlson, A.R., J. Letarte, J. Chen and K.J. Kasha, 2001. Visual screening of microspore-derived transgenic barley (*Hordeum vulgare* L.) with green-fluorescent protein. Plant Cell Rep. **20**: 331-337.

Jähne, A., D. Becker, R. Brettschneider, and H. Lörz, 1994. Regeneration of transgenic, microspore-derived, fertile barley. Theor.Appl.Genet. **89**: 525-533.

Kasha, K.J., E. Simion R. Oro, Q.A. Yao, T.C. Hu, and A.R. Carlson, 2001. An improved *in vitro* technique for isolated microspore culture of barley. Euphytica. **120**(3): 379-385.

Sanford, J.C., F.D. Smith and J.A. Russell, 1993. Optimizing the biolistic process for different biological applications. Methods in Enzymology **217**: 483-509.

Yao, Q.A., E. Simion, M. William, J. Krochko and K.J. Kasha, 1997. Biolistic transformation of haploid isolated microspores of barley (*Hordeum vulgare* L.). Genome **40**: 570-581.

4.4
Doubled haploids in genetic mapping and genomics

B.P. Forster and W.T.B. Thomas
Scottish Crop Research Institute, Invergowrie, Dundee, DD2 5DA, Scotland, UK

Genetic maps

Understanding gene organization in plants is a key goal of researchers seeking to unravel the complexities of a genotype that lead to a particular phenotype. Similarly, plant breeders need to know the location and relationships between genes so that they can determine how to manipulate linkage blocks in the best way to produce superior recombinants. Up to the 1980's, progress in mapping genes was limited to assembling information on morphological mutants, major disease resistance genes and some protein and isozyme markers. Generally, mapping studies were of relatively few genes in classical F_2, F_3 or backcross populations and results were amalgamated to produce composite morphological maps such as those in barley (Wettstein-Knowles, 1992). Considerable progress was made in this manner but the maps were of limited value as many of the markers were deleterious mutants not found in commercially relevant germplasm.

Early impact of doubled haploids

A small doubled haploid (DH) population was part of a study demonstrating that the radiation-induced dwarfing gene (*ari-e*.GP) found in the barley variety Golden Promise was located on chromosome 5H and linked to the gene *s* for short-haired rachilla (Thomas *et al.*, 1984). There is little other evidence of doubled haploids being used in similar mapping studies although another small (44 DH) population was used to study the segregation of a range of markers in barley (Schon *et al.*, 1990).

The fact that DH could be used to fix genetical variation after a single round of recombination meant that one could compare random inbred populations of F_1 derived DHs with, for example, random inbred populations derived by pedigree inbreeding or single seed descent. When such populations were segregating for a single major gene, the different major gene phenotypes could be compared for a range of other phenotypic traits, particularly economically important ones. If one could demonstrate a significant difference between the two major phenotype means for a trait, then there was positive evidence that the major gene either had a pleiotropic effect upon that trait or was linked to genes controlling the trait. If the association was due to the latter, then the amount of the genetic variation for the trait associated with the major gene should decrease with each round of recombination (Snape and Simpson, 1981). There will therefore be a difference between the amount detected in the F_1DH population and the pedigree inbred or single seed descent (SSD) population. This approach was used in studies of associations of some major genes found in commercial cultivars and has almost exclusively been applied in barley (Table 4.4-1).

Table 4.4-1. Genetic markers found to be associated with quantitative traits through the study of random DH populations from a cross

Crop	Cross	Marker	Trait association	Reference
Barley	Mona x Tysofte Prentice	*3q1*	agronomy, minerals	Kjaer *et al.*, 1991
	Mona x Tysofte Prentice	*6p1*	TGW, minerals	Kjaer *et al.*, 1991
	Mona x Tysofte Prentice	*7p3*	minerals	Kjaer *et al.*, 1991
	Mona x Tysofte Prentice	*Amy1*	minerals	Kjaer *et al.*, 1991
	BH4/143/2 x Ark Royal	*ari-e*.GP	agronomy, yield	Powell *et al.*, 1985b
	Golden Promise x Ark Royal	*ari-e*.GP	agronomy, yield	Powell *et al.*, 1985b
	Golden Promise x Mazurka	*ari-e*.GP	agronomy, yield	Powell *et al.*, 1985b
	TS42/3/5 x Apex	*ari-e*.GP	agronomy, yield, malting quality	Thomas *et al.*, 1991
	Mona x Tysofte Prentice	*Eam*1 (*Ea*)	agronomy, yield, minerals	Kjaer *et al.*, 1991
	Mona x Tysofte Prentice	*Eam*8 (*ea_k*)	agronomy, yield, minerals	Kjaer *et al.*, 1991
	Leger x CI9831	*Est5*	agronomy, yield, grain quality	Jui *et al.*, 1997
	Mona x Tysofte Prentice	*Mla9*	minerals	Kjaer *et al.*, 1991
	Mona x Tysofte Prentice	*Mlk*	minerals	Kjaer *et al.*, 1991
	H3003 x H354-295-2	*mlo5*	agronomy, yield	Bjornstad and Aastveit, 1990
	Riso 5678 x Foma	*mlo5*	agronomy, yield	Kjaer *et al.*, 1990
	Riso 6018 x Foma	*mlo6*	agronomy, yield	Kjaer *et al.*, 1990
	SR7 x Carlsberg II	*mlo10*	agronomy, yield	Kjaer *et al.*, 1990
	Kunlun no 1 x CIMMYT no 6	*nud*	height, yield, grain quality	Choo *et al.*, 2001
	Clipper x Ymer	*Ppd*	agronomy, yield	Powell *et al.*, 1985a
	Kunlun no 1 x CIMMYT no 6	*Raw*1	threshability	Choo *et al.*, 2001
	Leger x CI9831	*Raw*1	agronomy, yield, grain quality	Jui *et al.*, 1997

Crop	Cross	Marker	Trait association	Reference
	Blenheim x E224/3	*Rrn2*	grain quality	Powell *et al.*, 1992
	E224/3 x Blenheim	*Rrn2*	yield, grain quality	Powell *et al.*, 1992
	TS57/72/6 x Keg	*sdw*1	agronomy, yield, grain quality	Thomas *et al.*, 1991
	Universe x Mazurka	*sdw*1	agronomy	Powell *et al.*, 1985a
	Riso 5678 x Foma; Riso 6018 x Foma; SR7 x Carlsberg II	*srh*	TGW	Kjaer *et al.*, 1990
	Leger x CI9831	*srh*	agronomy, yield, grain quality	(Jui et al., 1997)
	Dissa x Sabarlis	*Vrs*1 (*vrs*)	agronomy, yield	Powell *et al.*, 1990
	Leger x CI9831	*Vrs*1 (*vrs*)	grain quality, agronomy	Jui *et al.*, 1997
Wheat	Chinese Spring x Norin 61	*Glu & Rht2*	agronomy	Inagaki and Egawa, 1994

Bulk Segregant Analysis
Bulk Segregant Analysis (BSA) is a form of marker/trait association analysis where deployment of DHs has a number of advantages. A population is phenotyped for a trait of interest and lines at the opposing extremes of the distribution are identified. Lines at high and low extremes are presumed to have all the increasing and decreasing alleles respectively so that a polymorphism between the two bulks of the extremes indicates linkage of the trait with that marker (Michelmore *et al.*, 1991). The technique relies upon accurate matching of genotype and phenotype so the deployment of doubled haploidy is of benefit here. If markers of known map location are used, genetic locations are quickly identified. This, however, is a lengthy process as such markers usually identify a single locus. BSA is much more powerful when combined with marker systems with a high multiplex ratio. Initially RAPDs were used in BSA studies but AFLPs and derivative systems are currently preferred. BSA is a quick method for identifying marker/trait associations and is often used as a prelude for more precise trait mapping in DH or other populations. Alternatively, it can be used to quickly identify markers of potential value in Marker-Assisted Selection schemes, where practical breeders can dispense with the need to know a map location. BSA is most powerful when a trait is controlled by relatively few genes but is of limited value when a trait is controlled by more genes and the population size is relatively small. It can be of value in map construction, however, as one can assemble bulks to try and identify markers that bridge fragmented linkage groups. Studies using DH populations in BSA published in refereed journals are listed in Table 4.4-2. It can be seen that BSA and DHs have been most frequently combined in oilseed rape, where eight crosses have been used to study five traits, and barley, where six crosses have been used to study eight traits. Overall, the list is comparatively short, reflecting the move of researchers to whole genome studies.

Doubled haploidy and molecular marker maps
The advent of molecular markers revolutionised the construction of genetic maps of plant genomes and doubled haploidy has played a major role in facilitating the generation of maps. The first genome wide map utilising a DH population was produced in the barley cross 'Proctor' x 'Nudinka' (Heun *et al.*, 1991), closely followed by another barley map in the cross 'Igri' x 'Franka' (Graner *et al.*, 1991). From the list of crosses that have been studied for QTLs (Table 4.4-5), it can be seen that doubled haploids have been produce to map 56 populations distributed across 8 species. Most (23) have been produced in barley followed by oilseed rape (9), wheat (9) and rice (8). This is an under-estimate as it does not consider DH populations that have been published in un-refereed journals nor DH populations that have not been used for QTL studies. For instance, no QTL studies in the second DH cross to be mapped (Igri x Franka) have been published in a refereed journal.

Table 4.4-2. List of crosses and traits where Bulked Segregant Analysis has been applied to DH populations to identify linked markers

Crop	Cross	Trait	Marker	Reference
Barley	Blenheim x E224/3	milling energy	RAPD	Chalmers et al., 1993
	Blenheim x E224/3	resistance to scald	RAPD, RFLP	Baura et al., 1993
	Harrington x TR306	resistance to loose smut	SSR	Li et al., 2001
	Igri x Franka	resistance to BaMMV and BaYMV	RAPD	Weyen et al., 1996
	Igri x Franka	resistance to BaMMV and BaYMV	RAPD	Ordon et al., 1995
	Ingrid x Pokko	resistance to powdery mildew	RAPD	Manninen et al., 1997
	Leger x CI9831	resistance to net blotch	RAPD	Molnar et al., 2000
	Q21861 x SM89010	resistance to leaf rust	RAPD	Borovkova et al., 199(Borovkova et al., 1997)
	Q21861 x SM89010	resistance to stem rust	RAPD, RFLP	Borovkova et al., 1995
Mustard	J90-4317 x J90-2733	resistance to white rust	RAPD	Prabhu et al., 1998
Oilseed rape	Apollo x YN90-1016	seed coat colour	RAPD, AFLP	Somers et al., 2001
	Apollo x YN90-1016	linoleic acid desaturation	RAPD	Somers et al., 1998
	Darmor x Yudal	erucic acid	RAPD	Jourdren et al., 1996b
	Maluka x Susceptible	resistance to blackleg	RAPD, RFLP	Mayerhofer et al., 1997
	Reston x LL09	linolenic and erucic acid	RAPD	Rajcan et al., 1999
	Shiralee x Susceptible	resistance to blackleg	RAPD, RFLP	Mayerhofer et al., 1997
	Stellar x Drakkar	linolenic acid content	RAPD	Jourdren et al., 1996a
	Westar T5 x Miyuki	T-DNA insertion	RAPD	Baranger et al., 1997
	WW989 x YO	seed colour	RFLP	Vandeynze et al., 1995
Pepper	Yolo Wonder x PM687	resistance to root-knot nematode	RAPD, AFLP	Djian-Caporalino et al., 2001
Rice	IR64 x Azucena	resistance to RYMV	RFLP	Ghesquiere et al., 1997
Tobacco	NC528 x Ky14	resistance to root-knot nematode	RAPD	Yi et al., 1998
Wheat	Ciano x Walter	androgenesis	AFLP	Torp et al., 2001
	Cranbrook x Halberd	dough quality	Range	Cornish et al., 2001

Doubled haploid populations are ideal for genetic mapping. As can be seen from Figure 4.4-1, one can have DH populations available for DNA extraction and mapping 1½ years after the initial cross, i.e. almost as quick as an F_2 or BC_1 population and definitely much faster than a pedigree inbred or SSD population. The advantage of using a DH population is that one can re-grow it and distribute it in seed form so that it is comparatively easy to screen it with a large number of markers. Current technology for screening markers means that one can generate a reasonably dense genetic map of a DH population within 3 years of making the initial cross. Map construction from a DH population derived from the F_1 of a cross is relatively simple because the expected segregation is that of a backcross (Snape, 1976). Given the number of markers necessary to construct a whole genome map, manual map construction is prohibitive but a range of computer software (Table 4.4-3) is available to speed up the process.

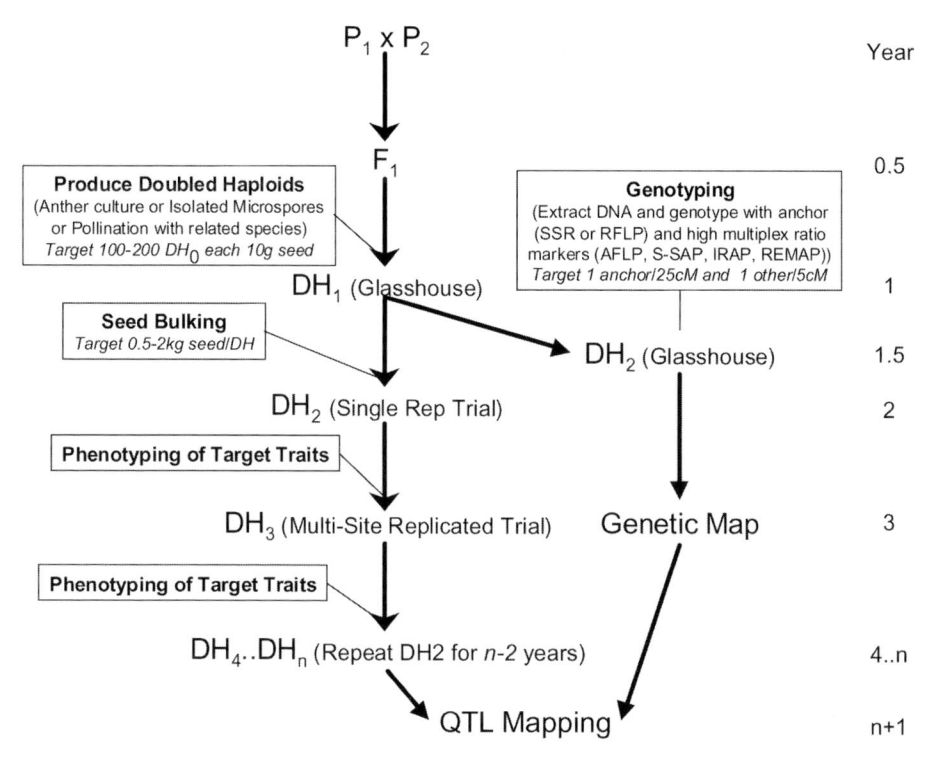

Figure 4.4-1. Outline scheme of the deployment of doubled haploids in marker and QTL mapping.

Table 4.4-3. Software for the construction of genetic maps

Software	Web address	Reference
JOINMAP	http://www.plant.wageningen-ur.nl/default.asp?section=products&page=/products/mapping/joinmap/jmintro.htm	Stam, 1993
GMENDEL	http://www.css.orst.edu/G-mendel/Default.htm	Liu and Knapp, 1990
MAPMAKER	http://www-genome.wi.mit.edu/ftp/distribution/software/mapmaker3/	Lander et al., 1987
MAPMANAGER/ QTX	http://mapmgr.roswellpark.org/mapmgr.html	Manly and Olson, 1999

All rely on examining the pair-wise linkage relationships between markers to form groupings with varying degrees of probability – the higher the probability, the closer the linkage between markers but the less the genome covered by a group. Once one is reasonably satisfied that all the markers within a group at a certain level of probability belong to the same region of the genome one can order the markers within a linkage group. Different software have different approaches to this problem but, provided one exercises some quality control of the mapping process, then reasonably robust maps can be produced by all of them. MAPMAKER and JOINMAP have tended to be the software of choice to construct the genetic maps appearing in refereed journals. JOINMAP has also been used to construct consensus maps from merging different maps of the same species when at least some of the markers on each chromosome are in common (e.g. Langridge et al., 1995; Qi et al., 1996).

Doubled haploidy and QTL mapping
Genes of relatively small effect that do not produce individually recognizable phenotypes control most economically important traits. Progress in locating such genes was limited to the association studies described above before the advent of molecular marker maps. When marker maps became available, it was soon possible to identify Quantitative Trait Loci (QTL) controlling any measurable character to defined regions of a genome. DH populations are again of immense benefit in such studies as one can grow multi-site replicated trials, to derive the most meaningful data, three years after the initial cross (Figure 4.4-1). This timescale would be impossible to achieve with any other population type, making the investment in deploying DHs well worth while. The benefit can be seen in the progress of studies of the Proctor x Nudinka barley population as it was not only the first DH population to be mapped with molecular markers but it was also the first genome wide QTL study of a DH population, just one year after the map was published (Heun, 1992). Similary, the extensive North American Barley Genome Mapping Project published its first map of the 'Steptoe' x 'Morex' DH population (Kleinhofs et al., 1993) shortly before publishing the results of an extensive QTL study with the same cross (Hayes et al., 1993b).

As with genetic map construction, there is a number of software packages for the detection of QTL, most of which can handle data from DH populations (Table 4.4-4). The first package to be widely used was MAPMAKER/QTL, which used an interval mapping technique. This method, later termed Simple Interval Mapping (SIM) was found to give misleading results in some circumstances (Haley and Knott, 1992; Martinez and Curnow, 1992).

Later software packages adopted the use of marker co-factors to account for variation in other portions of the genome in an attempt to refine QTL location and improve the sensitivity of detection. This approach, termed Composite or Compound Interval Mapping (CIM) is now the method of choice and different versions of it can be found in software such as MapQTL, MQTL, PLABQTL and QTL Cartographer.

Table 4.4-4. Software for detecting QTLs

Software	Web address	Reference
MAPL	http://peach.ab.a.u-tokyo.ac.jp/~ukai/	Ukai et al., 1995
MAPMAKER/QTL	http://www-genome.wi.mit.edu/ftp/distribution/software/mapmaker3/	Lander and Botstein, 1989
MAPMANAGER/QTX	http://mapmgr.roswellpark.org/mapmgr.html	Manly and Olson, 1999
MapQTL	http://www.plant.wageningen-ur.nl/default.asp?section=products&page=/products/mapping/mapqtl/mqintro.htm	Jansen et al., 1995
MCQTL	http://www.inra.fr/bia/T/MCQTL/MCQTL.html	Jourjon, 2000
MQTL	http://gnome.agrenv.mcgill.ca/dm/software.htm	Tinker and Mather, 1995
MultiQTL	http://esti.haifa.ac.il/~poptheor/MultiQtl/MultiQtl.htm	Korol et al., 1998
PLABQTL	http://www.uni-hohenheim.de/~ipspwww/soft.html	Bohn et al., 1999
Qgene	http://www.qgene.org/	Nelson, 1997
QTL Cafe	http://web.bham.ac.uk/g.g.seaton/	
QTL Cartographer	http://statgen.ncsu.edu/qtlcart/cartographer.html	Basten et al., 1994

Each package has its strengths and weaknesses and ones choice should be weighted by the type of problem one is trying to solve. For instance, if one wishes to detect the presence of QTL x Environment interactions of the cross-over type, then MQTL is superior to the others. If one still wishes to detect QTL x Environment interactions but wishes to consider just those QTLs that are consistent over environments then PLABQTL and QTL Cartographer have advantages. MapQTL has advantages for mapping data collected from just one environment and also has a routine for detecting QTL in non-parametric data.

We have surveyed refereed publications to determine the extent of the impact of DHs on QTL mapping (Table 4.4-5). We have grouped mapped traits under the following headings: yield, quality, agronomic, developmental, biotic stress, abiotic stress and others. It is clear that in species where an efficient DH protocol is available, most of the mapping populations have been produced by that means. For instance, some 23 different barley DH populations have been studied for QTLs (Table 4.4-5) whereas other populations have not been widely exploited. In contrast, DHs have been used in just one QTL study in maize but different population types have been widely used. In total, 56 DH mapping populations have been used for QTL detection and, after barley, 26 of the remaining 33 are almost equally divided between oilseed rape, rice and wheat. However, none of these crops has been as extensively phenotyped as many of the barley crosses. Only seven traits have been studied in oilseed rape, whereas many more traits have been mapped in the cereals. Again, barley is the dominant crop with over 50 mapped traits (single or groups) compared to nearly 40 for rice and just over 30 for wheat (Table 4.4-5). Many of the wheat studies have been published from 2000 onwards, reflecting the problems in generating maps for the crop, and we expect that the list for all crops will grow further. This growth will, however, depend upon improved DH protocols for many of the crops.

Whilst much information has been accumulated, it is disappointing that little use of it is being made in practical plant breeding programmes. Some examples can be found in the literature where QTLs that have been detected in a mapping population are then being exploited in developing new varieties e.g stripe rust resistance in barley (Toojinda *et al.*, 1998). In general, problems of QTL x Environment interactions and interactions with genetic background mean that plant breeders have not made much use of the information that has been collected. The movement from largely anonymous statistical effects to functional marker maps (see next section) may well lead to a greater take up of the information generated.

Doubled haploids and genomics

Despite all the QTL information that has been accumulated over the past decade, identification of genes affecting quantitative traits remains difficult. The resolution of QTLs is imprecise and the 1 LOD confidence interval is often in the order of 30cM (Kearsey and Farquhar, 1998), preventing map-based cloning. Utilisation of Recombinant Chromosome Substitution Lines (RCSLs) will enable more precise mapping of QTL and DHs can be used to speed up their development both as a genome wide coarse-focus set and a fine-focus set for a target region (Thomas *et al.*, 2000). We expect that the world-wide genomics initiatives currently being undertaken in many crop species around the world will have a major impact in gene identification as they will enable the transition from largely anonymous to functional molecular marker maps (Figure 4.4-2). Mapped ESTs will enable the identification of candidate genes for a trait, which can then be used to relate back to physical maps via BAC libraries. The complete homozygosity and immortality of DH populations will be of great advantage in facilitating this approach. The Igri x Franka DH population has been used to integrate translocation break-points on the genetic map and therefore allow a genome wide comparison of genetic and physical map distance in barley, highlighting large areas of low recombination (Künzel *et al.*, 2000). ESTs from rice have been found to be co-located with major genes and QTLs for resistance to rice blast, bacterial blight and sheath blight in the

IR64 x Azucena DH map (Wang *et al.*, 2001). In addition, anchoring the various BAC libraries that have been developed across the world for a range of crops will enable quite precise targeting of candidate regions for traits of interest. Ultimately, such strategies will lead to tools for allele mining and therefore a targeted means of exploiting germplasm resources.

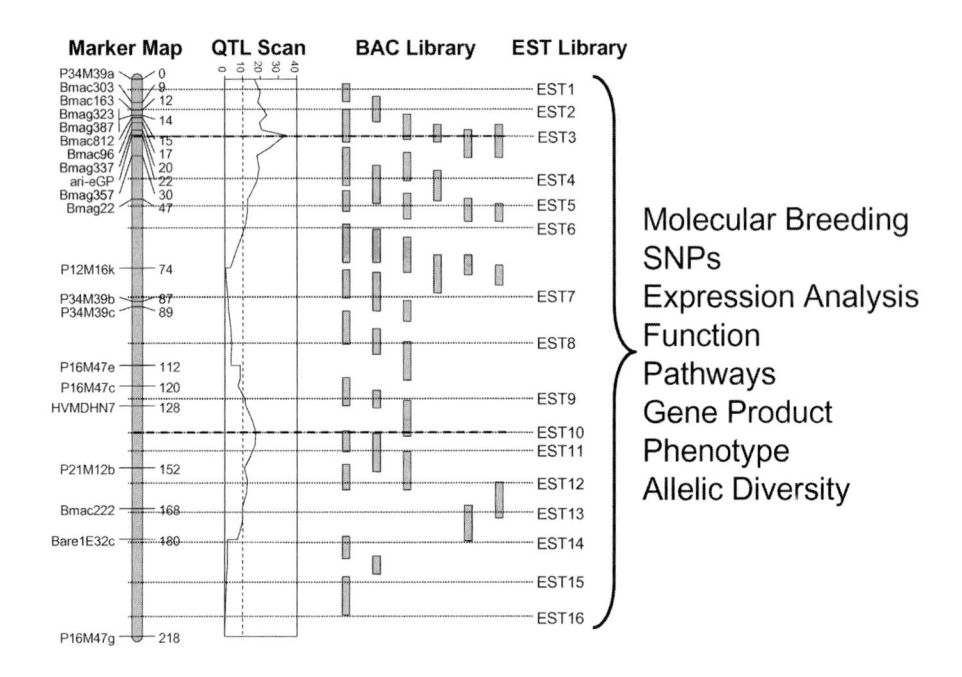

Figure 4.4-2. Moving from anonymous marker and QTL maps to functional maps with candidate genes.

Table 4.4-5. List of Crosses and Traits where DH populations have been used to detect QTLs

Crop	Cross	Main markers	Trait						References
			Yield	Quality	Agronomic	Developmental	Biotic stress	Abiotic stress / Other	
Barley	Blenheim x E224/3	AFLP, RAPD, RFLP	yield, TGW	hardness, malt extract, germination	grain size, specific weight	heading, height	mildew, scald, leaf and stripe rust		Powell et al., 1997; Thomas et al., 1995, 1996
	Blenheim x Kym	RFLP	Components	malt extract, grain nitrogen content		heading time, height			Bezant et al., 1996, 1997a,b
	Calicuchima sib x Bowman	RFLP					stripe rust resistance		Chen et al., 1994
	Chebec x Harrington	RFLP		malt enzymes activity, dormancy	seed length, width and ratio	heading time, habit, height, leaf number	Cereal Cyst Nematode resistance		Langridge et al., 1996
	Chevron x Stander	RFLP					Fusarium head blight resistance		Ma et al., 2000
	Clipper x Sahara3371	RFLP		malt enzyme activity			Cereal Cyst Nematode resistance	boron tolerance	Jefferies et al., 1999; Langridge et al., 1996
	Derkado x B83 -12/21/5	AFLP, SSR, S-SAPs	yield and yield components	malt extract, fermentability		heading time		salt tolerance	Ellis et al. 2002; Hackett et al. 2001; Swanston et al. 1999; Thomas et al. 1998
	Dicktoo x Morex	AFLP, RFLP		malt extract		heading time, photoperiod and vernalisation response, habit		winter hardiness	Hayes et al., 1993a; Karsai et al., 1997, 1999; Oziel et al., 1996; Pan et al., 1994; Zhu et al., 2000
	Galleon x Haruna Nino	RFLP		malt enzyme activity			net blotch resistance		Langridge et al., 1996; Williams et al., 1999
	Gobernadora x CMB643	RFLP				ear morphology, height	Fusarium head blight resistance		Zhu et al., 1999b
	Harrington x Morex	RFLP	Yield	malt extract, grain and malt nitrogen content, soluble	grain size, specific weight	heading time, height			Marquez-Cedillo et al., 2000, 2001

Crop	Cross	Main markers	Trait							References
			Yield	Quality	Agronomic	Developmental	Biotic stress	Abiotic stress	Other	
	Harrington x TR306	RFLP	yield, TGW	malt extract, grain and malt nitrogen content, wort B-glucan and viscosity, malt enzyme activity	lodging, grain size, specific weight	heading, height, maturity time	mildew, aphids, scald, leaf rust, net blotch resistance	deep sowing response		Falak et al., 1999; Kaneko et al. 2001; Mano and Takeda, 1997; Mather et al., 1997; Moharramipour et al., 1997; Sato et al. 2001; Spaner et al. 1999; Takahashi et al., 2001; Tinker et al., 1996
	HOR 1063 x Krona	RFLP	TGW			heading time, height	leaf rust resistance			Kicherer et al. 2000
	Igri x Danilo	RFLP	yield		brackling, ldging, shattering, hight, seed length, width and ratio		mildew, scald resistance			Backes et al., 1995, 1996
	Igri x Triumph	RFLP				heading time, photoperiod response				Laurie et al. 1994, 1995
	Lina x HS92 (Canada Park)	AFLP						salt tolerance		Ellis et al. 1997
	Post x Vixen	AFLP, RAPD, SSR					Barley Yellow Dwarf Virus resistance			Scheurer et al., 2001
	Proctor x Nudinka	RFLP	TGW				mildew, leaf stripe resistance			Heun, 1992; Pecchioni et al., 1996; 1999
	Rolfi x CI9819	IRAP, ISSR, RAPD, REMAP, RFLP, SSR					net blotch resistance			Manninen et al. 2000
	Shyri x Galena	AFLP, RFLP,					leaf and stripe rust, Barley			Toojinda et al., 2000

Crop	Cross	Main markers	Trait						References
			Yield	Quality	Agronomic	Developmental	Biotic stress	Abiotic stress	Other	
	SM73 x SM145	RGAP, SSR					Yellow Dwarf Virus resistance			Zhu et al., 1999a
	Steptoe x Morex	RFLP	yield	grain and malt extract, malt enzyme activity, seed dormancy, grain starch content	lodging, shattering	heading time, height	net and spot blotch, bacterial leaf streak resistance	salt tolerance	tissue culture response, crossability with wheat, callus and shoot growth in culture	Borem et al., 1999; Bregitzer and Campbell, 2001; El Attari et al., 1998; Han et al., 1995, 1996, 1999; Hayes et al., 1993b; Kandemir et al., 2000; Kaneko et al., 2001; Mano et al., 1996; Mano and Takeda, 1997; Romagosa et al., 1996, 1999; Steffenson et al., 1996; Takahashi et al., 2001; Taketa et al., 1998; Ullrich et al., 1997; Zwickert-Menteur et al., 1996 Kjaer et al., 1995; Kjaer and Jensen, 1995, 1996
	Tysofte Prentice x Vogelsanger Gold	RAPD, RFLP	yield components	grain and straw nitrogen and phosphorous content		heading time, height				
Brassicas (Brassica oleracea)	Alboglabra x Italica	RFLP				flowering time, vigour, germination				Bettey et al., 2000; Bohuon et al., 1998
	Cabbage (Bi) x Broccoli (Gr)	AFLP, RFLP, RGAP, SSR					club root resistance			Voorrips et al., 1997
Flax	Fz haploid x Glenelg	AFLP					Fusarium wilt resistance			Spielmeyer et al., 1998
Maize	A188 x DH7	RFLP							androgenesis	Murigneux et al., 1994
	DH5 x DH7	RFLP							androgenesis	Murigneux et al., 1994
	R6 x DH09	RFLP							androgenesis	Murigneux et al., 1994
Oilseed rape	Apollo x YN90	RAPD		linolenic acid content						Somers et al., 1998
	Cresor x Westar	RFLP					blackleg resistance			Dion et al., 1995
	Darmor x Samourai	RAPD, RFLP					blackleg resistance			Pilet et al., 2001

Crop	Cross	Main markers	Trait							References
			Yield	Quality	Agronomic	Developmental	Biotic stress	Abiotic stress	Other	
	Darmor-bzh x Yudal	RAPD, RFLP					blackleg, leaf spot, club root resistance			Manzanares-Dauleux et al., 2000; Pilet et al., 1998a,b, 2001
	Darmor x Yudal	RAPD, RFLP		erucic acid content						Jourdren et al., 1996b
	Major x Stellar	RFLP		erucic and linolenic acid content						Thormann et al., 1996
	Mansholt's HR x Samourai	RFLP, RAPD		glucosinolates, erucic acid content						Ecke et al., 1995; Uzunova et al., 1995
	R54 x Express	AFLP					Turnip Yellow Virus			Dreyer et al., 2001
	Stellar x Major	RFLP		gucosinolate, erucic and linolenic acid content						Thormann et al., 1996; Toroser et al., 1995
Pepper	Perennial x Yolo Wonder	DNA					Cucumber Mosaic Virus, Potato Virus Y, *Phytophora capisci* resistance			Caranta et al., 1997; Lefèbvre and Palloix, 1996
Rice	Akihikari x Koshohikari	RAPD, RFLP						cold tolerance		Takeuchi et al., 2001
	CT9993-5-10-1-M x IR62266-42-6-2	AFLP, RFLP			shoot biomass	root morphology and thickness		drought tolerance, cell membrane stability under drought stress		Kamoshita et al., 2002; Tripathy et al., 2000; Zhang et al., 2001
	Gui630 x 02428	RFLP	yield and yield components			heading time, height				Li et al., 1996
	Gui630 x Taiwanjing	RFLP			number of panicles	heading, growth components				Zhou et al., 2001
	IR64 x Azucena	EST, RAPD, RFLP,		aroma	seed length, width and ratio	panicle and tiller number, height, root morphology, penetration and	rice blast, bacterial and sheath blight,	drought tolerance, salt tolerance, low		Alam and Cohen, 1998; Albar et al., 1998; Cao et al., 2001; Courtois et al., 2000; Fang and

Crop	Cross	Main markers	Trait — Yield	Quality	Agronomic	Developmental	Biotic stress	Abiotic stress	Other	References
		STSs				thickness	Rice Yellow Mosaic Virus, brown planthopper resistance	potassium stress, ferrous ion toxicity		Wu, 2001; Ghesquiere et al., 1997; Hemamalini et al., 2000; Huang et al., 1996; 1997; Liao et al., 2001; Lorieux et al., 1996; Prasad et al., 2000; Wang et al., 2001; Wu et al., 1997; 1998; Yadav et al., 1997; Yan et al., 1998a,b, 1999; Zheng et al., 2000
	IRAT177 x Apura	RFLP					Rice Yellow Mosaic Virus resistance			Ghesquiere et al., 1997
	Miara x C6'	AFLP, SSR							domestication	Bres-Patry et al., 2001
	Zhai-Ye-Qing8 x Jing-Xi17	RFLP, SSR	yield components	paste viscosities, grain quality parameters	grain size and dimensions, specific weight	heading time, height, intermode lengths, morphological index		low temperature germination, cold, salt, low phosphorus and metal ion tolerance	anther culture response	Bao et al., 1999, 2000; Gong et al., 1999; He et al., 1998; He et al., 1999, 2001; Lu et al., 1997; Ming et al., 2000, 2001; Qian et al., 2000a,b; Teng et al., 2001; Yan et al., 2000
Wheat	AC Domain x Haruyutaka	RFLP			seed dormancy					Kato et al., 2001
	CD87 x Katepwa	SSR	milling yield	starch, flour clour			stripe rust resistance			Bariana et al., 2001; Batey et al., 2001a; Mares and Campbell, 2001; Smith et al., 2001
	Chinese Spring x SQ1	RFLP						drought induced ABA		Quarrie et al., 1994
	Ciano x Walter	AFLP							ather culture response	Torp et al., 2001
	Courtot x Chinese Spring	AFLP, RFLP, SSR		bread making; arabinoxylans	ear morphology	heading time, height, photoperiod response			rye cossability	Cadalen et al., 1998; Martinant et al., 1998; Perretant et al., 2000; Sourdille et al., 2000a,b; Tixier et al., 1998
	Cranbrook x Halberd	AFLP, RFLP	milling yield	grain protein, dormancy, grain α-	vgour, competition	coleoptile length, maturity date		boron toxicity		Bariana et al., 2001; Batey et al., 2001b; Coleman et al., 2001; Cornish et al., 2001;

Crop	Cross	Main markers	Trait						References
			Yield	Quality	Agronomic	Developmental	Biotic stress	Abiotic stress Other	
				amylase activity, grain starch content, flour colour, dough properties, grain hardness					Eckermann et al., 2001; Jefferies et al., 2000; Mares and Campbell, 2001; Mares and Mrva, 2001; Mrva and Mares, 2001; Osborne et al., 2001; Rebetzke et al., 2001; Smith et al., 2001
	Fukuho-komugi x Oligo Culm	RAPD					Fusarium head blight resistance		Ban, 2000
	Sicco(CS) x Highbury(CS)	RFLP	yield and yield components			heading time, height			Hyne et al., 1994
	Sunco x Tasman	SSR		noodle colour, grain hardness			stripe rust resistance		Bariana et al. 2001; Mares and Campbell, 2001; Osborne et al., 2001

Acknowledgements
SCRI recieves grant-in-aid from the Scottish Executive Environment and Rural Affairs Department.

References

Alam, S.N., and M.B. Cohen, 1998. Detection and analysis of QTLs for resistance to the brown planthopper, *Nilaparvata lugens*, in a doubled-haploid rice population. Theor. Appl. Genet. **97**: 1370-1379.

Albar, L., M. Lorieux, N. Ahmadi, I. Rimbault, A. Pinel, A.A. Sy, D. Fargette, and A. Ghesquiere, 1998. Genetic basis and mapping of the resistance to rice yellow mottle virus. I. QTLs identification and relationship between resistance and plant morphology. Theor. Appl. Genet. **97**: 1145-1154.

Bakes, G., A. Graner, B. Foroughi-Wehr, G. Fischbeck, G. Wenzel, and A. Jahoor, 1995. Localization of quantitative trait loci (QTL) for agronomic important characters by the use of a RFLP map in barley (*Hordeum vulgare* L.). Theor. Appl. Genet. **90**: 294-302.

Backes, G., G. Schwarz, G. Wenzel, and A. Jahoor, 1996. Comparison between QTL analysis of powdery mildew resistance in barley based on detached primary leaves and on field data. Plant Breed. **115**: 419-421.

Ban, T., 2000. Analysis of quantitative trait loci associated with resistance to fusarium head blight caused by *Fusarium graminearum* Schwabe and of resistance mechanisms in wheat (*Triticum aestivum* L.). Breed. Sci. **50**: 131-137.

Bao, J.S., P. He, Y.W. Xia, Y. Chen, and L.H. Zhu, 1999. Starch RVA profile parameters of rice are mainly controlled by Wx gene. Chinese Sci. Bull. **44**: 2047-2051.

Bao, J.S., X. W. Zheng, Y.W. Xia, P. He, Q.Y. Shu, X. Lu, Y. Chen, and L.H. Zhu, 2000. QTL mapping for the paste viscosity characteristics in rice (*Oryza sativa* L.). Theor. Appl. Genet. **100**: 280-284.

Baranger, A., R. Delourme, N. Foisset, F. Eber, P. Barret, P. Dupuis, M. Renard, and A.M. Chevre, 1997. Wide mapping of a T-DNA insertion site in oilseed rape using bulk segregant analysis and comparative mapping. Plant Breed. **116**: 553-560.

Bariana, H.S., M.J. Hayden, N.U. Ahmed, J.A. Bell, P.J. Sharp, and R.A. McIntosh, 2001. Mapping of durable adult plant and seedling resistances to stripe rust and stem rust diseases in wheat. Austral. J. Agric. Res. **52**: 1247-1255.

Barua, U.M., K.J. Chalmers, C.A. Hackett, W.T.B. Thomas, W. Powell, and R. Waugh, R., 1993. Identification of RAPD markers linked to a *Rhynchosporium secalis* resistance locus in barley using near-isogenic lines and bulked segregant analysis. Heredity. **71**: 177-184.

Basten, C.J., B.S. Weir and Z.-B. Zeng, 1994. Zmap-a QTL cartographer. In: Proceedings of the 5th World Congress on Genetics Applied to Livestock Production: Computing Strategies and Software. Smith, C., J.S. Gavora, B. Benkel, J. Chesnais, W. Fairfull, J.P. Gibson, B.W. Kennedy and E.B. Burnside (Eds). Volume **22**: 65-66. Published by the Organizing Committee, 5th World Congress on Genetics Applied to Livestock Production, Guelph, Ontario, Canada

Batey, I.L., M.J. Hayden, S. Cai, P.J. Sharp, G.B. Cornish, M.K. Morell, and R. Appels, 2001. Genetic mapping of commercially significant starch characteristics in wheat crosses. Austral. J. Agric. Res. **52**: 1287-1296.

Bettey, M., W.E. Finch-Savage, G.J. King, and J.R. Lynn, 2000. Quantitative genetic analysis of seed vigour and pre-emergence seedling growth traits in *Brassica oleracea*. New Phytol. **148**: 277-286.

Bezant, J., D. A. Laurie, N. Pratchett, J. Chojecki, and M. Kearsey, 1996. Marker regression mapping of QTL controlling flowering time and plant height in a spring barley (*Hordeum vulgare* L.) cross. Heredity. **77**: 64-73.

Bezant, J.H., D.A. Laurie, N. Pratchett, J. Chojecki, and M.J. Kearsey, 1997a. Mapping of QTL controlling NIR predicted hot water extract and grain nitrogen content in a spring barley cross using marker-regression. Plant Breed. **116**: 141-145.

Bezant, J., D.A. Laurie, N. Pratchett, J. Chojecki, and M. Kearsey, 1997b. Mapping QTL controlling yield and yield components in a spring barley (*Hordeum vulgare* L.) cross using marker regression. Molec. Breed. **3**: 29-38.

Bjornstad A. and K. Aastveit, 1990. Pleiotropic effects on the *ml-o* mildew resistance gene in barley in different genetic backgrounds. Euphytica **46**: 217-226.

Bohn, M., H.F. Utz, and A.E. Melchinger, 1999. Genetic similarities among winter wheat clutivars determined on the basis of RFLPs, AFLPs, and SSRs and their use for predicting progeny variance. Crop Sci. **39**: 228-237.

Bohuon, E.J.R., L.D. Ramsay, J.A. Craft, A.E. Arthur, D.F. Marshall, D.J. Lydiate, and M.J. Kearsey, 1998. The association of flowering time quantitative trait loci with duplicated regions and candidate loci in *Brassica oleracea*. Genetics. **150**: 393-401.

Borem, A., D.E. Mather, D.C. Rasmusson, R.G. Fulcher, and P.M. Hayes, 1999. Mapping quantitative trait loci for starch granule traits in barley. J. Cereal Sci. **29**: 153-160.

Borovkova, I.G., Y. Jin, B.J. Steffenson, A. Killian, T.K. Blake, and A. Kleinhofs, 1997. Identification and mapping of a leaf rust resistance gene in barley line Q21861. *Genome*. **40**: 236-241.

Borovkova, I.G., B.J. Steffenson, Y. Jin, J.B. Rasmussen, A. Killian, A. Kleinhofs, B.G. Rossnagel, and K. N. Kao, 1995. Identification of molecular markers linked to the stem rust resistance gene *rpg4* in barley. Phytopath. **85**: 181-185.

Bregitzer, P. and R.D. Campbell, 2001. Genetic markers associated with green and albino plant regeneration from embryogenic barley callus. Crop Sci. **41**: 173-179.

Bres-Patry, C., M. Lorieux, G. Clement, M. Bangratz, and A. Ghesquiere, 2001. Heredity and genetic mapping of domestication-related traits in a temperate japonica weedy rice. Theor. Appl. Genet. **102**: 118-126.

Cadalen, T., P. Sourdille, G. Charmet, M.H. Tixier, G. Gay, C. Boeuf, S. Bernard, P. Leroy, and M. Bernard, 1998. Molecular markers linked to genes affecting plant height in wheat using a doubled-haploid population. Theor. Appl. Genet. **96**: 933-940.

Cao, G., J. Zhu, C. He, Y. Gao, J. Yan, and P. Wu, 2001. Impact of epistasis and QTL x environment interaction on the developmental behavior of plant height in rice (*Oryza sativa* L.). Theor. Appl. Genet. **103**: 153-160.

Caranta, C., A. Palloix, V. Lefebvre, and A.M. Daubeze, 1997. QTLs for a component of partial resistance to cucumber mosaic virus in pepper: Restriction of virus installation in host- cells. Theor. Appl. Genet. **94**: 431-438.

Chalmers, K.J., U.M. Barua, C.A. Hackett, W.T.B. Thomas, R. Waugh, and W. Powell, 1993. Identification of RAPD markers linked to genetic factors controlling the milling energy requirement of barley. Theor. Appl. Genet. **87**: 314-320.

Chen, F.Q., D. Prehn, P.M. Hayes, D. Mulrooney, A. Corey, and H. Vivar, 1994. Mapping genes for resistance to barley stripe rust (*Puccinia striiformis* f. sp. *hordei*). Theor. Appl. Genet. **88**: 215-219.

Choo, T.M., K.M. Ho, and R.A. Martin, 2001. Genetic analysis of a hulless x covered cross of barley using doubled-haploid lines. Crop Sci. **41**: 1021-1026.

Coleman, R.D., G.S. Gill, and G.J. Rebetzke, 2001. Identification of quantitative trait loci for traits conferring weed competitiveness in wheat (*Triticum aestivum* L.). Austral. J. Agric. Res. **52**: 1235-1246.

Cornish, G.B., F. Bekes, H.M. Allen, and D.J. Martin, 2001. Flour proteins linked to quality traits in an Australian doubled haploid wheat population. Austral. J. Agric. Res. **52**: 1339-1348.

Courtois, B., G. McLaren, P.K. Sinha, K. Prasad, R. Yadav, and L. Shen, 2000. Mapping QTLs associated with drought avoidance in upland rice. Molec. Breed. **6**: 55-66.

Dion, Y., R.K. Gugel, G.F.W. Rakow, G. Seguin-Swartz, and B.S. Landry, 1995. RFLP mapping of resistance to the blackleg disease [causal agent, *Leptosphaeria maculans* (Desm) Ces, et de Not] in canola (*Brassica napus* L). Theor. Appl. Genet. **91**: 1190-1194.

Djian-Caporalino, C., L. Pijarowski, A. Fazari, M. Samson, L. Gaveau, C. O'Byrne, V. Lefebvre, C. Caranta, A. Palloix, and P. Abad, 2001. High-resolution genetic mapping of the pepper (*Capsicum annuum* L.) resistance loci Me-3 and Me-4 conferring heat-stable resistance to root-knot nematodes (*Meloidogyne* spp.). Theor. Appl. Genet. **103**: 592-600.

Dreyer, F., K. Graichen, and C. Jung, 2001. A major quantitative trait locus for resistance to Turnip Yellows Virus (TuYV, syn. beet western yellows virus, BWYV) in rapeseed. Plant Breed. **120**: 457-462.

Ecke, W., M. Uzunova, and K. Weissleder, 1995. Mapping the genome of rapeseed (*Brassica napus* L). 2. Localization of genes controlling erucic acid synthesis and seed oil content. Theor. Appl. Genet. **91**: 972-977.

Eckermann, P.J., A.P. Verbyla, B.R. Cullis, and R. Thompson, 2001. The analysis of quantitative traits in wheat mapping populations. Austral. J. Agric. Res. **52**: 1195-1206.

El Attari, .H., A. Rebai, P.M. Hayes, G. Barrault, G. Dechamp-Guillaume, and A. Sarrafi, 1998. Potential of doubled-haploid lines and localization of quantitative trait loci (QTL) for partial resistance to bacterial leaf streak (*Xanthomonas campestris* pv. *hordei*) in barley. Theor. Appl. Genet. **96:** 95-100.

Ellis, R.P., B.P. Forster, D.C. Gordon, L.L. Handley, R. Keith, P. Lawrence, R.C. Meyer, W. Powell, D. Robinson, C.M. Scrimgeour, G.R. Young, and W.T.B. Thomas, 2002. Phenotype/genotype associations of yield and salt tolerance in a barley mapping population segregating for two dwarfing genes. J. Ex. Bot. **53:** 1-14.

Ellis, R.P., B.P. Forster, R. Waugh, W.N. Bonar, L.L. Handley, D. Robinson, D.C. Gordon, and W. Powell, 1999. Mapping physiological traits in barley. New Phytol. **137:** 149-157.

Falak, I., D.E. Falk, N.A. Tinker, and D.E. Mather, 1999. Resistance to powdery mildew in a doubled haploid barley population and its association with marker loci. Euphytica. **107:** 185-192.

Fang, P., and P. Wu, 2001. QTL x N-level interaction for plant height in rice (*Oriza sativa* L). Plant and Soil, **236:** 237-242.

Ghesquiere, A., L. Albar, M. Lorieux, N. Ahmadi, D. Fargette, N. Huang, S.R. McCouch, and J.L. Notteghem, 1997. A major quantitative trait locus for rice yellow mottle virus resistance maps to a cluster of blast resistance genes on chromosome 12. Phytopath. **87:** 1243-1249.

Gong, J.M., P. He, Q.A. Qian, L.S. Shen, L.H. Zhu, and S.Y. Chen, 1999. Identification of salt-tolerance QTL in rice (*Oryza sativa* L.). Chinese Sci. Bull. **44:** 68-71.

Graner, A., A. Jahoor, J. Schondelmaier, H. Siedler, K. Pillen, G. Fischbeck, and G. Wenzel, 1991. Construction of an RFLP map of barley. Theor. Appl. Genet. **83:** 250-256.

Hackett, C.A., R.C. Meyer, and W.T.B. Thomas, 2001. Multi-trait QTL mapping in barley using multivariate regression. Genet. Res. Camb. **77:** 95-106.

Haley, C.S. and S.A. Knott, 1992. A simple regression method for mapping quantitative trait loci in line crosses using flanking markers. Heredity. **69:** 315-324.

Han, F., S.E. Ullrich, S. Chirat, S. Menteur, L. Jestin, A. Sarrafi, P.M. Hayes, B.L. Jones, T.K. Blake, D.M. Wesenberg, A. Kleinhofs, and A. Kilian, 1995. Mapping of a beta-glucan content and beta-glucanase activity loci in barley grain and malt. Theor. Appl. Genet. **91:** 921-927.

Han, F., S.E. Ullrich, J.A. Clancy, and I. Romagosa, 1999. Inheritance and fine mapping of a major barley seed dormancy QTL. Plant Sci. **143:** 113-118.

Han, F.S.E. Ullrich, A. Kleinhofs, B.L. Jones, P.M. Hayes and D.M. Wesenberg, 1997. Fine structure mapping of the barley chromosome-1 centromere region containing malting-quality QTLs. Theor. Appl. Genet. **95:** 903-910.

Hayes, P.M., T. Blake, T.H.H. Chen, S. Tragoonrung, F. Chen, A. Pan, and B. Liu, 1993a. Quantitative trait loci on barley (*Hordeum vulgare* L) chromosome 7 associated with components of winter hardiness. Genome. **36:** 66-71.

Hayes, P.M., B.H. Liu, S.J. Knapp, F. Chen, B. Jones, T.K. Blake, J.D. Franckowiak, D.C. Rasmusson, M. Sorrells, S.E. Ullrich, and D. Wesenberg, 1993b. Quantitative trait locus effects and environmental interaction in a sample of North American barley germplasm. Theor. Appl. Genet. **87:** 392-401.

He, P., S.G. Li, Q. Qian, Y.Q. Ma, J.Z. Li, W.M. Wang, Y. Chen, and L.H. Zhu, 1999. Genetic analysis of rice grain quality. Theor. Appl. Genet. **98:** 502-508.

He, P., J.Z. Li, X.W. Zhen, L.S. Shen, C.F. Lu, Y. Chen, and L.H. Zhu, 2001. Comparison of molecular linkage maps and agronomic trait loci between DH and RIL populations derived from the same rice cross. Crop Sci. **41:** 1240-1246.

He, P., L.H. Shen, C.F. Lu, Y. Chen, and L.H. Zhu, 1998. Analysis of quantitative trait loci which contribute to anther culturability in rice (*Oryza sativa* L.). Molec. Breed. **4:** 165-172.

Hemamalini, G.S., H.E. Shasshidhar, and S. Hittalmani, 2000. Molecular marker assisted tagging of morphological and physiological traits under two contrasting moisture regimes at peak vegetative stage in rice (*Oryza sativa* L.). Euphytica. **112:** 69-78.

Heun, M., 1992. Mapping quantitative powdery mildew resistance of barley using a restriction fragment length polymorphism map. Genome. **35:** 1019-1025.

Heun, M., A.E. Kennedy, J.A. Anderson, N.L.V. Lapitan, M.E. Sorrells, and S.D. Tanksley, 1991. Construction of a restriction fragment length polymorphism map for barley (*Horedum vulgare*). Genome. **34:** 437-447.

Huang, N., B. Courtois, G.S. Khush, H.X. Lin, G.L. Wang, P. Wu, and K.L. Zheng, 1996. Association of quantitative trait loci for plant height with major dwarfing genes in rice. Heredity. **77**: 130-137.

Huang, N., A. Parco, T. Mew, G. Magpantay, S. McCouch, E. Guiderdoni, J. C. Xu, P. Subudhi, E. R. Angeles, and G. S. Khush, 1997. RFLP mapping of isozymes, RAPD and QTLs for grain shape, brown plant hopper resistance in a doubled haploid rice population. Molec. Breed. **3**: 105-113.

Hyne, V., M.J. Kearsey, O. Martinez, W. Gang, and J.W. Snape, 1994. A partial genome assay for quantitative trait loci in wheat (*Triticum aestivum*) using different analytical techniques. Theor. Appl. Genet. **89**: 735-741.

Inagaki, M. and Y. Egawa, 1994. Association between high molecular weight subunit genotypes of seed storage glutenin and agronomic traits in a doubled haploid population of bread wheat. Breed. Sci. **44**: 41-45.

Jansen, R.C., J.W. Van Ooijen, P. Stam, C. Lister, and C. Dean, 1995. Genotype-by-environment interaction in genetic mapping of multiple quantitative trait loci. Theor. Appl. Genet. **91**: 33-37.

Jefferies, S.P., A.R. Barr, A. Karakousis, J.M. Kretschmer, S. Manning, K.J. Chalmers, J.C. Nelson, A. K. M. R. Islam, and P. Landridge, 1999. Mapping of chromosome regions conferring boron toxicity tolerance in barley (*Hordeum vulgare* L.). Theor. Appl. Genet. **98**: 1293-1303.

Jefferies, S.P., M.A. Pallotta, J.G. Paull, A. Karakousis, J.M. Kretschmer, S. Manning, A.K.M.R. Islam, P. Landridge, and K.J. Chalmers, 2000. Mapping and validation of chromosome regions conferring boron toxicity tolerance in wheat (*Triticum aestivum*). Theor. Appl. Genet. **101**: 767-777.

Jourdren, C., P. Barret, R. Horvais, R. Delourme, and M. Renard, 1996a. Identification of RAPD markers linked to linolenic acid genes in rapeseed. Euphytica **90**: 351-357.

Jourdren, C., P. Barret, R. Horvais, N. Foisset, R. Delmourne, and M. Renard, 1996b. Identification of RAPD markers linked to the loci controlling erucic acid level in rapeseed. Molec. Breed. **2**: 61-71.

Jourjon, M.F., 2000. MCQTL: Software For Mapping QTL in simple or composite design. *www.intl-pag.org/8/abstracts/pag8060.html*.

Jui, P.Y., T.M. Choo, K.M. Ho, T. Konishi, and R.A. Martin, 1997. Genetic analysis of a two-row x six-row cross of barley using doubled-haploid lines. Theor. Appl. Genet. **94**: 549-556.

Kamoshita, A., J.X. Zhang, J. Siopongco, S. Sarkarung, H.T. Nguyen, and L.J. Wade, 2002. Effects of phenotyping environment on identification of quantitative trait loci for rice root morphology under anaerobic conditions. Crop Sci. **42**: 255-265.

Kandemir, N., D.A. Kudrna, S.E. Ullrich, and A. Kleinhofs, 2000. Molecular marker assisted genetic analysis of head shattering in six-rowed barley. Theor. Appl. Genet. **101**: 203-210.

Kaneko, T., W.S. Zhang, H. Takahashi, K. Ito, and K. Takeda, 2001. QTL mapping for enzyme activity and thermostability of beta- amylase in barley (*Hordeum vulgare* L.). Breed. Sci. **51**: 99-105.

Karsai, I., K. Meszaros, Z. Bedo, P.M. Hayes, A. Pan, and F. Chen, 1997. Genetic analysis of the components of winterhardiness in barley (*Hordeum vulgare* L.). Acta Biol. Hung. **48**: 67-76.

Karsai, I., K. Meszaros, P. Szucs, P.M. Hayes, L. Lang, and Z. Bedo, 1999. Effects of loci determining photoperiod sensitivity (*Ppd-H1*) and vernalization response (*Sh2*) on agronomic traits in the 'Dicktoo' x 'Morex' barley mapping population. Plant Breed. **118**: 399-403.

Kato, K., W. Nakamura, T. Tabiki, H. Miura, and S. Sawada, 2001. Detection of loci controlling seed dormancy on group 4 chromosomes of wheat and comparative mapping with rice and barley genomes. Theor. Appl. Genet. **102**: 980-985.

Kearsey, M.J. and A.G.L. Farquhar, 1998. QTL analysis in plants; where are we now? Heredity, **80**: 137-142.

Kicherer, S., G. Backes, U. Walther, and A. Jahoor, 2000. Localising QTLs for leaf rust resistance and agronomic traits in barley (*Hordeum vulgare* L.). Theor. Appl. Genet. **100**: 881-888.

Kjaer, B., H.P. Jensen, J. Jensen and J.H. Jorgensen, 1990. Associations between three *ml-o* powdery mildew resistance genes and agronomic traits in barley. Euphytica **46**: 185-193.

Kjaer, B., V. Haahr, and J. Jensen, 1991. Associations between 23 quantitative traits and 10 genetic markers in a barley cross. Plant Breed. **106**: 261-274.

Kjaer, B. and J. Jensen, 1995. The inheritance of nitrogen and phosphorus content in barley analysed by genetic markers. Hereditas, **123**: 109-119.

Kjaer, B. and J. Jensen, 1996. Quantitative trait loci for grain yield and yield components in a cross between a six-rowed and a two-rowed barley. Euphytica, **90**: 39-48.

Kjaer, B., J. Jensen, and H. Giese, 1995. Quantitative trait loci for heading date and straw characters in barley. Genome. **38**: 1098-1104.

Kleinhofs, A., A. Kilian, M.A.S. Maroof, R.M. Biyashev, P. Hayes, F.Q. Chen, N. Lapitan, A. Fenwick, T.K. Blake, V. Kanazin, E. Ananiev, L. Dahleen, D. Kudrna, J. Bollinger, S.J. Knapp, B. Liu, M. Sorrells, M. Heun, J.D. Franckowiak, D. Hoffman, R. Skadsen, and B.J. Steffenson, 1993. A molecular, isozyme and morphological map of the barley (*Hordeum vulgare*) genome. Theor. Appl. Genet. **86**: 705-712.

Korol, A.B., Y.I. Ronin, E. Nevo, and P.M. Hayes, 1998. Multi-interval mapping of correlated trait complexes. Heredity. **80**: 273-284.

Künzel, G., L. Korzun, and A. Meister, 2000. Cytologically integrated physical restriction fragment length polymorphism maps for the barley genome based on translocation breakpoints. Genetics. **154**: 397-412.

Lander, E.S. and D. Botstein, 1989. Mapping Mendelian factors underlying quantitative traits using RFLP linkage maps. Genetics. **121**: 185-199.

Lander, E.S., P. Green, J. Abrahamson, A. Barlow, M.J. Daly, S.E. Lincoln, and L. Newburg, 1987. MAPMAKER: an interactive computer package for constructing primary genetic linkage maps of experimental and natural populations. Genomics. **1**: 174-181.

Langridge, P., A. Karakousis, N. Collins, J. Kretschmer, and S. Manning, 1995. A consensus linkage map of barley. Molec. Breed. **1**: 389-395.

Langridge, P., A. Karakousis, J. Kretschmer, S. Manning, K.J. Chalmers, R. Boyd, C.D. Li, R. Islam, S. Logue, R. Lance, and SARDI, 1996 RFLP and QTL analysis of barley mapping populations. *http://greengenes.cit.cornell.edu/WaiteQTL/*

Laurie, D.A., N. Pratchett, J.H. Bezant, and J.W. Snape, 1994. Genetic analysis of a photoperiod response gene on the short arm of chromosome 2(2H) of *Hordeum vulgare* (barley). Heredity. **72**: 619-627.

Laurie, D.A., N. Pratchett, J.H. Bezant, and J.W. Snape, 1995. RFLP mapping of five major genes and eight quantitative trait loci controlling flowering time in a winter x spring barley (*Hordeum vulgare* L.) cross. Genome. **38**: 575-585.

Lefebvre, V., and A. Palloix, 1996. Both epistatic and additive effects of QTLs are involved in polygenic induced resistance to disease: A case study, the interaction pepper - *Phytophthora capsici* Leonian. Theor. Appl. Genet. **93**: 503-511.

Li, C.D., P.E. Eckstein, M.Y. Lu, B.G. Rossnagel, and G.J. Scoles, 2001. Targeted development of a microsatellite marker associated with a true loose smut resistance gene in barley (*Hordeum vulgare* L.). Molec. Breed. **8**: 235-242.

Li, P., C.F. Lu, K.D. Zhou, Y. Chen, R.D. Li, and L.H. Zhu, 1996. RFLP mapping of major and minor genes for important agronomic characters in rice. Science in China Series C-Life Sciences, **39**: 440-448.

Liao, C.Y., P. Wu, B. Hu, and K.K. Yi, 2001. Effects of genetic background and environment on QTLs and epistasis for rice (*Oryza sativa* L.) panicle number. Theor. Appl. Genet. **103**: 104-111.

Liu, B.H. and S. Knapp, 1990. GMENDEL: a program for Mendelian segregation and linkage analysis of individual or multiple progeny populations using log-likelihood ratios. J. Heredity **81**: 407.

Lorieux, M., M. Petrov, N. Huang, E. Guiderdoni, and A. Ghesquiere, 1996. Aroma in rice: genetic analysis of a quantitative trait. Theor. Appl. Genet. **93**: 1145-1151.

Lu, C.F., L.S. Shen, Z.B. Tan, Y.B. Xu, P. He, Y. Chen, and L.H Zhu, 1997. Comparative mapping of QTLs for agronomic traits of rice across environments by using a doubled-haploid population. Theor. Appl. Genet. **94**: 145-150.

Ma, Z.P., B.J. Steffenson, L.K. Prom, and N.L. Lapitan, 2000. Mapping of quantitative trait loci for Fusarium head blight resistance in barley. Phytopath. **90**: 1079-1088.

Manly, K.F., and J.M. Olson, 1999. Overview of QTL mapping software and introduction to Map Manager QT. Mam. Genome. **10**: 327-334.

Manninen, O., R. Kalendar, J. Robinson, and A.H. Schulman, 2000. Application of BARE-1 retrotransposon markers to the mapping of a major resistance gene for net blotch in barley. Molec. Gen. Genet. **264**: 1-21.

Manninen, O.M., T. Turpeinen, and E. Nissila, 1997. Identification of RAPD markers closely linked to the *mlo*-locus in barley. Plant Breed. **116:** 461-464.

Mano, Y., H. Takahashi, K. Sato, and K. Takeda, 1996. Mapping genes for callus growth and shoot regeneration in barley (*Hordeum vulgare* L). Breed. Sci. **46:** 137-142.

Mano, Y. and K. Takeda, 1997. Mapping quantitative trait loci for salt tolerance at germination and the seedling stage in barley (*Hordeum vulgare* L). Euphytica. **94:** 263-272.

Manzanares-Dauleux, M.J., R. Delourme, F. Baron, and G. Thomas, 2000. Mapping of one major gene and of QTLs involved in resistance to clubroot in *Brassica napus*. Theor. Appl. Genet. **101,** 885-891.

Mares, D.J. and A.W. Campbell, 2001. Mapping components of flour and noodle colour in Australian wheat. Austral. J. Agric. Res. **52:** 1297-1309.

Mares, D.J. and K. Mrva, 2001. Mapping quantitative trait loci associated with variation in grain dormancy in Australian wheat. Austral. J. Agric. Res. **52:** 1257-1265.

Marquez-Cedillo, L.A., P.M. Hayes, B.L. Jones, A. Kleinhofs, W.G. Legge, B.G. Rossnagel, K. Sato, E. Ullrich, and D.M. Wesenberg, 2000. QTL analysis of malting quality in barley based on the doubled- haploid progeny of two elite North American varieties representing different germplasm groups. Theor. Appl. Genet. **101:** 173-184.

Marquez-Cedillo, L.A., P.M. Hayes, Kleinhofs, A., W.G. Legge, B.G. Rossnagel, K. Sato, S.E. Ullrich, and D.M. Wesenberg, 2001. QTL analysis of agronomic traits in barley based on the doubled haploid progeny of two elite North American varieties representing different germplasm groups. Theor. Appl. Genet. **103:** 625-637.

Martinant, J.P., T. Cadalen, A. Billot, S. Chartier, P. Leroy, M. Bernard, L. Saulnier, and G. Branlard, 1998. Genetic analysis of water extractable arabinoxylans in bread wheat endosperm. Theor. Appl. Genet. **97:** 1069-1075.

Martinez, O., and R.N. Curnow, 1992. Estimating the locations and the sizes of the effects of quantitative trait loci using flanking markers. Theor. Appl. Genet. **85:** 480-488.

Mather, D.E., N.A. Tinker, D.E. Laberge, M. Edney, B.L. Jones, B.G. Rossnagel, W.G. Legge, K.G. Briggs, R.B. Irvine, D.E. Falk, and K.J. Kasha, 1997. Regions of the genome that affect grain and malt quality in a North American two-row barley cross. Crop Sci. **37:** 544-554.

Mayerhofer, R., V.K. Bansal, M.R. Thiagarajah, G.R. Stringham, and A.G. Good, 1997. Molecular mapping of resistance to *Leptosphaeria maculans* in Australian cultivars of *Brassica napus*. Genome. **40:** 294-301.

Michelmore, R.W., I. Paran, and R.V. Kesseli, 1991. Identification of markers linked to disease resistance genes by bulked segregant analysis – a rapid method to detect markers in specific genomic regions by using segregating. PNAS. **88:** 9828-9832.

Ming, F., X.W. Zheng, G.H. Mi, P. He, L.H. Zhu, and F.S. Zhang, 2000. Identification of quantitative trait loci affecting tolerance to low phosphorus in rice (*Oryza Sativa* L.). Chinese Sci. Bull. **45:** 520-525.

Ming, F., X.W. Zheng, G.H. Mi, L.H. Zhu, and F.S. Zhang, 2001. Detection and verification of quantitative trait loci affecting tolerance to low phosphorus in rice. J. Plant Nutr. **24:** 1399-1408.

Moharramipour, S., H. Tsumuki, K. Sato, and H. Yoshida, 1997. Mapping resistance to cereal aphids in barley. Theor. Appl. Genet. **94:** 592-596.

Molnar, S.J., L.E. James, and K.J. Kasha, 2000. Inheritance and RAPD tagging of multiple genes for resistance to net blotch in barley. Genome. **43:** 224-231.

Mrva, K., and D.J. Mares, 2001. Quantitative trait locus analysis of late maturity alpha- amylase in wheat using the doubled haploid population Cranbrook x Halberd. Austral. J. Agric. Res. **52:** 1267-1273.

Murigneux, A., S. Bentolila, T. Hardy, S. Baud, C. Guitton, H. Jullien, S. Ben Tahar, G. Freyssinet, and M. Beckert, 1994. Genotypic variation of quantitative trait loci controlling *in vitro* androgenesis in maize. Genome. **37:** 970-976.

Nelson, J.C., 1997. QGENE: software for marker-based genomic analysis and breeding. Molec. Breed. **3:** 239-245.

Ordon, F., E. Bauer, W. Friedt, and A. Graner. 1995. Marker-based selection for the *ym4* BaMMV-resistance gene in barley using RAPDs. Agronomie. **15:** 481-485.

Osborne, B.G., K.M. Turnbull, R.S. Anderssen, S. Rahman, P.J. Sharp, and R. Appels, 2001. The hardness locus in Australian wheat lines. Austral. J. Agric. Res. **52:** 1275-1286.

Oziel, A., P.M. Hayes, F.Q. Chen, and B. Jones, 1996. Application of quantitative trait locus mapping to the development of winter-habit malting barley. Plant Breed. **115:** 43-51.
Pan, A., P.M. Hayes, F. Chen, T.H.H. Chen, T. Blake, S. Wright, I. Karsai, and Z. Bedo, 1994. Genetic analysis of the components of winterhardiness in barley (*Hordeum vulgare* L). Theor. Appl. Genet. **89:** 900-910.
Pecchioni, N., P. Faccioli, H. Toubiarahme, G. Vale, and V. Terzi, 1996. Quantitative resistance to barley leaf stripe (*Pyrenophora graminea*) is dominated by one major locus. Theor. Appl. Genet. **93:** 97-101.
Pecchioni, N., G. Vale, H. Toubia-Rahme, P. Faccioli, V. Terzi, and G. Delogu, 1999. Barley-*Pyrenophora graminea* interaction: QTL analysis and gene mapping. Plant Breed. **118:** 29-35.
Perretant, M.R., T. Cadalen, G. Charmet, P. Sourdille, P. Nicolas, C. Boeuf, M.H. Tixier, G. Branlard, S. Bernard, and M. Bernard, 2000. QTL analysis of bread-making quality in wheat using a doubled haploid population. Theor. Appl. Genet. **100:** 1167-1175.
Pilet, M.L., R. Delourme, N. Foisset, and M. Renard, 1998b. Identification of loci contributing to quantitative field resistance to blackleg disease, causal agent *Leptosphaeria maculans* (Desm.) Ces. et de Not., in Winter rapeseed (*Brassica napus* L.). Theor. Appl. Genet. **96:** 23-30.
Pilet, M.L., R. Delourme, N. Foisset, and M. Renard, 1998a. Identification of QTL involved in field resistance to light leaf spot (*Pyrenopeziza brassicae*) and blackleg resistance (*Leptosphaeria maculans*) in winter rapeseed (*Brassica napus* L.). Theor. Appl. Genet. **97:** 398-406.
Pilet, M.L., G. Duplan, H. Archipiano, P. Barret, C. Baron, R. Horvais, X. Tanguy, M.O. Lucas, M. Renard, and R. Delourme, 2001. Stability of QTL for field resistance to blackleg across two genetic backgrounds in oilseed rape. Crop Sci. **41:** 197-205.
Powell, W., P.D.S. Caligari, W.T.B. Thomas, and J.L. Jinks, 1985a. The effects of major genes on quantitatively varying characters in barley. 2. The *denso* and daylength response loci. Heredity. **54:** 349-352.
Powell, W., R.P. Ellis, and W.T.B. Thomas, 1990. The effects of major genes on quantitatively varying characters in barley. 3. The 2 row 6 locus (V-v). Heredity. **65:** 259-264.
Powell, W., W.T.B. Thomas, E. Baird, P. Lawrence, A. Booth, B. Harrower, J.W. McNicol, and R. Waugh, 1997. Analysis of quantitative traits in barley by the use of amplified fragment length polymorphisms. Heredity **79:** 48-59.
Powell, W., W.T.B. Thomas, P.D.S. Caligari, and J.L. Jinks, 1985b. The effects of major genes on quantitatively varying characters.1. The GP-ert locus. Heredity. **54:** 343-348.
Powell, W., W.T.B. Thomas, D.M. Thompson, J.S. Swanston, and R. Waugh, 1992. Association between rDNA alleles and quantitative traits in doubled haploid populations of barley. Genetics. **130:** 187-194.
Prabhu, K.V., D.J. Somers, G. Rakow, and R.K. Gugel, 1998. Molecular markers linked to white rust resistance in mustard *Brassica juncea*. Theor. Appl. Genet. **97:** 865-870.
Prasad, S.R., P.G. Bagali, S. Hittalmani, and H.E. Shashidhar, 2000. Molecular mapping of quantitative trait loci associated with seedling tolerance to salt stress in rice (*Oryza sativa* L.). Curr. Sci. **78:** 162-164.
Pressoir, G., L. Albar, N. Ahmadi, I. Rimbault, M. Lorieux, D. Fargette, and A. Ghesquiere, 1998. Genetic basis and mapping of the resistance to rice yellow mottle virus. II. Evidence of a completely epistasis between two QTLs. Theor. Appl. Genet. **97:** 1155-1161.
Qi, X.Q., P. Stam, and P. Lindhout, 1996. Comparison and integration of four barley genetic maps. Genome. **39:** 379-394.
Qian, Q., P. He, X.W. Zheng, Y. Chen, and L.H. Zhu, 2000a. Genetic analysis of morphological index and its related taxonomic traits for classification of indica/japonica rice. Science in China Series C-Life Sciences. **43:** 113-119.
Qian, Q., D.L. Zeng, P. He, X.W. Zheng, Y. Chen, and L.H. Zhu, 2000b. QTL analysis of the rice seedling cold tolerance in a double haploid population derived from anther culture of a hybrid between *indica* and *japonica* rice. Chinese Sci. Bull. **45:** 448-453.
Quarrie, S.A., M. Gulli, C. Calestani, A. Steed, and N. Marmiroli, 1994. Location of a gene regulating drought-induced abscisic acid production on the long arm of chromosome 5A of wheat. Theor. Appl. Genet. **89:** 794-800.
Rajcan, I., K.J. Kasha, L.S. Kott, and W.D. Beversdorf, 1999. Detection of molecular markers associated with linolenic and erucic acid levels in spring rapeseed (*Brassica napus* L.). Euphytica. **105:** 173-181.

Rebetzke, G.J., R. Appels, A.D. Morrison, R.A. Richards, G. McDonald, M.H. Ellis, W. Spielmeyer, and D. G. Bonnett, 2001. Quantitative trait loci on chromosome 4B for coleoptile length and early vigour in wheat (*Triticum aestivum* L.). Austral. J. Agric. Res. **52**: 1221-1234.

Romagosa, I., F. Han, J.A. Clancy, and S.E. Ullrich, 1999. Individual locus effects on dormancy during seed development and after ripening in barley. Crop Sci. **39**: 74-79.

Romagosa, I., S.E. Ullrich, F. Han, and P.M. Hayes, 1996. Use of the additive main effects and multiplicative interaction model in QTL mapping for adaptation in barley. Theor. Appl. Genet. **93**: 30-37.

Sato, K., T. Inukai, and P.M. Hayes, 2001. QTL analysis of resistance to the rice blast pathogen in barley (*Hordeum vulgare*). Theor. Appl. Genet. **102**: 916-920.

Scheurer, K.S., W. Friedt, W. Huth, R. Waugh, and F. Ordon, 2001. QTL analysis of tolerance to a German strain of BYDV-PAV in barley (*Hordeum vulgare* L.). Theor. Appl. Genet. **103**: 1074-1083.

Schon, C., M. Sanchez, T. Blake, and P.M. Hayes, 1990. Segregation of Mendelian markers in doubled haploid and F_2 progeny of a barley cross. Hereditas. **113**: 69-72.

Smith, A.B., B.R. Cullis, R. Appels, A.W. Campbell, G.B. Cornish, D. Martin, and H.M. Allen, 2001. The statistical analysis of quality traits in plant improvement programs with application to the mapping of milling yield in wheat. Austral. J. Agric. Res. **52**: 1207-1219.

Snape, J.W., 1976. A theoretical comparison of diploidised haploid and single seed descent populations. Heredity. **36**: 275-277.

Snape, J.W. and E. Simpson, 1981. The genetic expectations of doubled haploid lines derived from different filial generations. Theor. Appl. Genet. **60**: 123-128.

Somers, D.J., K.R.D. Friesen, and G. Rakow, 1998. Identification of molecular markers associated with linoleic acid desaturation in *Brassica napus*. Theor. Appl. Genet. **96**: 897-903.

Somers, D.J., G. Rakow, V.K. Prabhu, and K. R.D. Friesen, 2001. Identification of a major gene and RAPD markers for yellow seed coat colour in *Brassica napus*. Genome. **44**: 1077-1082.

Sourdille, P., J.W. Snape, T. Cadalen, G. Charmet, N. Nakata, S. Bernard, and M. Bernard, 2000a. Detection of QTLs for heading time and photoperiod response in wheat using a doubled-haploid population. Genome. **43**: 487-494.

Sourdille, P., M.H. Tixier, G. Charmet, G. Gay, T. Cadalen, S. Bernard, and M. Bernard, 2000b. Location of genes involved in ear compactness in wheat (*Triticum aestivum*) by means of molecular markers. Molec. Breed. **6**: 247-255.

Spaner, D., B.G. Rossnagel, W.G. Legge, G.J. Scoles, P.E. Eckstein, G.A. Penner, N.A. Tinker, K.G. Briggs, D.E. Falk, J.C. Afele, P.M. Hayes, and D.E. Mather, 1999. Verification of a quantitative trait locus affecting agronomic traits in two-row barley. Crop Sci. **39**: 248-252.

Spielmeyer, W., A.G. Green, D. Bittisnich, N. Mendham, and E.S. Lagudah, 1998. Identification of quantitative trait loci contributing to Fusarium wilt resistance on an AFLP linkage map of flax (*Linum usitatissimum*). Theor. Appl. Genet. **97**: 633-641.

Stam, P., 1993. Construction of integrated genetic linkage maps by means of a new computer package -Joinmap. Plant J. **3**: 739-744.

Steffenson, B.J., P.M. Hayes, and A. Kleinhofs, 1996. Genetics of seedling and adult plant resistance to net blotch (*Pyrenophora teres f teres*) and spot blotch (*Cochliobolus sativus*) in barley. Theor. Appl. Genet. **92**: 552-558.

Swanston, J.S., W.T.B. Thomas, W. Powell, G.R. Young, P.E. Lawrence, L. Ramsay, and R. Waugh, 1999. Using molecular markers to determine barleys most suitable for malt whisky distilling. Molec. Breed. **5**: 103-109.

Takahashi, H., K. Sato, and K. Takeda, 20 01. Mapping genes for deep-seeding tolerance in barley. Euphytica. **122**: 37-43.

Taketa, S., H. Takahashi, and K. Takeda, 1998. Genetic variation in barley of crossability with wheat and its quantitative trait loci analysis. Euphytica. **103**: 187-193.

Takeuchi, Y., H. Hayasaka, B. Chiba, I. Tanaka, T. Shimano, M. Yamagishi, K. Nagano, T. Sasaki, and M. Yano, 2001. Mapping quantitative trait loci controlling cool-temperature tolerance at booting stage in temperate japonica rice. Breed. Sci. **51**: 191-197.

Teng, S., D.L. Zeng, Q. Qian, K. Yasufumi, D.I. Huang, and L.H. Zhu, 2001. QTL analysis of rice low temperature germinability. Chinese Sci. Bull. **46**: 1800-1804.

Thomas, W.T.B., E. Baird, J.D. Fuller, P. Lawrence, G.R. Young, J. Russell, L. Ramsay, R. Waugh, and W. Powell, 1998. Identification of a QTL decreasing yield in barley linked to *Mlo* powdery mildew resistance. Molec. Breed. **4**: 381-393.

Thomas, W.T.B., A.C. Newton, A. Wilson, A. Booth, M. Macaulay, and R. Keith, 2000. Development of recombinant chromosome substitution lines - a barley resource. SCRI Ann. Rep. 1999/2000, 99-100.

Thomas, W.T.B., W. Powell, and J.S. Swanston, 1991. The effects of major genes on quantitatively varying characters in barley. 4. The *GPert* and *denso* loci and quality characters. Heredity. **66**: 381-389.

Thomas, W.T.B., W. Powell, J.S. Swanston, R.P. Ellis, K.J. Chalmers, U.M. Barua, P. Jack, V. Lea, B.P. Forster, R. Waugh, and D.B. Smith, 1996. Quantitative trait loci for germination and malting quality characters in a spring barley cross. Crop Sci. **36**: 265-273.

Thomas, W.T.B., W. Powell, R. Waugh, K.J. Chalmers, U.M. Barua, P. Jack, V. Lea, B.P. Forster, J.S. Swanston, R.P. Ellis, P.R. Hanson, and R.C.M. Lance, 1995. Detection of quantitative traits loci for agronomic, yield, grain and disease characters in spring barley (*Hordeum vulgare* L.). Theor. Appl. Genet. **91**: 1037-1047.

Thomas, W.T.B., W. Powell, and W. Wood, 1984. The chromosomal location of the dwarfing gene present in the spring barley variety Golden Promise. Heredity. **53**: 177-183.

Thormann, C.E., J. Romero, J. Mantet, and T.C. Osborn, 1996. Mapping loci controlling the concentrations of erucic and linolenic acids in seed oil of *Brassica napus* L. Theor. Appl. Genet. **93**, 282-286.

Tinker, N.A., and D.E. Mather, 1995. MQTL: software for simplified composite interval mapping of QTL in multiple environments. J. Agric. Genomics. **1** http://www.ncgr.org/ag/jag/papers95/paper295/indexp295.html

Tinker, N.A., D.E. Mather, B.G. Rossnagel, K J. Kasha, A. Kleinhofs, P.M. Hayes, D.E. Falk, T. Ferguson, L.P. Shugar, W.G. Legge, R.B. Irvine, T.M. Choo, K.G. Briggs, S.E. Ullrich, J.D. Franckowiak, T.K. Blake, R.J. GRraf, S.M. Dofing, M.A.S. Maroof, G.J. Scoles, D. Hoffman, L.S. Dahleen, A. Kilian, F. Chen, R.M. Biyashev, D.A. Kudrna, and B.J. Steffenson, 1996. Regions of the genome that affect agronomic performance in two-row barley. Crop Sci. **36**: 1053-1062.

Tixier, M.H., P. Sourdille, G. Charmet, G. Gay, C. Jaby, T. Cadalen, S. Bernard, P. Nicolas, and M. Bernard, 1998. Detection of QTLs for crossability in wheat using a doubled haploid population. Theor. Appl. Genet. **97**: 1076-1082.

Toojinda, T., E. Baird, A. Booth, L. Broers, P. Hayes, W. Powell, W. Thomas, H. Vivar, and G. Young, 1998. Introgression of quantitative trait loci (QTLs) determining stripe rust resistance in barley: an example of marker-assisted line development. Theor. Appl. Genet. **96**: 123-131.

Toojinda, T., L.H. Broers, X.M. Chen, P.M. Hayes, A. Kleinhofs, J. Korte, D. Kudrna, H. Leung, R.F. Line, W. Powell, L. Ramsay, H. Vivar, and R. Waugh, 2000. Mapping quantitative and qualitative disease resistance genes in a doubled haploid population of barley (*Hordeum vulgare*). Theor. Appl. Genet. **101**: 580-589.

Toroser, D., C.E. Thormann, T.C. Osborn, and R. Mithen, 1995. RFLP mapping of mapping of quantitative trait loci controlling seed aliphatic-glucosinolate content in oilseed rape (*Brassica napus* L). Theor. Appl. Genet. **91**: 802-808.

Torp, A.M., A.L. Hansen, and S.B. Andersen, 2001. Chromosomal regions associated with green plant regeneration in wheat (*Triticum aestivum* L.) anther culture. Euphytica. **119**: 377-387.

Tripathy, J.N., J. Zhang, S. Robin, T.T. Nguyen, and H.T. Nguyen, 2000. QTLs for cell-membrane stability mapped in rice (*Oryza sativa* L.) under drought stress. Theor. Appl. Genet. **100**: 1197-1202.

Ukai, Y., R. Ohsawa, A. Saito, and T. Hayashi, 1995. MAPL - A package of computer programs for construction of DNA polymorphism linkage maps and analysis of QTL. Breed. Sci. **45**: 139-142.

Ullrich, S.E., F. Han, and B.L. Jones, 1997. Genetic complexity of the malt extract trait in barley suggested by QTL analysis. J. Amer. Soc. Brew. Chem. **55**: 1-4.

Uzunova, M., W. Ecke, K. Weissleder, and G. Robbelen, 1995. Mapping the genome of rapeseed (*Brassica napus* L). 1. Construction of an RFLP linkage map and localization of QTLs for seed glucosinolate content. Theor. Appl. Genet. **90**: 194-204.

Vandeynze, A.E., B.S. Landry, and K.P. Pauls, 1995. The identification of restriction-fragment-length-polymorphisms linked to seed color genes in *Brassica napus*. Genome. **38**: 534-542.

Voorrips, R.E., M.C. Jongerius, and H.J. Kanne, 1997. Mapping of two genes for resistance to clubroot (*Plasmodiophora brassicae*) in a population of doubled haploid lines of *Brassica oleracea* by means of RFLP and AFLP markers. Theor. Appl. Genet. **94**: 75-82.

Wang, Z., G. Taramino, D. Yang, G. Liu, S.V. Tingey, G. H. Miao, and G. L. Wang, 2001. Rice ESTs with disease-resistance gene- or defense-response gene-like sequences mapped to regions containing major resistance genes or QTLs. Molec. Genet. Genom. **265**: 302-310.

Wettstein-Knowles, P.V., 1992. Cloned and mapped genes: current status. In: Barley: genetics, biochemistry, molecular biology and biotechnology. Shewry, P.R. (Ed.), 73-98.

Weyen, J., E. Bauer, E., A. Graner, W. Friedt, and F. Ordon, 1996. RAPD-mapping of the distal portion of chromosome 3 of barley, including the BaMMV/BaYMV resistance gene *ym4*. Plant Breed. **115**: 285-287.

Williams, K.J., A. Lichon, P. Gianquitto, J.M. Kretscmer, A. Karakousis, S. Manning, P. Langridge, and H. Wallwork, 1999. Identification and mapping of a gene conferring resistance to the spot form of net blotch (*Pyrenophora teres f maculata*) in barley. Theor. Appl. Genet. **99**: 323-327.

Wu, P., A. Luo, J. Zhu, J. Yang, N. Huang, and D. Senadhira, 1997. Molecular markers linked to genes underlying seedling tolerance for ferrous iron toxicity. Plant Soil. **196**: 317-320.

Wu, P., J.J. Ni, and A.C. Luo, 1998. QTLs underlying rice tolerance to low-potassium stress in rice seedlings. Crop Sci. **38**: 1458-1462.

Yadav, R., B. Courtois, N. Huang, and G. McLaren, 1997. Mapping genes controlling root morphology and root distribution in a doubled-haploid population of rice. Theoret. Appl. Genet. **94**, 619-632.

Yan, J.Q., J. Zhu, C.X. He, M. Benmoussa, and P. Wu, 1998a. Molecular dissection of developmental behavior of plant height in rice (*Oryza sativa* L.). Genetics **150**: 1257-1265.

Yan, J.Q., J. Zhu, C.X. He, M. Benmoussa, and P. Wu, 1998b. Quantitative trait loci analysis for the developmental behavior of tiller number in rice (*Oryza sativa* L.). Theor. Appl. Genet. **97**: 267-274.

Yan, J.Q., J. Zhu, C.X. He, M. Benmoussa, and P. Wu, 1999. Molecular marker-assisted dissection of genotype x environment interaction for plant type traits in rice (*Oryza sativa* L.). Crop Sci. **39**: 538-544.

Yan, L.J., C.G. Yi, and Q. Wei, 2000. Effects of metal ions on recombinant calcineurin A subunit. Chinese Sci. Bull. **45**: 453-456.

Yi, H.Y., R.C. Rufty, E. A. Wernsman, and M. C. Conkling, 1998. Mapping the root-knot nematode resistance gene (*Rk*) in tobacco with RAPD markers. Plant Dis. **82**: 1319-1322.

Zhang, J., H.G. Zheng, A. Aarti, G. Pantuwan, T.T. Nguyen, J.N. Tripathy, A.K. Sarial, S. Robin, R.C. Babu, B.D. Nguyen, S. Sarkarung, A. Blum, and H.T. Nguyen, 2001. Locating genomic regions associated with components of drought resistance in rice: comparative mapping within and across species. Theor. Appl. Genet. **103**: 19-29.

Zheng, H.G., R.C. Babu, M.S. Pathan, L. Ali, N. Huang, B. Courtois, and H.T. Nguyen, 2000. Quantitative trait loci for root-penetration ability and root thickness in rice: Comparison of genetic backgrounds. Genome. **43**: 53-61.

Zhou, Y., W. Li, W. Wu, Q. Chen, D, Mao, and A.J. Worland, 2001. Genetic dissection of heading time and its components in rice. Theor. Appl. Genet. **102**: 1236-1242.

Zhu, B., D.W. Choi, R. Fenton, and T.J. Close, 2000. Expression of the barley dehydrin multigene family and the development of freezing tolerance. Molec. Gen. Genet. **264**: 145-153.

Zhu, H., G. Briceno, R. Dovel, P.M. Hayes, B.H. Liu, C.T. Liu, and S.E. Ullrich, 1999a. Molecular breeding for grain yield in barley: an evaluation of QTL effects in a spring barley cross. Theor. Appl. Genet. **98**: 772-779.

Zhu, H., L. Gilchrist, P. Hayes, A. Kleinhofs, D. Kudrna, Z. Liu, L. Prom, B. Steffenson, T. Toojinda, and H. Vivar, 1999b. Does function follow form? Principal QTLs for Fusarium head blight (FHB) resistance are coincident with QTLs for inflorescence traits and plant height in a doubled-haploid population of barley. Theor. Appl. Genet. **99**: 1221-1232.

Zwickertmenteur, S., L. Jestin, and G. Branlard, 1996. Amy2 polymorphism as a possible marker of beta-glucanase activity in barley (*Hordeum vulgare* L). J. Cer. Sci. **24**: 55-63.

5.1
Cytogenetic tests for ploidy level analyses - chromosome counting

J. Maluszynska
Department of Plant Anatomy and Cytology, University of Silesia, Jagiellonska 28, 40-032 Katowice, Poland

Introduction

The chromosome number can be established in cells during mitotic or meiotic cell division. Counting of mitotic chromosomes is easier and faster. Root tips are the most convenient source of mitotic cells. When roots are not available, young buds, leaves or callus can be used. Cytological procedures of chromosome preparation and staining are modified depending upon plant species. Nevertheless, basic principles for handling the mitotic chromosomes of all plant species are similar and consist of collection of material, fixation and chromosome staining. Critical for chromosome counting is proper chromosome preparation. It is important to obtain sufficient number of well spread metaphase plates and good physical separation of the chromosomes. There are two types of chromosome preparation: one is staining of intact root meristem or other tissue prior to squashing and other is preparation of chromosome samples before staining. Method of chromosome staining applied for ploidy level analysis depends on plant species and chromosome size. Generally, methods presented here can be used for all species. For example Feulgen staining, which is based on reaction of Schiff's reagent with the aldehyde groups of DNA, previously exposed by acid hydrolysis, is effective in each species but in plants with small genome size (low amount of nuclear DNA) can give weak staining result. For species with small chromosomes such as *Arabidopsis, Beta, Brassica, Oryza, Solanum* and others, fluorescent staining with DAPI is specially recommended. The method is simple and quick, can be applied for each species but required a fluorescent microscope. Both, Feulgen method and DAPI stain only chromosomes while cytoplasm remains clear. Detail modifications for some plant species are included in protocols for DH.

Protocols

Plant material
- meristematic tissue of root-tips
- meristematic tissue of shoot-tips
- young leaves
- young flower buds
- callus.

Pretreatment
The aim of pretreatment is to increase the number of metaphase cells and chromosome condensation. Method of pretreatment should be experimentally established for each species (Table 5.3.1-1).

Table 5.3.1-1. Pretreatments methods most often used for plant material

Agent	Concentration	Time (h)	Conditions (temperature/ light)	Recommended for following species
Colchicine	0.5-1.0%	2-3	Room temperature/dark	Most species
8-Hydroxyquinoline	0.002 M	2 and 2	4°C and room temperature	Dicots e.g. *Arabidopsis, Brassica, Musa, Solanum*
α-Bromonaphthalene	saturated	2-4	Room temperature/dark	*Triticum, Hordeum, Nicotiana*
Ice-cold water		12-24	0°C	Cereals, temperature grasses, *Arabidopsis*

Fixation
Carnoy's fixative
    methanol or ethanol (95-100%)   1 part
    glacial acetic acid   3 parts

Procedure
    Transfer pretreated material into cold freshly prepared fixative for 2-4 h at room temperature and then store in the deep freezer (-20°C) until use. For longer storage, fixative should be replaced with 70% ethanol and material kept at 4°C.

Chromosome staining

Staining of chromosomes prior to squashing
For the staining of plant chromosomes prior to squashing usually aceto-orcein, aceto-carmine or Feulgen stains are used.

Aceto-orcein and aceto-carmine staining
Preparation of reagents
- Aceto-orcein – 2% orcein (Sigma, O7380) in 45% acetic acid

Dissolve 2 g orcein powder in 100 ml hot 45% acetic acid, shake well and filter (stock solution). Mix 9 parts of orcein solution and 1 part 1N HCl
- Aceto-carmine – 0.5% carmine (Sigma, C1022) in 45% acetic acid
  Add 0.5g carmine powder to 100 ml boiling 45% acetic acid, boil carmine solution for 30 min. Cool down to room temperature and filter.

– Acetic acid-gelatine
  Dissolve 5 g gelatine in 50 ml 45% acetic acid at 60°C.

Staining procedure
1. Place fixed material in a drop of aceto-orcein or aceto-carmine on the microscope slide warmed gently over a spirit lamp for a few minutes. The temperature should never exceed 60°C that can be burned on the back of the hand.

2. Remove stain with a piece of filter paper and add a drop of 45% acetic acid, cover material with cover slip and squash.

3. To prevent the chromosome from drying out, seal the edges of the cover slip with acetic acid-gelatine or nail varnish for temporal observation or make a permanent slide.

Permanent slides
To make the chromosome preparation permanent, the cover slip must be removed carefully, avoiding the cell loss from the slide. Quick-freezing (-70°C) provides a rapid and useful method for the removal of cover slips.
1. Place slide with squash preparations on a block of dry ice, with cover slip uppermost or dip slide in liquid nitrogen, until they are thoroughly frozen. The minimum time required is about 1 min. Flip off the cover slip with the sharp scalpel and place cold slide immediately in ethanol- acetic acid (3:1) for 1 min. followed by 3 washes in 96% ethanol.

2. Mount in Euparal (Roth, 7356) or DPX (Fluka, 44581).

Feulgen staining
Preparation of reagents
- Feulgen stain (Leuco-basic-fuchsin = Schiff's reagent)
  Dissolve 1 g basic fuchsin in 200 ml boiling distilled water, cool to about 50°C and filter. Add 20 ml 1 N HCl and 1 g potassium metabisulphate. Leave at least for 12 hours in the dark. Then add 0.5 g activated charcoal. Shake well and filter quickly. Store in dark bottle at 4°C.
  Ready-to-use Feulgen stain is also available: Sigma, Schiff's regent, S5133.

Staining procedure
1. Rinse fixed material for 10 minutes in distilled water.

2. Hydrolize in 5 N HCl at 20°C for 40 min or in 1 N HCl at 60°C 6-10 min.

3. Wash in distilled water.

4. Transfer material to Feulgen stain for 1-2 h.

5. Wash in distilled water.

6. Place stained material on a microscope slide, add a drop of 45% acetic acid, cover with cover slip and squash. The squash is ready for immediate microscope examination but is not permanent and cannot be stored.
7. Remove cover slip from the squash after freezing and make permanent slide.

Post-squashing staining of chromosomes

Chromosome preparation with enzymatic maceration prior the staining
Preparation of reagents
- Enzyme buffer - citric acid-sodium citrate buffer, pH 4.8
    Stock A. 0.1 M citric acid-monohydrate (21.01 g/L)
    Stock B. 0.1 M trisodium citrate dihydrate (29.41 g/L)
    Mix 40 ml A and 60 ml B.
    Use 1:10 dilution in distilled water

- Enzyme solution
    2% (w/v) cellulase from *Aspergillus niger* (Calbiochem, 21947) and 3% (v/v) pectinase from *A. niger* (solution in 40% glycerol; Sigma P4716)
    or
    1.8% (w/v) cellulase from *Aspergillus niger* (Calbiochem, 21947) and 0.2% 'Onozuka' RS cellulase (Serva, 16419) and 3% (v/v) pectinase from *A. niger* (solution in 40% glycerol; Sigma, P4716)
    Make up in enzyme buffer. Store in aliquots at -20°C.

Staining procedure
1. Rinse fixed tissue for 10 min in enzyme buffer to remove the fixative.

2. Transfer material into enzyme solution for 1-2 h at 37°C.

3. Remove the enzyme solution carefully with a Pasteur pipette and wash material with enzyme buffer for 30 min at room temperature.

4. Place enough material for one preparation (one root tip, small bud or anther) in 45% acetic acid in embryo dish or small Petri dish for a few minutes.

5. Transfer material under the stereo microscope in 1-2 drops (10-20 µl) of 45% acetic acid on a clean slide, tease the tissue to fragments with fine needle and remove unneeded parts e.g. root cap. Apply a cover slip and gently squash the material.

6. Check preparation and number of well spread metaphases under phase contrast microscope.

7. Remove cover slip from the squash after freezing and air dry.

8. The slides can be stored at 4°C until required.

9. The slides can be used for different staining with Giemsa or fluorochromes, as well as for *in situ* hybridization.

Non-fluorescent staining with Giemsa
Preparation of reagents
- Giemsa stain

4% Giemsa (GURR, R66, 35086 or Fluka, 48900) solution in Söerensen phosphate buffer.

- Söerensen phosphate buffer pH 6.8
  0.05 M Na_2HPO_4 and 0.05 M KH_2PO_4.

Staining procedure
1. Place slides in Giemsa solution for 15-30 min.

2. Rinse slides briefly in distilled water and air dry.

3. Mount in Euparal/DPX or chromosomes can be destained with three changes of 96% ethanol for 1-2 h and used for other staining or *in situ* hybridization.

Fluorescent staining with DAPI (4'-6-diamino-2phenylindole)
Preparation of reagents
- DAPI solutions
  Stock solution - dissolve 100 µg DAPI (Sigma, D9542) in 1 ml of distilled water. Store in aliquot at -20°C (it is stable for years).
  Working solution - dilute stock solution in McIlvaine buffer to final concentration 2–4 µl/ml (w/v).

- McIlvaine buffer pH 7.0
  Mix 0.1 M citric acid (18 ml) and 0.2 M Na_2HPO_4 x $2H_2O$ (82 ml).

Staining procedure
1. Add 50 µl of DAPI working solution to the slide for 15-30 min at room temperature in dark.

2. Wash off with distilled water and air dry.

3. Mount slides in antifade buffer Citifluor (AFI, Pelco) or Vectashield (Vector Laboratories).

4. After analysis, DAPI stained chromosomes can be destained with methanol-acetic acid (3:1) for 6-12 h, then rinsed with three changes of methanol and air dry. Destained slides can be restained by any other method or used for *in situ* hybridization.

Chromosome number analysis
Chromosome counting analysis can be done directly on slides with a proper microscope or indirectly on photographs. In the case of small and numerous plant chromosomes, it is sometimes necessary to enlarge them by taking photographs or capturing mataphase plate image for computer analysis and saving data. For each sample minimum of 5 complete mataphase plates should be analyzed.

Recommended references

Fukui, K. and S. Nakayama, 1996. Plant Chromosomes: Laboratory Methods. CRC Press, Tokyo.
Singh, R.J., 1993. Plant Cytogenetics. CRC Press, Tokyo.
Sharma, A.K., and A. Sharma, 1999. Plant Chromosomes. Harwood Academic Publishers, Amsterdam.

5.2
Ploidy determination using flow cytometry

B. Bohanec
University of Ljubljana, Biotechnical Faculty, Centre for Plant Biotechnology and Breeding, Jamnikarjeva 101, 1111 Ljubljana, Slovenia

Introduction

Measuring ploidy level is of highest importance in the final stages of haploid induction programs. For breeding purposes, a large number of haploids usually need to be tested, so an efficient ploidy analysis is a prerequisite for successful application. Methods to determine ploidy level may be direct (chromosome counting) or indirect (flow cytometry, stomatal size, chloroplast number of the guard cells and morphological observations). Various opinions about the usefulness of the mentioned techniques can be found. Sari *et al.* (1999) for instance, comparing various ploidy measurement methods in haploid watermelons concluded that "counting chromosomes is cumbersome, producing plants for morphological observations requires a long time and flow cytometry is expensive and labour intensive." They proposed that "measurement of stomata and chloroplast counting methods are simple to use and less labor intensive, and hence can be considered a practical alternative to the others." The author's experiences are based on flow cytometric measurements of a large number of haploid regenerants obtained mainly by androgenesis (cabbage), gynogenesis (onion) or distant fertilization (potato), as well as on genome size analysis of various species. Several points will be made to explain why, in the author's opinion, for ploidy analysis of haploid regenerants, flow cytometry is of much higher relevance than any other proposed method. The main aim of this article is to explain the basic features of flow cytometry and to propose optimized protocols for rapid and simple flow cytometric evaluations using either a HBO lamp or laser equipped flow cytometers.

Principles of flow cytometry

A flow cytometer is a device that measures light fluoresced or scattered from particles. For ploidy analysis, the particles used are interphase nuclei obtained from somatic tissues. Excitation light can be emitted from either a high pressure mercury lamp (HBO lamp) with broad spectral characteristic or laser light of very narrow wavelength. Nuclei are stained with DNA specific fluorochromes that after absorption of excitation light emit fluorescence of longer wavelength. The emitted light passes through specific optical filters and dichroic mirrors (which reflect some wavelengths and allow others to pass through) enabling only emitted light to reach the appropriate photomultiplier.

On photomultipliers, signals are detected and measured for each particle (nucleus) separately. More than 1000 nuclei per second can be measured. Acquired data are processed by computer and results are displayed on screen in real time distributed to linear or logarithmic scales, usually divided into 1024 channels. The position of signals is achieved by adjustment of "gain" which is related to the voltage on the photomultiplier. After acquisition, data can be stored and, if necessary, further analyzed using specialized software, normally provided by the producer of the flow cytometer. For ploidy analysis, only a single parameter, the fluorescence excited by the DNA specific fluorochrome, is usually measured, although a number of parameters can be measured for each particle simultaneously. This means that, for instance, only blue or red light signals can be measured, or up to 5 different color emissions including scattered light and time can be measured for each particle in a multiparameter function.

The term flow cytometry is based on the fluidic flow of particles passing a quartz chamber (flow cuvette). Particles from a sample are pumped through the flow cuvette in a very narrow stream flowing inside a larger "sheath" stream of water or saline, which is used to focus the sample stream into the center.

When nuclei are measured, undesired small particles are also often stained, representing »debris« consisting of broken nuclei or other smaller particles. It is useful to set a left threshold level to exclude such particles from analysis, allowing higher precision of measurement of nuclei. A similar improvement can be achieved using a side or forward scatter function on laser based flow cytometers. In such case, two parameter dot plot analysis and setting a gating region are used, allowing elimination of both small particles and nuclear doublets or similar larger particles that represent noise in the measured interval.

How is ploidy interpreted in flow cytometry

In eukaryotes, cell growth and division is a cyclical process including mitosis (M) followed by an interphase period starting with the first "gap" (G_1), a synthetic (S) phase and a final "gap" (G_2). Non-proliferating cells may leave the cell cycle either in G_1 or G_2. In somatic diploid tissues, nuclei arrested in both G_1 and G_2 stages are usually found representing 2n and 4n nuclei. The terms used to describe ploidy are C-values rather than chromosome number n values, 1C value representing the quantity of DNA corresponding to the haploid complement. C value may be presented in arbitrary units or expressed as picograms/nucleus or number of nucleic base pairs - Mbp/nucleus.

For determination of genome size, a standard species with known genome size is mixed with the studied species, and the mean values of histograms of G_1 of both species are used for calculation of the genome size of the measured plant. Typically, such analysis requires staining of nuclei with intercalating stain (mainly propidium iodide), slow measurement (about 20 nuclei/second) resulting in low variation, 10,000 nuclei measured per sample and several repetitions of measurement. Detailed explanations have been given elsewhere (Dolezel, 1991; Marie and Brown, 1993).

For determination and interpretation of the haploid status of regenerants, a greatly simplified procedure can be used. Firstly, tissue of known ploidy of the studied species, usually diploid tissue, is analyzed and gain values are set to the appropriate channel, for instance in such a way that the first peak, representing 2C nuclei (G_1 stage), appears around channel 200 (Fig. 5.2-1b) so the second peak (G_2 stage) appears around channel 400, assuming that a linear scale is used. The apparatus is thus adjusted using diploid tissue of known origin as external standard. The next samples, of unknown ploidy, are then observed under the same settings. For haploid samples, the first peak (Fig. 5.2-1a) would be expected to appear around channel 100 and the second peak around channel 200.

Measurements can be performed at a much faster speed since low CV values of histograms are not needed. A much lower number of nuclei need to be measured, 1000 (depending on CV value and noise ratio) often being sufficient. Visual observation of the position of peaks and their relative size can be used instead of calculation of mean values. All these simplifications result in much faster measurement compared to exact genome size analysis; one sample can be measured as fast in a minute as compared to at least ten minutes needed to obtain data for genome size measurements.

More complex interpretations
Nuclei in somatic tissues of plants often pass one or more additional divisions without apparent mitosis, this process of endomitosis is termed polysomaty or endopolyploidy. In such case, flow cytometric measurement of nuclei would reveal additional peaks, each following peak with doubled genome size. This phenomenon is usually expressed in some plant species in tissues at a particular developmental stage. For a diploid plant, 8C and 16C nuclei are relatively common in addition to 2C and 4C nuclei. Very similar multiple histograms are obtained when haploid regenerants are measured, even in species that do not tend to produce endopolyploid nuclei. In analyses of haploid regenerants, a 4C additional peak, in particular, in addition to 1C and 2C peaks, is also frequent (Fig. 5.2-1c). However, the origin of this additional peak is not endopolyploidy but represents a mixoploid character of the analyzed tissue. Mixoploids vary considerably in that the proportion of the original haploid nuclei (1C peak) might be predominant with only a minor proportion of already doubled nuclei or it might also be present in minimal proportion. In the latter case, the plant might be considered almost diploidized, although in some species a tendency towards haploidy might still be preserved in apex cells. Accurate measurement of the proportion is difficult, since the second peak (2C) represents mixed cells that might be a G_2 stage of the 1C nuclei or G_1 stage of the 2C nuclei. The proportion of the first and the third peak can be used as an indirect estimation of the second peak constitution. It is of course very important to distinguish between the two phenomena, endopolyploidy and mixoploidy, since an endoplyploid tissue always produces gametes with basic ploidy, while mixoploid plants can produce some branches or flower parts with haploid or doubled haploid cells.

When haploid regenerants are treated with chromosome doubling agents frequently some proportion of lines are not just doubled but even tetraploidized. On very rare occasions, spontaneous triploids have been reported as a result of either androgenesis or gynogenesis. Such events are easily detected with flow cytometry using the described comparison to the diploid standard.

Advantages of flow cytometric measurements over other ploidy analysis methods
As explained in the previous paragraph, the result of haploid induction, including chromosome doubling treatments, is often highly variable. It should be emphasized that flow cytometry is the only technique among direct and indirect protocols for ploidy determination that gives detailed information about the existence of mixoploid tissues and their proportions. In our experience, mixoploids are usually induced in at least 25% of lines submitted to chromosome doubling treatments.

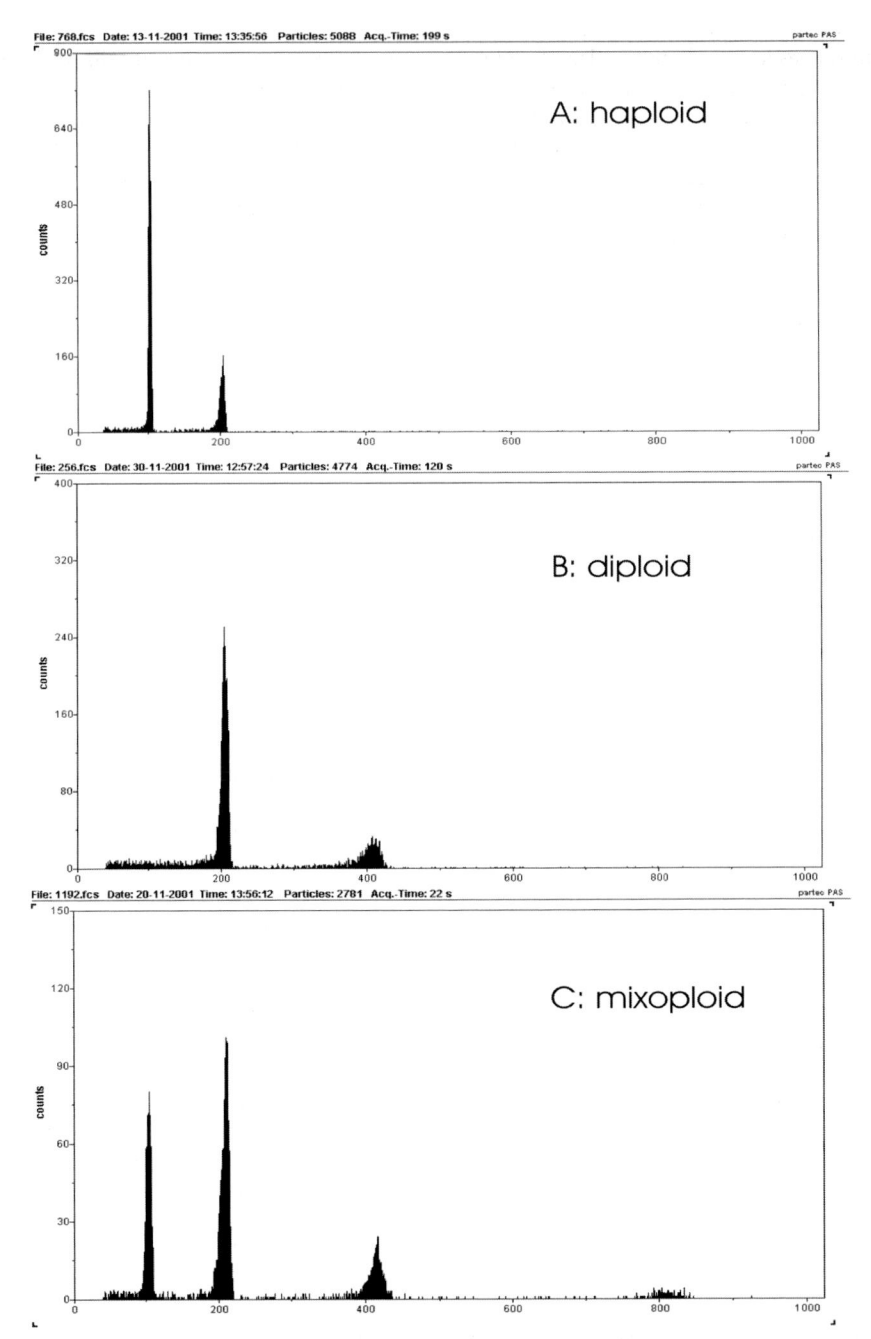

Figure. 5.2-1. Histograms of haploid, diploid and mixoploid samples showing typical fluorescence intensities proportional to nuclear DNA content.

The chromosome counting method is not only laborious and prone to errors when chromosomes are small, but is also often limited to a relatively low number of cells analyzed. Chromosome counting is usually determined from the root apices. In haploid regenerants, root tissues do not necessarily reflect the ploidy in shoot apices. For instance, in cabbage haploid regenerants, flow cytometric measurements revealed a tetraploid leaf structure, while chromosome counting in root tissues exhibited a mixture of haploid and diploid cells (Bohanec, unpublished). When optimized protocols are used, a high number of samples, 200 or more per day, can be accurately analyzed.

Procedures for sample preparations and staining

To obtain C-value information, interphase nuclei must be isolated and stained. Various tissues can be used, but often the highest value for analysis of haploids is given by the analysis of leaf tissue. In general, tissues from young leaves are superior to older ones. Numerous isolation and staining procedures have been published, differing in the choice of fluorochromes, composition of buffers, isolation and staining procedure and other features. Some procedures use one step isolation and staining buffers, while others apply two buffers - the first being used for the release of nuclei and the second for staining. Nuclei can be isolated and concentrated by centrifugation or measured immediately after filtration. In most cases, nuclei are analyzed soon after isolation, but can also be fixed and analyzed later.

We shall describe only two protocols, one based on 4',6'-diamidino-2-phenylindole (DAPI) and the other on propidium iodide (PI) staining. In fact, either fluorochrome can be used in both protocols but we found the described combination to be optimal. In general, for determination of ploidy, DAPI staining has several advantages – histograms with lower coefficients of variation are obtained, staining is almost immediate, and treatments can be done at room temperature. DAPI is also used in much lower quantities, usually 2-5 µg/ml compared to 50-100 µg/ml PI. As described earlier, DAPI binds to AT base pairs and is therefore of lower value for absolute genome size analysis. For determination of haploid status, absolute values are not needed, making DAPI the recommended fluorochrome for DNA staining in haploid research. The only reason that PI (optimal excitation wavelength at 495 nm) is often used, also in determination of ploidy, is the prevalence of flow cytometers equipped with 488 nm argon laser as the light source. For measurement using DAPI with optimal excitation wavelength at 372 nm, flow cytometers equipped with a HBO lamp are usually used. It should be noted that flow cytometers equipped with a lamp as the only light source are so far the cheapest, so if a flow cytometer is purchased only for ploidy analysis, such a model can be recommended.

Protocol for DAPI staining in Otto buffers (modified from Otto, 1988)

1. Chop an appropriate amount of plant material with a new razor blade (discard after 5-10 choppings) in ca. 0.5 ml of Otto I buffer in a 5 cm plastic Petri dish. The quantity of chopping buffer should be adjusted to the amount of plant material chopped. Pieces of young leaves are usually used for ploidy analysis, but occasionally when necessary, calli of androgenic origin can be frozen to -20 or $-80^{\circ}C$ and chopped in the same buffer while still frozen.

2. Chopped particles are mixed well or pipetted a few times with a Pasteur pipette to accelerate the release of nuclei. The suspension is filtered through a 30 µm nylon mesh into test tubes, the brand and item being defined by the producer of the flow cytometer. Filters can be placed in funnels or, preferably, purchased already mounted (Celltrics, Partec –www.partec.de). Note that various mesh sizes are available; we have found a 30 µm size efficient with various plant species.

3. Depending on the quantity of filtered suspension, add 3-5 volumes of Otto II buffer. Typically 1-2 ml of final quantity is sufficient for determination of ploidy.

4. Store at room temperature and analyze within 1 h.

Composition of buffers (for 100 ml of solution)
Otto chopping buffer (I):
0.1 M citric acid monohydrate	2.1 g
0.5% (v/v) Tween 20	0.5 ml

Otto staining buffer (II)*
0.4 M $Na_2HPO_4 \times 12H_2O$	7.1 g
DAPI	0.4 mg**

*do not refrigerate
**0.4 ml from stock solution in concentration 10 mg in 10 ml.

Protocol for PI staining in LB01 buffer (Dolezel *et al.*, 1989)
Procedures for chopping and filtering are the same as described for DAPI staining, the same lysis buffer is also used for staining.

1. Chop an appropriate amount of plant material with a new razor blade, as described for the DAPI staining technique, in 2-3 ml ice-cold LB01 buffer. PI and RNase are included in the buffer prior to chopping or - alternatively - are added after filtration. In the first case, add 5 ml of each PI and RNase from stock solutions to 100 ml LB01 buffer.

2. Filter suspension and place on ice for 15 min prior to analysis.

Composition of LB01 buffer (for 100 ml of solution):
15 mM Tris	181.7 mg
2 mM Na_2EDTA	74.5 mg
0.5 mM spermine tetrahydrochloride	17.4 mg
80 mM KCl	597 mg
20 mM NaCl	116.9 mg
0.1% (v/v) Triton X-100	100 µl

adjust volume to 100 ml
adjust to pH 7.5 with 1N HCl
add 110 µl ß-mercaptoethanol (15 mM)
add 5 ml of each PI and RNase from stock solutions:

PI and RNase stock solutions
1 mg/ml propidium iodide	25 mg
1 mg/ml RNase	25 mg

disolve separately in 25 ml H_2O.

For preparation of solutions, it is advisable to use highly purified water filtered through 0.22 µm filters to remove small particles.

References

Dolezel, J., 1991. Flow cytometric analysis of nuclear DNA content in higher plants. Phytochem.Anal. **2**: 143-154

Dolezel, J., P. Binarova and S. Lucretti, 1989. Analysis of nuclear DNA content in plant cells by flow cytometry. Biol.Plant. **31**: 113-120

Marie, D. and S. Brown, 1993. A cytometric exercise in plant DNA histograms, with 2C values for 70 species. Biol.Cell. **78**: 41-51

Otto, F., 1988. High resolution DNA-flow cytometry using DAPI. Partec, protocol 1. Partec, Münster, Arlesheim, pp. 6.

Sari, N., K. Abak, M. Pitrat, J.C. Rode and R.D. de Vaulx, 1994. Induction of parthenogenetic haploid embryos after pollination by irradiated pollen in watermelon. HortSci. **29**(10): 1189-1190.

6.1
Major media composition

Compiled by I. Szarejko
Department of Genetics, University of Silesia, Jagiellonska 28, 40-032 Katowice, Poland

Table 6.2-1. Composition of MS medium after Murashige and Skoog (1962) - growth regulators omitted

Medium components	Concentration (mg/L)
Major elements	
NH_4NO_3	1,650
KNO_3	1,900
$CaCl_2 \times 2H_2O$	440
$MgSO_4 \times 7H_2O$	370
KH_2PO_4	170
Na_2EDTA	37.3*
$FeSO_4 \times 7H_2O$	27.8*
Minor elements	
H_3BO_3	6.2
$MnSO_4 \times 4H_2O$	22.3
$ZnSO_4 \times 4H_2O$	8.6
KI	0.83
$Na_2MoO_4 \times 2H_2O$	0.25
$CoCl_2 \times 6H_2O$	0.025
$CuSO_4 \times 5H_2O$	0.025
Organic constituents	
Sucrose	30,000
Glycine	2
Agar	10,000
myo-Inositol	100
Nicotinic acid	0.5
Pyridoxine HCl	0.5
Thiamine HCl	0.4
pH	5.7-5.8

*/5 ml/L of a stock solution containing 5.57 g $FeSO_4 \times 7H_2O$ and 7.45 g Na_2EDTA per litre of H_2O

Table 6.2-2. Composition of FHG medium after Hunter (1988)

Medium components	Concentration (mg/L)
Macro elements	
KNO_3	1,900
NH_4NO_3	165
KH_2PO_4	170
$MgSO_4 \times 7H_2O$	370
$CaCl_2 \times 2H_2O$	440
FeNaEDTA	40
Micro elements	
$MnSO_4 \times 5H_2O$	22.3
H_3BO_3	6.2
$ZnSO_4 \times 7H_2O$	8.6
$CoCl_2 \times 5H_2O$	0.025
$CuSO_4 \times 5H_2O$	0.025
$Na_2MoO_4 \times 2H_2O$	0.25
Organic	
Maltose	62,000
myo-Inositol	100
Thiamine HCl	0.4
Glutamine	730
Hormones	
BAP	1
pH	5.6

Table 6.2-3. Composition of N6 medium after Chu (1978) - growth regulators omitted

Medium components	Concentration (mg/L)
$(NH_4)_2SO_4$	463
KNO_3	2,830
KH_2PO_4	400
$MgSO_4 \times 7H_2O$	185
$CaCl_2 \times 2H_2O$	166
$MnSO_4 \times 4H_2O$	4.4
$ZnSO_4 \times 7H_2O$	1.5
H_3BO_3	1.6
KI	0.8
Iron: 5 ml of solution obtained by dissolving 5.57 g $FeSO_4 \times 7H_2O$ and 7.45 g Na_2EDTA in 1 L distilled water	
Glycine	2
Thiamine HCl	1
Pyridoxine HCl	0.5
Nicotinic acid	0.5
Sucrose	50,000-120,000
Agar	8,000
pH	5.8

Table 6.2-4. Composition of B5 medium after Gamborg *et al.* (1968) - growth regulators omitted

Medium components	Concentration (mg/L)
$NaH_2PO_4 \times H_2O$	150
KNO_3	2,525
$(NH_4)_2SO_4$	134
$MgSO_4 \times 7H_2O$	246
$CaCl_2 \times 2H_2O$	150
Sequestrene 330 Fe	28
Nicotinic acid	1
Thiamine HCl	10
Pyridoxine HCl	1
myo-Inositol	100
$MnSO_4 \times H_2O$	10
H_3BO_3	3
$ZnSO_4 \times 7H_2O$	2
$NaMoO_4 \times 2H_2O$	0.25
$CuSO_4$	0.025
$CoCl_2 \times 6H_2O$	0.025
KI	0.75
Sucrose	20,000
pH	5.5

Table 6.2-5. Composition of Potato II medium after Wang and Hu (1984) - growth regulators omitted

Medium components	Concentration (mg/L)
KNO_3	1,000
$(NH_4)_2SO_4$	100
KH_2PO_4	200
$Ca(NO_3)_2 \times 4H_2O$	100
$MgSO_4 \times 7H_2O$	125
KCl	35
$FeSO_4 \times 7H_2O$	27.5
Na_2EDTA	37.3
Thiamine HCl	1
Potato extract*	100,000
Sucrose	90,000

*/ According to the authors, for medium containing 10% potato extract, 100 g potato tubers are boiled and the liquid used for making 1,000 ml medium.

Table 6.2-6. Composition of 190-2 medium after Wang and Hu (1984) - growth regulators omitted

Medium components	Concentration (mg/L)
KNO_3	1,000
$(NH_4)_2SO_4$	200
KH_2PO_4	300
$Ca(NO_3)_2 \times 4H_2O$	100
$MgSO_4 \times 7H_2O$	200
KCl	40
$MnSO_4 \times 4H_2O$	8
KI	0.5
$ZnSO_4 \times 7H_2O$	3
H_3BO_3	3
$FeSO_4 \times 7H_2O$	27.5
$Na_2EDTA \times 2H_2O$	37.3
Glycine	2
Thiamine HCl	1
Pyridoxine HCl	0.5
Nicotinic acid	0.5
myo-Inositol	100
Sucrose	30,000

Table 6.2-7. Composition of Nitsch & Nitsch medium after Nitsch and Nitsch (1969) - growth regulators omitted

Medium components	Concentration (mg/L)
Mineral salts	
KNO_3	950
NH_4NO_3	720
$MgSO_4 \times 7H_2O$	185
$CaCl_2$	166
KH_2PO_4	68
$MnSO_4 \times 4H_2O$	25
H_3BO_3	10
$ZnSO_4 \times 7H_2O$	10
$Na_2MoO_4 \times 2H_2O$	0.25
$CuSO_4 \times 5H_2O$	0.025
Iron: 5 ml of a solution of 7.45 g Na_2EDTA and 5.57 g of $FeSO_4 \times 7H_2O$ in 1 L of distilled water	
Organic addenda	
myo-Inositol	100
Glycine	2
Nicotinic acid	5
Pyridoxine HCl	0.5
Thiamine HCl	0.5
Folic acid	0.5
Biotin	0.05
Sucrose	20,000
Difco Bacto agar	8,000
pH	5.5

Table 6.2-8. Composition of NLN medium after Nitsch and Nitsch (1967) and Lichter (1981) - growth regulators omitted

Medium components	Concentration (mg/L)
$Ca(NO_3)_2 \times 4H_2O$	500
KNO_3	125
$MgSO_4 \times 7H_2O$	125
KH_2PO_4	125
$MnSO_4 \times 4H_2O$	25
H_3BO_3	10
$ZnSO_4 \times 7H_2O$	10
$Na_2MoO_4 \times 2H_2O$	0.25
$CuSO_4 \times 5H_2O$	0.025
$CoCl_2 \times 6H_2O$	0.025
Iron: 5 ml of a solution of 7.45 g Na_2EDTA and 5.57 g of $FeSO_4 \times 7H_2O$ in 1 L of distilled water	
Glycine	2
myo-Inositol	100
Nicotinic acid	5
Pyridoxine HCl	0.5
Thiamine HCl	0.5
Folic acid	0.5
Biotin	0.05
Glutathione	30
L-Glutamine	800
L-Serine	100
Sucrose	40,000-120,000
Difco potato extract	2,500

Reference

Chu, C.C., 1978. The N6 medium and its applications to anther culture of cereal crops. In: Proc. Symp.Plant Tissue Cult. Science Press, Beijing. pp.45-50.

Gamborg, O.L., R.A. Miller and K. Ojima, 1968. Nutrient requirements of suspension cultures of soybean root cells. Exp.Cell Res. **50**: 151-158.

Hunter, C.P. (1988) Plant regeneration from microspores of barley, *Hordeum vulgare* L. Ph.D. Thesis. Wye College, Univ. of London, London.

Lichter, R., 1981. Anther culture of *Brassica napus* in a liquid culture medium. Z.Pflanzenphysiol. **103**: 229-237.

Murashige, T. and F. Skoog, 1962. A revised medium for rapid growth and bioassays with tobacco tissue cultures. Physiol.Plant. **15**: 473-497.

Nitsch, C. and J.P. Nitsch, 1967. The induction of flowering *in vitro* in stem segments of *Plumbago indica* L. I. The production of vegetative buds. Planta. **72**: 355-370.

Nitsch, J.P. and C. Nitsch, 1969. Haploid plants from pollen grains. Science. **163**: 85-87.

Wang, X.Z. and H. Hu, 1984. The effect of potato II medium for triticale anther culture. Plant Sci.Lett. **36**: 237-239.

6.2
Basic equipment for maize microspore culture laboratory

Compiled by M.Y. Zheng[1], Y. Weng[2], R. Sahibzada and C.F. Konzak
Northwest Plant Breeding Co., 2001 Country Club Rd., Pullman, WA 99163, USA
[1] *Department of Biology, Gordon College, 255 Grapevine Rd., Wenham, MA 01985, USA*
[2] *USDA-ARS, 215 Johnson Hall, Washington State University, Pullman, WA 99164, USA*

1. Laminar flow hood for sterile operation. One 6 ft (sits two persons) hood is recommended. *Microzone Corporation*, or through *Fisher Scientific, VWR Scientific Products* or other.

2. Incubators with automatic temperature setting and control: Temperature range: 4-60C. *Fisher Scientific or VWR Scientific Products* or other.

3. Swinging bucket (arm) centrifuge with rotors that hold 15 to 50 ml *Falcon/Maizeing Conical* tubes: Automatic speed control and timer. *Fisher Scientific or VWR Scientific Products* or other.

4. MC-2 Waring blender with variable speeds. *Fisher Scientific*.

5. MC-2 Waring blender cups. *Fisher Scientific*.

6. MC-2 Waring blender cup lids (autoclavable). *Fisher Scientific*.

7. Micropipetter: 5 ml (1+)   Tips (1000+)
   1 ml (1+)   Tips (1000+)
   100 µl (1+) Tips (400+). *Fisher Scientific, VWR Scientific Products or various suppliers.*

8. Filters with mesh: Stainless mesh from *Universal Wire Cloth:*

Mesh	Type	Opening
150x150	304/316	0.0041" (100 µm)
325x325	304/316	0.002" (50 µm)
400x400	TW316	0.0015" (38 µm)

9. Petri dishes: 60x15 mm (200+)
   35x10 mm (200+)
   100x20 mm (200+). *Becton Dickinson & Co.* or other.

10. Electronic pH meter (1). *VWR Scientific Products* or other.

11. Balance: Accuracy to 0.001g. *VWR Scientific Products* or other.

12. Graduate cylinders: 1000 ml (1+)
    500 ml (1+)
    100 ml (4+)
    25 ml (2+). *Fisher Scientific or VWR Scientific Products* or other.

13. Sterile Conical Tube with cap: 15 ml (200+)
    50 ml (100, optional). *Becton Dickinson & Co.* or other.

14. Glassware: flasks, beakers, bottles, etc.

15. Autoclave. *Fisher Scientific or VWR Scientific Products* or other.

16. Magnetic hot/stirring plate (1). *Fisher Scientific or VWR Scientific Products* or other.

17. Magnetic stirring bars (10 of variable size). *Fisher Scientific or VWR Scientific Products* or other.

18. Surgical scissors (1 pair for each person), forceps (2 pairs for each person). *Fisher Scientific* or other.

19. Inverted microscope. (*Zeiss, Olympus* or any other brands).

20. Haemocytometer. *Fisher Scientific or VWR Scientific Products* or other.

22. Filter unit (0.22 µm pore size) for filter sterilization. *Millipore Corporation.*

23. Syringe filter. *Nalge Nunc International.*

24. Burner-alcohol lamps (1 for each person). *VWR Scientific Products* or other.

25. Alcohol, paper towels, Kimavipes, Clorox bleach. *Any local vendor(s).*

26. Chemicals including all mineral nutrients, vitamins, plant growth regulators, maltose, mannitol, sucrose, etc. *Sigma.*

List of Contributors

Andersen, S.B.
The Royal Veterinary and Agricultural University
Department of Agricultural Sciences
Section Plant Breeding and Crop Science
Thorvaldsensvej 40
DK-1871 Frederiksberg C, **Denmark**
Phone: +45-35-283-444
Fax: +45-35-283-468
E-mail: sven.b.anderson@agsci.kvl.da

Barnabás, B.
Agricultural Research Institute of the Hungarian Academy of Sciences
2462 Martonvásár, **Hungary**
Phone: +36-22-569-526/569-500
Fax: +36-22-569-576
E-mail: bea@penguin.mgki.hu

Bohanec, B.
Biotechnical Faculty
Centre for Plant Biotechnology and Breeding
Jamnikarjeva 101
1000 Ljubljana, **Slovenia**
Phone: +386-1-423-1161
Fax: +386-1-423-1088
E-mail: borut.bohanec@uni-lj.si

Bueno, M.A.
INIA, Carretera de la Coruña
Km 7.5
E-28040 Madrid, **Spain**
Phone: +34-1-347-6863/22
Fax: +34-1-357-2293
E-mail: bueno@inia.es

Castillo, A.
Departamento de Genética y Producción Vegetal
Estación Experimental de Aula Dei
Apartado 202
50080 Zaragoza, **Spain**
Phone: +34-9-7671-6072
Fax: +34-976-716-145
E-mail: amcast@eead.csic.es

Cistué, L.
Departamento de Genética y Producción Vegetal
Estación Experimental de Aula Dei
Apartado 202
50080 Zaragoza, **Spain**
Fax: +34-976-716-145
E-mail: lcistue@eead.csic.es

Clément, C.
Université de Reims Champagne Ardenne
UFR Sciences
Laboratoire de Biologie et Physiologie Végétales
BP 1039
51687 Reims Cedex 2, **France**
Phone: +33-3-2691-3339
Fax: +33-3-2691-3339
E-mail: christophe.clement@univ-reims.fr

Corey, A.
Department of Crop and Soil Science
Oregon State University
Corvallis, OR 97331, **USA**
Phone: +1-541-737-5878
Fax: +1-541-737-1589

Custers, J.B.M.
Business Unit "Plant Development and Reproduction"
Cluster "Seed and Reproduction Strategies"
Plant Research International B.V.
PO Box 16
6700 AA Wageningen, **The Netherlands**
Phone: +31-317-477-047
Fax: +31-317-418-094
E-mail: j.b.m.custers@plant.wag-ur.nl

da Silva Dias, J.C.
Instituto Superior de Agronomia DPAA
Technical University of Lisbon
Tapada da Ajuda
1349-017 Lisboa, **Portugal**
E-mail: mop06572@mail.telepac.pt

Davies, P.A.
SARDI, Field Crops Pathology Unit
GPO Box 397
Adelaide SA 5001, **Australia**
Phone: +61-8-8303-9494
Fax: +61-8-8303-7321
E-mail: davies.phil@saugov.sa.gov.au

DeMaine, M.J.
Scottish Crop Research Institute
Invergowrie, Dundee
DD2 5DA, **UK**
Phone: +44-1382-562-731
Fax: +44-1382-562-426
E-mail: mdemai@scri.sari.ac.uk

LIST OF CONTRIBUTORS

DeNoma, J.
Department of Crop and Soil Science
Oregon State University
Corvallis, OR 97331, **USA**
Phone: +1-541-737-5878
Fax: +1-541-737-1589

Devaux, P.
Florimond Desprez
Biotechnology Laboratory
P.O. Box 41
59242 Cappelle en Pevele, **France**
Fax: +33-3-2059-6601
E-mail: pierre.devaux@florimond-desprez.fr

Echávarri, B.
Departamento de Genética y Producción Vegetal
Estación Experimental de Aula Dei
Apartado 202
50080 Zaragoza, **Spain**
Phone: +34-9-7671-6071
Fax: +34-976-716-145

Ferrie, A.
Cell Technologies
Plant Biotechnology Institute / NRC
110 Gymnasium Place
Saskatoon, SK
S7N OW9, **Canada**
Phone: +306-975-5993
Fax: +306-975-4465
E-mail: alison.ferrie@nrc.ca

Forster, B.P.
Scottish Crop Research Institute
Invergowrie
Dundee DD2 5DA
Scotland, **UK**
Phone: +44-1382-562-731
Fax: +44-1382-562-426
E-mail: bforst@scri.sari.ac.uk

Germanà, M.A.
Dipartimento di Colture Arboree
Facoltà di Agraria
Università degli Studi di Palermo
Viale Delle Scienze 11
90128 Palermo, **Italy**
E-mail: agermana@unipa.it

Gertsson, B.
Svalöf Weibull AB
Oil Crop Department
SE-26881, Svalöv, **Sweden**
Phone: +46-418-667-046
Fax: +46-418-667-104
E-mail: bo.gertsson@swseed.se

Guo, Y.-D. *(current address)*
Laboratory of Grass Breeding
National Agricultural Research Centre for Hokkaido Region
National Agricultural Research Organization
Hitsujigaoka
Sapporo 062-8555, **Japan**
Phone: +81-11-857-9273
Fax: +81-11-859-2178
E-mail: yangdongguo@hotmail.com
         yaguo@affrc.go.jp

Hansen, M.
Agricultural University of Norway
Department of Horticulture and Crop Sciences
P.O. Box 5022
N-1432 As, **Norway**
Phone: +47-64-947-800
Fax: +47-64-947-802
E-mail: magnor.hansen@ipf.nlh.no

Hayes, P.M.
Barley Project
Department of Crop and Soil Science
Oregon State University
Corvallis, OR 97331, **USA**
Phone: +1-541-737-5878
Fax: +1-541-737-1589
E-mail: patrick.m.hayes@orst.edu

Heberle-Bors, E.
Vienna Biocenter
Institute for Microbiology and Genetics
University of Vienna
Dr. Bohrgasse 9
A-1030 Vienna, **Austria**
Phone: +43-1-4277-54603
Fax: +43-1-4277-9546
E-mail: erwin@gem.univie.ac.at

Höfer, M.
Federal Centre for Breeding Research on Cultivated Plants
Institute for Fruit Breeding
Pillnitzer Platz 2
D-01326 Dresden, **Germany**
Fax: +49-351-261-6213
E-mail: m.hoefer@bafz.de

LIST OF CONTRIBUTORS

Hu, T.C.
Monsanto Co.
700 Chesterfield Pkwy. N.
St. Louis, MO 63198, **USA**
Phone: +1-636-737-6685
Fax: +1-636-737-6567

Immonen, S.
FAO, SDRC, C-626
Viale delle Terme di Caracalla
00100 Rome, **Italy**
Phone: +39-06-5705-4861
Fax: +39-06-5705-3298
E-mail: sirkka.immonen@fao.org

Inagaki, M.N.
Japan International Research Center for Agricultural Sciences
Ministry of Agriculture, Forestry and Fisheries
1-2 Ohwashi, Tsukuba
Ibaraki 305-8686, **Japan**
Fax: +81-298-386-337
E-mail: minagaki@jircas.affrc.go.jp

Jacquard, C.
Université de Reims Champagne Ardenne
UFR Sciences
Laboratoire de Biologie et Physiologie Végétales
BP 1039
51687 Reims Cedex 2, **France**
Phone: +33-3-2691-3427
Fax: +33-3-2691-3427/2691-3339

Jakse, M.
Biotechnical Faculty
Centre for Plant Biotechnology and Breeding
Jamnikarjeva 101
1000 Ljubljana, **Slovenia**
Phone: +386-1-423-1161
Fax: +386-1-423-1088

Jauhar, P.P.
USDA-ARS Northern Crop Science Laboratory
State University Station
Fargo, ND 58105-5677, **USA**
Phone: +1-701-239-1309
Fax: +1-701-239-1369
E-mail: prem_jauhar@ndsu.nodak.edu

Kasha, K.J.
University of Guelph
Department of Plant Agriculture
Crop Science Building
Guelph, ON
N1G 2W1, **Canada**
Phone: +1-519-824-4120 x2507
Fax: +1-519-763-8933
E-mail: kkasha@uoguelph.ca

Konzak, C.
Northwest Plant Breeding Co.
2001 Country Club Rd.
Pullman, WA 99163, **USA**
Phone: +1-509-334-4404
Fax: +1-509-334-5320
E-mail: npb@completebbs.com

Letarte, J.
University of Guelph
Department of Plant Agriculture
Crop Science Building
Guelph, ON
N1G 2W1, **Canada**
Phone: +1-519-824-4120
Fax: +1-519-763-8933

Levall, M.
Syngenta Seeds AB
P.O. Box 302
S-261 23 Landeskrona, **Sweden**
Phone: +46-418-437-298
Fax: +46-418-437-283
E-mail: mats.levall@syngenta.com

Liu, W.
Northwest Plant Breeding Co.
2001 Country Club Rd.
Pullman, WA 99163, **USA**
Phone: +1-509-334-4404
Fax: +1-509-334-5320

Ljungberg, A.
SW Laboratory
Svalöf Weibull AB
SE-268 81 Svalöv, **Sweden**
Phone: +46-418-667-000/667-145
Fax: +46-418-667-219
E-mail: anita.ljungberg@swseed.se

LIST OF CONTRIBUTORS

Maluszynska, J.
Department of Plant Anatomy and Cytology
University of Silesia
Jagiellonska 28
40-032 Katowice, **Poland**
Phone: +48-32-200-9443
Fax: +48-32-200-9396
E-mail: maluszyn@us.edu.pl

Maluszynski, M.
Plant Breeding and Genetics Section
Joint FAO/IAEA Division
P.O. Box 100
Vienna, **Austria**
Phone: +43-1-2600-21619
Fax: +43-1-26007
E-mail: m.maluszynski@iaea.org

Manzanera, J.A. *(current address)*
UPM, Ciudad Universitaria
s/n 28040 Madrid, **Spain**
Phone: +34-1-336-7113
Fax: +34-1-243-9557

Martínez, L.E.
Catedra de Fisiologia Vegetal
Facultad de Ciencias Agrarias
CC No. 7, 5505,Chacras de Coria
Mendoza, **Argentina**
Fax: +54-261-496-0469
E-mail: lmartinez@fca.uncu.edu.ar

Mihály, R.
Cereal Research Non-Profit Company
Wheat Cell and Tissue Culture Laboratory
Szeged H-6701
P.O. Box 391, **Hungary**
Phone: +36-62-435-235
Fax: +36-62-434-163

Miner, M.
University of Guelph
Department of Plant Agriculture
Crop Science Building
Guelph, ON
N1G 2W1, **Canada**
Phone: +1-519-824-4120
Fax: +1-519-763-8933

Monostori, T.
Cereal Research Non-Profit Company
Wheat Cell and Tissue Culture Laboratory
Szeged H-6701
P.O. Box 391, **Hungary**
Phone:	+36-62-435-235
Fax:	+36-62-434-163

Nichterlein, K.
Plant Breeding and Genetics Section
Joint FAO/IAEA Division
P.O. Box 100
Vienna, **Austria**
Phone:	+43-1-2600-21617
Fax:	+43-1-26007
E-mail:	k.nichterlein@iaea.org

Nichols, B.
Department of Plant Agriculture
Bovey Building
University of Guelph
Guelph, ON N1G 2W1, **Canada**
Phone:	+1-519-824-4120
Fax:	+1-519-767-0755
E-mail:	rnichols@uoguelph.ca

Oro, R.
University of Guelph
Department of Plant Agriculture
Crop Science Building
Guelph, ON
N1G 2W1, **Canada**
Phone:	+1-519-824-4120
Fax:	+1-519-763-8933

Pauk, J.
Cereal Research Non-Profit Company
Wheat Cell and Tissue Culture Laboratory
Szeged H-6701
P.O. Box 391, **Hungary**
Phone:	+36-62-435-235
Fax:	+36-62-434-163
E-mail:	janos.pauk@gk-szeged.hu

Polle, E.
Northwest Plant Breeding Co.
2001 Country Club Rd.
Pullman, WA 99163, **USA**
Phone:	+1-509-334-4404
Fax:	+1-509-334-5320

LIST OF CONTRIBUTORS

Pulli, S.
Laboratory of Plant Physiology and Molecular Biology
Department of Biology
University of Turku
20014 Turku, **Finland**
Fax: +358-2-333-5549
E-mail: seppo.pulli@utu.fi

Puolimatka, M.
Plant Production Inspection Centre
Seed Testing Department
Tampereentie 51
P.O. Box 111
FIN-32201, Loimaa, **Finland**
Phone: +358-2-760-561
Fax: +358-2-7650-6222

Rines, H.W.
U.S. Department of Agriculture
Agricultural Research Service
Dept. Agronomy and Plant Genetics
University of Minnesota
411 Borlaug Hall
1991 Upper Buford Circle
St. Paul, MN 55108, **USA**
Fax: +1-651-649-5058
E-mail: rines001@tc.umn.edu

Rokka, V.M.
Agricultural Research Centre of Finland
Plant Production Research
Crops and Soil
Myllytie 10
FIN-31600 Jokioinen, **Finland**
Fax: +358-34-188-2496
E-mail: veli-matti.rokka@mtt.fi
vrokka@scri.sari.ac.uk

Sahibzada, R.
Northwest Plant Breeding Co.
2001 Country Club Rd.
Pullman, WA 99163, **USA**
Phone: +1-509-334-4404
Fax: +1-509-334-5320

Sanz, J.M.
Departamento de Genética y Producción Vegetal
Estación Experimental de Aula Dei
Apartado 202
50059 Zaragoza, **Spain**
Phone: +34-9-7671-6088
Fax: +34-976-716-145
E-mail: jmsanz@eead.csic.es

Shim, Y.S.
University of Guelph
Department of Plant Agriculture
Crop Science Building
Guelph, ON
N1G 2W1, **Canada**
Phone: +1-519-824-4120 x2507
Fax: +1-519-763-8933

Simion, E.
University of Guelph
Department of Plant Agriculture
Crop Science Building
Guelph, ON
N1G 2W1, **Canada**
Phone: +1-519-824-4120 x2507
Fax: +1-519-763-8933

Szarejko, I.
Department of Genetics
University of Silesia
Jagiellonska 28
40032 Katowice, **Poland**
Phone: +48-32-251-5446
Fax: +48-32-200-9396
E-mail: szarejko@us.edu.pl

Tai, G.C.C.
Potato Research Centre
Agriculture and Agri-Food Canada
P.O. Box 20280
Fredericton, N.B., E3B 4Z7, **Canada**
E-mail: taig@em.agr.ca

Tenhola-Roininen, T.
MTT Agrifood Research Finland
Plant Production Research
Crops and Biotechnology
Myllytie 10
31600 Jokioinen, **Finland**
Phone: +358-3-4188-2515
Fax: +358-3-4188-2496
E-mail: teija.tenhola-roininen@mtt.fi

Thomas, W.T.B.
Genome Dynamics
Scottish Crop Research Institute
Dundee, DD2 5DA
Scotland, **UK**
Tel. +1382-562-731
Fax: +1382-562-426
E-mail: wthoma@scri.sari.ac.uk

LIST OF CONTRIBUTORS

Touraev, A.
Vienna Biocenter
Institute of Microbiology and Genetics
University of Vienna
Dr. Bohr-Gasse 9
A-1030 Vienna, **Austria**
Tel.   +43-1-4277-54681
Fax: +43-1-4277-9546
E-mail: alisher@gem.univie.ac.at

Tuvesson, S.
SW Laboratory
Svalöf Weibull AB
SE-268 81 Svalöv, **Sweden**
Phone: +46-418-667-000/667-221
Fax: +46-418-667-219
E-mail: stine.tuvesson@swseed.se

Vallés, M.P.
Departamento de Genética y Producción Vegetal
Estación Experimental de Aula Dei
Apartado 202
50080 Zaragoza, **Spain**
Phone: +34-9-7671-6084
Fax: +34-976-716-145
E-mail: valles@eead.csic.es

von Post, R.
The Nilsson-Ehle Laboratory
Svalöf Weibull AB
26881 Svalöv, **Sweden**
Phone: +46-418-667-000/667-126
Fax: +46-418-667-219
E-mail: rebecka.v.post@sweed.se

Wedzony, M.
Department of Plant Physiology
Polish Academy of Sciences
Podluzna 3
30-239 Krakow, **Poland**
Phone: +48-12-425-3301
Fax: +48-12-425-3202
E-mail: niwedzon@cyf-kr.edu.pl

Weng, Y.
USDA-ARS
215 Johnson Hall
Washington State University
Pullman, WA 99164, **USA**
Phone: +1-509-335-3453
Fax: +1-509-335-3842

Wojnarowiez, G.
Université de Reims Champagne Ardenne
UFR Sciences
Laboratoire de Biologie et Physiologie Végétales
BP 1039
51687 Reims Cedex 2, **France**
Phone: +33-3-2691-3427
Fax: +33-3-2691-3427/2691-3339

Wolyn, D.J.
Department of Horticultural Science
University of Guelph
Ontario, N1G 2W1, **Canada**
Phone: +1-519-824-4120 ext. 3092
Fax: +1-519-767-0755
E-mail: dwolyn@evbhort.uoguelph.ca
dwolyn@uoguelph.ca

Wremerth-Weich, E.
Syngenta Seeds AB
P.O. Box 302
S-261 23 Landeskrona, **Sweden**
Phone: +46-418-437-271
Fax: +46-418-437-283
E-mail: Elisabeth.wremerthweich@syngenta.com

Xiong, X.Y.
Potato Research Centre
Agriculture and Agri-Food Canada
P.O. Box 20280
Fredericton, N.B., E3B 4Z7, **Canada**
Phone: +1-506-452-3316
Fax: +1-506-452-3260

Zapata Arias, J.F.
Plant Breeding Unit
FAO/IAEA Agriculture and Biotechnology Laboratory
Seibersdorf
P.O. Box 100
Vienna, **Austria**
Phone: +43-1-2600-28279
Fax: +43-1-26007
E-mail: f.zapata-arias@iaea.org

Zheng, M.Y.
Department of Biology
Gordon College
255 Grapevine Rd.
Wenham, MA 01985, **USA**
Phone: +1-978-867-4388
Fax: +1-978-867-4666
E-mail: mzheng@gordon.edu